ISBN 978-1-5280-1212-6
PIBN 10909527

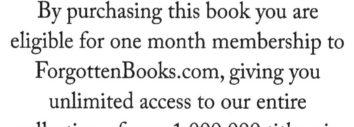

H. O. No. 168

AUSTRALIAN PILOT

VOLUME II

SOUTH AND EAST COASTS OF AUSTRALIA

FROM

CAPE NORTHUMBERLAND TO PORT JACKSON, INCLUDING BASS STRAIT AND TASMANIA

FIRST EDITION

1920

PUBLISHED AND SOLD BY THE HYDROGRAPHIC OFFICE
UNDER THE AUTHORITY OF THE
SECRETARY OF THE NAVY

PRICE, 90 CENTS

WASHINGTON
GOVERNMENT PRINTING OFFICE
1920

H. O. 168.

A summary of the Notices to Mariners affecting this publication, published during the year 1920, will be sent free of expense upon the receipt of this coupon at the United States Hydrographic Office, Washington, D. C.

Name _____

Address _____

H. O. 168.

A summary of the Notices to Mariners affecting this publication, published during the year 1921, will be sent free of expense upon the receipt of this coupon at the United States Hydrographic Office, Washington, D. C.

Name _____

Address _____

H. O. 168.

A summary of the Notices to Mariners affecting this publication, published during the year 1922, will be sent free of expense upon the receipt of this coupon at the United States Hydrographic Office, Washington, D. C.

Name _____

Address _____

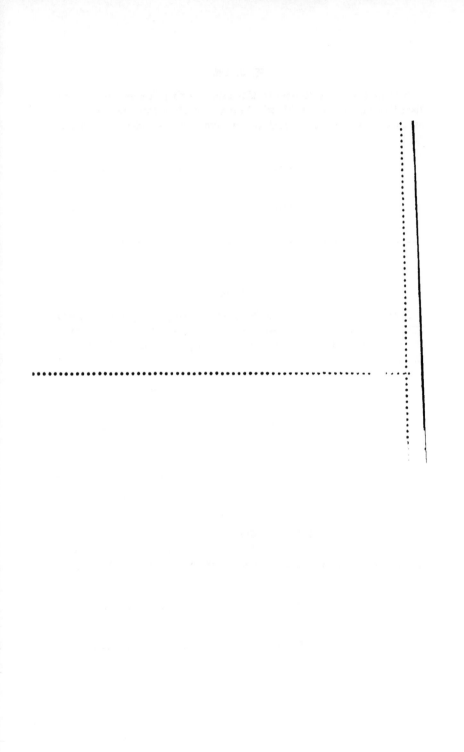

H. O. 168.

A summary of the Notices to Mariners affecting this publication, published during the year 1923, will be sent free of expense upon the receipt of this coupon at the United States Hydrographic Office, Washington, D. C.

Name _____

Address _____

H. O. 168.

A summary of the Notices to Mariners affecting this publication, published during the year 1924, will be sent free of expense upon the receipt of this coupon at the United States Hydrographic Office, Washington, D. C.

Name _____

Address _____

H. O. 168.

A summary of the Notices to Mariners affecting this publication, published during the year 1925, will be sent free of expense upon the receipt of this coupon at the United States Hydrographic Office, Washington, D. C.

Name _____

Address _____

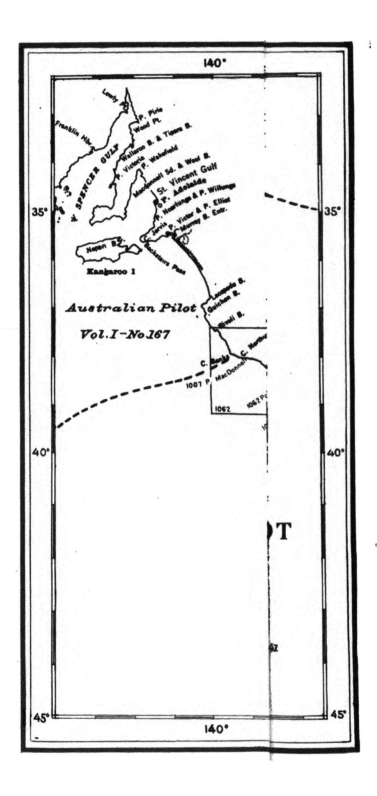

PREFACE.

This publication comprises descriptions of and sailing directions for the south and east coasts of Australia, from Cape Northumberland to Port Jackson, including Bass Strait and Tasmania. It is based primarily on information obtained from British Admiralty publication, Australia Pilot Vol. II.

The book contains the latest available information from reliable sources, and is corrected to April 10, 1920.

The bearings and courses are true, in degrees from 0° (North) to 360° (clockwise).

Bearings limiting the sectors of lights are toward the light.

The directions of winds refer to the points from which the blow; of currents, the points toward which they set. These directions are true.

Variations, with the annual rate of change, may be obtained from H. O. Chart No. 2406, " Variation of the Compass."

Distances are expressed in nautical miles, the mile being approximately 2,000 yards.

Sounding are referred to low water ordinary springs unless otherwise stated.

Heighths are referred to mean high water, spring tides.

The latest information regarding lights, their characteristics, sectors, fog signals, and submarine bells should always be sought in the Light Lists, as all the details are not given in this volume and changes are likely to occur.

Summary of Notices to Mariners.—While it is the intention of the Hyprographic Office to public about the first of each year a Summary of Notices to Mariners of the preceding year affecting the volume, it must be understood that these summaries are intended to include only important changes and corrections and that their publications may be discontinued at any time, especially when a new edition of the book is issued.

Masters of vessels should keep complete files of weekly Notices to Mariners and supply themselves with the latest List of Lights, and seek from local authorities, pilots, and harbor masters the latest information relative to any special regulations in force in the particular locality visited.

Mariners are requested to notify the United States Hydrographic Office, Washington, D. C., or one of its branch offices of errors they may discover in this publication, or of additional matter which they think should be inserted.

CONTENTS.

INFORMATION RELATING TO NAVIGATIONAL AIDS AND GENERAL NAVIGATION.

Publications.—The principal publications of the United States Hydrographic Office for the use of navigators are: Charts, Sailing Directions, American Practical Navigator, Altitude and Azimuth Tables, International Code of Signals, Light List, Notices to Mariners, Pilot Charts, and Hydrographic Bulletins. Of these the Notices to Mariners and the Hydrographic Bulletins are free to mariners and others interested in shipping. The Pilot Charts are free to contributors of professional information, but are sold to the general public at 10 cents a copy; other publications of the office are sold under the law at cost price, and can be purchased directly from the office or through its sales agencies, but are not sold by branch hydrographic offices.

Charts when issued are corrected to date.

The dates on which extensive corrections are made are noted on the chart on the right of the middle of the lower edge; those of the smaller corrections at the left lower corner.

The edition, and corresponding date, of the chart will be found in the right lower corner, outside the outer neat line.

Planes of reference.—The plane of reference for soundings on Hydrographic Office charts made from United States Government surveys and on Coast and Geodetic Survey charts of the Atlantic coast of the United States is mean low water; on the Pacific coast of the United States as far as the Strait of Juan de Fuca, it is the mean of the lower low waters; and from Puget Sound to Alaska, the plane employed on Hydrographic Office charts is low water ordinary springs.

On most of the British Admiralty charts the plane of reference is the low water of ordinary springs; on French charts, the low water of equinoctial springs.

In the case of many charts compiled from old or various sources the plane of reference may be in doubt. In such case, or whenever not stated on the chart, the assumption that the reference plane is low water ordinary springs gives a larger margin of safety than mean low water.

Whichever plane of reference may be used for a chart it must be remembered that there are times when the tide falls below it. Low water is lower than mean low water about half the time, and when a new or full moon occurs at perigee the low water is lower than the average low water of springs. At the equinoxes the spring range is also increased on the coasts of Europe, but in some other parts of the world, and especially in the Tropics, such periodic low tides may coincide more frequently with the solstices.

Wind or a high barometer may at times cause the water to fall below even a very low plane of reference.

On coasts where there is much diurnal inequality in the tides, the amount of rise and fall can not be depended upon and additional caution is necessary.

Mean sea level.—The important fact should be remembered that the depths at half tide are practically the same for all tides whether neaps or springs. Half tide therefore corresponds with mean sea level. This makes a very exact plane of reference, easily found, to which it would be well to refer all high and low waters.

If called on to take special soundings for the chart at a place where there is no tidal bench mark, mean sea level should be found and the plane for reductions established at the proper distance below it, as ascertained by the Tide Tables, or by observations, or in some cases, if the time be short, by estimation, the data used being made a part of the record.

Accuracy of chart.—The value of a chart must manifestly depend upon the character and accurancy of the survey on which it is based, and the larger the scale of the chart the more important do these become.

To judge a survey, its source and date, which are generally given in the title, are good guides. Besides the changes that may have taken place since the date of the survey in waters where sand or mud prevails, the earlier surveys were mostly made under circumstances that precluded great accuracy of detail; until a chart founded on such a survey is tested it should be used with caution. It may, indeed, be said that, except in well frequented harbors and their approaches, no surveys yet made have been so thorough as to make it certain that all dangers have been found. The number of the soundings is another method of estimating the completeness of the survey, remembering, however, that the chart is not expected to show all soundings that were obtained. When the soundings are sparse or unevenly distributed it may be taken for granted that the survey was not in great detail.

Large or irregular blank spaces among soundings mean that no soundings were obtained in these spots. When the surrounding soundings are deep it may fairly be assumed that in the blanks the water is also deep; but when they are shallow, or it can be seen from the rest of the chart that reefs or banks are present, such banks should be regarded with suspicion. This is especially the case in coral regions and off rocky coasts, and it should be remembered that in waters where rocks abound it is always possible that a survey, however complete and detailed, may have failed to find every small patch or pinnacle rock.

Fathom curves a caution.—Except in charts of harbors that have been surveyed in detail, the 5-fathom curve on most charts may be considered as a danger line, or caution against unnecessarily approaching the shore or bank within that curve on account of the possible existence of undiscovered inequalities of the bottom, which only an elaborate detailed survey could reveal. In general surveys of coasts or of little frequented anchorages the necessities of navi-gation do not demand the great expenditure of time required for so detailed a survey. It is not contemplated that ships will approach the shores in such localities without taking special precautions.

The 10-fathom curves on rocky shores is another warning, especially for ships of heavy draft.

A useful danger curve will be obtained by tracing out with a colored pencil, or ink, the line of depth next greater than the draft of the ship using the chart. For vessels drawing less than 18 feet the edge of the sanding serves as a well-marked danger line.

Charts on which no fathom curves are marked must especially be regarded with caution, as indicating that soundings were too scanty and the bottom too uneven to enable the curves to be drawn with accuracy.

Isolated soundings, shoaler than surrounding depths, should always be avoided, especially if ringed around, as it is doubtful how closely the spot may have been examined and whether the least depth has been found.

The chart on largest scale should always be used on account of its greater detail and the greater accuracy with which positions may be plotted on it.

Caution in using small-scale charts.—In approaching the land or dangerous banks, regard must always be had to the scale of the chart used. A small error in plotting a position means only yards on a large-scale chart, whereas on one of small scale the same amount of displacement means a large fraction of a mile.

Mercator chart.—Observed bearings are not identical with those measured on the Mercator chart (excepting only the bearings north and south, and east and west on the equator) because the line of sight, except as affected by refraction, is a straight line and lies in the plane of the great circle, while the straight line on the chart (except the meridian line) represents not the arc of a great circle but the loxodromic curve, or rhumb line, which on the globe is a spiral approaching but never in theory reaching the pole, or, if the direction be east or west, a circle of latitude.

The difference is not appreciable with near objects, and in ordinary navigation may be neglected. But in high latitudes, when the objects are very distant, and especially when lying near east or west, the bearings must be corrected for the convergence of the meridians in order to be accurately placed on the Mercator chart, which represents the meridians as parallel.

Polyconic chart.—On the polyconic chart, since a straight line represents (within the limits of 15 or 20 degrees of longitude) the arc of a great circle or the shortest distance between two points, bearings of the chart are identical with observed bearings.

The Mercator projection is unsuited to surveying, for which purpose the polyconic projection is used by the Hydrographic Office and the Coast and Geodetic Survey.

Notes on charts should always be read with care, as they may give important information that can not be graphically represented.

Current arrows on charts show only the most usual or the mean direction of a current; it must not be assumed that the direction of a current will not vary from that indicated by the arrow. The velocities, also, of currents vary with circumstances, and those given on the charts are merely the mean of those determined, possibly from very few observations.

Compass roses on charts.—The gradual change in the variation must not be forgotten in laying down on the chart courses and bearings from the magnetic compass roses, which become in time slightly in error, and in some cases, such as with small scales or when the lines are long, the displacement of position from neglect of this change may be of importance. The date of the variation and the annual change, as given on the compass rose, facilitate corrections when the change has been considerable. It is better to reduce all magnetic bearings and courses to true and then use the true compass rose.

The change in the variation for a change of position is in some parts of the world so rapid as to need careful consideration, requiring a frequent change of the course. For instance, in approaching Halifax from Newfoundland the variation changes 10° in less than 500 miles.

Local magnetic disturbance of the compass on board ship.—The term "local magnetic disturbance" has reference only to the effects on the compass of natural magnetic masses external to the ship. Observation shows that such disturbance of the compass in a ship afloat is experienced in many places on the globe.

Magnetic laws do not permit of the supposition that the visible land causes such disturbance, because the effect of a magnetic force diminishes so rapidly with distance that it would require a local center of magnetic force of an amount absolutely unknown to affect a compass ½ mile distant.

Such deflections of the compass are due to magnetic minerals in the bed of the sea under the ship, and when the water is shallow and the force strong, the compass may be temporarily deflected when passing over such a spot; but the area of disturbances will be small unless there are many centers near together.

Aids—Buoys.—Too much reliance should not be placed on buoys always maintaining their exact positions. They should therefore be regarded as warnings and not as infallible navigational marks, especially when in exposed places and in the wintertime; and a ship's position should always, when possible, be checked by bearings or angles of fixed objects on shore.

The light shown by a lightbuoy can not be implicitly relied on; it may be altogether extinguished, or, if periodic, the apparatus may get out of order.

Whistle and bell buoys are sounded only by the action of the sea; therefore, in calm weather, they are less effective or may not sound.

Lights.—All the distances given in the Light Lists and on the charts for the visibility of lights are calculated for a height of 15 feet for the observer's eye. The effect of a greater or less height of eye can be ascertained by means of the table of distances of possible visibility due to height, published in the Light Lists.

The loom of a powerful light is often seen far beyond the limit of visibility of the actual rays of the light, and this must not be confounded with the true range. Refraction, too, may often cause a light to be seen farther than under ordinary circumstances.

When looking out for a light, the fact may be forgotten that aloft the range of vision is much increased. By noting a star immediately over the light a very correct bearing may be obtained from the standard compass when you lay down from aloft.

On first making a light from the bridge, by at once lowering the eye several feet and noting whether the light is made to dip, it may be determined whether the ship is on the circle of visibility corresponding with the usual height of the eye, or unexpectedly nearer the light.

The intrinsic power of a light should always be considered when expecting to make it in thick weather. A weak light is easily obscured by haze, and no dependence can be placed on its being seen.

When a light is sighted, it should be identified at once by checking its characteristics. This is particularly necessary when approaching well-lighted coasts where lights with similar characteristics are often found close together.

The power of a light can be estimated by its candlepower or order, as given in the Light Lists, and in some cases by noting how much its visibility in clear weather falls short of the range corresponding to its height. Thus, a light standing 200 feet above the sea and recorded as visible only 10 miles in clear weather, is manifestly of little brilliancy, as its height would permit it to be seen over 20 miles if of sufficient power.

Sailing Directions or Pilots are books treating of certain sections or divisions of the navigable waters of the globe. They contain descriptions of coast lines, dangers and harbors, information of winds, currents, and tides, and directions for approaching and entering harbors, and much other general information of interest to mariners.

The Sailing Directions are corrected, as far as practicable, to the date of issue from the office; they can not, from their nature and the infrequency of their revision, be so fully corrected as charts and Light Lists, and for that reason, when they differ the one of the most recent issue should be accepted as correct.

Light Lists, published about once a year, are corrected before issue, and changes affecting them are published in the weekly Notices to Mariners.

The navigator should make notations of corrections in the tabular form in the Light Lists and paste in at the appropriate places the slips from the Notices to Mariners.

Notices to Mariners, containing newly acquired information pertaining to various parts of the world, are published weekly and mailed to all United States ships in commission, Branch Hydrographic offices and agencies, and United States consulates. Copies are furnished free by the main office or by any of the branch offices on application.

With each Notice to naval vessels is sent also a separate sheet, giving the items relative to lights contained in the latest Notice, intended especially for use in correcting the Light Lists.

Pilot Charts of the North Atlantic, Central American Waters, and North Pacific and Indian Oceans are published each month, and of the South Atlantic and South Pacific Oceans each quarter. These charts give the average conditions of wind and weather, barometer, percentage of fog and gales, routes for steam and sailing vessels, ice, derelicts, ocean currents, storm tracks, and other useful information. They are furnished free only in exchange for marine data or observations.

Hydrographic Bulletins, published weekly are supplemental to the Pilot Charts, and contain the latest reports of obstructions and dangers along the coast and principal ocean routes and other information for mariners. They are to be had free upon application.

The bulletins are supplemented by the Daily Memorandum published daily, Sundays and holidays excepted, in order that the information relating to dangers and aids to navigation received may be disseminated as quickly as possible.

Tides.—A knowledge of the times of high and low water and of the amount of vertical rise and fall of the tide is of great importance in the case of vessels entering or leaving port, especially when the low water is less than or near their draft. Such knowledge is also useful at times to vessels running close along a coast, in enabling them to anticipate the effect of the tidal currents in setting them on or off shore. This is especially important in fog or thick weather.

Tidal currents.—In navigating along coasts where the tidal range is considerable, special caution is necessary. It should be remembered that there are generally indrafts and corresponding outdrafts abreast of all large bays and bights, although the general run of the current may be nearly parallel with the shore outside the entrances.

The turn of the tidal currents, offshore is seldom coincident with the times of high and low water along the shore. In some channels the tidal current may overrun the turn of the vertical movement of the tide by three hours, the effect of which is that at high and low water by the shore the current is running at its greatest velocity.

The effect of the tidal wave in causing currents may be illustrated by two simple cases:

(1) Where there is a small tidal basin connected with the sea by a large opening.

(2) Where there is a large tidal basin connected with the sea by a small opening.

In the first case the velocity of the current in the opening will have its maximum value when the height of the tide within is changing most rapidly—i. e., at a time about midway between high and low water. The water in the basin keeps at approximately the same level as the water outside. The flood current corresponds with the rising and the ebb with the falling of the tide.

In the second case the velocity of the current in the opening will have its maximum value when it is high water or low water without, for then there is the greatest head of water for producing motion. The flood current begins about three hours after low water, and the ebb current about three hours after high water, slack water thus occurring about midway between the tides.

Along most shores not much affected by bays, tidal rivers, etc., the current usually turns soon after high water and low water.

The swiftest current in straight portions of tidal rivers is usually in the middle of the current, but in curved portions the most rapid current is toward the outer edge of the curve, and here the deepest water will generally be found. The pilot rule for best water is to follow the ebb-tide reaches.

Countercurrents and eddies may occur near the shores of straits, especially in bights and near points. A knowledge of them is useful in order that they may be taken advantage of or avoided.

A swift current often occurs in a narrow passage connecting two large bodies of water, owing to their considerable difference of level at the same instant. The several passages between Vineyard Sound and Buzzards Bay are cases in point.

Tide rips are made by a rapid current setting over an irregular bottom, as at the edges of banks where the change of depth is considerable.

The Tide Tables, which are published annually by the United States Coast and Geodetic Survey, give the predicted times and heights of the high and low waters for every day in the year at 81 of the principal ports of the world, and, through the medium of these by means of tidal differences and ratios, at a very large number of subordinate ports. The tables for the Atlantic and the Pacific coasts ports of the United States are also published separately.

It should be remembered that these tables aim to give the times of high and low water, and not the times of turning of the current or of slack water, which may be quite different.

The distinction between "rise" and "range" of the tide should be understood. The former expression refers to the height attained above the datum plane for soundings, differing with the different planes of reference; the latter, to the difference of level between successive high and low waters.

Full explanations and directions for their use are given in the Tide Tables.

Fog signals.—Sound is conveyed in a very capricious way through the atmosphere. Apart from the influence of the wind large areas of silence have been found in different directions and at different distances from the origin of sound, even in clear weather; therefore, too much confidence should not be felt as to hearing a fog signal. The apparatus, moreover, for sounding the signal often requires some time before it is in readiness to act. A fog often creeps imperceptibly toward the land, and may not be observed by the lighthouse keepers until upon them; a ship may have been for many hours in it, and approaching the land in confidence, depending on the signal, which is not sounded. When sound travels against the wind, it may be thrown upward; a man aloft might hear it though inaudible on deck.

The submarine bell system of fogsignals is much more reliable than systems transmitting sound through the air, as sound traveling in water is not subject to the same disturbing influences; the fallibility of the lighthouse keeper is, however, about the same in all systems, so that caution should be observed even by vessels equipped with submarine bell receiving apparatus.

Submarine bells have an effective range of audibility greater than signals sounded in air, and a vessel equipped with receiving apparatus may determine the approximate bearing of the signal. These signals may be heard also on vessels not equipped with receiving apparatus by observers below the water line, but the bearing of the signal can not then be readily determined.

Vessels equipped with radio apparatus and submarine bell receivers may fix their distance from a light vessel having radio and submarine bell, utilizing the difference in volecity of sound waves of the radio and the bell. Sound travels 4,794 feet per second at 66° F. in water, and the travel of radio sound waves for practicable distances may be taken as instantaneous.

All vessels should observe the utmost caution in closing the land in fogs. The lead is very often the safest guide and should be faithfully used.

Radio compass stations.—Most valuable aids to navigation in a fog are the radio compass stations, which will fix a ship's position by two or more bearings from a single radio station, or by simultaneous bearings from two or more stations.

In localities where only one radio station is available, mariners may use the single bearing like a Sumner's line of position, or a single bearing of any object whose position is known.

All reports from mariners indicate great accuracy in the bearings given by the radio station, and they should be used whenever available.

PILOTING—FIXING POSITION.

Piloting, in the sense given the word by modern and popular usage, is the art of conducting a vessel in channels and harbors and along coasts, where landmarks and aids to navigation are available for fixing the position, and where the depth of water and dangers to navigation are such as to require a constant watch to be kept upon the vessel's course and frequent changes to be made therein.

Piloting is the most important part of navigation and the part requiring the most experience and nicest judgment. An error in position on the high seas may be rectified by later observations, but an error in position while piloting often results in disaster. Therefore the navigator should make every effort to be proficient in this important branch, bearing in mind that a modern vessel is usually

safe on the high seas, and in danger when approaching the land and
making the harbor.

The navigator, in making his plan for entering a strange port,
should give very careful previous study to the chart and sailing
directions, and should select what appear to be the most suitable
marks for use, also providing himself with substitutes to use in case
those selected as most suitable should prove unreliable in not being
recognized with absolute certainty. Channel buoys seen from a distance are difficult to identify, because their color is sometimes not
easily distinguished and they may appear equally distant from the
observer even though they be at widely varying distances. Ranges
should be noted, if possible, and the lines drawn, both for leading
through the best water in channels, and also for guarding against
particular dangers; for the latter purpose safety bearings should
in all cases be laid down where no suitable ranges appear to offer.
The courses to be steered in entering should also be laid down and
distances marked thereon. If intending to use the sextant and
danger angle in passing dangers, and especially in passing between
dangers, the danger circles should be plotted and regular courses
planned rather than to run haphazard by the indications of the
angle alone, with the possible trouble from bad steering at critical
points.

The ship's position should not be allowed to be in doubt at any
time, even in entering ports considered safe and easy of access, and
should be constantly checked, continuing to use for this purpose
those marks concerning which there can be no doubt until others
are unmistakably identified.

The ship should ordinarily steer exact courses and follow an exact
line, as planned from the chart, changing course at precise points,
and, where the distances are considerable, her position on the line
should be checked at frequent intervals. This is desirable even
where it may seem unnecessary for safety, because if running by the
eye alone and the ship's exact position be immediately required, as
in a sudden fog or squall, fixing at that particular moment may be
attended with difficulty.

The habit of running exact courses with precise changes of course
will be found most useful when it is desired to enter port or pass
through inclosed waters during fog by means of the buoys; here
safety demands that the buoys be made successively, to do which
requires, if the fog be dense, very accurate courses and careful attention to the times, the speed of the ship, and the set of the current;
failure to make a buoy as expected leaves, as a rule, no safe alternative but to anchor at once, with perhaps a consequent serious loss
of time.

In passing between dangers where there are no suitable leading marks, as, for instance, between two islands or an island and the main shore when the conformations of the shore line are very similar, with dangers extending from both, a mid-channel course may be steered by the eye alone with great accuracy, as the eye is able to estimate very closely the direction midway between.

In piloting among coral reefs or banks, a time should be chosen when the sun will be astern, conning the vessel from aloft or from an elevated position forward. The line of demarcation between the deep water and the edges of the shoals is indicated with surprising clearness. This method is of frequent application in the numerous passages of the Florida Keys.

Changes of course should in general be made by exact amounts, naming the new course or the amount of the change desired, rather than by ordering the rudder to be put over and then steadying when on the desired heading, with the possibility of the attention being diverted and so of forgetting in the meantime, as may happen, that the ship is still swinging. The steersman, knowing just what is desired and the amount of the change to be made, is thus enabled to act more intelligently and to avoid bad steering, which in narrow channels is a very positive source of danger.

Coast piloting involves the same principles and requires that the ship's position be continuously determined or checked as the landmarks are passed. On well-surveyed coasts there is a great advantage in keeping near the land, thus holding on to the marks and the soundings, and thereby knowing at all times the position rather than keeping offshore and losing the marks, with the necessity of again making the land from a vague position, and perhaps the added inconvenience of fog or bad weather, involving a serious loss of time and fuel.

The route should be planned for normal conditions of weather, with suitable variations where necessary in case of fog or bad weather or making points at night, the courses and distances, in case of regular runs over the same route, being entered in a notebook for ready reference, as well as laid down on the chart. The danger circles for either the horizontal or the vertical danger angles should be plotted, wherever the method can be usefully employed, and the angles marked thereon; many a mile may thus be saved in rounding dangerous points with no sacrifice in safety. Ranges should also be marked in, where useful for position or for safety, and also to use in checking the deviation of the compass by comparing in crossing, the compass bearing of the range with its magnetic bearing, as given by the chart.

A continuous record of the progress of the ship should be kept by the officer of the watch, the time and patent log reading of all changes of course and of all bearings, especially the two and four point bearings, with distance of object when abeam, being noted in a book kept in the pilot house for this special purpose. The ship's reckoning is thus continuously cared for as a matter of routine and without the presence or particular order of the captain or navigating officer. The value of thus keeping the reckoning always fresh and exact will be especially appreciated in cases of sudden fog or when making points at night.

Where the coastwise trip must be made against a strong offshore or head wind it may be desirable with trustworthy charts to skirt the shore as closely as possible in order to avoid the heavier seas and adverse currents that prevail farther out. In some cases, with small ships, a passage can be made only in this way. The important saving of coal and of time, which is even more precious, thus effected by skillful coast piloting makes this subject one of prime importance to the navigator. However, many vessels have gotten into serious trouble by attempting to save time and cut down distances by rounding too closely dangers and aids, and navigators should always bear in mind that the safety of the vessels is the first consideration.

Fixing position.—A navigator in sight of objects whose positions are shown on the chart and which he can recognize may locate his vessel by ane one of the following methods:

1. Sextant angles between three known objects.
2. The bearing of a known object and angle between two known objects.
3. Cross bearings of two known objects.
4. Two bearings of a known object, separated by an interval of time, with the run during that interval.
5. The bearing and distance of a known object.

Besides the foregoing there are two mothods by which, without obtaining the precise position, the navigator may assure himself that he is clear of any particular danger.

1. The danger angle.
2. The danger bearing.

These various methods are fully explained in most textbooks on navigation and in Bowditch's American Practical Navigator, a copy of which should be in the navigator's outfit.

The first method of fixing the position, by the "three-point problem," is the most accurate of all methods, but requires expertness in the use of the sextant and protractor. However, the choice of the method should be governed by circumstances, depending upon which is best adapted to prevailing conditions.

Soundings are of very great advantage when approaching land or shoal banks in determining the position, and the convenience in the use of modern sounding machines renders any neglect to take soundings inexcusable.

Soundings taken at random are of little value in fixing or checking position, and may at times be misleading. In thick weather, when near or closing the land, soundings should be taken continuously and at regular intervals and, with the character of the bottom, systematically recorded. By laying the soundings on tracing paper, according to the scale of the chart, along a line representing the track of the ship, and then moving the paper over the chart, keeping the line representing the track parallel with the course until the observed soundings agree with those of the chart, the ship's position will in general be quite well determined.

At sea the only methods of determining the position of the vessel are by "dead reckoning" and by observations of heavenly bodies, though observations may be made use of by various methods. (See American Practical Navigator and textbooks on navigation.)

The one which should be best understood and put to the most constant use is that employing position or Sumner lines. These lines give the most comprehensive information to the navigator with the least expenditure of labor and time. The knowledge gained is that the vessel must be somewhere on the line, provided the data used is accurate and the chronometer correct. As the information given by one line of position is not sufficient to determine the definite location of the vessel, it is necessary to cross this line by another similarly obtained, and the vessel, being somewhere on both, must be at their intersection. However, a single line at times will furnish the mariner with invaluable information: For instance, if it is directed toward the coast, it marks the bearing of a definite point on the shore, or, if parallel to the coast, it clearly indicates the distance off, and so will often be found useful as a course. A sounding taken at the same time with the observation will in certain conditions prove of great value in giving an approximate position on the line.

The easiest and quickest way to establish a line of position is by employing the method of Marcq St. Hilaire, as modified by the use of tables of altitude.

A very accurate position can be obtained by observing two or more stars at morning or evening twilight, at which time the horizon is well defined. The position lines thus obtained will, if the bearings of the stars differ three points or more, give an excellent result. A star or planet at twilight and the sun afterwards or before may be combined; also two observations of the sun, with sufficient interval to admit of a considerable change of bearing; in these cases one of the lines must be moved for the run of the ship.

Use of oil for modifying the effect of breaking waves.— Many experiences of late years have shown that the utility of oil for this purpose is undoubted and the application simple.

The following may serve for the guidance of seamen, whose attention is called to the fact that a very small quantity of oil, skillfully applied, may prevent much damage both to ships, especially of the smaller classes, and to boats by modifying the action of breaking seas.

The principal facts as to the use of oil are as follows:

1. On free waves, i. e., waves in deep water, the effect is greatest.

2. In a surf, or waves breaking on a bar, where a mass of liquid is in actual motion in shallow water, the effect of the oil is uncertain, as nothing can prevent the larger waves from breaking under such circumstances; but even here it is of some service.

3. The heaviest and thickest oils are most effectual. Refined kerosene is of little use; crude petroleum is serviceable when no other oil is obtainable, or it may be mixed with other oils; all animal and vegetable oils, such as waste oil from the engines, have great effect.

4. In cold water, the oil, being thickened by the low temperature and not being able to spread freely, will have its effect much reduced. A rapid spreading oil should be used.

5. A small quantity of oil suffices, if applied in such a manner as to spread to windward.

6. It is useful in a ship or boat either when running, or lying-to, or in wearing.

7. When lowering and hoisting boats in a heavy sea the use of oil has been found greatly to facilitate the operation.

8. For a ship at sea the best method of application appears to be to hang over the side, in such a manner as to be in the water, small canvas bags, capable of holding from 1 to 2 gallons of oil, the bags being pricked with a said needle to permit leakage. The waste pipes forward are also very useful for this purpose.

9. Cross a bar with a flood current, to pour oil overboard and allow it to float in ahead of the boat, which would follow with a bag towing astern, would appear to be the best plan.

On a bar, with the ebb current running, it would seem to be useless to try oil for the purpose of entering.

10. For boarding a wreck, it is recommended to pour oil overboard to windward of her before going alongside, bearing in mind that her natural tendency is always to forge ahead. If she is aground, the effect of oil will depend upon attending circumstances.

11. For a boat riding in bad weather to a sea anchor, it is recommended to fasten the bag to an endless line rove through a block on the sea anchor, by which means the oil can be diffused well ahead of the boat, and the bag readily hauled on board for refilling, if necessary.

CHAPTER I.

GENERAL REMARKS—VICTORIA—TASMANIA—NEW SOUTH
WALES—PHYSICAL FEATURES—PRODUCTIONS—STATISTICS—
COMMUNICATION—CLIMATE—UNIFORM BUOYAGE—SIGNALS—
WINDS AND WEATHER—CURRENTS—PASSAGES.

GENERAL REMARKS.

This volume treats of the coasts and harbors of Victoria and New South Wales, from Cape Northumberland to Port Jackson, inclusive; also of Bass Strait and the coasts and harbors of Tasmania.

Standard time.—In all parts of Victoria, New South Wales, and Tasmania the mean time of the meridian of 150° E., or 10 hours fast on Greenwich mean time, has been adopted as the standard time.

Target practice—Regulations.—The commonwealth defense act contains the following sections:

71. The officer in charge of any artillery or rifle range may stop all traffic during artillery or rifle practice on any road or waterway crossing the line of fire, or in dangerous proximity thereto.

72. No ships, boats, or persons shall come, or remain within the prescribed distance of any ship, battery, gun, or person engaged in artillery or rifle practice, or shall remain in any position so as to obstruct such practice.

The defense authorities have also given notice that all forts from which practice is carried on will fly a red flag from the masthead of the flagstaff. All ships and boats should be kept at a distance of at least 800 yards to the left and 1 mile to the right of the line of fire from a distance of 6,000 yards from the battery, in accordance with part 6, No. 20, of the above act.

GENERAL INFORMATION.

Steam trials—Caution.—Flag A, International Code, when flown by H. M. A. or other ships in the vicinity of any measured distance, indicates that such vessels are running speed trials. For mutual safety vessels, both steaming and sailing, should endeavor to keep out of the way whilst these trials are in progress.

Naval dockyard.—The only naval establishment is at Garden Island, Port Jackson.

15

Docks.—At Sydney and Melbourne are wet and dry docks and patent slips suitable for all classes of vessels. (*See* Appendices I and II.)

Lloyd's signal stations are established at the following places mentioned in this volume, viz: Cape Northumberland, Cape Nelson, Cape Otway, Point Lonsdale, Queenscliff, Cape Schanck, Wilson Promontory, Gabo Island, Table Cape, Mersey Bluff, Low Head, Eddystone Point, Cape Sorell, Currie Harbor, Cape Bruny, and Deal Island.

Morse signal stations.—The following lighthouses are prepared to communicate by Morse signals at night with passing vessels: Cape Nelson, Cape Otway, Eagle Nest (Split) Point, Point Lonsdale, Cape Schanck, Wilson Promontory, and Gabo Island./ In communicating follow rules given in H. O. Publication No. 87.

Life boats and life-saving stations are inserted only on plans of harbors and anchorages.

Radio.—Radio stations are established at the following places mentioned in this volume, viz: Mount Gambier, Melbourne, Flinders Island (Bass Strait), Hobart, and Sydney. The station at Sydney is capable of communication with New Zealand and the radio stations in the Pacific.

Radio stations are shown on H. O. Chart No. 2180.

Signals to be made by vessels approaching ports when inconvenienced by searchlights.—Any vessel approaching a port in the British Empire when searchlights are being worked and finding that they interfere with her safe navigation may make use of the following signals, either singly or combined:

(*a*) By flashing lamp, four short flashes, followed by one long flash.

(*b*) By whistle, siren, or fog horn, four short blasts, followed by one long blast.

Whenever possible, both flashing lamp signals and sound signals should be used.

On these signals being made, the searchlights will be worked, as far as circumstances will permit, so as to cause the least inconvenience, being either extinguished, raised, or their direction altered.

The signals should not be used without real necessity, as unless the vessel is actually in the rays of a searchlight it is impossible to know which searchlight is affected. These signals should be repeated until the inconvenience is removed.

These signals are designed to assist mariners, and do not render the Government liable in any way.

Lights carried by pilot vessels.—Pilot vessels, when engaged on their station on pilotage duty, carry the lights required by the International Rules of the Road.

Trawling grounds.—The principal trawling ground along the east coast extends from near Port Stephens in New South Wales, southward past Sydney to Gabo Island; it continues across the eastern slope of Bass Strait, past Flinders Island to Tasmania. It covers approximately 6,000 square miles, and is within easy access of the two principal markets (Sydney and Melbourne). A central port is at Eden, in Twofold Bay, where, if a cool store were established, a base station could with advantage be established, and smokehouses, etc., kept in continuous operation.

In Bass Strait, only limited areas have been found to carry fish in paying quantity.

So far there has been located approximately 10,000 square miles of trawable ground carrying fish in paying quantity, and all this area is within reasonable distance of principal centers.

For the purpose of giving a general view as to the relative abundance of fish within the various localities, the fishing grounds have been divided into sections as follows:

1. **Gabo to Sydney.**—This section has been examined during the whole or part of 16 different cruises at varying intervals between April, 1909, and August, 1913. During this period the trawl was on the bottom for 228¼ hours, and produced a total catch of 84,721 pounds of marketable fish, or at the rate of 371 pounds per hour of fishing.

2. **The grounds to the south and west from Gabo Island, extending eastward from Flinders Island to Tasmania.**—This section has been visited during 23 different cruises between April, 1909, and August, 1913; trawling 432¼ hours, marketable fish caught 81,715 pounds, or an average of 189 pounds per hour of fishing.

3. **The deeper section, extending beyond 200 fathoms, was** examined during three cruises in May and June, 1913. Trawling was for 122 hours; marketable fish taken, 13,939 pounds; average, 118 pounds an hour.

The fishing grounds can accommodate about 120 trawlers. Investigations are not completed, but enough has been done to show that there is an available supply of fish on the trawling grounds. See H. O. Publication No. 167.

Barometer.—The graduation of barometric scales in millibars having now been largely introduced, the accompanying diagram is inserted to enable the mariner to convert millibars into inches and vice versa.

Victoria.—The State of Victoria is situated between the parallels of 34° and 39° of south latitude, and the meridians of 141° and 150° of east longitude. Its northern boundary is an imaginary line from Cape Howe in a northwest direction to the source of the Murray

River, thence by that river downward to its intersection by the 141st meridian of east longitude. Its area is 87,884 square miles.

Physical features.—A broad and irregular chain of mountains, ranging in height from 4,000 to 6,500 feet, extends through the State from east to west, dividing it into two unequal parts, the rivers having their sources on the southern side of the watershed flowing to the sea, and those rising on the northern slopes empty themselves into the Murray.

There is no connected system of coast ranges.

The larger portion of Victoria is mountainous or hilly; only in the northwest are vast sandy and sparsely grassed plains intersected with belts of scrub and forests.

Rivers.—With the exception of the Murray, there are few navigable rivers. In winter most of the rivers become angry torrents, carrying devastation over much fertile country, while in summer many of them dwindle down to small streams or detached pools.

Lakes.—There are many shallow lakes, both fresh and saline, some of which dry up in the summer heat. The largest of these is Lake Corangamite, covering 57,700 acres; this lake, with a number of small ones in its neighborhood, has no visible outlet, and is extremely saline; Lakes Colac and Burrumbeet are 6,560 and 5,100 acres in extent and are fresh-water lakes; Lakes Albacutya 13,000 acres, Bolac 3,500 acres, Connewarre 3,880 acres, Hindmarsh 30,000 acres, and Tynell 42,600 acres, are salt.

The Gipps Land lakes, Victoria, King, Reeves, and Wellington, are approachable from the sea at certain times, a belt of sand only separating them therefrom, through which there is a narrow artificial entrance; the largest of them, Lake Wellington, covers an area of 34,500 acres, and Lake Victoria occupies 28,500 acres.

Geology.—The geological formation of Victoria is very varied, which helps to give the country much of its beautiful scenery. It possesses a large proportion of Paleozoic and volcanic rocks to which it owes its extreme fertility. The Dividing range running east and west through the country consists of granitic and Silurian rocks. South of the mountains there was once an enormous deposit of Upper Paleozoic or Secondary rocks which have since been greatly denuded. Devonian sandstones, slates, and limestones occur in Gipps Land. Secondary rocks occur in the Cape Otway country and in the region east of Port Western. .

Tertiary deposits cover one-third of the surface of the State. The calcareous or desert sandstone of Pliocene age, which is so largely developed in Western and South Australia, enters Victoria in the west and northwest, and forms much of the poor arid pastures of that country. To the south of the mountains small patches of it are found at Port Phillip heads and Port Western.

The quartz, gravel, clay, sandstone, and conglomerate, in which alluvial gold is found, are Older Pliocene, while the fresh-water sandstones of Geelong and of the Loddon Valley are Newer Pliocene. The colored clays of Warrnambool on the southwestern coast are Post-Pliocene. Miocene beds occur in the Moorabool Valley west of Geelong and in the Cape Otway region; while the sandstones of Portland in the west, and the rough limestone of the Gipps Land lakes are of the same age.

Flora.—The largest trees indigenous to the State of Victoria belong to the eucalyptus or gum-tree tribe; the casuarina or Shea oak, the banksia or honeysuckle, the native cherry tree, the titree, the acacia, the myrtle, and the beech comprise the most widely distributed of the native trees. Ferns are numerous.

Fauna.—The largest indigenous animal to be met with in Victoria is the kangaroo, some reaching 6 feet high, known by the distinctive names of kangaroo, wallaby, and paddymelon. Rat kangaroos and mice are plentiful; opossums are numerous, and so are the wombat, porcupine or ant-eating echidna, bandicoot, native bear, native cat, and others. There are several species of the bat tribe, the largest of which is known as the flying fox.

The feathered tribe is largely represented; there are several kinds of eagle and hawk, also the emu, crane, black swan, wild bustard, pigeon, duck, teal, water-hen, quail, snipe, plover, magpie, laughing jackass, and many other birds; the parrot tribe are very numerous and of the most brilliant plumage.

Snakes are to be found in most parts of the State; many of them are of a venomous nature, the black and brown snake and the death adder especially; other, mostly of brighter and more variegated colors, are not so dangerous. The insect world is, perhaps, in no other country so variously and widely distributed.

The paucity of animals suitable for game and food, and the scarcity of song and other birds, have led to the introduction of numerous well-known English animals and birls. Hares, rabbits, white swans, foreign ducks, deer, thrushes, and also several kinds of birds are now becoming quite plentiful.

Products.—The principal product of Victoria was formerly gold, to which it owes its extraordinarily rapid progress; considerable quantities of silver, tin, copper, antimony, lead, coal, lignite, and kaolin have also been worked; diamonds, sapphires, turquoise, and other precious stones have been found. Wheat is grown in such quantities as not only to support the population, but to leave a large surplus for export. Potatoes, oats, barley, and maize are cultivated. In 1917–18, 800,000 gallons of wine were produced. The rearing of cattle and sheep forms one of the staple industries.

Population.—The population in 1918 was estimated to be 1,417,000.

Trade.—In 1917–18 the value of the imports was £20,688,039 and of the exports £18,805,160. The chief articles of import were wool (from across the border), arms, ammunition, explosives, bags, sacks, bicycles, coal, gold, hides, skins, oil, paper, silks, spirits, tea, tobacco, sugar, cottons, woolens, live stock, timber, iron, and steel; of export, wool, gold, wine, tallow, sugar, cheese, butter, hay, oil, potatoes, preserved and frozen meats, wheat, flour, biscuits, oats, leather, skins, hides, and live stock.

Ports and harbors.—The principal ports and harbors are Port Phillip, Portland Bay, Port Fairy, Lady Bay, Port Western, Venus Bay, Port Albert, and Gipps Land Lakes.

Communication—Steam vessels.—The Peninsular & Oriental Steam Navigation Co. and the Orient Royal Mail Line, each provide service to and from England and Melbourne. There are also steamers to Melbourne; of the Messageries Maritimes from Marseilles; of the Anglo-Australasian Steam Navigation Co., and of Lund's and other lines from the United Kingdom. Local companies' steam vessels connect Melbourne with all the Australian and New Zealand ports.

Railroads.—In 1918 there were 4,205 miles of railroad completed, all of which belong to the Government, and which have a gauge of 5 feet 3 inches, except 122 miles which have a 2 foot 6 inch gauge. Lines radiate in every direction from Melbourne; by the North-Eastern line to Wodonga, through railroad communication is effected with Sydney (which can be reached in 19 hours from Melbourne); and the Western line to Serviceton give through communication with Adelaide. Other extreme points reached are Portland, Belfast, on the southwest coast; Swan Hill and Echuca on the Murray River; Warracknabeal, Donald, and Wychproof to the northwest; and eastward to Bairnside, on the Mitchell River, and Port Albert.

Coal, coke, liquid fuel, and lubricating oil can be obtained at Melbourne.

Explosives.—Ships with explosives on board when entering any port of Victoria shall specially report the same to the pilot and at the time of making entry at the customhouse. All vessels entering, or in, the ports of Victoria shall hoist a red burgee at the main. Explosives may be landed only between sunrise and sunset. No boat shall be used for the conveyance of explosives, either to or from any ship or wharf or other place, unless duly licensed for that purpose, and no explosives shall be landed or conveyed from the ship until notice shall have been given to the water police (if there be any) at the port or place where the ship shall lie, in sufficient time to enable the police to give such directions as may be necessary to prevent danger, which directions the person in charge of such explosives shall obey. Boats licensed to convey explosives are subject to all the regulations for the management of hulks containing explosives, and no boat with explosives on board shall be towed by a high-pressure open-decked steamboat whose furnaces are exposed, or by any steamer with less towline than 60 feet in length, and no steamer shall approach within 200 yards of any hulk, lighter, or boat containing explosives, unless the explosives are stowed in the hold and the hatches are closed and covered with tarpaulin. No explosives shall be removed from any ship for conveyance to the magazine except between sunrise and sunset, and explosives shall only be permitted to be deposited in the magazine between those hours.

Vessels receiving explosives must be anchored beyond the limits within which ships having explosives on board are not permitted to anchor, and it may only be put on board between sunrise and sunset.

No vessel having explosives on board arriving in or off any of the ports of Victoria shall go alongside any wharf or jetty within these ports or be at anchor otherwise than as directed for each port.

General signals.—The following signals are in use at the ports of Victoria:

Mails on board_____ { White flag at the fore, to be kept flying till the mails are out of the ship.

Explosives on board_____ Red burgee at the main.

Government emigrants on board__ Ensign at the mizzenhead.

Sea pilot_____ Pilot jack at the foremasthead.

Harbor pilot_____ Ensign at the foremasthead.

Boarding officer_____ Blue flag at the main.

Medical assistance_____ Letter B at the peak.

Water police_____ { Day signal, ensign at the mainmasthead; by night, two lights vertical 5 feet apart, at any masthead or peak.

Customs boat_____ Pilot jack at the peak.

Steam tug_____ Rendezvous flag at the peak or mizzenmast.

Clearing officer outwards_____ { White flag at the main when the ship is ready for sea.

Water_____ Letter M at mizzen.

Immigration officer_____ { Ensign at mainmasthead with blue flag underneath.

Pratique required_____ Yellow flag at the main.

Quarantinable or suspicious disease_____ { Yellow and black flag at the main.

Night signals_____ { Three lights, two red and one white, 6 feet apart, in the form of an equilateral triangle, with the white light at the apex, upward.

Launching vessels from patent slips or building yards_____ { Red flag on flagstaff one hour before launching.

Ballast_____ Letter S at mizzen.

Blasting operations in progress___ A square red flag hoisted on the works.

And from the signal station:

A ball at the yardarm_____	Sailing vessel in sight making for port from eastward or westward, as ball is hoisted at east or west yardarm. Hauled down when anchored.
A ball at the masthead and at yardarm_____	Steamer in sight making for port from eastward or westward, as ball is hoisted at east or west yardarm.
Flag P at the masthead_____	Bad weather signal, can not work in the bay.

Wind warning signals, Victoria:

A white pennant over a red flag with square black center_____	Northwesterly winds.
A red pennant over a red flag with square black center_____	Northeasterly winds.
A red flag with square black center over a white pennant_t____	Southwesterly winds.
A red flag with square black center over a red pennant_____	Southeasterly winds.
Two red flags with square black centers hoisted one over another.	Very severe gales.

The above signals are also exhibited from the yardarm at Gellibrand Lighthouse.

Uniform system of buoyage.—The following rules have been adopted in Victorian waters:

The term " starboard hand " denotes that side which would be on the right hand of the mariner, either going with the main stream of flood or entering a harbor, river, or estuary from seaward.

The term " port hand " denotes the left hand of the mariner under the same circumstances.

Starboard-hand buoys are conical, painted red with even numbers, and show the painted top of a cone above the water.

Port hand buoys are can, painted black with odd numbers, and show a flat top above the water.

Spherical buoys show a domed top above the water, and when used to mark middle ground are distinguished by horizontal stripes of white color.

Surmounting beacons, such as staff and globe, are painted of one dark color.

Staff and globe are only used on starboard hand buoys, staff and cage on port hand; diamonds at the outer ends of middle grounds, and triangles at the inner ends.

Mooring buoys are painted red, and are either barrel, can, or spherical buoys.

Cable buoys, for marking submarine telegraph cables, are painted green, with the word " Telegraph " painted thereon in white letters.

Buoying and marking of wrecks:
- (*a*) All buoys and topsides of vessels used for marking wrecks are painted green, with the word "Wreck" painted thereon in white letters and moored, where practicable, on the side of the wreck which is next to midchannel.
- (*b*) When a wreck-marking vessel is used, it exhibits by day one ball on the side nearest the wreck, and two placed vertically on the other side at a height of 20 feet above the sea. By night three fixed white lights similarly arranged, but not the ordinary white riding light, shall be shown from sunset to sunrise.

Climate.—Although the maximum summer temperature rises to 105° F. in the shade, there is a considerable amount of fine clear weather not oppressively warm, and, excepting when the hot northerly winds blow, the climate is exceeding agreeable. The mean temperature at Melbourne is 58°, and the minimum is 27°. In the low, lands frost is almost unknown, but in the mountainous districts it often freezes at night, though the days may be hot. The average rainfall in Melbourne is 25½ inches; it is very variable. In the northwest interior it is very dry, the rainfall being seldom more than 14 inches.

Tasmania, formerly known as Van Diemen's Land, is situated southward of the southeastern portion of Australia, from which it is separated by Bass Strait, 150 miles wide. The state includes numerous dependent islands, of which King Island and Flinders Island are the largest. It contains an area of 26,215 square miles, including Macquarie Island.

Physical features.—Probably Tasmania is the most thoroughly mauntainous island on the globe; being one continuous series of mountains, valleys, peaks, and glens. The highest mountains just exceed 5,000 feet. The southern and western parts of the island are particularly remarkable for bold and commanding scenery. The coast, which is rocky and bold in its outlines, is broken by numerous inlets, many of which constitute good natural harbors. Rivers are numerous, and a few of them are navigable for a portion of their course. There are also several mountain lakes near the sources of the rivers, the largest, Great Lake, 3,820 feet above the sea, is about 12 miles long, and has an area of 28,000 acres.

Geology.—Formations of the ancient and Paleozoic and metamorphic rocks, and abundance of granite, constitute almost the whole of the table lands and lofty peaks. Mesozoic rocks occur in the lower hills and are more prevalent than in Australia.

Sandstone, supposed to be of Triassic age, occurs near Hobart, forming hills capped with basalt. Tertiary beds occupy much of the larger valleys and plains. Igneous and volcanic rocks abound. Porphyries and greenstones occur on most of the plateau, and form parts of many of the highest mountains. Basalts occur of every age down to the Plicoene Tertiary. The islands in Bass Strait are granite, as are also the northeastern corner of Tasmania and Wilson promontory in Victoria. Tasman Peninsula presents grand vertical precipices of basaltic columns. The Secondary sandstones produce fine building material. Limestone occurs in the Derwent Valley and on the north coast, where are extensive caves. Coal and lignite occur in many localities and are believed to be both of Paleozoic and Mesozoic age. Some of the coal is of good quality and it is becoming more largely worked. Rich iron ore occurs on the north coast and in many other places. Tin, lead, antimony, manganese, and plumbago also occur, but only the tin has been worked to any extent. There are also some quarries of good roofing slates.

Flora.—There are about 1,100 known species of plants in Tasmania, of which about 1,000 are indigenous. The larger timber is comprised in the blue gum, Huon pine, myrtle, wattle, blackwood, and King William pine; in the mountain gullies and ravines tree ferns of large size are very plentiful.

Fauna.—In all about 230 species are indigenous to Tasmania, comprising 26 mammals (of which 12 are peculiar to the island), 156 birds, 7 lizards, 3 snakes, 7 frogs, and 21 fresh-water fishes. Of the birds, not more than 15 kinds are peculiar to Tasmania. The larger animals indigenous to the island are kangaroo, wallaby, native hyena, native devil, wombat, platypus, opossum, and tiger cat; smaller ones are the kangaroo rat, bandicoot, and native cat. Among reptiles are snakes, tiger, copperhead, and whip, all venomous; lizards, iguanas, scorpions, centipedes, and tarantulas. Insect life is very prevalent; as many as 1,460 species have been described. The feathered tribes are abundantly represented both in land and aquatic birds. Among the former, black and white cockatoos, jays, magpies, white hawks, crows, eagles, sparrow hawks, owls, moreporks, miner, quail, pigeons, parrots, paroquets, thrushes, robins, diamond birds, larks, honeysuckers, wrens, firetails, redbills, and many others. Among the latter are black swans, snipe, herons, bitterns, teal, duck, tern, penguins, petrel, pelicans, gulls, cormorants, native hen, baldcoot, and divers.

The bays contain excellent fish, particularly the trumpeter, found on the south side of the island. The species include the trumpeter up to 60 pounds weight, the salmon, the flathead, trevally, garfish, barracouta and kingfish, perch, flounder, gurnard, and bream. The anchovy is migratory. English mackerel have been seen off the east coast and some of the herrings are like the English. Rock and bull kelp cod are favorites. English trout and the true salmon are found;

the efforts to acclimatize the latter have been crowned with success. Among fresh-water fish are a so-called fresh-water herring, various kinds of trout, eel, lamprey, black fish, and fine perch.

There are various species of whales, seals, and porpoises in the neighboring seas. The production of oysters upon the east coast is of some importance.

Products.—The chief products of the State are wool, gold, silver, tin, fruit, and sheep. The wool is much esteemed and commands a high price as well as the apples. The live stock is celebrated for its general excellence, especially the stud sheep. The woods are scarcely yet fully appreciated; the sources of supply are practically inexhaustible, abounding with the most beautiful cabinet woods and the largest-sized timbers adapted for every variety of purpose. Much beer is brewed for exportation to the neighboring States.

Population.—The population in 1918 was estimated at 202,850.

Harbors.—The principal harbors are Port Davey, Macquarie Harbor, Stanley, Emu Bay, Devonport, Port Sorell, Port Dalrymple, George Bay, Spring Bay, Port Arthur, Storm Bay, Norfolk Bay, Frederick Henry Bay, D'Entrecasteaux Channel, Port Esperance, Southport, and Recherche Bay.

Trade.—In 1917–18 the total value of the imports was about £489,249, the exports £961,123. This does not include interstate trade or indirect foreign trade.

The imports are chiefly railway material, apparel and haberdashery, cottons and woolens, wrought and unwrought iron; and the exports, wool, gold, silver, copper, tin, timber, jam, hops, grain, hides, bark, fruit, and sheep.

Communication—Steam vessels.—The steam vessels of Shaw, Savill & Albion Co., and of the New Zealand Shipping Co., from London for New Zealand, via the Cape, call at Hobart, also large cargo steamers.

The Union Steamship Co. maintain frequent and speedy communication between Hobart, Sydney, and Launceston, and between Hobart and Melbourne, Launceston and Melbourne, Hobart and Strahan, and Strahan and Melbourne. They also run vessels regularly to Stanley, Emu Bay, and Devonport, and their steamers leave Hobart at regular intervals for New Zealand ports.

Huddart, Parker & Co.'s steamers ply in the interstate trade, and many steamers call at Launceston.

Railroads.—The railroads in Tasmania are 767 miles in length and consist of the main line from Hobart to Launceston and the western line from Launceston to Burnie. There are also branches to Glenora and Apsley, from Corners to St. Marys, from Parattah to Oatlands, and from Launceston to Scottsdale; and a line connects

Bellerive with Sorrell. A line runs from Strahan in Macquarie Harbor to Zeehan, Dundas, and Burnie.

The railroads of Tasmania are constructed on the 3 foot 6 inch gauge.

Telegraphs.—Electric telegraphs extend along the railroads and to all the chief towns. A submarine line connects Tasmania with Victoria and thus with the universal telegraph system.

Coal.—Coal can be obtained at Hobart.

The chief coal fields at present worked in the State are those of the Fingal district on the east coast. The principal collieries are the Cornwall and Mount Nicholas. The coal is bituminous and of an excellent quality. Coal of a poorer quality is found on the rivers Don and Mersey on the northwest coast, Hamilton, in the center, and in many other places in the island. In 1917 coal to the value of £38,697 was mined.

Climate.—The climate, though differing in the eastern and western portions—the former being dry and the latter very wet—is very fine and salubrious, and well suited to European constitutions, and although the hot northern winds of Australia do occasionally reach the island, they are greatly subdued in temperature by their passage over Bass Strait. It possesses the full summer heat due to its latitude, but the nights are always cool and refreshing. The mean temperature at Hobart is 54° F. The maximum temperature of Hobart in summer is 105°, which is rarely reached; in winter it seldom falls below 29°, though on the uplands it often sinks to 18° below freezing, producing ice of a considerable thickness. The rainfall at Hobart averages 23½ inches annually, at Launceston 28 inches, and at Macquarie Harbor on the west coast over 100 inches. But it varies greatly at the same place. There is abundance of wind, often violent. Thunderstorms are rare.

New South Wales.—The first authentic account of the discovery of New South Wales is from Capt. Cook, in the year 1770, who bestowed upon it the name it still bears.

It is divided from Queensland by the Macpherson Range, west to the Great Dividing Range, thence southwesterly to the source of the Dumaresq River, thence by that river and Mackintyre and Barwon Rivers west to the one hundred and forty-ninth meridian of longitude; from this meridian west to the one hundred and forty-first meridian or the border of South Australia the parallel of 29° has been adopted as the dividing line.

On the seaboard the northern limit is Danger Point in latitude 28° 10′, and on the south, Cape Howe in latitude 37° 31′, embracing a coast of 683 miles.

The State is divided from that of Victoria by an imaginary line from Cape Howe in a northwest direction to the source of the Murray

River, thence by that river downwards to its intersection by the one hundred forty-first meridian of longitude.

It became a British possession in 1788. In 1855 responsible government was granted. The legislative power is vested in a parliament of two houses, the first called the Legislative Council and the second the Legislative Assembly. The executive is in the hands of a governor, appointed by the Imperial Government, assisted by a cabinet of ministers.

It comprises an area of 309,460 square miles.

Physical features.—The surface of the country may be divided into three divisions—the coast district, the table-lands, and the plains of the interior. The coast district is a comparatively narrow strip of undulating land, its average width being 30 miles, and extending back to the Great Dividing Range. It is fertile and well watered, with many navigable rivers, which are generally barred, and vary much in volume according to the season. Most of these streams periodically overflow their banks and cover vast tracts of low lands.

The table-lands are a high plateau traversing the entire length of the colony, furrowed by precipitous valleys, and frequently presenting on the seaward side nearly perpendicular escarpments. This high plateau extends to about the one hundred and forty-first meridian; westward of this line there is a gradual fall to the great central plains.

The southern portion of the coast line generally shows as abrupt rocky cliffs, alternating with stretches of beach; farther north, low sandy shores will be more frequently found, with many bold headlands.

Rivers.—All the rivers of the State, with three exceptions, have their sources in the Great Dividing Range, and flow thence into the sea by the eastern or western watershed. The great rivers of the western watershed are the Darling, Lachlan, Murrumbidgee and Murray with their affluents.

All these rivers unite their streams with the Murray, which flows into Lake Alexandrina in South Australia, and thence into the ocean. The rivers of the eastern watershed fall into the Pacific Ocean, the principal ones being the Hawkesbury, Hunter, Shoalhaven, Clarence, Macleay, Richmond, and Manning; they are all partially navigable by light-draft steamers, but have bar entrances that are more or less difficult or dangerous.

Lakes.—The largest lake in New South Wales is Lake George, about 25 miles southwestward of Goulburn; it is 25 miles long and 8 wide, and situated on the top of the table-land of the Dividing Range, 2,129 feet above the level of the sea; there is no outlet for the water, which is consequently saline, from the accumulation of salt held in solution. Other lakes are Lake Bathurst, with an area of 8 square miles; Tarrago Lake, Burra Burra Lake, Lake Macquarie, and Lake Illawarra, the two latter being connected with the sea.

Geology.—The mountain ranges and table-lands of New South Wales consist mainly of the older Paleozoic formations, pierced and rent by intrusive igneous rocks of various ages. The settled districts of the east coast lie mostly on rocks of the carboniferous formation, or on newer deposits of Mesozic age, while the great western plains and valleys are almost wholly Tertiary sandstone, or more recent deposits, with intervening areas covered by overflows of igneous trap rock. Granitic rocks of various kinds are abundant, and syenite forms the summit of Kosciusko in latitude $36\frac{1}{2}°$ (7,305 feet), the highest mountain in Australia. Gold occurs in granite, in quartz veins and in beds of iron pyrites. The Carboniferous rocks cover an immense area and are largely coal bearing, so that the coal fields are among the most extensive in the world. These deposits are of Paleozoic formation and correspond to the coal of England. No active volcano exists in the State.

Flora.—Among the indigenous trees, shrubs, and plants of New South Wales are the acacia, the eucalyptus or gum tree, cedar, casuarina or Sheaoak, honeysuckle, figtree, cabbage-tree palm, ferns of large size, salt bush, and mallee scrub; of fruits, the orange, shaddock, banana, loquot, and pineapple are grown; wild flowers of great variety and beauty adorn the landscape in the spring, of these the waratah or native tulip, the Christmas bush, and varities of rock lily are some of the most striking.

Fauna.—The native animals comprise the kangaroo, wallaby, paddymelon, bandicoot, wombat, opossum, koala or native bear, native cat, platypus, and native hedgehog. Bats are very numerous, from the flying fox of large sixe to the flying mouse. Snakes are very numerous, the death adder and yellow snake are the most dangerous, their bite being frequently fatal. The lizard tribes are well represented, and a peculiar long-necked tortoise is found in the swamps of some of the rivers. The feathered tribe is well represented, among them being the parrot (of which there are over 60 species), eagle, owl, great kingfisher or laughing jackass, magpie, quail, native pheasant, bush turkey, wild turkey, emu, lyre bird, bower bird, native companion, and several other species of crane and heron, and black swan. Leeches abound in most of the creeks and lagoons, and insects are prolific almost beyond belief. Fish swarm in the rivers that intersect, and seas that fringe the state, among them are the bream, mullet, whitting, snapper, jewfish, flathead, garfish, Murray cod, and perch; crustaceans abound and the Sydney oysters are much prized.

Numerous animals have been at times imported from Great Britain and elsewhere, and have been found to thrive in their new home; among these are deer, hares, rabbits, and several of the familiar songsters of the English hedgerows and fields, besides certain feathered pests like the starling and the sparrow.

Fisheries have been described under Trawling grounds.

Products.—The great staple productions of New South Wales are wool, gold, wheat, coal, and timber. The wool is of fine quality. Accessory products are tallow, skins, and preserved meats. The country is rich in minerals; gold, copper, platinum, silver, tin, lead, zinc, iron, coal, and antimony are found; also small quantities of diamonds, rubies, opals, and other precious stones. Agriculture is one of the principal industries; maize, potatoes, tobacco, and sugar cane are grown. Large quantities of wine are also produced.

Population.—The population on June 30, 1918, was estimated to be 1,897,000.

Harbors.—The principal harbors of New South Wales described in this volume are Twofold Bay, Bateman Bay, Jervis Bay, and Port Jackson.

Docks.—At Sydney are docks and patent slips suitable for all classes of vessels. (See Appendix I.)

Trade.—The oversea imports comprise nearly every article in ordinary use, the value in the year ending June, 1918, being £27,975,582. The principal oversea exports are wool, tallow, leather, timber, coal, preserved and frozen meats, butter, skins, fruits, wheat, flour, tin, silver, and gold, valued at £36,216,779.

Communication — Steamers. — Frequent communication with Europe and America, also with New Zealand, India, China, and Japan, is maintained by steamers.

The steamers of the Peninsula & Oriental Steam Navigation Co. and of the Orient Royal Mail Line have regular sailings between London and Sydney.

Those of the British India Co. run via Queensland and Torres Strait to India and Great Britain.

Those of the Austral-Canadian Line run to and from Vancouver Island, calling at Suva, in Fiji, and Honolulu.

Steamers of the Union Steamship Co., of New Zealand, run from Sydney to San Francisco, calling at Auckland and Honolulu. There are also steamers of the Messageries Maritimes from Marseilles. Local companies' steamers connect Sydney with Australian, New Zealand, and Tasmanian ports, Fiji, etc.

Railroads.—The main railroad systems are the Southern, Southwestern, Western, and Northern, all connected with each other, Sydney being their starting point. The Southern system embraces the main line, with branches, from Sydney through Goulburn to Wagga Wagga, at the head of the Murrumbidgee navigation, with a continuation to Albury on the Victoria frontier, which now completes the communication between Sydney and Melbourne. The Southwestern leaves the main line at Junee Junction and extends westward to Hay, with branches to other places. The Western line runs from Sydney across the Blue Mountains to Bourke on the Darling River, with several branches. A railroad runs into the well-known Illawarra district; it passes along and near the coast to the southward of Sydney, the present terminus being at Nowra. The Northern system runs, via Newcastle, to Tenterfield and on to Wallangarra in Queensland, passing through a fine pastoral, agricultural, and mining district. There is a branch line from Werris Creek to Moree. The gauge of the lines is 4 feet 8½ inches.

Hence although there is through railroad communication between Adelaide and Brisbane, a break of gauge occurs both on entering and leaving the State of New South Wales.

There were in 1918, 4,679 miles of Government and 142 miles of private railroad lines in the State.

Telegraphs.—The telegraph system extends over all the settled parts of Australia, and submarine cables join it with the United Kingdom, New Zealand, Tasmania, and New Caledonia.

Radio.—The radio station at Sydney is capable of communication with New Zealand and stations in the Pacific.

Coal.—Coal for steaming purposes in any quantity may be obtained at Sydney.

The amount of coal produced in the State in 1916 was 8,127,161 tons.

Port regulations.—Vessels on arrival in any port of New South Wales should obtain a copy of the port regulations.

The pilot service of New South Wales is under the control of the department of navigation.

Pilotage signals.—For the ports of New South Wales the day signal for a pilot is the pilot jack at the fore; and the night signal, burning a *blue* light. Vessels exempt from pilotage fly a white flag at the main. Naval vessels, cable-laying vessels, whalers, and pleasure yachts are exempt from pilotage and harbor rates, and vessels in distress are charged half rates.

Pilotage rates.—The rates are reasonable.

Every steamship whilst proceeding in any harbor in New South Wales, between sunset and sunrise, shall sound a prolonged blast from a steam-whistle at intervals of not more than two minutes.

Pilot Signals.—The following are the signals in use at all the bar harbors on the coast of New South Wales, excepting Newcastle:

No. 1. Pennant on right yardarm.
No. 2. Pennant on left yardarm.
No. 3. Two pennants on right yardarm.
No. 4. Two pennants on left yardarm.
No. 5. Pennant over ball on right yardarm.
No. 6. Pennant over ball on left yardarm.
No. 7. Ball over two pennants on right yard arm.
No. 8. Two balls over pennant on left yardarm.
No. 9. Ball on right yardarm.
No. 10. Ball on left yardarm.
No. 11. Two balls on right yardarm.
No. 12. Ball on each yardarm.

The significations are as follows:

No. 1. You may approach with safety.
No. 2. Stand in.
No. 3. Stand in; the pilot has left to board you.
No. 4. If the pilot can not board you, the boat will be inside the bar; steer for her.
No. 5. A boat or tug will be sent off immediately.
No. 6. A boat or tug will be sent off when practicable.
No. 7. The flood tide has commenced.
No. 8. The ebb tide has commenced.
No. 9. There is too much sea on the bar to send a boat or tug.
No. 10. Stand off.
No. 11. It blows too hard to send a boat or tug.
No. 12. The pilot can not board you; stand off and on till the morning.

NOTE.—To be a red flag in each case.

Any other communications will be made by the International Code.

Care must be taken to prevent anything liable to be mistaken for a white flag being displayed, as a white flag is the signal that a pilot is not required and that the vessel is exempt from the necessity of taking a pilot.

Storm signals.—The existence of gales which are likely to endanger shipping will be signaled at the principal telegraph stations on the coast of New South Wales, in the following manner, viz.:

The signal staffs will support two yards, which cross each other at right angles in the direction of the cardinal points of the compass, the yardarms denoting respectively north, south, east, and west; midway between north and east will denote northeast, etc.

A violent squall will be represented by a conspicuous diamond-shaped signal.

A heavy sea by a drum-shaped signal.

Gale, with clear weather, by a diamond-shaped signal over a drum.

Gale, with thick weather and rain, by a drum over a diamond-shaped signal.

The direction from which the gale is blowing will be indicated by the particular yardarm between which and the mast-head the signal is suspended.

Place where squall or gale is blowing will be shown by the numerical flag at the masthead.

Gales that are general over a large portion of the coast will be indicated by the geometrical figures without the masthead flags.

At Port Jackson the signals will be shown from South Head and from Fort Phillip signal stations. The latter is near the observatory.

Numerical flags.—The following flags or pendants are used at the signal stations of New South Wales to indicate the place from which a vessel arrives, and, in connection with storm signals, the place where a gale is blowing:

1. Red.
2. Yellow and blue, horizontal, 2 divisions.
3. Blue, yellow, red, vertical.
4. Red and white, 4 divisions.
5. White, with 5 blue crosses.
6. Blue and yellow, 6 horizontal stripes.
7. Blue, with 7 white crosses.
8. Blue and white, 8 triangles.
9. Red and white, 9 vertical stripes.
0. Blue, white ball in center.

Substitute, white.

Numeral pendant, yellow and red, vertical.

Ports represented by numerical flags:

10. Torres Strait.	50. Portland Bay.	76. Circular Head.
11. Cleveland Bay.	51. South Australia.	80. Keppel Bay.
37. Wilson Promontory.	52. King George Sound.	81. Port Denison.
40. Sydney.	53. Western Australia.	82. Wollongong.
41. Moreton Bay.	54. Launceston.	83. Wide Bay.
42. Clarence River.	55. Hobart.	84. Port Curtis.
43. Port Macquarie.	56. Gulf of Carpentaria.	88. Port Fairy or Warr-
44. Port Stephens.	61. Shoalhaven.	nambool.
45. Newcastle.	68. Richmond River.	97. Hawke's Bay.
46. Jervis Bay.	70. Macleay River.	98. Kiama.
47. Twofold Bay.	72. Gabo Island.	99. Wallaroo.
48. Corner Inlet.	75. Manning River.	101. Port Mackay.
49. Port Philip.		

Note.—Other numbers signify ports outside Australia from which a vessel arrives; they are not inserted, as they would not be used for storm signals.

Uniform system of buoyage.—In New South Wales a uniform system of buoyage is maintained in all the ports and harbors of the State, and is as follows: When entering, all red buoys must be left on the starboard hand and black buoys on the port hand.

Signal stations.—Nearly all the signal stations mentioned in this volume are connected by telegraph with Sydney. Vessels can communicate by the International Code.

Climate.—As New South Wales is divided naturally into three distinct geographical districts, so the climate is separated into three meteorological regions. Over the coast district, which attains a maximum elevation of about 600 feet, the extreme shade temperatures range between 16° and 117° F., but the mean annual temperature is 63°. The mean annual rainfall is 37 inches, the greatest, at Port Macquarie, on the coast, being 63 inches, and the least 24 inches on Jerry's plains, 53 miles inland.

On the tablelands the temperature range between 8° and 118°, the mean being 58°. The rainfall over this region is from 19 inches to 63 inches and the mean 32 inches.

On the dry plains to the westward the lowest reading of the thermometer is 17° and the highest 127°, with a mean of 64°. The rainfall is from 11 to 31 inches and the mean 19 inches.

WINDS AND WEATHER.

South coast of Australia.—Within 100 miles of the south coast of Australia the most settled weather prevails during January, February, March, and April, the wind being generally southeasterly, and partaking of the nature of land and sea breezes, being more easterly during the night and early morning and more southerly during the day and afternoon. With the above winds the barometer is usually very high, often above 30.5 inches.

The easterly wind, in this season, falls light after sunrise, freshening in the forenoon from south-southeast to a force of 5 to 6, and often bringing up a haze if the morning has been hot; the sea breeze attains its greatest force during the afternoon, becoming lighter nearer sunset as its direction changes toward the land.

Should the barometer fall, the wind, instead of shifting to the southward in the morning, may turn to the northward and blow from that direction a very hot, dry wind for one to three days. When this northerly wind falls light a moderate gale from west to southwest usually springs up, seldom lasting more than 24 hours, after which a period of fine weather again ensues with southeasterly winds as above.

At the end of April the southeasterly winds almost entirely cease, though sometimes they blow at intervals during the whole of May;

at this time there are occasionally fresh northeast winds not followed by any change to the westward.

From the middle of May until the end of October westerly winds prevail, the gales from that direction quickly raising a heavy sea, and blowing with as much strength near the land as further seaward.

Gales.—The signs of the approach of a westerly gale on the south coast of Australia are so well marked that no vessel need encounter one unprepared. From May to October, if the barometer falls rapidly when below 30 inches with a fresh and gusty northerly wind, whilst heavy clouds with lightning gather to the northwestward, a westerly gale is certainly approaching. The northerly wind usually falls light as the bank of clouds to the northwestward rapidly rises, and the wind then shifts to northwest in a heavy squall, with rain and lightning. In the lull before this squall St. Elmo's fire is often seen on the ironwork of the masts and yards.

When the barometer rises the wind soon shifts to the westward and southwest, the weather clearing up when the wind becomes well southward of southwest. Frequently the barometer remains nearly stationary, or falls after the gale begins, and the wind continues to blow hard from northwest to southwest for a week or 10 days, though the average of these winter gales is from 3 to 4 days. The rising of the barometer to above 30 inches, and the entire clearing of the clouds from the western horizon are sure signs of the gale having passed. A short interval of fine weather then ensues, and the wind turns again to the northward on the approach of the next gale.

The three months of October, November, and December have sometimes settled weather, with a preponderance of southeasterly winds, but the westerly gales of October are frequently as severe as those experienced at any time during the winter, and an occasional gale from that quarter is likely to be experienced in November, sometimes in December and sometimes, but more rarely, in the early months of the year.

The force of the southeasterly wind in summer has been known to increase to a strong gale lasting about 48 hours and raising a very irregular cross sea, as the constant southwesterly swell does not subside with the southeasterly wind. These gales are accompanied by a red haze, the barometer being steady but below the average height for the time of the year. The wind continues from the southeastward while the gale lasts, not changing its direction seaward and landward as in fine weather. Several years sometimes elapse without the recurrence of one of these gales.

Fogs.—Fogs are extremely rare on the south coast of Australia; the haze which comes up with the sea breeze in the summer is occasionally sufficiently thick to render objects indistinct at a distance of 3 or 4 miles.

Bass Strait.—In Bass Strait the winds are similar to those which are met with along the whole of south coast of Australia, except toward its eastern part, where they partake of the nature of those on the east coast. The strongest gales blow frequently from between south and southeast, accompanied by thick weather and often by heavy rain. In Bass Strait northerly winds are common both in summer and winter, and preponderate over all others in frequency and force, more especially during the winter months; these winds being off the land are not so much felt or dreaded.

In fine weather a light northerly wind is frequently found near the shore though light southerly outside.

Next in force come southwesterly and southerly winds. The north wind of the Victorian coast is generally a northwest wind in the vicinity of Cape Howe. In January, February, and March easterly winds with fine weather seem to be not uncommon, but no dependence is to be placed on them at any other season. At the eastern side of the strait and of Tasmania it is not unusual to meet a northeast wind, though it seldom blows strong.

As the western part of the Ninety-mile beach is approached, easterly gales are not so generally felt; Wilson promontory appears to mark the dividing line.

January and February are the best months for making a passage to the westward through Bass Strait, although easterly winds blow on some rare occasions at other times, but these are mostly gales and generally terminate in a breeze from the opposite quarter, having much the character of a rotatory gale. The gales that prevail in the strait begin at north-northwest and gradually draw around by west to southwest, at which point they subside; if, however, the wind, before it has so much southing, shifts again to the northward of west, the gale will continue. It is seldom fine when the barometer is lower than 29.95 inches, and bad weather is certain if it falls to 29.70.

Thick weather, accompanying a breeze from the southeastward, especially from May to September, is generally the precursor of a gale and should be regarded accordingly.

West coast of Tasmania.—The prevailing winds are from southwest and bring much bad weather, especially in the months of June, July, and August. Northwest and westerly gales are frequent.

Coast of New South Wales.—The prevailing winds between Cape Howe and Port Jackson are from the northeast from October to April and from the westward from May to September. There are occasional gales from southwest, as well as strong breezes from between north and east-northeast, bringing rain, with thunder and lightning; these, however, are usually of short duration. Very oppressive hot winds from northwest sometimes blow fiercely from

November to February, and are usually followed by a sudden shift
from between southeast and south-southwest, and against which vessels near the coast should be particularly guarded, as the first gust
is generally very violent and apt to occasion damage unless due
precautions have been taken.

Besides the sudden change from northwest to the southward, a
similar change from northeast to south is very frequent from September to February, and generally happens after some days of northeast wind. These changes, as a rule, may be foreseen by clouds rising
in the southward, with lightning; sometimes, however, very little
warning is given, as the shift of wind may happen with a cloudless
sky; they are the well-known southerly bursters. The effect of these
sudden changes is so great that the thermometer at Port Jackson
sometimes descends from 100° to 64° in less than half an hour.
These storms may last only a few hours or for several days. They
average 32 in number during the season.

The barometer rises slowly for several hours before the arrival of
the storm. The proximity of a relatively low pressure is considered
an indication of an approaching southerly burster, as is also a foggy
morning following a hot day.

Southerly winds are more frequent from April to October than
from October to April, yet they occasionally blow for three or four
successive days in the latter period; the southerly wind usually draws
off the land at night, from a southwest or even a west-southwest
direction, especially from April to October, and with more westing
in it the nearer the land.

From May to September cold westerly winds are prevalent and are
generally accompanied with fine weather and a dry atmosphere;
gales from between northeast and south bring rain with them; indeed, there is no settled weather during these months with any winds
from the sea, and even with northwest and north winds, which are
usually light, there is frequent rain.

Land and sea breezes are frequent from November to February;
the northeast wind springs up from a calm early in the forenoon and
subsides at about midnight, a slight draft off the land being occasionally felt close inshore between these intervals.

A heavy dew in the night is an indication of a northeast or sea
breeze the following day.

The northeast breeze sometimes blows a steady gale for three or
four days, shifting from north to northeast in squals. When likely
to be of this duration it sets in with thick overcast weather, and increasing in strength is accompanied with gloomy dense clouds and
heavy rain and an atmosphere so thick that during the squalls objects
are not distinguishable at a distance of a quarter of a mile. These

gales are locally known as black northeasters; the barometer gives no indication of them and is not affected during their continuance.

If at any time during the months of June, July, and August the weather is unsettled, with the wind unsteady, and with gloomy weather and occasional rain, an easterly gale may be expected which will last for two or three days, shifting from northeast to east-southeast, accompanied with heavy leaden clouds and sheets of incessant blinding rain. The barometer is not in any way affected by the approach or continuance of these gales and stands steady at 30.12 to 30.18 inches.

A heavy sea rolls in on the cost, which is a dead lee shore, and there is little chance for small craft, caught close, being able to gain an offing. There is nothing, however, to prevent a well-appointed ship having an offing, from holding it by watching the shifts of wind and keeping as long as prudent on the starboard tack, thus bringing the prevailing current setting to the southward under her lee.

The barometer.—If the weather be tolerably fine and the mercury does not stand above 30 inches, there is no probability of danger; but when the mercury is much higher and begins to fall, with the weather becoming thick, a gale from southeast to east is to be apprehended, and a proper offing should be immediately obtained. With respect to the rise and fall of the barometer, it may be taken as a general rule upon this coast that a rise denotes either a fresher wind in the quarter where it then may be, or that it will shift more to seaward; and a fall denotes less wind, or a breeze more off the land. Too much faith, however, should not be put in the barometer, unless the observer can combine local experience with the use of the instrument.

On this coast the barometer is at its greatest mean height in August and September with southeast and southerly winds, and at its lowest mean height in December, January, and February with northwest winds. It ranges between 30.92 and 29.26 inches. From April to October a marked fall in the barometer is certain to be followed by westerly winds and fine weather, whatever may be the quarter or the conditions under which the wind may be blowing when it commences to fall.

From October to April it may be similarly and as surely depended on as the forerunner of a northwest hot wind.

From November to February the barometer generally falls on the approach of a southerly gale whilst the northeast wind is blowing, but this fall must not be implicitly relied on, as southerly gales have occurred without its showing any perceptible change. After the strength of a southwest or southerly gale is over the barometer rises to about 30 inches, when fine weather and a gradual change of wind to northeast may be expected.

Fogs rarely occur except in the summer months, and then soldom last longer than from day dawn to 10 a. m.

In the vicinity of Port Jackson.—From the early part of October to April, in the vicinity of Port Jackson tolerably regular land and sea breezes prevail. The sea breeze generally begins at 10 a. m. and subsides after sunset; the land breeze commences at about midnight and continues to 8 a. m. North and south winds and also the northwest hot winds occasionally interfere with the regularity of the land and sea breezes. The northwest hot winds, after blowing for a period of from a half to three days, are usually succeeded by sudden violent gusts from south-southeast to south-southwest, which generally settle into a gale from those quarters accompanied with rain. The greatest vigilance of the masters of vessels possessing local experience is frequently insufficient to prepare for these gusts, owing to the suddenness with which they come; mariners, therefore, should be careful to be ready for the change during the time the hot wind is blowing; the calm which sometimes intervenes is brief.

From May to September the wind prevails strong from the westward, between northwest and southwest, with fine clear weather and occasional gales from the north and south, with rain.

The wind rarely blows on-shore with sufficient violence to endanger the safety of a well-appointed vessel, but sometimes in September and October gales set in from southeast to east, accompanied with heavy rain and a high barometer; they blow with considerable fury from one to two days and finish with a long, slowly declining gale from south to southwest.

Barometer.—As a general rule the barometer stands low with westing in the wind, lowest with a northwest wind, high with easting in the wind, and highest with southeast gales.

Fogs rarely occur except in the summer months, and then seldom last longer than from dawn to 10 a. m. When the sea breeze blows it is accompanied with a thin haze, which envelopes the land and renders it indistinct; this haze disperses with the land breeze.

CURRENTS.

From Cape Northumberland to Bass Strait the currents appear to depend chiefly on the winds and can not be allowed for with any certainty. For the greater part of the year they run, more or less, in an easterly direction; in January, February, and March, when the easterly winds prevail, a westerly set may be expected.

These currents are stronger as the coast is approached and strongest off the headlands such as Cape Nelson and Moonlight Head, and more particularly off Cape Wickam (King Island), though they are scarcely felt at a distance from it of 6 miles.

Vessels making the land about Cape Otway during the continuance of strong westerly winds should be prepared for a southerly set, though sometimes the current is found to set toward the land.

At the western entrance of Bass Strait, from April to December, a current sets to the southeast at rates varying from ½ to 2½ knots an hour, according to the strength, direction, and duration of the wind. It sets strongly on to King Island at times. Many wrecks have occurred on this island, apparently from errors in reckoning.

In the strait between King Island and Tasmania the streams are tidal. On the west coast of King Island the current often sets to the northwest. Southwestward of King Island the currents and tidal currents are irregular; they are sometimes very strong.

Through Bass Strait an easterly current of ½ to 1½ knots per hour will be found during the greater part of the year, but with easterly winds, which prevail from January to the beginning of April, the current is reversed.

Coasts of Tasmania.—In the bight of the north coast, between Circular head and Cape Portland, there is almost a constant current running to the eastward during the greater part of the year.

On the east coast the set is generally to the southward.

On the west coast the current generally sets to the northward, particularly during the prevalence of southwest and southerly winds.

Coast of New South Wales.—The current almost constantly sets to the southward along this coast, in a broad serpentine belt, extending 20 to 60 miles from the land, at a rate varying from half a knot to 3 knots, the greatest strength being at about the 100-fathom curve, near which it has been observed on the parallel of about 31°, running at a rate of about 4½ knots an hour. To the eastward of this southgoing belt the currents seem to be variable; and close to the land, especially in the bights, there is commonly an eddy, setting to the northward, from a quarter of a knot to one knot. Along the southern part of this coast the current runs strongest; and toward Cape Howe, the southeast corner of Australia; it may run in either direction at a rate of 1 to 1½ knots an hour.

Sailing vessels bound northward avoid the current by keeping within about 2 miles off the land, though the wind may be puffy. See H. O. Pilot charts of the Indian Ocean and South Pacific Ocean.

OCEAN PASSAGES.

Ocean passages.—In all ocean passages mariners should consult H. O. Pilot charts, H. O. Chart No. 1262, and H. O. Chart, No. 5300, and lay courses accordingly, remembering that in the high latitudes of the shortest routes bad weather is almost certain to be encountered.

Full-powered steamers.—From the Cape of Good Hope steer to cross the meridians of 30° E., in lat. 39½° S., of 40° E. in lat 42° S., and of 60° E. in lat. 45° S. Proceed to the eastward in about lat. 45° S. until in long. 90° E., then steer on a great circle direct for Cape Otway in Bass Strait.

Low-powered steamers.—From the Cape of Good Hope steer to the southward to cross the meridian of 20° E. in about lat. 39° S. and proceed to the eastward about the parallels 39° S. and 40° S. St. Paul and Amsterdam Islands may be seen from a distance of 60 miles in clear weather. Vessels may sometimes make quicker passages by going further south, but better weather will, as a rule, be found on the parallels mentioned.

Sailing vessels.—Sailing vessels should proceed to the southward until the westerly winds are reached and then as directed for low-powered steamers. Those not touching at the cape should enter the Indian Ocean in about latitude 39° S. to 40° S.

For Cape Otway to Port Phillip and Melbourne, see Chapter III.

For Bass Strait to Sydney, see later.

Approaching Bass Strait, see later.

To Hobart—Full-powered steamers.—Proceed as above, but continue in about latitude 45° S. until longitude 130° E. is reached, then steer direct for 10 miles south of southwest Cape of Tasmania.

The Shaw, Savill, and Albion Co.'s steamers from the Cape of Good Hope cross longitude 40° E. in latitude 44° S., 60° E. in 45° 25′ S., 80° E. in 45° 55′ S., 100° E. in 47° S., and 120° E. in 46° 20′ S., and thence proceed south of Tasmania to Hobart.

Low-powered steamers.—Proceed as directed for low-powered steamers to Melbourne until in longitude 123° E., then steer for 10 miles south of southwest Cape of Tasmania.

All steamers make the Maatsuyker Islets or light, pass north of the Mewstone, 3 miles southward of South Cape and to Hobart by the d'Entrecasteaux Channel. See Chapter VII.

Sailing vessels follow the route of low-powered steamers, but when nearing Tasmania should be far enough to the south to avoid falling in with its rocky western coast in the night through any error in the reckoning, or being caught on a lee shore by a southwest gale. In fine weather, from 10 miles southward of the southwest cape of Tasmania, pass between Maatsuyker Islets and the Mewstone, then steer to pass 3 miles southward of South Cape. When blowing heavily from the southwest or southward, especially if unable to obtain observations before making the land, it is desirable to keep more to the southward, passing south of the Mewstone and on either side of Piedra Blanca and the Eddystone, taking care to avoid Sidmouth Rock. Proceed to Hobart through Storm Bay. See Chapter VII.

Great circle and composite tracks—Caution.—The routes recommended above are not the shortest in point of distance. The distance from the Atlantic or Cape of Good Hope to Australian ports is diminished as higher parallels of latitudes are adopted until the great circle is reached. Thus, for example, the distance from the Cape of Good Hope to Melbourne by the great circle, which reaches the latitude 58° 19′ S., is 5,592 miles; by composite tracks, with the maximum latitude 50° S., it is 5,666 miles, with the maximum latitude 45° S., it is 5,790 miles; with the maximum latitude 40° S., it is 5,988 miles; and by Mercator's tract it is 6,156 miles. So the advantage of the composite track with the maximum latitude of 50° S. over that with a maximum latitude of 40° S. is a distance of 322 miles. But the disadvantages attending the selection of any route in high latitudes should be clearly understood by the seaman, especially for passenger ships proceeding at a high speed or small, ill-found, or deeply laden vessels. The steadiness and comparatively moderate strength of the winds, with the smoother seas and more genial climate north of 40° S., compensate by comfort and security for the time presumed to be saved by taking a shorter route, with tempestuous gales, sudden violent and fitful shifts of wind, accompanied by hail or snow and the terrific and irregular seas, which have been so often encountered in the higher latitudes; moreover, the islands in the higher latitudes are so frequently shrouded in fog that often the first sign of their vicinity is the surf beating against them.

Icebergs.—Independently of the severity of the climate occasionally experienced in high latitudes, there exists the lurking danger of disrupted masses of ice and icebergs of large dimensions. The absence or approximate positions of these dangers can not be depended on for any season of the year; they are, however, rarely encountered north of latitude 40° S. Nevertheless there are instances of icebergs being seen off the cape, and north of latitude 40° S. as far as longitude 60° E., and it is therefore desirable to keep a good lookout for them.

Between 40° and 45° S. they have been occasionally met with as far as 65° E., on the forty-fifth parallel to 135° E., and on the fiftieth parallel to 140° E.

Icebergs are seldom sighted between the meridians and 130° E. and 170° W., along the usual tracks adopted by shipping, and more especially from April to October.

Southeastward of the Cape of Good Hope, midway between Kerguelen Island and the meridian of Cape Leeuwin, midway between New Zealand and Cape Horn, and northeastward of Cape

Horn, icebergs are most numerous. The periods of maximum and minimum frequency vary greatly. There were 12 reports in 1905, 305 in 1906, and 36 in 1907. It may happen that while ships are passing ice in lower latitudes, others, in higher latitudes, find the ocean free of ice.

The lengths of many of the southern ocean icebergs are remarkable; bergs of 5 to 20 miles in length are frequently sighted south of the fortieth parallel, and bergs of from 20 to 50 miles in length are far from uncommon.

It may be gathered from numerous observations that bergs may, in places, be fallen in with anywhere south of the thirtieth parallel, that as many as 4,500 bergs have been observed in a run of 2,000 miles, that estimated heights of 800 to 1,700 feet are not uncommon, and that bergs of from 6 to 82 miles in length are numerous.

Approaching Bass Strait.—Steam vessels approaching Bass Strait from the westward make the land at Moonlight Head or the light at Cape Otway. The high bold promontory of Cape Otway is easily distinguished by the white lighthouse on it, and the signal station to which all passing vessels are recommended to show their numbers and communicate what public intelligence they may have.

It is desirable to round Cape Otway at a distance of not less than 3 or 4 miles.

In approaching Bass Strait the winds and currents must be carefully attended to, particularly during the prevalence of southwest or southerly gales. The 100-fathom curve is 35 miles to the southwestward of Moonlight Head, and the depth of 40 fathoms at 10 miles from that headland. When approaching Bass Strait in thick weather, or when uncertain of the vessel's position, do no reduce the soundings to less than 40 fathoms. Soundings of 60 or 70 fathoms will be found at 25 or 30 miles westward of King Island. Outside this limit the soundings deepen rapidly to over 100 fathoms. Inshore of 60 fathoms the depths are irregular, but there are 30 fathoms at a distance of 4 miles to the northwest of Cape Wickham.

Sailing vessels.—Follow the same directions.

Caution.—In approaching King Island from the westward, especially during thick or hazy weather, caution is required on account of the variable strength of the current, which sets to the southeast with a force varying from a half to $2\frac{1}{2}$ knots an hour, according to the strength and duration of the westerly winds, and the use of the lead is enjoined. Many vessels have been wrecked on this island in consequence of not making the land near Cape Otway, and from errors in reckoning.

Commanders of iron ships, especially of those newly built, are cautioned as to the necessity of ascertaining the deviation of their compasses on approaching the Australia coast.

To Port Phillip—Steam vessels.—See Chapter III.

Should Cape Otway be rounded early in the evening, with a fresh southerly wind, beware of overrunning the distance, as a strong current after a prevalence of southerly gales, often sets northeastward along the land. Bearings of Eagle Nest Light give a good check.

Sailing vessels.—Follow the directions for steamers. When abreast of Eagle Nest Point Lighthouse, if there is not sufficient daylight to get into pilot waters, a sailing vessel should stand off and on shore till daylight, not shoaling the water to less than 20 fathoms. Do not heave to. *See* Chapter III.

Bass Strait to Sydney—Steam vessels.—From a position off Cape Otway steer to pass about 2 miles south of the Anser Islands, 3 miles north of Rodondo, and 2 miles south of Southeast Point, Wilson Promotory. Then steer to pass about 5 miles southeastward of Rame Head and Gabo Island. Occasionally and especially during and after easterly gales the current sets strongly toward the land; in thick weather the lead must not be neglected. From a position east of Cape Howe steer to the northward along the land to Port Jackson, in fine weather at a distance of about 2 miles, passing inside of Montagu Island, to avoid the southerly current.

Sailing vessels.—When the position off Cape Otway is ascertained, shape an easterly course as desired or for Rodondo Island, which is visible in clear weather from a distance of 30 miles. Having passed Rodondo Island, or the Kent group, steer for a position about 20 miles to the southeast of Rame Head and make Gabo Island Light or the land in the vicinity of Cape Howe; but should it blow hard from the southward, a more easterly course should be steered to avoid Ninety-mile Beach, extending from Corner Inlet for 150 miles, or nearly to Cape Howe, which would then be a dangerous lee shore. From a position east of Cape Howe, steer to the northward along the east coast for Port Jackson at such distance from the land as the wind and weather would suggest, bearing in mind that the current generally sets to the southward at a distance of 20 to 60 miles from the land. Having made the Outer South Head Light enter the port as directed later.

The soundings off this coast to the 100-fathom curve have been carefully obtained, and being accompanied with the nature of the ground at various depths, a mariner making the coast in thick weather, and uncertain of his position, can by sounding estimate the distance from the land.

Southwestern entrance to Bass Strait.—The entrance to Bass Strait between King Island and the Hunter group is not recommended, on account of Bell Reef and Reid Rocks which lie in it. If from necessity or choice entering Bass Strait by this passage, keep to the southward of Reid Rocks and Bell Reef, the latter being passed at the distance of 2½ miles to the southward of it by steering for Black Pyramid on a, 98° bearing. With steam or a commanding breeze the passage between King Island and Reid Rocks may be taken without danger by paying attention to the tidal currents, which set somewhat across the channel at times.

From Black Pyramid pass about 1 mile north of Albatross Islet, thence to Port Dalyrymple, round the sunken danger Mermaid Rock, off Three Hummock Island, and then make a direct course.

From south of Tasmania to Sydney—Steam vessels.—From off South Cape steer to pass about 5 miles off the Friar Rocks and 1 mile off Tasman Island, then make a direct course to Cape Howe, but not closing the coast of Tasmania within 5 miles. From Cape Howe proceed to the northward as previously directed.

Sailing vessels.—After rounding South Cape, give a berth of 20 or 30 miles to Cape Pillar and the east coast of Tasmania, to escape the baffling winds and calms which frequently perplex vessels in-shore, while a steady breeze is blowing in the offing. This is more desirable from December to March, when easterly winds prevail, and a current is said to be experienced on the southeast coast at 20 to 60 miles offshore, running northward at the rate of three-quarters of a knot, while in-shore it is running in the opposite direction, with nearly double that rate. From a position about 30 miles eastward of Cape Pillar, 350 miles on a 12° course will take a vessel to a position 15 miles eastward of Cape Howe, whence proceed as previously directed.

Sydney through Bass Strait to Cape Otway—Steam vessels.—It is usual to keep about 15 miles off the east coast of Australia, along the 100-fathom curve, in order to take advantage of the southerly current closing the land in the vicinity of Cape Howe or Gabo Island. From off Gabo Island steer for Southeast Point Lighthouse, Wilson Promontory; pass about 2 miles southward of the lighthouse and the Anser Islands, thence shape a course for Cape Otway, which should not be approached nearer than 3 miles.

Sailing vessels.—Proceeding from Sydney to Bass Strait, in order to take advantage of the current as far as Cape Howe, which appears to run strongest from November to March, keep along the outer edge of the 100-fathom curve, or a distance of 15 to 18 miles from the coast, where the current runs stronger and with more regularity than elsewhere.

From about 15 miles eastward of Cape Howe, if the wind is southerly do not steer a more westerly course than 212° until in latitude 39° 30′ on account of the danger to be apprehended from southeasterly or southerly gales upon the Ninety-mile Beach, between Cape Howe and Corner Inlet. On reaching the parallel of 39° 30′, steer to pass about 3 miles northward of Wright Rock, and the same distance southward of the south point of Deal Island, the south-easternmost of the Kent Group. Having passed the Kent Group, steer to pass 2 or 3 miles south of the Sugarloaf Rock, leaving the Judgment Rocks on the starboard hand.

Local experience has shown that with westerly and southwesterly winds smoother water is found near the coast between Shallow Inlet and Conran Point, known as the Ninety-mile Beach; and as south-westerly winds are the prevailing ones, mariners bound westward may often take advantage of the smoother water and an absence of danger to approach the beach instead of avoiding it.

A vessel inshore when an easterly gale threatened should at once get an offing; these gales give signs of warning.

As westerly gales veer to the southward from December to March, it is advisable to stand toward the Tasmanian coast, and so be ready to take advantage of the shift of wind.

In the other months, and more particularly in September, October, and November, the same course can not be recommended; then the wind does not shift for a continuance, but is constantly shifting to the west-northwestward.

From the Sugarloaf steer 15 or 20 miles to the northward of King Island, if the winds permit; but should the wind hang to the west-ward of north, a course may be safely directed for the northern ex-tremity of Three Hummock Island, taking care to avoid the Mermaid and the Taniwha Rocks, passing afterwards north or south of King Island, as may be most favorable; the former is preferable.

Bass Strait to St. Vincent or Spencer Gulfs—Full-powered steam vessels.—In fine weather from off Cape Otway steer to pass about 5 miles southward of Cape Nelson, 10 miles southwestward of Capes Northumberland and Banks, thence make a direct course to Cape Willoughby. Care must at all times be taken to guard against a set toward the land, but with southerly and westerly winds the coast should be given a much greater berth, as a current of a knot an hour sometimes sets toward it between Cape Otway and Cape Willoughby. From Cape Northumberland to Spencer Gulf give a good berth to the southwest Young Rock, which is only 5 feet high; and, except with strong southeasterly winds, make allowance for the easterly set which usually prevails. From December to March, with southeasterly winds, a current runs about 1 knot an hour to the northwest. See H. O. Publication No. 167.

Low-powered steam vessels.—Follow the same track as full-powered vessels, but in the event of threatening weather from the south and westward, care must be taken to secure a good offing.

Sailing vessels.—Follow as nearly as possible the same track, giving the coast a wider berth with westerly winds.

To the westward, south of Australia.—**Full-powered steam vessels** make a direct course from Cape Otway or Investigator Strait to the westward at all seasons and keep at a distance of about 10 miles off the land from King George Sound to Cape Leeuwin, giving a sufficient berth to the White-topped Rocks.

Low-powered steam vessels.—Follow the directions for sailing vessels, using favorable opportunities to get to the westward.

Sailing vessels bound westward from Sydney may, from December to March, proceed through Bass Strait, or round Tasmania, easterly winds prevailing in the strait and along the south coast of Australia at that season, when vessels have made good passages, by keeping to the northward of latitude 40° south, and have passed round Cape Leeuwin into the southeast trade wind, which then extends well to the southward. A vessel from Bass Strait bound round Cape Leeuwin is recommended, with a favorable wind, to shape a course which will lead about 150 miles south of that cape. In adopting this route advantage must be taken of every favorable change of wind, in order to make westing; and it is advisable not to approach too near the land, as it would become with southwest gales, which are often experienced, even from December to March, a most dangerous lee shore, and the contrary currents run strongest near the land. The prevalence of strong westerly gales renders the southern route very difficult, indeed, generally impracticable, for sailing vessels, from April to November. The northern route, through Torres Strait, is then preferred, directions for which are given in H. O. Publication 169.

The worst months for making a passage to the westward are September, October, and November, for westerly gales are then of frequent occurrence, the wind sometimes being from west-southwest to west-northwest for more than a week at a time, and blowing very strong. From December to August northerly winds are very common.

AUSTRALIA AND TASMANIA TO AND FROM NEW ZEALAND.

Full-powered steamers.—To New Zealand, as direct as possible; proceeding round North Cape to Auckland, through Cook Strait for Wellington, round the southern end of South Island, and through Foveaux Strait for Otago or Port Lyttelton.

Low-powered steamers and sailing vessels follow the same routes as above.

From New Zealand to Australia and Tasmania reverse the above routes, observing that between Australia and New Zealand a southern route is usually more favorable than a northern one when making easting, and vice versa.

From Port Phillip.—**Full-powered steam vessels** bound round Cape Horn, on leaving Port Phillip, proceed through Bass Strait and then steer to a position in latitude 49° south, longitude 165° east, between the Snares and Auckland Isles, south of New Zealand.

Low-powered steam and sailing vessels with a westerly wind follow the route of full-powered steam vessels, but if on leaving Port Phillip the wind should blow from east or northeast, it may be desirable to run to the southwestward, pass between Cape Otway and King Island, and then proceed along the west coast of Tasmania, being prepared for the prevailing westerly or southwesterly winds, when this coast becomes a dangerous lee shore. Having rounded the outlying dangers off the south coast of Tasmania, proceed for the position before mentioned, between the Snares and Auckland Isles.

From Sydney.—**Full-powered steam vessels** proceed to the the position in latitude 49° S., longitude 165° E., south of New Zealand.

Low-powered steam and sailing vessels.—At all seasons, and from whatever quarter the wind may blow, it is advisable on leaving Port Jackson to proceed to the southward rather than to the northward of New Zealand. Advantage therefore should be taken of the most favorable winds for either reaching the before-mentioned position, between the Snares and Auckland Isles, or, if baffled by southerly winds and favored by fine weather, the passage through Cook Strait may be taken with advantage, especially from October to February.

Eastward to Cape Horn—Full-powered steam vessels.— From the position south of the Snares proceed eastward (passing between the Antipodes and Bounty Islands) in about latitude 49° S. until in longitude 150° W., and thence make a direct course to Cape Horn.

The routes adopted in the South Pacific Ocean by the steamships of the Shaw, Savill & Albion Co., the New Zealand Shipping Co., and the White Star Line cross the respective meridians in the undermentioned latitudes:

Longitude 165° W. in latitude 48° 30′ S.
Longitude 160° W. in latitude 49° 00′ S.
Longitude 150° W. in latitude 49° 30′ S.
Longitude 140° W. in latitude 50° 00′ S.
Longitude 130° W. in latitude 50° 50′ S.
Longitude 120° W. in latitude 51° 30′ S.

Longitude 110° W. in latitude 52° 10′ S.
Longitude 100° W. in latitude 52° 45′ S.
Longitude 90° W. in latitude 54° 00′ S.
Longitude 80° W. in latitude 55° 00′ S.

Low-powered steam and sailing vessels, from the position south of the Snares, proceed eastward between the Antipodes and Bounty Islands, keeping the parallel of 49° S. to about the meridian of 115° W., and then gradually incline to the southward, to round Diego Ramirez and Cape Horn. Or having passed through Cook Strait, steer to the southeast between the Chatham and Bounty Islands until in the parallel of 49° S.

Caution.—The course frequently pursued between the fiftieth and sixtieth parallels, and even in higher latitudes in this great extent of ocean would, with a clear sea and favorable weather, doubtless insure the quickest passage, as being the shorter distance, but experience has proved that at nearly all seasons of the year so much time is lost at night and in thick weather, and even serious danger incurred in avoiding the great quantities of ice met with in these higher latitudes, that a parallel even as far north as 47° has been adopted with advantage. Between this latter parallel and that of 50°, it is believed the mariner will experience steadier winds, smoother water, absence of ice, and will probably make as short a passage, and certainly one in a more genial climate and with more security than in a higher latitude.

In navigating this wide expanse of ocean, and also for rounding Cape Horn, H. O. Pilot Charts should be studied; and reference should be made to previous descriptions in this chapter of the gales, heavy seas, and icebergs which are apt to be encountered in the higher latitudes.

Cape Horn—Winds and weather.—In the neighborhood of Cape Horn, March and September are, generally speaking, the worst months in the year; heavy gales then prevail. March is usually the most boisterous month. In April, May and June the finest weather is experienced. Bad weather often occurs during these months, but not so much as at other times. Easterly winds are frequent, with fine, clear, settled weather. June and July are much alike, but easterly gales blow more during July. In August, September, and October westerly winds and cold weather prevail. December, January, and February are the warmest months, but westerly winds, which often increase to very strong gales with much rain, are frequent.

The barometer is lowest with northwest winds, and highest with southeast; if it falls to 29 inches, or 28.9, a southwest gale may be expected, but the gale does not commence until the barometer has ceased to fall.

CHAPTER II.

Cape Northumberland (lat. 38° 04', long. 140° 40'), rugged and cliffy, is about 100 feet high; a hill behind it rises to 136 feet. It may be easily distinguished by the lighthouse; Mounts Gambier and Schanck are also excellent marks by which to recognize it. Several detached rocks lie close to it.

Light.—A flashing white light, 150 feet high, visible 18 miles, is shown from a white lighthouse, 42 feet high, painted with three bands, white, red, and white, on the southern extremity of Cape Northumberland.

The light-keepers are provided with a gun, to warn vessels, if observed standing into danger.

Lloyd's signal station.—There is a signal station at the lighthouse, and communication can be made by the International code of signals, and at night by Morse lamp. This station is connected by telegraph.

All vessels passing Cape Northumberland Lighthouse during the day, and wishing to be reported, will, on showing their numbers, be telegraphed to Adelaide and Port Adelaide free of expense.

In consequence of the difficulty in making out the answering pendant, a round ball with the answering pendant underneath is used at the signal station, instead of the answering pendant only.

The station is connected with Port Macdonnell by a telephone.

The storm signal is a blue swallow-tailed flag under a red ball.

Meteorological observations.—The mean annual height of the barometer at Cape Northumberland is 30.01 inches; the maximum 30.70 inches; and the minimum 29.04.

The mean annual temperature is 56°.4 F., the maximum being 106° in January, and the minimum 32° in June.

The mean annual rainfall for 35 years was 26.47 inches.

Caution.—In bad weather, with the wind from the southward, the lead should be carefully attended to. Several vessels have been wrecked between Cape Northumberland and Cape Buffon from neglecting this precaution.

The coast northwest of Cape Northumberland soon becomes low, and owing to the heavy ocean swell which sets directly on it should be very carefully avoided.

Kelp.—Westward of Cape Northumberland, and from 1 to 4 miles offshore, there are forests of kelp, the tops of the plant trailing a long distance on the surface of the water; it does not appear to grow where the depth is greater than 15 fathoms. Steam vessels have been obliged to stop to clear their screws of the accumulated weed.

Fish.—Barracouta are very plentiful in the waters between Jaffa and Cape Northumberland; they are easily caught when in more than 20 fathoms, and with the vessel going from 4 to 6 knots.

Breaksea Reef, the south end of which is 129°, 2 miles from Cape Northumberland, is a dangerous rocky reef, extending 1½ miles eastward from the cape and the same distance offshore. There are less than 2 fathoms on most of it, and the sea generally breaks all over it with great violence.

Clearing marks.—Coming from the westward the sand hill on the beach between Middle and Douglas Points well open of Cape Northumberland 313°, until the customhouse at Port Macdonnell bears westward of 6° leads southwestward of the reef; and from the eastward the customhouse should be kept westward of 6° until the sand hill is well open of Cape Northumberland.

There are 11 fathoms water ½ mile southwest from Cape Northumberland, and 5 to 6 fathoms close to the southwest edge of the Breaksea Reef.

Macdonnell Bay (lat. 38° 04′, long. 140° 41′) is a very slight indentation of the coast, extending about 4 miles eastward from Cape Northumberland, and affords shelter from northwesterly and northerly winds, within Breaksea Reef.

Port Macdonnell being near to the most fertile portion of the state, and connected with it by good roads, is one of the principal trading places of the southeast districts of South Australia, and is situated on the coast, 2 miles to the eastward of Cape Northumberland. There is a considerable export of wheat, flour, wool, potatoes, ground bark, and dairy produce. The population is about 400.

Communication.—There is telegraphic communication, and there are six mails a week from Adelaide. It is 14 miles from the railway at Mount Gumbier.

A steam vessel trading between Adelaide and Melbourne calls weekly, and there is a large trade by small vessels between the port and Melbourne.

Supplies.—Provisions, water, and ships' stores can be procured.

Jetty (lat. 38° 04′, long. 140° 41′).—There is a convenient jetty, 353 yards in length, in the most sheltered part of the bay, with trucks and cranes, and having at its outer end 5½ feet at low water. It is used only by lighters. The lighters are fine sailing boats, carrying about 15 tons of cargo.

Moorings.—There are five sets of moorings with anchors of 80 hundredweight in 2¼ to 3 fathoms at low water; these represent the total possible accommodation in from 14 to 17 feet water for shipping at Port Macdonnell, for with a southwesterly swell coming in it breaks everywhere else for miles around. There are also three sets of moorings situated about a mile southeastward from the jetty, in a depth of 15 to 18 feet at low water; but no vessel coming to load should draw more than 13 feet, in consequence of the range in bad weather.

Although the moorings now laid down at this port are of the heaviest description, and fully competent to hold any vessel that can enter Macdonnell Bay, it must be remembered that during and directly after heavy southwest gales, the sea rolls in over the outlying reefs, breaking heavily in the bay and in 7 to 9 fathoms to the southward of the port.

Vessels parting from the moorings usually run on the shore north from them, and generally get off uninjured when the water has smoothed down.

Pilots.—Before approaching the coast strangers should hoist the signal for a pilot, who will come off in favorable weather. Should the pilot not be able to board, it is recommended to maintain an offing until the weather moderates.

Directions.—It is necessary to have daylight to enter the bay.

From the westward, with Cape Northumberland bearing 6°, 2 miles distant, steer 90° until the peak of Mount Gambier, is seen over the right or eastern fall of Mount Schanck, a truncated cone, bearing 6°, the depth will be then about 5 fathoms. Steer in on the above range, on, which leads directly to the moorings, the water gradually shoaling; 13 feet will be the least passed over, which depth is 500 yards, 192°, from the outer moorings. This may be avoided by hauling a little to the eastward when half a mile from the moorings, and steering for them when they bear 344°.

From the eastward do not approach the shore nearer than 3 miles until on the range given above, then proceed as directed.

In leaving the anchorage, if the wind is from the southward make the first tack to the eastward, if the vessel will lie 118°, or to the southward of it. If obliged to cast to the westward, do not stand in a southwesterly or westerly direction for more than ½ mile. A 208° course made good, clears Breaksea reef, and leads in safety to sea. The best course, if practicable, is to go out with the range, astern, for 3 miles, by so doing passing through the smoothest water obtainable; a vessel will then be in 10 fathoms, clear of all breaks and dangers, and may proceed as desired. The chart is a good guide.

In the event of all the moorings being occupied, vessels entering the bay must anchor, and be kept in such a condition, as to ballast and trim, as will enable them to seek an offing should bad weather come on.

It is obvious, from the nature of the bottom, that no vessel is safe in bad weather from the westward, if at her own anchors.

The harbor master has coir springs for the use of vessels in bad weather.

At night.—Do not enter the bay at night without a pilot, but keep Cape Northumberland Light bearing from 344° to 28°, taking care not to come under 25 fathoms water, or about 5 or 6 miles from the cape.

Tides.—It is high water, full and change at Port Macdonnell at 0 h. 2 m.; springs rise 4 feet.

Signals.—The flagstaff at which the signals are shown is situated near the inner end of the jetty.

A blue flag is hoisted by the harbor master at the flagstaff, when he deems it unsafe for vessels in the offing to come in and moor; or for boats to land from vessels at the moorings.

A lifeboat and rocket apparatus are in readiness in case of accident, and there is a pilot boat, with coir springs, available for vessels at Port Macdonnell. In the event of shipwreck near and the lives of the crew being in danger, assistance will, if possible, be rendered.

Mounts Gambier and Schanck are two isolated conspicuous hills inland from this part of the coast. Mount Gambier, 14 miles northward of Cape Northumberland, is a peak 630 feet high, with table land attached, which extends to the eastward of it. It is an extinct volcano, and there are four lakes in the crater; the eastern, known as the Blue Lake, is 160 fathoms deep and about half a mile in diameter.

Mount Schanck, 8 miles north-northeast from Cape Northumberland, is a truncated cone 380 feet high; it is also an extinct volcano, and the crater is dry.

Radio.—A radio station has been established at Mount Gambier. It is open to the public from 6 a. m. to 8 p. m. Weather forecasts are furnished to vessels on request.

The coast.—Flint Point (lat. 38° 04′, long. 140° 46′), 5 miles eastward of Cape Northumberland, is very low and fronted by rocks and heaps of stones dry at low water. There are 3 fathoms more than a mile south of it. From Cape Northumberland to this point the coast is low, a sandy beach with a bank behind, and, except from Port Macdonnell jetty to 2 miles to the eastward of it, fronted by extensive rocky ledges dry at low water. A low wooded range runs in a north-easterly direction 1½ miles from Cape Northumberland; elsewhere the country at the back of Port Macdonnell is swampy for more than a mile inland. The swamps discharge themselves into the sea by Cress Creek, the mouth of which is nearly 1 mile east of the jetty.

Danger Point, 1¾ miles, east-northeastward of Flint Point, is also low, with fresh-water swamps at the back. The indentation between Danger and Flint Points, called Brown Bay, is shallow; a rocky reef, with 3 fathoms on its extremity, extends southward 1½ miles from Danger Point. A range of wooded hills, which continues to the Glenelg River, commences 3 miles northwestward from Danger Point, with an elevation at that spot of 125 feet. Moorak Creek, a fine, fresh-water creek, discharges itself into the sea close to Danger Point; when there is no ocean swell on a boat may run into the mouth of the creek and fill water casks from alongside.

Green Point, 50 feet high, and 3 miles eastward of Danger Point, is named from its verdant appearance. There is a sandy beach between it and Danger point, forming Riddoch Bay, and a range of sand hills, the highest 70 feet, commences in the bight of Riddoch Bay and extends to Green Point.

Landing.—Butte Reef, with 2 feet on it, having deeper water inside, makes landing practicable on Green Point in ordinary fine weather, when there is a swell outside.

Ruby Rock, situated 98°, 12 miles distant from Cape Northumberland, and nearly 2 miles offshore, has 3 feet on it at low water, and during southeast and easterly winds seldom breaks; with the sun ahead, there is no indication of the rock, attention must be given to the lead and bearings. There are 2 to 3 fathoms in an east-south-easterly direction 600 yards from the rock, 8 fathoms close-to seaward and 4 fathoms directly inshore of it; no clearing mark can be given for clearing it. There are 16 fathoms 2 miles, and 20 fathoms 4 miles south of it.

At night, should the light on Cape Northumberland be obscured, Ruby Rock may be avoided by keeping in more than 10 fathoms water.

VICTORIA.

Boundary.—**Mount Ruskin** (lat. 38° 04′, long. 141° 00′), 150 feet high, 1½ miles northwestward from Glenelg River mouth, is situated on the boundary of South Australia and Victoria.

Glenelg River, which discharges itself into the sea at 1½ miles eastward of the boundary is situated 15¼ miles eastward of Cape Northumberland. The coast between it and Green Point is a sandy beach with low sandhills behind. There is a sandy bar at the mouth, which is fordable at low water when the sea is smooth.

The coast.—Eastward of Glenelg River the coast in the bight is a succession of hummocks about 150 feet high, partly covered with bushes, the sand in many places showing and reaching the summits. At 2 or 3 miles inland there are densely timbered tracts of rising ground about 300 feet high.

A heavy swell constantly rolls on this coast, rendering a wide berth necessary.

At a distance of about 12 miles to the northwestward of Cape Bridgewater a range of hills 500 feet high, and heavily timbered, lies at the back of the coast hummocks, about 2 miles from the coast. At the western extremity of this range, between it and the coast, is a group of high bare sand hummocks, and a large tract of bare sand is situated at a distance of 4 to 7 miles from the cape.

Mount Kincaid, 692 feet high, lies about 12 miles northward from Cape Bridgewater, and about 4 miles from the coast. It is scarcely visible from seaward, its position being only indicated by a few trees slightly elevated above the surrounding country.

Mount Richmond, 711 feet high, is conspicuous, and has a broad, flat top. It lies 7¼ miles northward of Cape Bridgewater.

Cape Bridgewater is situated about 40 miles east-southeastward from Cape Northumberland, has a flat summit 441 feet above the level of the sea, and falls gradually to the cliffy coast south and west of it, and to the cultivated land to the northward, the latter at its lowest part being about 200 feet high. The cape may be seen from a distance of 25 miles.

Anchorage.—West of Cape Bridgewater there is slight shelter from easterly winds, but the bay is exposed to the prevailing winds. With discretion steam vessels may use it, but a heavy swell almost constantly rolls into the bay.

Bridgewater Bay.—East of Cape Bridgewater is a bight known as Bridgewater Bay, but which, like the bay to the westward, can not be recommended as an anchorage. A heavy swell rolls in during southerly and southwesterly breezes, and, except under favorable circumstances, vessels ride uneasily. The swell threatens to break in 20 fathoms, on a line between Capes Bridgewater and Nelson, and does actually break at nearly a mile off shore. The current often sets outward along the cape.

In the bight between Capes Bridgewater and Nelson, but nearer the latter, there is a large conspicuous body of drift sand, just eastward of which is Mount Chaucer, a small peaked hill 405 feet high.

Cape Nelson (lat. 38° 26′, long. 141° 33′) lies 7 miles eastward of Cape Bridgewater, and is an irregular cape of jagged cliffs, 200 feet high, rising, at the back and center to lightly timbered and grassy hummocks, the highest of which, Picnic Hill, is 459 feet high; it is bold to the southeast. From Cape Nelson the land trends northerly for nearly 3 miles, and thence east for 2 miles, where it suddenly turns to the southeast, forming a promontory named Cape Sir William Grant; this coast is composed of limestone cliffs from 100 to 200 feet in height.

Light.—A fixed white light with red sectors, 250 feet above water, the white visible 22 miles, the red 12 miles, is shown from a white stone lighthouse, 79 feet high on Cape Nelson.

Lloyd's signal station.—There is a signal station at the lighthouse, and communication can be made by the International Code of signals, and at night by Morse lamp. Signals are telephoned to Portland, which is connected by telegraph.

Life-saving apparatus.—A life-saving rocket apparatus is kept at the lighthouse.

Cape Sir William Grant.—A well-defined point, projecting 1 mile to eastward, lies east northeastward 4 miles from Cape Nelson, and has a table summit, the highest part of which is 222 feet high; the cape on all sides has precipitous cliffs, about 150 feet in height.

Danger Point lies northeastward from Cape Sir William Grant, forming a bight between, outside of which, at a distance of 1,200 yards south of the point, is a reef, with only 16 feet water, upon which the sea breaks heavily. A reef, with 17 feet water, also extends from the point half a mile in an easterly direction.

Lawrence Rocks (lat. 38° 24′, long. 141° 40′), lying eastward from Cape Sir William Grant and 1 mile southeastward from Danger Point, consist of two small but conspicuous islets of limestone, the larger having two summits, the higher of which is 132 feet above water. The passage between Danger Point and Lawrence Rocks is not safe. The red sector of Cape Nelson light shows over Lawrence Rocks.

With strong winds from seaward a current sets out through this channel sometimes at the rate of 3 knots.

Aspect.—In clear weather, when off Portland Bay, Mount Napier, 1,449 feet high, distant about 32 miles, is visible, and with Mount Clay, 612 feet high, 2¼ miles northward of the mouth of Surrey River, will enable a stranger to identify the land in the vicinity. The appearance of Mount Clay is that of a flat-topped hill with a notch in the center; but for the notch it would closely resemble Mount Richmond, which is 14 miles westward from it.

Portland Bay may be said to extend from Danger Point 12½ miles northeastward to Fitzroy River, and is the natural outlet of many millions of acres of agricultural and pastoral country. The port of Portland consists of that portion of the bay contained within a line running due north from Danger Point to the opposite shore. In the depth of the bay off the town of Portland there is good anchorage, sheltered from all both southeasterly gales, which seldom occur, and still more rarely with strength to do any damage to shipping.

The holding ground is good, being limestone ledges full of holes generally filled with sand, but occasionally with blue clay and small bowlders, apparently of volcanic origin.

From Danger Point the coast trends northwestward 1 mile to Blacknose Point, and thence nearly 2 miles to Observatory Hill.

The coast about Danger and Blacknose Points is low, being only from 60 to 70 feet in height. Blacknose Point has a reef extending from it nearly 400 yards, at which distance the depth is 3 fathoms.

From Observatory Hill the coast trends west-northwestward nearly ¼ mile to the entrance of Wattle Hill Creek, which winds westward by the southern end of the town of Portland; from the entrance of the creek the coast curves along the front of the town for nearly 1 miles to Whaler Point.

The coast from Observatory Hill to Whaler Point, or what may be termed Portland Bay proper, is bordered by a sand bank; the edge of the sand bank in 3 fathoms water is ¼ mile from the shore; the water then deepens more suddenly, and at about 600 yards there is a depth of 5 fathoms. The water then deepens gradually until at a distance of nearly 2 miles there is 10 fathoms.

From Whaler point the land trends northwestward for about 1½ miles, whence it turns suddenly to the northward, and at ¼ mile again suddenly to northeastward. At a distance of 5¼ miles is the mouth of Surrey River, near which is the village of Narrawong. This coast from its turn to the northeastward is low, being only from 6 to 12 feet above high water. At a short distance from the beach it rises, but the whole coast is so densely timbered as to make it uncertain where the elevation takes place. A sandy beach fringes the coast described, and off it is Minerva Reef.

From the mouth of Surrey River the land trends with a slight curve in an easterly direction nearly 8 miles to the mouth of Fitzroy River.

The whole coast from Surrey River is a succession of sand hummocks about 30 feet high, nearly destitute of vegetation, having perpendicular or cliffy faces.

Whaler Point (lat. 38° 20′, long. 141° 36′) is a limestone cliff 107 feet high, off which a reef of rocks extends ¼ mile, with 7 feet water on its outer and shoalest part. There is no channel over this reef.

Light.—A group flashing white light, 135 feet above water, visible 12 miles, is shown from a white stone tower on Whaler Point.

Buoy.—Eastward of the point, on the tail of the reef, a conical chequered black and white buoy is moored in 4 fathoms.

Minerva Reef extends almost the whole distance between Surrey River and Whaler Point. Its shoal water of 1½ fathoms does not lie more than ½ mile from the shore, but there are depths of 3½ fathoms at the distance of a mile; the whole forms a large area of uneven bottom on which the sea breaks at times heavily.

Portland (lat. 38° 20′, long. 141° 36′) is the oldest settlement in Victoria and the outlet of a large area of agricultural and pastoral country. The chief exports are fish, wool, hides, bark, butter, wheat, corn, tallow, hardwood, fruit, bluestone, lime, and agricultural produce.

Portland is in communication with Melbourne by rail. There is also telegraphic communication. Steam vessels frequently call. The population is about 3,000.

A church with a conspicuous spire is situated in the town near the end of the new pier.

Piers.—The railroad jetty in the southwestern corner of the bay is 40 feet wide and projects 400 yards from the shore in an easterly direction into 17 feet of water. The new pier, about 200 yards northward of the railway jetty, projects 85°, ¼ mile from the shore into 31 feet of water; at the outer end it is 40 yards wide for a length of 400 yards, with a least depth of 29 feet alongside. The depth alongside is to be dredged to 40 feet at low water.

This pier and railway pier are fitted with a system of spring piles.

Jetty lights.—A fixed red light, visible 3 miles, is exhibited from the end of the railroad jetty.

A flashing red light, visible 2 miles, is exhibited from a lamppost on the end of the new pier.

A breakwater called Fisherman's Pier is built immediately to the southeastward of the railway pier. It affords shelter to small craft from the swell that usually rolls into the bay.

Berthage.—Vessels may berth at either side of the railway jetty for 250 feet from the outer end if drawing 15 feet, and for 500 feet from the outer end if drawing 12 feet. The new pier is also available for the accommodation of vessels, but it is considered unsafe for any person unacquainted with all conditions of the port to take oversea deep draft vessels alongside unless under the supervision of the pilot thereat. Vessels intending to make use of the pier or jetty must be supplied with all necessary moorings, warps, and springs, or they may be hired, through the harbor master.

Alongside the piers the swell and undertow cause vessels to surge backwards and forwards. This is worst after westerly winds; southeasterly winds cause a good amount of surface commotion, but the surging is not so bad.

A ship berthing alongside the piers must have a clear side. There are no floating wooden pontoon fenders, and a vessel's side takes against the horizontal planking on the side of the pier.

Meteorological observations.—At Portland the mean annual height of the barometer is 30.02 inches. The highest temperature was 108° F. in January, the lowest 27° in July, the mean 58.2°. The mean annual rainfall is 33.38 inches.

Supplies.—Fresh provisions, vegetables, and water can be obtained. There is no regular stock of coal at Portland, but a small quantity, about 40 tons, can usually be obtained. If water is required, special notice must be given, as it has to be obtained from the country and sent down by rail.

The inhabitants use rainwater. The water obtained from local springs contains a large proportion of chemicals, and would be injurious if used in boilers.

Trade.—During the freezing season, from October to March, the oversea shipings call for frozen mutton, etc.

A life-boat with life-saving rocket apparatus is stationed at the railway jetty.

Anchorage.—The best anchorage is with the light on the end of the railway jetty bearing 228°, and according to draft. At night vessels must not anchor outside the limits of the light at the end of the railway jetty.

Explosives anchorage.—The explosives anchorage is eastward of the line bearing 6° from Observatory Point and beyond a distance of 1,500 yards from the shore.

Pilots.—There is a pilot stationed at Portland.

Directions.—From the westward to Portland Bay endeavor to sight the high land of Cape Bridgewater, which, when seen from the distance of 12 or 15 miles from the southwestward, appears covered with white sand patches. Then, after making Cape Nelson, steer for a prudent distance outside Lawrence Rocks. When Whaler Point Lighthouse bears 315°, steer for it and anchor when the railway-pier light bears 228°, or as convenient. As the vessel proceeds northward, the houses of Portland open out from Observatory Hill. Should the wind be scant, the vessel may pass to the northward of the town until it bears 231°, and then tack for the anchorage.

At night.—Entering Portland Bay from the westward do not round the Lawrence Rocks until the light on Whaler Point becomes visible, when shape a course for the light. The outer end of the New Pier is marked by a light. Pass to the northeastward of this light and anchor when the railway-pier light bears 228°.

From the eastward.—In proceeding to Portland Bay from the eastward, Lady Julia Percy Island should be sighted. It lies 17 miles eastward from Cape Sir William Grant, and may be passed at the distance of ¼ mile. Then shape a course for the bay.

At night.—From the eastward shape a course for Whaler Point Light until the railway-jetty light becomes visible, when steer for the jetty or anchorage. Care must be taken not to lose sight of the light on the railway jetty. The southern edge of the light on Whaler Point leads about ¼ mile off shoal water. A boat is always in readiness to afford assistance when required.

Tides.—The tide in Portland Bay as regards its rise and fall, is greatly dependent on the winds. It is high water, full and change, at 0h. 30m.; springs rise about 3 feet.

The coast.—From Fitzroy River, which is 12 miles to the northeastward of Portland, the coast trends with a curve east-southeastward 10 miles to the entrance of Lake Yambuk.

From Lake Yambuk, Boulder Point lies east-southeastward about 6 miles. The coast for the first half of the distance is sandy, having bare sand and grassy hummocks ‚immᵉdiᵃᵗᵉly over it; the highest, Mount Hummock, 213 feet high, forms one of the points in the triangulation of the State. The remaining half of the distance is of a rocky character.

From Boulder Point to the south point of Griffith Island, 4¾ miles, the general direction of the coast is easterly; it is strewn with bowlders of various sizes, some uncovered at high water, and a few sunken rocks lie at about ¼ mile off it.

Mills Reef lies 1 mile eastward of the entrance to Lake Yambuk, and 1,500 yards from the shore, abreast of Lady Julia Percy Island; it consists of several rocks awash at high water, and marked by kelp. See caution below.

Lady Julia Percy Island (lat. 38° 25′, long. 142° 00′), lying 21 miles eastward from Cape Nelson, and 4½ miles from the mainland, is of a triangular form, 155 feet high, flat topped and cliffy on all sides. The island presents the same appearance from all directions, with the exception that the southern end is a few feet higher than the other parts, toward which the island has a small decline. There is indifferent landing on the north side in a small bay.

Caution.—Between Lady Julia Percy Island and the mainland is a passage 3 miles wide; but it is not advisable for a sailing vessel to go through it, as a heavy swell from the southwest generally rolls in upon the coast, and frequent calms in summer make it unsafe; the whole coast being fronted by a border of dangerous rocks extending for 1,500 yards offshore, with a breaking sea even farther off. Steam vessels and other trading vessels using this passage are therefore cautioned against approaching the land in this vicinity.

Port Fairy (lat. 38° 24′, long. 142° 15′).—For 7 miles on either side of Port Fairy the coast is low, that to the westward having grassy slopes with a few scattered trees, whilst that to the eastward is composed for the most part of bare sand hummocks about 60 feet in height. In making this port from the southward the most remarkable land seen is Tower Hill, lying 6½ miles northeastward from Griffith Island, which extends from off the land in a northeasterly direction, and forms Port Fairy.

Tower Hill, 300 feet high, presents the appearance of a table-land, but that part more particularly named Tower Hill is a peak thrown up by volcanic agency in the center of a fresh-water lake.

From the westward, Tower Hill itself is not usually visible, as it then appears in line with the higher table-land which lies 1 mile east of it. When Tower Hill begins to bear northerly it opens out west of the table-land, and continues to be visible as a single conical peak. The table-land falls to the westward, and appears to join Tower Hill; eastward it falls to the same elevation as the western land. The land in the vicinity, both eastward and westward, is higher than the general coast.

This hill is not only a good mark for Port Fairy, but also for the adjoining port of Warrnambool, it being situated midway between the two places. After making Tower Hill, Griffith Island is the next conspicuous land seen.

Griffith Island is conspicuous on this coast, and has two or three hummocks, the highest of which is 74 feet above high water; it is ¾ mile long and ½ mile broad, tapering away to the northeastern point, where it is only 15 feet above high water; in it is included what was formerly Rabbit Island; these two islands were united to seaward by artificial means, since which the sand has heaped up inside.

From the southern end of Griffith Island, which is composed of large volcanic bowlders, the same description of coast extends to the eastward ¼ mile, terminating in a hillock 10 feet high, known as Dusty Miller Island, there being a channel at high water between it and Griffith Island.

Sunken Rocks extend 200 yards from the southern coast of Griffith and Dusty Miller Island, and continue 200 yards off the eastern point of Griffith Island, upon which the lighthouse stands.

Lights.—A group flashing white light, 41 feet above water, visible 11 miles, is shown from a red circular stone lighthouse on the eastern point of Griffith Island.

A fixed white light with red sector, 24 feet above water, visible 3 miles as shown from the end of the southern training wall, the entrance to Moyne River.

A fixed green light, 40 feet above water, visible 3 miles, is shown from a house on Lookout Hill near the inner end of Port Fairy Jetty.

Reef—Buoy.—From the eastern point of Griffith Island to the eastward and northeastward a reef, dry at low water, extends 200 yards off. Also from the same point, rocky ground extends more than 600 yards in a northerly direction, with depths of 5 to 10 feet near its north part, and with not more than 15 feet anywhere on it. At the northern end of this rocky ground a black conical buoy is moored in 17 feet. Vessels either steering for the anchorage or entering the Moyne River must leave this buoy on the port hand.

Back Pass is a narrow channel between Griffith Island and the mainland, which boats occasionally use in very calm weather. The bottom is rocky and uneven, the depths varying from 4 to 12 feet. Just outside the line of sunken rocks at either side of the entrance to the pass the water suddenly deepens to 7 fathoms.

The coast westward of Back Pass is bordered by Sunken Rocks, which extend from 300 to 400 yards from it, and is formed principally of large volcanic bowlders.

Moyne River flows into Port Fairy, and on Lookout Hill, on the eastern bank, 38 feet high, and close to the river's mouth, stands the signal flagstaff.

The entrance to the river has been improved by two stone training walls, which extend 500 yards from the shore in a northeasterly direction. Between these walls a channel has been excavated to depths of 10 to 12 feet. From the entrance to the Belfast wharves, the depth in mid-channel is about 10 feet at low water, but owing to the silt brought down and deposited during floods, this depth can not always be relied on. The width between the ends of the training walls at the entrance to the river is 350 feet, decreasing to 200 feet at the shore end of the works.

Belfast (Port Fairy) (lat. 38° 24', long. 142° 15'), situated on the western side of the Moyne River, and 1,500 yards from its mouth, is the post and telegraph town. It is a railroad terminus 187 miles from Melbourne, and steamers ply to and from it at frequent intervals. Vessels drawing 10 feet water can be loaded and discharged at the wharf stores in the middle of the town. It is the principal shipping port of the western district, and there is a large trade in wool, grain, potatoes, and general produce.

Supplies of all sorts are good and easily obtained.

The population is about 2,500.

Port limits.—The port of Port Fairy includes all inlets, rivers, bays, harbors, and navigable waters northwestward of and within a line bearing 52° from the western end of Griffith Island to the opposite shore.

Tidal signals.—The following signals, indicating the depth of water in the Moyne River, are shown from the flagstaff on Lookout Hill.

> 10 feet of water in river=low water.
> 10½ feet of water in river=a pendant.
> 11 feet of water in river=a square flag.
> 11½ feet of water in river=a square flag and pendant.
> 12 feet of water in river=two square flags.
> 12½ feet of water in river=two square flags and pendant.
> 13 feet of water in river=three square flags.

Signal station.—There is a signal station on Lookout Hill, and communication can be made by the International Code. It is connected by telegraph.

Jetty.—At 300 yards north of Moyne River entrance a jetty extends 400 yards into 7 feet water, but it is now in ruins. At 200 and 400 yards northward of the jetty, and 300 yards from the shore, are two patches of sunken rocks; being in shoal water they do not interfere with shipping.

Wharves.—The depth of the water at Belfast wharves is about 10 feet. There is berthing accommodation for two vessels of 10-foot draft.

A lifeboat with life-saving rocket apparatus is kept on the east side of the Moyne River opposite the government wharf.

Directions.—After making out the hill on Griffith Island steer so as to clear the reef which extends from the lighthouse, then haul up for the anchorage, for which the flagstaff on Lookout Hill in line with the jetty is a good mark; or if bound to Belfast, round the black buoy on the port hand, then steer for the Moyne River entrance, keeping in midstream to the Belfast wharves.

At night do not enter Port Fairy until the light on Lookout Hill is opened out, when steer for it and anchor as convenient.

In thick weather vessels should not attempt to enter the port.

Caution.—Vessels working inshore to the westward of Port Fairy must not bring the lighthouse on Griffith Island to bear eastward of 68°, nor should it be approached nearer than 1 mile until it bears 265°, when a 310° course may be steered for the anchorage.

Tides.—It is high water, full and change, at Port Fairy at 0 h. 31 m.; ordinary springs rise 3 feet.

Anchorage.—The best anchorage for small vessels is in about 3 fathoms water, 200 yards northwestward of the black buoy of the foul ground northward of the east point of Griffith Island, with Griffith Island lighthouse bearing 164°. The anchorage for large ships is in 5 to 6 fathoms off the tail of the reef extending from the northeast point of Griffith Island, with the lighthouse bearing 195° and the flagstaff in range with or a little open north of the end of the jetty 240°.

Vessels trading to Port Fairy generally pick up an anchorage in about 15 feet water between the black buoy and the jetty; vessels making use of the port only during the continuance of a southwesterly gale may get as close in as their draft of water will permit.

The anchorage is bad with easterly winds, and vessels are recommended not to try and ride out a southeasterly gale, except as a matter of necessity, and then all precautions should be taken and springs placed on the cable.

Explosives anchorage.—The explosives anchorage is outside the limits of the port, namely, eastward of an imaginary line bearing 322° from the eastern end of Griffith Island.

Reef Point.—From Moyne River entrance the coast trends northward and thence eastward to Reef Point, which is 2¼ miles northeastward of Griffith Island lighthouse. All this coast has a sandy beach with grassy sand-hummocks until within a mile of Reef Point, when the hummocks are all of bare sand 50 to 65 feet in height.

Off Reef Point volcanic bowlders from 9 to 2 feet above high water extend a distance of 400 yards, and sunken rocks extend 100 yards farther.

Sisters Point (lat. 38° 21′, long. 142° 20′) is conspicuous from its having immediately over it two hummocks 65 feet high, so like each other as to have obtained the name of the Sisters. Bowlders 4 feet above high water lie ¼ mile seaward of this point.

The coast.—From Reef Point, Sisters Point lies east-northeastward 1¼ miles. A point lies midway between forming a sandy bight on either side, but the whole distance between them from ¼ to ½ mile from the land is filled with high water, half tide, and sunken rocks. The coast between Reef and Sisters Points is a succession of bare sand hummocks about 50 feet high.

From Armstrong Bay the coast is a sandy bight with grassy hummocks over it, from 100 to 160 feet in height, trending to the southeastward for about 6 miles to Middle Island, at the western part of Lady Bay. Between 1 and 2 miles from Middle Island is a tract of bare sand.

Armstrong Bay.—One mile eastward of Sisters Point is a small sandy point fringed with bowlders, forming a small bay known by the name of Armstrong. Sunken rocks are numerous and nearly fill it up. This bay is used by fishing boats.

Helen Rock, with 1 fathom on it, lies eastward, distant 2¼ miles from Sisters Point. The rock is 1 mile from the shore, has 8 or 10 fathoms close to on all sides, and is of so pinnacle a form that a lead will not rest upon its summit. It rarely breaks, and is much in the way of coasters.

Mount Warrnambool.—In clear weather, and if more than 5 miles from the land, between Port Fairy and Lady Bay, Mount Warrnambool is visible; it has a round but not very even summit 707 feet above water. It lies 13 miles east-northeastward from Warrnambool Lighthouse. A low spur of the same hill lies about 3 miles west of it.

Lady Bay (lat. 38° 24′, long. 14° 28′), in which is included the port of Warrnambool, is an indentation of the mainland situated between the reef extending southward of Warrnambool Breakwater and Hopkins River, 1½ miles east-northeastward, off which Hopkins Reef, which is awash at high water, projects ⅛ mile. From Pickering Point, at the west side of the Merri River mouth, the land trends west-northwestward for about 1,500 yards, this coast being composed of sandstone cliffs, having numerous indentations with half tide and sunken rocks lying off it, in some places to a distance of 600 yards. Immediately over the cliffy coast are numerous sand hummocks, in some cases grassed but generally bare; the western and highest is 115 feet high, the others vary from 60 to 80 feet. The northern shore of Lady Bay consists of low bare sand ridges with higher and well-wooded land at the back, on which is the town of Warrnambool with its surrounding cultivated land. To the eastward of Hopkins River the country is open, rising gradually from the coast, and terminating in a high grass down 1½ miles inland.

Warrnambool Harbor, on the western side of Lady Bay, is formed by several outlying islands and rocks, nearly connected with each other, which extend from Pickering Point in a southeasterly direction. The largest of these is named Middle Island; between Middle Island and Pickering Point is Merri Island; and outside to the eastward is Breakwater Rock.

Merri Island, lying 100 yards southward from Pickering Point, to which it is all but attached by half-tide rocks, is 47 feet high, and very small, being about 120 yards in extent.

Middle Island, the central and largest of the three islets which protect Warrnambool Harbor, is 250 yards long northwest and southeast and 100 yards broad; it is 70 feet high, and on its summit is the old lighthouse; it lies southeastward of Merri Island, to which it is almost joined by rocks of various heights.

Datum mark.—On one of these, Datum Rock, a datum mark for tides, has been placed, on which is written, " The bottom of this is 5 feet 6 inches above ordinary low water."

From Middle Island several half-tide rocks extend in a southerly direction for a distance of 200 yards, and at a distance of 600 yards in a south-southeasterly direction is a dangerous rocky patch of 17 feet, upon which the sea breaks heavily, the intervening space between it and the island being uneven and rocky.

Breakwater Rock, a small islet 18 feet high and encircled ly sandstone ledges which uncover at half tide, lies about 200 yards eastward of Middle Island. Between it and Middle Island is a small rocky passage with from 2 to 12 feet water. Fronting Breakwater Rock to the south and southeastward are several half-tide ledges nearly joined to one another, and distant from Breakwater Rock nearly 200 yards. Off these again to the southeastward at a farther distance of 200 yards is another half-tide ledge, with two small patches, each a foot above high water. Rocks awash at low water extend 200 yards from the last-mentioned ledge.

Breakwater Pier, built of concrete blocks, extends about 420 yards in a 69° direction from Breakwater Rock, and is connected with the shore at the east side of Merri River entrance by a timber viaduct. The depth of water within the pier is subject to continual alteration, especially at the outer end. A depth of 15 feet can generally be obtained for a length of 400 feet, but at times sand patches with only 14 feet over them form for about 150 feet along from the outer end, and extend nearly 100 feet off the pier. The pier is connected by rail with the town of Warrnambool, and there is every facility for loading and unloading vessels.

Two red cask warping buoys are moored northward of the breakwater pier; the outer buoy is moored in about 4¼ fathoms, 150 yards north-northeastward, the inner buoy in 2 fathoms, 100 yards northwestward from the end of the pier.

Range lights.—A flashing white light, 109 feet above water, visible 14 miles, is shown from a white obelisk immediately in front of the town.

A fixed white light with red sectors, 87 feet above water, visible 5 miles, is shown from another obelisk situated 140 yards 8° from the rear light.

These lights in line bearing 8° lead into the harbor.

A flashing red light with green sectors, 30 feet above water, visible 3 miles, is shown from a post at the end of Warrnambool breakwater pier.

Warrnambool (lat. 38° 24′, long. 142° 28′).—The main part of the town is situated about ½ mile northward from Lady Bay. A large trade is done from the port; the principal exports are wool, potatoes, pigs, bacon, and dairy produce. Steam vessels ply three times a week to Melbourne. There is a railroad in connection with the system of Victoria. The population in 1917 was 7,400 persons.

There is telegraphic communication.

Port limits.—The port of Warrnambool includes all the inlets, rivers, bays, harbors, and navigable waters north of and within a line bearing 258° from the entrance of Hopkins River to the outer end of the reef southward of the breakwater pier.

Merri River.—Immediately behind Pickering Point is the mouth of the Merri River, which ordinarily may be stepped across, but floods wash the sand from its mouth, allowing the discharge of a large body of water.

From Merri Point, at the eastern side of the entrance to the Merri River, the land trends in a northerly direction for about ⅓ mile, whence it trends in an easterly direction for about ½ mile, and then southeastward to the mouth of the Hopkins River, the heads of which form the eastern side of Lady Bay.

The bar.—Warrnambool Harbor·is protected to the southeastward by a bar of $3\frac{1}{2}$ to 5 fathoms water, which adjoins and extends westward from a rocky patch awash at high water, lying 600 yards southward from Hopkins River heads.

The 5-fathom western extremity of this bar is only 600 yards from the low water rocks extending southeasterly from the islands off Pickering Point, and this distance, in which are patches of $4\frac{1}{2}$ to 5 fathoms, forms the main or south channel into Warrnambool Harbor.

Between the bar and about a quarter of a mile from the shore the soundings are from 7 to 4 fathoms, the bottom being generally sand over rock or sandstone rock.

Landmark.—A tall water tower with a flat top, situated on a summit northward of the town, forms a good landmark. When the high lighthouse can not be seen the water tower, open its own breadth to the westward of the low lighthouse, may be taken as the range for the harbor.

Directions.—The range into Warrnambool Harbor is marked by the two light towers, 140 yards apart, bearing 8°, erected on the north shore, the summits being, respectively, 109 and 87 feet above high water.

The south channel, which is the best entrance into Warrnambool Harbor, has, on the range, two rocky patches of $4\frac{1}{2}$ and $4\frac{3}{4}$ fathoms. The bottom of the whole channel is rocky and uneven, varying from 9 fathoms to $4\frac{1}{2}$ fathoms, but in which a depth of $5\frac{1}{4}$ fathoms might be maintained by keeping the upper lighthouse just open westward of the lower one, bearing 10°.

A stranger bound to Warrnambool Harbor from the westward or southward will be greatly guided as to his relative position by Tower Hill, which is only 3 miles from the coast, and 7 miles west of Warrnambool.

From the westward, having taken care to avoid the $2\frac{3}{4}$ fathoms patch which lies nearly $\frac{1}{4}$ mile south-southeastward from Middle Island, bring the upper light tower just open to the westward of the lower light tower bearing 10°, which mark leads in $5\frac{1}{4}$ fathoms water between the 5-fathom bank and the foul ground to the southeast of Breakwater Rock. When the lighthouse on the breakwater pier bears about 333°, a vessel will be within the dangerous rocks to the westward, and may steer for the harbor, giving the end of the breakwater a fair berth. Sailing vessels can not do better than hug the western rocks, as by getting under their lee they are enabled, without danger of shipping a heavy sea, to haul up for the anchorage.

From the eastward, Tower Hill is the best guide to the locality. Mount Warrnambool is hidden by the land if within 4 miles, and being upwards of 10 miles inland is often obscured by mist.

Having made out Warrnambool, either cross the eastern part of the bar, to about 4 fathoms, or if the sea is at all rough, haul off and stand to the westward until the coast in that direction be opened clear of the islands, then proceed to get on the range as before directed. Crossing the bar must depend entirely on the weather. The great disadvantage of crossing it is that vessels have to proceed broadside to the swell.

Vessels are recommended not to approach too near the mouth of Hopkins River. In bad weather, or with a heavy southerly swell, the sea breaks a mile off the land. In fine weather, however, vessels may, and do, cross in all directions, the bar extending from Hopkins Reef.

Warrnambool is the only one of the three western ports of Victoria which may be considered safe in southeasterly gales. This is in consequence of the outer swell being broken on the bar fronting Lady Bay and the harbor.

At night.—To enter Warrnambool Harbor from the westward or southward (having avoided the $2\frac{3}{4}$-fathoms patch which lies half a mile south-southeastward from Middle Island by giving it a wide berth), the vessel should bring the range lights in line bearing 8° and steer for them, which leads in over $4\frac{1}{2}$ fathoms water between the 4-fathoms bank and the foul ground southeast of Breakwater Rock, until abreast the breakwater light, when steer in for the anchorage, giving the breakwater pier end, on which is a flashing light, showing red over the harbor and green to seaward, a fair berth in passing.

From the the eastward, either bring the marks above described on or cross the bar to the southeastward, taking care not to shut the rear range light in when standing toward the mouth of Hopkins River.

In bad weather, or with a heavy southerly swell, the sea breaks 1 mile off the land.

On the approach of a heavy southwest gale with night coming on Portland bay is easy of access, and affords good shelter until the gale abates. This is considered of great importance, as it would be dangerous to take Lady Bay in a gale from southwest or south, the sea then breaking with great violence across the southeast entrance.

Caution.—It is not safe to enter or leave the harbor in southwesterly or southerly gales.

Anchorage.—Warrnambool Harbor is small and not adapted for large vessels, the outer anchorage being in $3\frac{1}{2}$ to 4 fathoms, with a swell sometimes which causes a diminution of the depth. The best anchorage is in about 15 feet water, about 200 yards north-northwestward from the end of the breakwater.

Vessels having entered the harbor must pick up an anchorage where most convenient, according to their draft of water, endeavoring to take as much advantage of the shelter afforded by the breakwater pier as their own safety and convenience will permit.

All vessels using this port should be provided with good springs for their cables, as even in the finest weather there is a heavy swell.

Explosives anchorage.—The explosives anchorage is outside an imaginary line bearing 7° from the outer end of the breakwater.

Tides (lat. 38° 24′, long. 142° 28′).—It is high water, full and change, in Warrnambool Harbor at 0 h. 37 m.; springs rise about 3 feet. Southwesterly winds cause the highest and easterly winds the lowest tides.

Pilot.—There is a pilot stationed at Warrnambool.

Signal station.—There is a signal station at Warrnambool pilot station, and communication can be made by the International Code. It is connected by telegraph.

Danger signals.—A black ball hoisted at the masthead of the pilot flagstaff on the breakwater indicates that it is unsafe for vessels to enter the harbor.

When the sea is braking in the fairway, rendering the entrance to Warrnambool Harbor unsafe for boats, a chequered black and white cone is shown from the eastern yardarm at the signal station, between sunrise and sunset.

Lifeboat.—A lifeboat with life-saving rocket apparatus is stationed at the inner end of the Breakwater Pier.

The coast.—From 4 miles eastward of Warrnambool to Moonlight Head, which is 38 miles farther to the southeastward, the coast is of a cliffy character and presents an almost unbroken appearance, the only break to its uniformity being a broad-topped cultivated hill, 221 feet high, over the east bank of Hopkins River, and a fall in the land 9 miles east of Warrnambool. The cliffs are higher as Moonlight Head is approached.

The coast from Hopkins River, at the eastern part of Lady Bay, to Flaxman Hill is nearly straight and apparently bold, but a heavy swell constantly rolls in and breaks in about 5 fathoms water; the coast thence continues to trend in the same direction and is of the same character for a further distance of about 3 miles, and is locally known as the Bold Projection. Sunken rocks here exist at a distance of ¼ mile from the coast.

Flaxman Hill (lat. 38° 33′, long. 142° 45′), 262 feet high, lies east-southeastward 14½ miles from the mouth of Hopkins River; ¼ mile northwest of Flaxman Hill is a second hill not quite so high, but sometimes more conspicuous, in consequence of its sandy appearance. The two hills together are a good guide to the locality of a part of the cost which otherwise presents a great sameness of appearance, overhanging cliffs forming the principal feature. About midway between Hopkins River and Flaxman Hill the coast range immediately over the cliffs is rather higher than the adjacent land, being there elevated 242 feet above the level of the sea. A large pile of stones has been built upon the summit of Flaxman Hill.

Bay of Islands.—The western land of the Bay of Islands lies close southeastward of the Bold Projection and east-southeastward 18 miles from Warrnambool. The bay may be identified by its white cliffy appearance, varied by numerous small islands all of the same character, the whole presenting a pleasing and striking appearance.

From the western part of the Bay of Islands to Curdie Inlet, distant 4 miles east-southeastward, the coast is cut by bays and studded with small islands. The sea breaks heavily ½ mile from the shore, and it is probable that sunken rocks fringe the whole distance. It was not safe at the time of the survey to sound close off this part of the coast, and therefore it should be given a berth.

Curdie Inlet is conspicuous from the sandy nature of the entrance and is often barred across. The mouth is low and interspersed with low water rocks. At the western point of the inlet, on the highest part of the coast, there is a conspicuous sandpatch, and eastward there are other sandhills or patches; these are more conspicuous from their contrast with the cliffy coast on either side.

From the immediate mouth where the fresh water discharges itself there is a widening of the entrance to a second or outer mouth, and at the points which form the outer mouth are several limestone rocks, those about the western point being more numerous; those off the eastern point are about ⅓ mile from it, and are joined to it by a narrow neck of sand, the central portion of which is washed over by the sea.

The highest of the eastern rocks, Schomberg Rock, is about 17 feet in height, and a ledge extends from it in a northwesterly and southeasterly direction. The sea breaks violently to the eastward and southward, and across the mouth from the ledge to the rocks off the western point there is also a heavy break.

From Curdie Inlet, Hesse Point lies 3 miles to the eastward; the coast between is irregular and cliffy. At Curdie Inlet the appearance of the coast begins to change in consequence of the cliffs being backed by higher ground. From Hesse Point the coast trends eastward for 2 miles to the mouth of Port Campbell.

Port Campbell (lat. 38° 37′, long. 142° 59′) is the only anchorage between Warrnambool and Cape Otway, but it is directly open to the southwest. The entrance to Port Campbell, at the mouth of Campbell Creek, is marked by two headlands, and is easily dis-

tinguished by Hesse Point, 2¼ miles to the westward, and by a remarkable islet, about 200 feet high and 400 yards off the shore, 1¼ miles
to the eastward.

A reef off the eastern head reaches in a southwesterly direction to
nearly 1,500 yards and breaks heavily; whilst the reef off the western
head extends in the same direction only a little more than ¼ mile,
and on which there is very little break.

The channel between these reefs is from 600 feet wide at the seaward end of the reefs to 200 feet just inside the western head, with
depths decreasing from 8 to 3½ fathoms.

Anchorage.—There are about 12 acres of anchorage ground, with
depths of from 1 to 3 fathoms, gradually increasing to 4 fathoms
when abreast of the eastern head and to 5 and 6 fathoms to a point in
line with the western head. The bottom is sandy, with patches of
limestone rock. In heavy weather there is a great drawback off the
beach, which causes vessels to surge considerably at their anchors,
necessitating a spring being run out to the shore. It is reported
that during the summer months (from December to March, inclusive) there is smooth water at the anchorage. The port may be
considered an anchorage for small craft of 30 to 40 tons and for steam
vessels of the Murray River class, drawing from 6 to 8 feet water.
After northerly winds there is good landing.

The town contains a post, telegraph, and telephone office, and has
a population of about 150.

Communication is by coach to Timboon, thence by rail.

The district is agricultural and pastoral, with exceptionally rich
soil.

Jetty.—There is a jetty at Port Campbell about 50 yards in
length, alongside the end of which there is a depth of 9 feet at low
water.

Buoys.—There is a red cask buoy for warping purposes off the
jetty.

Life-saving apparatus.—There is a life-saving rocket apparatus
at Port Campbell.

Directions.—Upon approaching Port Campbell from either side
the sea appears to break right across the entrance, but when the
sandy beach becomes well open, a passage will be seen between the
breakers, and can with confidence be taken in moderate weather on
the fairway marks. The eastern break is very defined, as it is one
continuous break from its outer extremity to the shore, with bold
water immediately clear of the break on its west side. The left extremity of the eastern head in line with beacon on hummocks 49°
leads between the east and west breakers, until the two poles on
Napier Bluff, on the western side, are in line 21°, which line keep

until the beacon on the eastern head bears about 100°, when steer for the jetty.

A 6-foot rock lies about 130 yards westward from the beacon near the inner end of the jetty, and eastward and southward of this rock there is foul ground, with a depth of 7 to 8 feet over it at low water; this foul ground is a great protection, as the sea generally breaks upon it, making the waters inside of it comparatively smooth.

The tides and the tidal currents are influenced greatly by the wind. The set is principally southeasterly, or outward across the eastern breakers. Mean rise of tide about 4 feet.

The coast.—From Port Campbell the coast trends east-southeastward 3 miles to the Sherbrook River, and thence with a slight curve southeastward 11 miles to Moonlight Head. At 1½ miles eastward of Port Campbell is a remarkable rock, 236 feet high, 400 yards off the shore, and at 1 and 2 miles eastward of the Sherbrook River are a few islets and rocks known as the Sow and Pigs. At a distance of 1 to 3 miles west of Moonlight Head there are several ledges which cover and uncover and are skirted by a few sunken rocks, distant about ¼ mile from the shore.

Ronald Point (lat. 38° 42′, long. 143° 10′), lying midway between Sherbrook River and Moonlight Head, is a bluff point 257 feet high, made conspicuous by a large body of drift sand to the eastward; the point forms the western head of the entrance to the Gellibrand River. This river, though draining a rather large tract of country, is similar to Curdie Inlet, Campbell Creek, and Sherbrook River, having a small mouth never very broad and barred across in dry seasons.

Life-saving apparatus.—There is a life-saving rocket apparatus kept at the adjacent village of Princetown. (Population about 150.)

Moonlight Head is bold, rounded, and densely timbered, not only over the cliffs, but wherever it is possible for vegetation to cling; the undergrowth is almost impenetrable. The hills immediately over the coast are about 500 feet high, the highest being 546 feet; these hills form spurs of the Otway Ranges, which rise gradually at the back, until at the distance of 2 to 3 miles inland they attain an elevation of over 1,000 feet.

The highest hill of the Otway ranges west of Cape Otway is 1,800 feet high, and has a rounded summit; it is situated about 10 miles northeastward from Moonlight Head.

Several rocks above water closely skirt Moonlight Head.

The coast from Moonlight Head trends to the northeastward and forms a bight to Lion Headland, which is 3¼ miles distant.

Northeastward of Moonlight Head, distant ½ mile, is Reginald Point, with a small islet close to.

Lion Headland is formed of bold high cliffs, perhaps the highest on the coast of Victoria; here too the Otway Ranges have the greatest elevation when near the coast.

Rotten Point lies 4 miles eastward from Lion Headland. Between the two points a bight is formed, in the depth of which, at 3 miles from the headland, is the mouth of Joanna River, with a sand island at its mouth. Rotten Point is rocky, and there is a rock awash at high water ¼ mile southward of it. Three miles farther, to the southeastward, is the mouth of Ayr River. There are several conspicuous sandpatches about the mouth of Joanna River and Rotten Point, and there is one very large body of drift sand just to the eastward of Ayr River,

The coast between Rottan Point and Cape Otway is rocky, and the sea generally breaks in 5 fathoms of water; back from the shore are numerous sandhills about 350 feet high covered with stunted bush.

A conspicuous conical peak 1,650 feet high, with a range of about the same elevation near it to the northward, lies 10 miles northward from Cape Otway.

Cape Otway (38° 51′, long. 143° 31′), on the north side of the western approach to Bass Strait, is a bluff cliffy projection 250 feet high, of a dark brown color, with patches of coarse sandstone rising to openly timbered grassy hummocks, not exceeding 350 feet in height. A rocky ledge, with 10 feet water on its shoalest part, extends south-southeastward 1,500 yards from the cape; and a very heavy ripple extends nearly 2 miles from the land.

The cape should not be approached within a mile on a northwesterly to north-northeasterly direction and to the westward nearer than 2 miles; it should be rounded at a distance of not less than 3 miles.

Light.—A group flashing white light, with red sectors, 300 feet above water, visible 24 miles, is shown from a white circular lighthouse, 62 feet high on the southwestern extremity of Cape Otway.

Danger light.—A fixed red danger light is exhibited from Cape Otway lighthouse, 48 feet below the main light.

This light is so screened as to be obscured when approaching it from seaward, until 3 miles distant, on a 358° bearing; and is visible (in clear weather) to a vessel proceeding on a 266° or 86° course, until 8 miles distant.

It is exhibited to warn mariners of the proximity of danger, and when seen, the course should be altered off the land to run it out of sight. In thick or foggy weather mariners should not rely on sighting the fixed red light, but should keep a good offing.

Fog signal.—A fog signal is made from the lighthouse.

Lloyd's signal station.—There is a signal station at Cape Otway Lighthouse, and communication can be made by the International code, and at night by Morse lamp. It is connected by telegraph.

Life-saving apparatus.—There is a life-saving rocket apparatus at Cape Otway Lighthouse.

Meteorological observations.—The mean annual height of the barometer is 30.01 inches. The maximum recorded temperature in the shade is 109° F., the minimum 30°, and the mean 55° The mean annual rainfall is 34.45 inches.

Soundings.—The 50-fathom curve, distant 3 miles south of Cape Nelson, increases its distance from the land rapidly until south of Lady Julia Percy Island, where it is distant 23 miles from the main shore. South of Moonlight Head, it is distant 30 miles; it then takes a gradual sweep in toward the mouth of Bass Strait, and at Cape Otway is distant only 8 miles.

Inshore of the 50-fathom curve the soundings shoal very gradually.

The 100-fathom curve is found at about 15 miles distant from Cape Northumberland, 17 miles from Capes Bridgewater and Nelson, and thence it increases its distance from the land until southwest of Moonlight Head it is 40 miles off. It is about 50 miles from a line joining Capes Otway and Wickham, and 30 miles from the west coast of King Island. At the depth of 100 fathoms the bank of soundings appears to drop very suddenly. Seaward of this depth no bottom was obtained at 165 fathoms and 175 fathoms.

Tidal currents.—The tides and tidal currents are much affected by the winds, and are uncertain. A southwesterly or westerly breeze keeps up the flood or east-going tidal current, and increases its rate; an easterly breeze has an opposite effect. While the tides were observed in Surprise Bay, an easterly gale had the effect of doing away entirely with one flood tide.

Currents.—In October, November, and December, when southwesterly breezes mostly prevail, a current may be expected to run to the eastward. In January, February, and March, a westerly current may be expected, but as these currents do not appear to be at any time continuous, they can not with certainty be allowed for. They are stronger as the coast is approached, and strongest off the various headlands, such as Capes Bridgewater and Nelson, Moonlight Head, and particularly near Cape Wickham.

Bass Strait separates Australia from Tasmania. It is about 200 miles long, nearly east and west, and 120 miles wide. The west end between Cape Otway and Cape Grim, the northwest point of Tasmania, is 120 miles wide, but King Island, which lies midway, occupies nearly 36 miles of this space. The safest entrance, 47 miles wide, is to the northwest, and the other entrance, 37 miles wide, to the southeast of the island; the latter entrance, however, being much impeded by numerous dangers, is only recommended to the general navigator in cases of emergency.

The east end of Bass Strait is still more crowded with islands and rocks, more than 50 miles of the southern portion of the entrance being occupied by Flinders and Barren Islands, the latter being separated from the northeast part of Tasmania by Banks Strait.

As the north portion of Bass Strait contains the approach to Port Phillip, and the most frequented route between the south and east coast of Australia, it will be described first. The coast from Cape Otway to Port Phillip and Port Phillip itself will be described in Chapter III; the southern portion, with the north coast of Tasmania, being described in Chapters IV and V.

King Island (lat. 39° 35', long. 143° 57'), the north end of which forms the southeast side of the safest entrance into Bass Strait from the westward, is 36 miles long north and south, and 13 miles broad at the center. It is in the Government of Tasmania.

Caution.—In approaching King Island from the westward, especially during thick or hazy weather, caution will be required on account of the variable strength of the current which sets to the southeast with a force varying from ½ to 2½ knots an hour, according to the strength and duration of the westerly winds, and the use of the lead is enjoined. Many fatal wrecks have occurred on this island from errors in reckoning and in consequence of not making the land near Cape Otway. Commanders of iron ships, especially of those newly built, are warned as to the necessity of ascertaining the deviation of their compasses on approaching the Australian coast.

Soundings of 60 or 70 fathoms are found at 25 to 30 miles westward of King Island. Outside this limit the soundings deepen rapidly to no bottom at 100 fathoms. Inshore of 60 fathoms soundings the depths are irregular, but there are 30 fathoms at a distance of 4 miles to the northwestward of Cape Wickham. Further description of depths around King Island are given later.

Cape Wickham (lat. 39° 35', long. 143° 57'), the north point of King Island, is formed of gray granite, and lies 48 miles southeastward from Cape Otway. A few sunken rocks fringe it at the distance of 200 yards. Northward of the cape the unevenness of the bottom and the strong tidal currents often cause a break at a much greater distance than the rocks extend.

Light.—A group flashing white light, 280 feet above water, visible 24 miles, is shown from a white circular tower 145 feet high, 1,200 yards southwestward of the cape and to the northwest of a round hill 300 feet high.

Harbinger Rocks (lat. 39° 33', long. 143° 53').—East Harbinger Rock lies 306°, 3¾ miles from Cape Wickham Lighthouse, and consists of a group of sunken rocks about 200 yards in extent. In heavy weather or when there is a swell this reef breaks much more heavily than the West Harbinger, but there are times when it will only occasionally break.

West Harbinger, lying 289°, 4½ miles from Cape Wickham Lighthouse, has the appearance of a small, flat-topped bowlder about a foot above high water. A sunken rock, which does not always break, lies 300 yards to the southwestward.

The Harbingers are 1¼ miles apart. There is deep water between them and from 9 to 14 fathoms all round. Irregular depths, varying from 15 to 28 fathoms, are found between them and the shore.

Navarin Reef lies 36° 2¼ miles from Cape Wickham Lighthouse and 1¼ miles from the shore. The principal part is a rock awash at high water, northeastward of which, at the distance of 200 yards, is another rock occasionally dry. The body of the reef is nearly ½ mile long in an east-northeast and west-southwest direction. The sea always breaks on this reef.

Victoria Cove.—At 1¼ miles from Cape Wickham, in a southwest direction, is Cape Farewell. Between the two capes is Victoria Cove. It has a small, sandy beach, on which the sea breaks. This cove, however, being in the vicinity of the lighthouse, is used as a landing places for stores. The lighthouse keeper has a large surfboat, which lessens the danger of landing, but no ordinary boat should attempt to land without a thorough understanding with the keeper that it is safe.

The following signals are adopted:

A ball at the south yardarm of the flagstaff, in addition to the ensign at the head of the staff, and then lowered a little signifies that a boat can land at the cove.

Ensign at south yardarm. Vessels should anchor at New Year Islands.

Ensign at north yardarm. Vessels should anchor on the east side of King Island.

Two fires on the point is a signal to a vessel waiting at New Year Islands that there is safe landing at the cove.

Phoques Bay is the name of the bight between Cape Farewell and the New Year Islands.

The coast—Rocks.—From Cape Farewell the coast is of the same nature for 2 miles as Cape Wickham, and trends with a curve south-southwestward 8 miles to Whistler Point. At 2 miles distant from Cape Farewell some sunken rocks extend to a distance of nearly 1,500 yards from the land; and a sandy beach commences and continues to 1,500 yards of Whistler Point. At the southwestern end of the sandy beach there is good landing in nearly all weathers. A dilapidated hut formerly pointed out the landing place. At 1,500 yards, north-northeastward, from Whistler Point is Elizabeth Rock, dry at low water. Numerous other rocks above water, as well as sunken, lie off the point in all directions. At 2 miles northeastward of Whistler Point a fresh water creek empties itself. At 1 mile southeastward of the point the land rises to a height of 265 feet.

New Year Islands and Franklin Road.—North New Year Island lies 7 miles southwestward from Cape Wickham Lighthouse, is curved in form, and about 1 mile long northeast and southwest; its highest part, near the southwestern end, is 117 feet above high water. A channel ¼ mile broad divides North from South New Year Island. The latter island is 1,500 yards long, in a north-northwest and south-southeast direction, and less than 100 feet high. East of these islands is Franklin Road, an anchorage for small craft protected from all weathers, known locally as New Year Islands anchorage.

Between South New Year Island and King Island there is a distance of over a mile. Several rocks, some above water, others sunken, occupy at nearly equal distance the whole of this space, leaving, however, channels of deep water between. As the sea breaks upon the various dangers the channels may be used in a case of necessity, such as a sailing vessel happening to get upon a lee shore.

The anchorage (lat. 39° 40′, long. 143° 50′) in Franklin Road is in 5 or 6 fathoms water, with the east point of North New Year Island bearing 356°, and a remarkable rock at the northern extremity of South New Year Island, know as the Asses ears, bearing about 232°. The best guide for the anchorage is the absence of kelp. Kelp grows everywhere except in the tidal gutter setting between the islands; here only is the bottom comparatively free from rocks. The anchorage ground being small in extent it is necessary to moor, unless in a small craft, for which there would be room nearer the shore. A moderate sized vessel must either moor, or anchor further out and be exposed to the swell, which, more than the wind, has to be guarded against at this anchorage. Immediately a swell sets in, a spring should be placed on the cable, and care taken that the cable does not foul any sunken boulders, but this is not likely to happen in the position recommended. A small rock, generally above water, but sometimes covered, occupies what would otherwise be the best anchorage.

Though the anchorage may be considered quite safe if the above precautions are taken, yet mariners unacquainted with the place are not advised to use it. Independently of the foul bottom and the small extent of the anchorage ground, which will only accommodate one vessel, the tidal currents often run too strong to enable a ship to pick up a berth as wished.

The principal use of Franklin road anchorage is as a place of waiting for the vessel bringing stores for, or wishing to communicate with the lighthouse.

Tides and tidal currents.—It is high water, full and change, at New Year Islands at 0 h. 48 m.; springs rise 3 feet. The current turns, in fine weather, at high or low water, but is greatly affected by prevailing winds.

Supplies.—Crayfish are plentiful here, and occasionally other fish abound. The mutton bird, the flesh of which is eaten and the oil used for tanning purposes, has a breeding place on New Year Islands, and arrives regularly every year, towards the end of November, to deposit its eggs.

There is a watering place in the southeastern corner of North New Year Island facing the anchorage.

Snakes are numerous on the islands.

The coast.—Netherby Point (lat. 39° 55', long. 143° 50') lies 12 miles southward from Whistler Point, the southern extremity of Franklin Road. The intervening coast presents a very uniform appearance; the coast ranges are densely timbered and about 300 feet in height. The coast is broken up into small bays, with offlying rocks generally above high water, but sometimes sunken. The sunken rocks in some cases extend 1,500 yards from the shore, and outside of these there is much foul ground, which, with tidal currents and a westerly swell, often make a breaking sea, leading anyone unacquainted with the coast to imagine rocks everywhere. At 1½ miles southward of Whistler Point there is a small sandpatch, and at 7½ miles there is a very conspicuous long and bare sandhill, at the foot of which there is a sandy beach.

Bank.—At 8½ miles south-southwestward from Whistler Point is a patch of foul ground which often breaks, but not less than 6 fathoms water has been found.

At 2 miles northwestward from Netherby point is a rock awash at low water which breaks heavily.

Coast—Waterwitch Point.—From Netherby Point the land trends southward for nearly 2 miles to Waterwitch Point; between is Currie Harbor.

From Waterwitch Point the coast trends southeasterly for 2 miles to a conspicuous long sandhill similar to that to the northward, and thence the coast, of the same broken and rocky character, trends 5½ miles southward to Fitzmaurice Bay.

Currie Harbor lies just to the southward of Netherby Point and affords shelter from all winds. It is only adapted for very small craft locally acquainted.

Signals.—The following shipping and tidal signals are shown from the flagstaff at Currie Harbor Lighthouse:

Signal.	Signification.
(a) Red pennant at masthead............	Steamer in sight to the northward.
(b) Blue pennant at masthead.............	Steamer in sight to the southward.
(c) Red flag at masthead.................	Low water.
(d) Black ball at masthead...............	Half flood.
(e) Blue flag at masthead................	High water.
(f) Black ball at masthead and another black ball at half-mast.	Half ebb.

The tidal signals are only shown when vessels are in sight.

Light.—A flashing white light, 150 feet above water, visible 17 miles, is shown from a white iron tower on six columns 70 feet high.

Lloyds signal station.—There is a Lloyds signal station near the lighthouse.

Waterwitch Reef (lat. 39° 58′, long. 143° 50′).—At a distance of 2 miles southward from Waterwitch Point is Waterwitch Reef, nearly awash.

This reef, with the foul ground adjacent, is nearly 1 mile in extent, but the center is the only part which continuously breaks. Midway between Waterwitch Reef and the shore is a rock which uncovers, and between it and Waterwitch Point it is all foul ground.

Fitzmaurice Bay affords good shelter in easterly winds in about 10 fathoms, sand, off the sandy beach in the depth of the bay. A sand patch is a good guide to the locality. As the wind usually shifts from east round northerly to northwest and west and as the westerly change is often very sudden, this bay can only be used with caution.

Water.—There is a good fresh-water stream near the northern corner of the sandy beach, but a heavy surf will nearly always be found on the beach.

Cataraque Point forms the southwestern extremity of Fitzmaurice Bay, and lies south-southeast distant 8¾ miles from Netherby Point. At 200 yards northwestward there are a few sunken rocks, some of which are awash at low water.

From Cataraque Point the coast, which has an elevation of about 300 feet and is here bold and cliffy, trends south-southeastward for 3½ miles to Surprise Point, eastward of which is the bay of the same name.

Surprise Point.—Rocks above water extend ¼ mile south of this point, and between it and the opposite point of Surprise Bay is a rock just above high water, with a group of sunken rocks lying around it. South of Surprise Point the land falls suddenly to 100 feet in height.

Surprise Bay is much used by sealers and small craft visiting the island. It affords good protection in all weather for this class of vessel, the sea being broken upon the group of rocks in the center of the bay. In strong westerly winds the bay should not be entered.

Tides.—It is high water, full and change, in Surprise Bay at 0 h. 43 m.; springs rise 3 feet. During the period tidal observations were being taken an easterly gale had the effect of doing away with one flood tide, showing how the tides are influenced by the winds.

Stokes Point (lat. 40° 10′, long. 143° 56′), the southern end of King Island, lies 3½ miles southeastward from Surprise Point; it is only a few feet above high water, and has the appearance of a group of bowlders, over and outside which the sea is constantly breaking; there are a few sunken rocks south of the point at 300 yards from the high-water line. At 1 mile north of the point the land has an elevation of 144 feet, and falls gradually on the opposite side to about 100 feet.

In rounding Stokes Point care must be taken to give it a good wide berth; the low point and the rocks lying off it appear more distant than they are in reality, in consequence of the gradually rising hill to the northward.

Bank.—At 6 miles westward from Stokes Point there is a rocky bank on which a least depth of 10 fathoms was found; it is probable that it breaks in bad weather.

Seal Bay.—From Stokes Point the east coast of King Island trends northward for about a mile, and then northwestward ¼ mile to the sandy beach of Seal Bay. Off Middle point in the center of the bay, half tide and sunken rocks extend in an east southeasterly direction for ¼ mile.

Anchorage.—Seal Bay, though seemingly protected from the prevailing winds, is actually exposed, for easterly winds are of more frequent occurrence here than on the Victorian coast; the bay has a bleak and warning appearance, and sealers never use it, as they prefer the safer anchorage upon the opposite side of the island in Surprise Bay. The anchorage in Seal Bay is near the center, in 7 or 8 fathoms water, over coarse sand of a loose nature, with the eastern part of Stokes Point just open of the next point to the northward, bearing 183°. A sailing vessel is recommended to anchor farther off in about 10 fathoms water. A swell setting into the bay, or indications of an easterly wind, should be the signal for a vessel to get under way.

Seal and Stanley Rocks.—Seal Rock, 12 feet high, lies 1¼ miles east-northeastward from Stokes Point; at 300 yards southward from

Seal Rock is a smaller rock which uncovers at low water; a few sunken rocks lie near it.

Eastward of the Seal Rock, at a distance of 1½ miles, are the Stanley Rocks, consisting of several rocky patches, with less than 1 fathom upon them; between these and Seal Rock the general depth is about 7 fathoms, but there is one patch of 3 fathoms at 800 yards from the rock, and another of 5 fathoms at 1,500 yards. No shoaler water could be found, but in stormy weather the sea breaks the whole distance from Seal Rock to the outer rocky patches.

Reid Rocks, about 10 miles southeastward of Stanley Rocks, are described later.

The coast.—Black Point lies about 3 miles north-northeastward from Stokes Point, and may almost be considered the northeastern point of Seal Bay. Over the point there is a hummock 113 feet high, and to the northward over the coast there is a higher range of conspicuous sandy hummocks. The point itself is a black rock about 30 feet high; and ½ mile eastward of it, but only ¼ mile from the nearest shore, is a rock above water, with sunken rocks between it and the land.

Two miles northeastward from Black Point is another small point, at the back of which the land rises, and off which to the southeastward, at a distance of 400 yards, is a small rock above water. At 1 mile east-northeastward from the latter point is a smaller point, off which, in a south-southeasterly direction, and at a distance of 800 to 1,400 yards, respectively, are the Brig and South Brig Rocks. At the back of this land King Island attains its greatest elevation, namely, 700 feet.

Brig Rock, so called from its resemblance to a brig under sail, is 45 feet high; there is deep water between it and the shore, and between it and South Brig Rock.

South Brig Rock is 40 feet high, and of much greater extent than Brig Rock; it has no resemblance to a vessel under sail, but is more easily seen, from its black appearance. A few detached rocks lie off it to the southward, and the sea breaks 200 yards off its south side. South Brig Rock bears from Seal Rock northeastward, distant 4½ miles. The coast abreast should not be closely approached in light winds on account of the swell which usually breaks upon the rocks fronting it.

Bold Point (lat. 40° 03′, long. 144° 06′) is situated 5 miles northeastward from South Brig Rock. Several small points and bays occupy the space between; the first half of the distance has several rocks, most of them above high water, lying about 600 yards off the shore. Over the point the coast range has an elevation of 630 feet and is densely timbered. Three-quarters of a mile south from Bold

Point there is a point with a small detached rock forming its southern extremity; at 200 yards off this are a few sunken rocks.

From Bold Point the coast trends northward 7 miles to the south point of Sea Elephant Bay. This coast is broken and almost steep-to. Small sandy beaches vary its rocky character, and over it are densely timbered ranges about 500 feet in height, which at the south point of Sea Elephant Bay trend away to the northwestward.

Sea Elephant Bay (lat. 39° 50′, long. 144° 10′), nearly 6 miles broad, and 1½ miles deep, is open to the eastward.

Off its northern point, and distant 1½ miles to the eastward, is Sea Elephant Rock, 76 feet high and nearly ¼ mile in extent. Between the point and the rock is a channel of about 3 fathoms water. At 1 mile northward from Sea Elephant Rock is Sea Elephant Reef, which at very low tides is uncovered about 2 feet; there is foul ground round it ½ mile in extent. Two hundred yards to the southwestward of Sea Elephant Rock is a rock above water, near which are a few sunken rocks.

Anchorage.—Sea Elephant Bay affords a safe anchorage during westerly gales, and the wind generally, when the weather is clearing, shifts to the southward. The bottom throughout the bay is sand, or sand and shells, and there is anchorage anywhere in about 9 fathoms; but the center of the bay, in a line between its southern point and Sea Elephant Rock, is the most convenient, as there is nothing in the way of a sailing vessel getting to sea on the first appearance of a fresh breeze from the eastward. In the summer months there is much easterly weather, and a swell rolls in.

Water.—In the southern part of the bay there is a good fresh-water stream. Also an abundance of firewood.

Elephant Shoal.—Eastward of Sea Elephant Bay, at nearly 7 miles from the shore, there is a shoal with 3¼ fathoms least water upon it; the shoal generally has a depth of 4½ and 5 fathoms, sand, and at this depth is 3 miles long, in a north and south direction. The northern or shoalest part lies 106° from Sea Elephant Rock, about 4½ miles. Midway between the shoal and the shore the water deepens to 12 and 14 fathoms, and thence shoals gradually again, until ½ mile from the shore there are 5 fathoms. As the sea breaks heavily on the shoal in strong winds, it should be given a wide berth.

Tides and tidal currents.—It is high water, full and change, in Sea Elephant Bay at 0 h. 50 m.; springs rise 3 feet. The flood current runs to the northward and the ebb to the southward, at springs 1½ knots. The turn of the current is influenced by the wind; in fine weather it occurs at high and at low water.

The coast.—From Cowper Point, the northern point of Sea Elephant Bay, the coast consisting of low sand hummocks, trends northward 9¼ miles to Lavinia Point. At 1 mile distant from the northern point of the bay is Sea Elephant River, a small stream accessible at high water to small craft drawing 3 feet water; at the back is a swamp. Midway, and at 2 miles inland, is a double-topped hill, densely timbered, 338 feet in height, know as Sea Elephant Hill.

Lavinia Point (lat. 39° 40′, long. 144° 07′), the northeastern extremity of King Island, is low and sandy; thence the coast, which continues sandy, trends northwestward 3¾ miles to Boulder Point, so named from a large granite boulder which forms it. At 1¼ miles to the northwestward of Lavinia Point is a conspicuous sand patch. A few sunken rocks lie off Boulder Point, and a shoal with 10 feet water extends northward from the point 1,500 yards.

The meeting of the tidal currents has caused a heaping up of the sand in the vicinity of Lavinia Point, and it is not uncommon for coasters to anchor in westerly gales in about 9 fathoms upon the bank thus formed. If the gale should have settled into a westerly one, this anchorage is as safe as Sea Elephant Bay, and it is handier for proceeding westward when the weather clears.

At 1 mile to the northwestward of Boulder Point is a large and conspicuous sand patch, much more conspicuous than that between Lavinia and Boulder Points.

Doughboy Rock.—The coast from Boulder Point trends west-northwestward for 3 miles to another point, off which at 200 yards is a rock awash; 1¾ miles west-northwestward from the latter point lies a rock above water known as the Doughboy, and a reef dry at low water connects it with the shore, from which it is distant 600 yards. Doughboy Rock lies 1 mile eastward from Cape Wickham. There is a passage of deep water between it and Navarin Reef, but the tidal current often runs strong and causes a rip. The passage is not recommended.

Depths around King Island.—The 30-fathom curve, commencing at about 4 miles northwestward of Cape Wickham, just outside Harbinger Rocks, follows the curve of the land, and passes New Year Islands at a mile distant, thence down the west coast of King Island at a distance of about 3 miles until at 5 miles northwestward of Netherby Point it is distant 5 miles from the adjacent coast. Here are depths of 21 and 22 fathoms with much foul ground leading to the rocky patch of 6 fathoms already described. Thence the 40-fathom curve approaches to within 3 miles of Netherby Point, increasing its distance from the land to 4 miles, but again nearing the land at Cataraque Point, where it is distant only 1 mile. At Surprise Point it is distant only ½ mile, and at Stokes Point nearly 1 mile, whence it becomes a very irregular line trending first easterly and then toward Reid Rocks.

At 6 miles westward from Stokes Point there is a rocky bank, referred to previously, on which not less than 10 fathoms were found; it is probable that the sea breaks here in bad weather.

On the eastern side of King Island the depths are less than 30 fathoms. About 20 miles to the eastward of Sea Elephant Bay is a depth of 25 fathoms, sand, and at about 23 miles to the eastward of Lavinia Point 24 fathoms, fine white sand. From these positions toward the island the water appears to shoal very gradually, while eastward it appears to deepen as gradually. Northward of the island the 30-fathom curve is 2 miles northward from Cape Wickham; it passes Navarin Reef at 1 mile distant and trends easterly.

Currents and tidal currents.—Off Cape Wickham (lat. 39° 35′, long. 143° 57′) there is occasionally a very strong current, which may be more correctly termed a tidal current accelerated by the wind. Close to the cape it is said to run occasionally as much as 5 knots, but something like 2 knots is the ordinary velocity at spring tides.

The current loses in force as its distance from the shore increases. It is probable that a westerly gale keeps up the flood currents, which here sets to eastward, and an easterly gale has an opposite effect.

Southward and westward of King Island the currents or tidal currents are irregular; they are known at times to be very strong, but they were never experienced of any strength during the survey of the island.

Sealers have reported that in the strait between King Island and Tasmania a current sets eastward during easterly weather; if this be so in the center of the strait it is likely that in-shore on both sides there is a current setting in an opposite direction.

CHAPTER III.

CAPE OTWAY TO PORT PHILLIP—PORT PHILLIP, GEELONG, AND
MELBOURNE.

Parker River.—From Cape Otway the coast trends eastward 2
miles to Franklin Point, which is low and sandy, with some rocks
lying near it. At 1,500 yards north of this point is the mouth of
the small river Parker. As there is usually a heavy surf at the mouth
of the river, it is hazardous to attempt a landing there.

The coast.—From Franklin Point the coast trends northeastward
43 miles to Addis Point. It begins with high dark-colored cliffs,
backed by densely wooded hills, rising to the height of 2,297 feet at
a distance of 25 miles north-northeastward from Cape Otway, and
extends to within 5 miles of Addis Point. At about 8 miles northeastward of Addis Point the coast changes to sand hummocks, backed
by undulating hills, with patches of wood, and farmhouses.

From Blanket Bay, a small bight 1½ miles northeast of Parker
River, where the Cape Otway lighthouse stores are landed, the coast
trends northeast and eastward for 4 miles to Storm Point, and thence
north-northeastward 2¼ miles to Bunbury Point. Hayley Reef, with
rocks, just above high water, projects ½ mile from the shore between
the two points.

Henty Shoal, lying 1¾ miles off the shore, is situated 9¾ miles
northeastward from Cape Otway Lighthouse. It has 19 feet water
over it, and the sea breaks heavily in moderate weather. It is steep-
to, with 8 to 10 fathoms all round within 200 yards of its shoalest
part.

Beacons.—The position of Henty Shoal is shown by the intersection of two lines drawn through two pairs of pillar beacons, each
surmounted by a ball. Two of these beacons, which are 200 yards
apart and in range when bearing 277°, the inshore one being white
and the outer black, are situated 670 yards southwestward of Hayley
Point.

The other two beacons are about 300 yards apart, when in range
bear 325°, the northwestern one being white, the other red, and are
situated on Bunbury Point.

Directions.—If bound to the northeast, keep the black beacon
near Hayley Point well open north of the white one until the white
beacon on Bunbury Point opens well to the northeast of the red

beacon. In proceeding to the southwest, keep the outer or red beacon on Bunbury Point well open south of the white one until the white beacon near Hayley Point is well open southward of the black beacon.

Apollo Bay (lat. 38° 46′, long. 143° 41′), on the northeastern side of Bunbury Point, lies just under a high part of the Otway Range and may be known by the beacons on the point and the houses of the township of Kambruk along the shore of the bay.

The population of Apollo Bay is about 600.

A reef, on which the sea breaks heavily at times, extends off Bunbury point for 670 yards.

Port limits.—The port of Apollo Bay includes all inlets, rivers, bays, harbors, and navigable waters westward of and within a line bearing 27° from the eastern end of Hayley Reef to the mouth of Wild Dog Creek.

Kambruk (Krambruk) is a postal township with telegraph station and is a favorite watering place; communication is by coach to Forrest, thence by rail; also by steamer twice a week.

Jetties.—The old jetty, situated 150 yards northward of the south (red) beacon on Bunbury Point, is no longer available.

A new jetty, the inner end of which is situated 475 yards west-northwestward from the north (white) beacon on Bunbury Point, extends thence 563 yards in a direction 81° into a depth of 12 feet at low water.

Light.—A fixed red light, 25 feet above water, visible 2 miles, is shown from a lamp-post at the head of a new jetty.

Anchorage.—There is anchorage during west or southwest gales in Apollo Bay 1,600 yards off the shore in 6 fathoms water, with the white beacon on Bunbury point bearing 227°. There is generally a swell in the bay, which is especially heavy during easterly or southerly winds. The holding ground is good. Vessels must be prepared for a change of wind to the south or southeast.

Buoy.—A mooring buoy is moored about 100 yards northwestward from the outer end of the jetty.

Life-saving apparatus.—There is a life-saving rocket apparatus at Apollo Bay.

Cape Patton (lat. 38° 42′, long. 143° 51′), situated 8½ miles northeastward of Bunbury Point, is a bold, dark looking, wooded head. At 1½ miles southwestward from it, a shoal with a depth of 1¼ fathoms projects ½ mile from the shore. At 1½ miles northeastward from the cape, a 2-fathoms spit extends ½ mile from Hawdon Point. From this point the coast extends 9½ miles northeastward to Grey Point, a low grassy projection, with a reef dry at low water extending 400 yards from it, and forming the southern side of Louttit Bay.

Landings.—Kennett River Landing, between the mouth of the river and Hawdon Point, is 420 feet long, 8 feet wide, with a depth of 1¼ feet at the outer end. Wys River Landing, immediately north-ward of Sturt Point is 320 feet long, 8 feet wide, with a depth of 5 feet at the outer end.

Tides.—The rise of tide at both landings is 5¼ feet.

Louttit Bay, at the head of which is the township of Lorne, lies about midway between Cape Otway and Port Phillip heads.

Port limits.—The port of Lorne, Louttit Bay, includes all inlets, rivers, bays, harbors, and navigable waters westward of and within a line bearing 148° from the eastern side of Erspine River to abreast the jetty.

Louttit Bank, with a least known depth of 6 fathoms, extends eastward 1¼ miles off Grey Point.

Lorne, a watering place and the settlement in Louttit Bay, has a population of about 400. There are three mails a week from Melbourne, and there is a telegraph office.

The mean temperature at Lorne in winter is 12° higher than Melbourne, and in summer 10° lower; annual rainfall 36 inches.

Jetty.—There is a jetty 193 yards long at Lorne, which projects from the shore in a northeast direction, with a depth of 9 feet at low water on the outer end.

Light.—A fixed green light, 27 feet above water, visible 2 miles, is shown on the end of the jetty.

Buoy.—A mooring buoy is moored off the jetty.

Anchorage, during southwest or westerly gales, may be obtained in Louttit Bay to the northwestward of Louttit Bank, in 5 fathoms water about 1,600 yards off the shore. The anchorage in this bay is preferable to that in Apollo Bay, there being less swell. Sailing vessels anchoring in this bay, with westerly gales, must prepare for a change of wind, as it often chops round to south, and sometimes to southeast.

Eagle Nest (Split) Point (lat. 38° 28′, long. 144° 06′), 7½ miles northeastward from Grey Point, is of a reddish-brown color, and appears like three cliffs close together, divided by dark ravines. Eagle Nest Reef, which is awash, projects ½ mile from the shore at 1,340 yards northeastward of Eagle Nest Point.

Light.—A group flashing white light with red sectors, 218 feet above water, the white visible 20 miles, the red 9 miles, is shown from a white concrete tower 83 feet high on Eagle Nest Point.

Danger light.—An auxiliary light, visible from a distance of 3 miles and through an arc of 180° seaward, is also shown from this lighthouse, 50 feet below the principal light. It is invisible, from a height of 14 feet above the sea, until within the distance of about 3 miles from it.

The danger light is to warn mariners of too close approach to the land, and when seen the course should be altered to run it out of sight.

Signal station.—There is a signal station at Eagle Nest Point lighthouse; communication can be made by the International code, and at night by the Morse lamp. Signals are telephoned to Lorne, which is connected by telegraph.

Life-saving apparatus.—There is a life-saving rocket apparatus at Eagle Nest Point lighthouse.

Demons Bay.—Between Eagle Nest Reef and Addis Point, at 7¼ miles northeast from it, the coast forms two bights, separated by Roadnight Point, the northeastern being Demons Bay. Northward 1 mile from Roadnight Point is a creek, with a sunken rock close off it, between which and Addis Point there are two rocks above water surrounded by sunken rocks, outside which there are 5 to 7 fathoms water; these are the Ingoldsby Reefs, and they break heavily.

Addis Point.—From Addis Point the coast trends northeastward nearly 5 miles to Zealey Point, whence it curves eastward 9¼ miles to Barwon Head.

Zealey Point is fronted by a reef extending ⅛ mile from the shore partially dry at low water. Close to this point to the southwestward is the outlet of Spring Creek, and to the northwestward is the small watering place of Torquay.

Victoria Reef, on which there is a depth of 2¼ fathoms, lies 1¼ miles eastward from Zealey Point, with which it is connected by a bank that continues along the coast to Barwon Head; at midway it only extends ¼ mile from the shore.

Ant Spit, on which there is a depth of 2 fathoms, and which breaks heavily, projects from this bank to 2½ miles westward of the head.

Barwon Head (lat. 38° 18′, long. 144° 30′) is a saddle-shaped scrubby hummock 122 feet high terminating seaward in Flinder's Point and appearing from seaward like an island, on account of the low land in its rear. This head forms the south side of Barwon River, which boats can only enter at high water when it is smooth, owing to the rocky ground at its mouth. On the north bank of this river, at about 10 miles to the northwestward of its mouth, is situated the important town of Geelong.

Charlemont Reef, 1 mile southwestward from Barwon Head, is a detached 1½ fathoms patch, with deep water about it, which breaks heavily at times.

The coast.—From Barwon River the low sandy coast curves eastward nearly 6 miles to Point Lonsdale, the outer point on the west side of the entrance to Port Phillip. A spit having from 1½ to 2½ fathoms on it projects 1 mile eastward from the mouth of the river, whence a continuous rocky shoal, nearly ½ mile broad, with from 1 to 3 fathoms, extends to Point Lonsdale. From the outer edge of this shoal the soundings gradually increase to 10 fathoms at 1¾ miles and 20 fathoms at 2¾ miles off the shore.

Port Phillip includes all inlets, rivers, bays, harbors, and navigable waters contained within a line drawn from Point Lonsdale to Point Nepean, and not included in the ports of Melbourne, the metropolis of the State of Victoria, and Geelong. It is situated at the head of the extensive bight which lies between Cape Otway on the west, and Wilson Promontory, 130 miles to the eastward of the cape.

In approaching the port from the westward, the entrance is not easily distinguished until Point Nepean, the eastern entrance head, bears 30°, when Shortland Bluff, on which the highest and leading lighthouses are erected, shows out, and the estuary becomes visible. If Barwon Head is previously seen, the entrance to Port Phillip is easily found by its relative position with that head.

Port Phillip is an extensive bay about 31 miles in length, north and south, and 20 miles in breadth at the middle, where on the western side it forms an arm which trends 15 miles in a west-south-westerly direction to Geelong. At the northern end of the bay the waters contract, forming the portion known as Hobson Bay.

Depth in channels.—The dredged channel over the rocky flat seaward of the heads has a depth of 40 feet at low water, with a width of 435 yards. The width of the channel carrying not less than 37 feet water is 900 yards. South channel has a depth of 33 feet. West channel has 17 feet at low water.

Caution.—Except under the most favorable conditions of sea and tide, it is considered dangerous to navigate any vessel having a greater draft than 32 feet through the entrance. Owing to the varying conditions of sea and tide the limit of draft is left to the discretion of the sea pilots.

During rough seas, vessels may scend or dip from 8 to 10 feet below the ordinary water level, and great caution should be used in the case of heavy draft vessels outward bound, which are recommended, under these conditions, to leave only on the last quarter of the flood or at slack water flood.

Prohibited entrance when deep draft vessels are leaving.— Outward bound vessels drawing over 29 feet will, when off Portsea, hoist a ball or shape 2 feet in diameter, where it can best be seen. A similar signal will then be hoisted on the examination steamer outside the Heads, as an indication that the channel is closed to inward traffic.

Caution with regard to inward bound vessels.—A black ball or shape, 2 feet in diameter, displayed from Point Lonsdale signal station, indicates that the entrance to Port Phillip is temporarily closed to inward-bound traffic, and while this signal is flying no inward-bound vessel will be permitted to enter.

Pilots.—The pilot service at Port Phillip Heads is conducted by means of steam pilot vessels. The steamer is kept constantly cruising outside the heads within the limits of the cruising ground.

The cruising ground extends seawards from Port Phillip Heads to a point distant 15 miles 222° from the center of Port Phillip Heads. From this point the western boundary extends 334°, toward the shore at Zealey Point, and the southern boundary 92°, toward Cape Schanck until Arthur's Seat bears 50°, which line towards the shore forms the eastern boundary.

Vessels which miss the pilot vessel will be boarded by a pilot from a whaleboat, when they are inside Point Lonsdale. But no stranger should attempt to enter without taking a pilot; although the channels are so carefully lighted and buoyed that it is quite possible to do so.

Signals.—Vessels steering for Port Phillip are bound to show the usual signal for a pilot when within pilot waters, and if the pilot vessel be in sight they must allow a reasonable time for a pilot to board.

At night, vessels requiring a pilot should show a blue light every 15 minutes; or a bright white light flashed or shown at short or frequent intervals, just above the bulwarks, for about 1 minute at a time.

In thick or foggy weather, the steam pilot vessel shall, in addition to any signal required by law, sound two blasts every 5 minutes on her steam whistle or siren, the first blast short, the second long. Vessels requiring the services of a pilot should, in fog, make the reverse signal, viz.,—two blasts, the first long, the second short; steamers on their whistle or siren, sailing vessels on the fog horn.

Vessels which are exempt from pilotage must, on arriving within pilot waters, have a large white flag flying at the main masthead until past Swan Point, 6 miles within the entrance, under a heavy penalty, to prevent the pilots' time being unnecessarily taken up running after vessels which do not require their services.

The employment of pilots is optional with all vessels belonging to the British Government.

Tidal signals are shown at Point Lonsdale (lat. 38° 18′, long. 144° 37′) denoting the quarter of the tide with reference to the current.

The flood or in-going current—

During the first quarter is denoted by a blue flag half-mast.

During the second quarter is denoted by a blue flag at mast-head.

During the third quarter is denoted by a red flag half-mast.

During the fourth quarter is denoted by a red flag at mast-head.

The ebb or outgoing current.—The same signals are used for the four quarters of this stream with a ball below the flag.

At night during the in-going current one green fixed light, and during the outgoing current two green fixed lights will be exhibited at Point Lonsdale Lighthouse below the main light.

By attention to these signals a mariner will know the state of the tidal current, which can not be always ascertained by the usual process of finding the time of high water, its strength and duration being much influenced by the wind and weather.

The signal keeper has instructions if he sees vessels approaching the heads and running into danger to warn them by the International code of signals; strangers should therefore watch these signals.

Tides and tidal currents are described later.

Point Lonsdale (lat. 38° 18′, long. 144° 37′), the western head of the entrance to Port Phillip, is low and juts out from a dark, rocky cliff, it being neither so high nor so well marked in outline as Point Nepean, the eastern head, but can be easily distinguished by the lighthouse, lookout house, and a tidal signal flagstaff near its southern extremity.

Lloyd's signal station.—There is a signal station at Point Lonsdale Lighthouse, and communication can be made by the International code and at night by Morse lamp. It is connected by telephone with Queenscliff telegraph station.

Inward-bound vessels will be reported from Point Lonsdale lighthouse station from sunrise to sunset.

Lonsdale Reef, the greater part of which dries at low water, projects ¼ mile southeastward from Point Lonsdale, and is about 200 yards in extent, having dangerous rocky patches extending nearly 400 yards farther to the southeast, with 5 fathoms water close outside them.

At about 200 yards from Point Lonsdale the reef is intersected by a channel with a depth of 5 feet, which may be used by boats in fine weather.

Lonsdale Rock, 135°, a little more than ¼ mile from Point Lonsdale Lighthouse, with 18 feet water on it, lies on the west side of the fairway. Depths of 4 fathoms extend about 200 yards to the south and east of Lonsdale Rock.

Light.—A group flashing white light with red sectors, 120 feet above water the white visible 17 miles, the red 8 miles, is exhibited from a white concrete tower with black roof and gallery 70 feet high, on Lonsdale Point.

Fogsignal.—A fogsignal is made from the lighthouse.

Lifeboat.—A lifeboat and life-saving rocket apparatus are maintained at a jetty at Point Lonsdale in case of shipwreck.

Point Nepean (lat. 38°, 18′, long. 144° 39′).—The eastern head of the entrance to Port Phillip is the narrow western termination of a peninsula, which extends 15 miles in a westerly direction from Arthur's seat, and consists of a series of sand hummocks slightly covered with low bushes.

A beacon consisting of a white triangle, 20 feet high, with a square on top, stands on the extremity of Point Nepean.

Life-saving apparatus.—There is a life-saving rocket apparatus kept at Point Nepean.

Nepean Reef and Rock.—Nepean Reef projects westward nearly 400 yards from Point Nepean to Nepean Rock, a small dry rock; thence the reef and several pinnacle rocks outside it extend 700 yards westward to Corsair Rock. Nepean Reef dries at low water out for 700 yards from the point.

Rock Beacon.—A red triangular-shaped beacon, 25 feet high, with a ball on top stands on the dry reef, 400 yards from Point Nepean; the beacon is known as Rock Beacon.

The northern edge of the rocky ledge along Nepean Reef trends from Corsair Rock eastward to 200 yards northward of Point Nepean. The coast outside Point Nepean is bordered by a continuation of this reef and numerous rocks; but they do not extend more than 200 to 300 yards, and the coast may be approached to ¼ mile in 5 fathoms.

Corsair Rock, the outer end of Nepean Reef, is 20 feet in diameter, having 8 feet water over it with 3 to 5 fathoms close-to on the channel side; this rock lies with the red Rock Beacon in line with the white beacon on Point Nepean, bearing 101°, the red beacon being distant 700 yards.

Entrance to Port Phillip (lat. 38° 18′, long. 144° 38′).—The entrance between Points Lonsdale and Nepean is 1¾ miles wide, but the navigable channel is contracted to a little less than 1 mile in width between the reefs that project from these points, just without the heads.

Range marks.—The line of the Queenscliff range lights, bearing 42°, marks the western side of the 40-foot channel. The Obelisk and Queenscliff high light in line, bearing 38°, marks the eastern side. Between these lines is the 40-foot channel.

Lonsdale Bight Beacons.—These beacons are near the shore in the bight between Point Lonsdale and Shortland Bluff. The front beacon is a white triangle with disk as topmark. The rear beacon is an inverted triangle painted black with a vertical white stripe.

These beacons in range 343° lead in through the heads in about 6 fathoms. The beacons do not define the line of deepest water through the heads, but may be used as a range for entering from or leaving for the eastward.

Swan Beacon, about 75 feet high, on Swan Point, is white, with a red top, and staff and globe.

The Rip.—The rocky flat through which the newly dredged channel has been made lies about ½ mile outside Port Phillip heads with from 5 to 9 fathoms over it. The water deepens outside this flat to 12 and 15 fathoms, and inside the heads to as much as 15 to 47 fathoms. This great inequality of depth, combined with tidal currents, at times running 5 to 7 knots, causes the well-known " Race," or " Rip," which during or immediately after a southwesterly gale breaks so furiously as to be dangerous to small vessels.

Tides.—It is slack water, on the flood, in the Rip at full and change at about 2 h.

Blasting operations.—When blasting or sweeping operations are in progress off the Heads the following signals will be shown from the masthead of the steamer engaged on the work:

(a) Letter " B " when on blasting duty.

(b) International Code signal " X. H. C.," when on sweeping duty.

Mariners and boatmen are warned not to approach nearer than ¼ mile to the boats engaged in blasting work; warning will be given by prolonged whistle blasts when a charge is about to be exploded.

Mariners are specially requested to keep clear of the steamer when she is sweeping, as the apparatus will render her movements slow and uncertain.

The western shore of Port Phillip from Point Lonsdale curves northward and eastward, forming a bay 1,500 yards deep; it is mostly occupied by shoals with irregular depths between them, extending from the shore to a line from Point Lonsdale to Shortland Bluff. The only part of the bay which appears free from shoals and has tolerably regular depths is within about one mile of Shortland Bluff; even here anchorage is not recommended.

Victory Shoal lies nearly in the center of the above bay, its outer edge, on which there are 11 to 14 feet water, being in line between Point Lonsdale and Shortland Bluff, the least depth of water on the shoal is about 6 feet.

Queenscliff, at the entrance of Port Phillip, is 2¼ miles northeastward of Point Lonsdale. Two trains run daily to and from Geelong, 20 miles distant; and a steam vessel plies daily to and from

Melbourne in summer. During the winer months a special coach runs daily via Portarlington, in connection with a line of steamers. When the weather permits the interstate steam vessels embark and disembark passengers. All vessels arriving from foreign ports are boarded here by the health officer. It is much used by visitors as a watering and bathing place. The population is about 2,000.

Shortland Bluff (lat. 36° 16′, long. 149° 39′), about 50 feet high, on which are two lighthouses, lookout, and telegraph station, flagstaff, and a red and white stone obelisk 50 feet high, with the town of Queenscliff in their rear, is the southeastern extremity of a peninsula projecting nearly 2 miles in a northeasterly direction from the line of coast, with which it is connected by an isthmus little more than 200 yards in breadth. The peninsula is about $\frac{1}{4}$ mile broad at Shortland Bluff, from whence it gradually contracts to the northeastward, where it terminates in a low, narrow point.

Lloyd's signal station.—Communication can be made by the International Code of Signals. Queenscliff is connected by telegraph. Inward-bound vessels will be reported from Queenscliff signal station from sunset to sunrise. The station will be unattended from sunrise to sunset.

Time signal.—A flag is dipped daily, except Sundays and public holidays, at the signal station at 1 h. 00 m. 00 s. standard time of Victoria, equivalent to 15 h. 00 m. 00 s. Greenwich mean time.

Lights.—A fixed white light, 130 feet above water, visible 17 miles, is shown from a blue stone tower, 81 feet high, on Shortland Bluff, Queenscliff.

An occulting white light, with red sectors, 90 feet above water, the white visible 14 miles, the red 10 miles, is shown from a white tower with black roof and gallery, 69 feet high. It is 222°, 352 yards from the high lighthouse.

Another occulting white light, with red sectors, is shown from the gallery of the same tower. For details, see Light List.

An occulting green light, 75 feet high, is shown from the top of the obelisk situated near Queenscliff low lighthouse.

Queenscliff Jetties.—There are two jetties at Queenscliff, the northern one projects about 750 feet from the shore, with a face 70 feet long and 30 feet wide pointing south-southwestward with a depth of 10 feet alongside and in line with the face of the south jetty. About 400 feet from the face is an arm extending 200 feet to the northwestward, with a depth of 6 feet alongside.

The south jetty, about 330 yards southward of the north jetty, is 26 feet wide, projects from the shore in a southeasterly direction about 400 yards, and has a face 300 feet long and 50 feet wide, outside of which is a depth of 11 feet at low water.

A breakwater, 160 feet long, and in line with the face lies northward of the south jetty, with an opening of 100 feet between the jetty and breakwater.

Lights.—A fixed red light, visible 2 miles, is on the outer end of the northern jetty.

A fixed green light, visible 3 miles, is shown from a lamp-post on the outer end of the southern jetty.

A fixed green light, visible 3 miles, is shown from a lamp-post on the inner end of the southern jetty.

Buoys.—A red cask buoy is moored in 10 feet of water 350 yards southward from the south end of the southern jetty. Shoal water of 8 to 9 feet extends for about 200 yards northeastward of this buoy. Vessels going to the southern jetty should leave the red buoy on the starboard hand, and steer straight for the jetty in not less than 10½ feet at low water.

A red cask is moored in 12 feet of water, 300 yards eastward of the north end of the southern jetty, and marks the port side of the North Channel leading to Queenscliff southern jetty. Vessels approaching the Queenscliff northern jetty may pass on either side of this buoy.

A black cask buoy is moored in 14 feet of water, 800 yards eastnortheastward from the north end of the southern jetty. This buoy lies about 100 yards shoreward of the north side of the fairway.

All vessels should keep to the southward of this buoy to avoid the foul ground to the northward of it, on which there are patches with only 6 to 7 feet of water over them at low tide.

Lifeboat.—A lifeboat, with life-saving rocket apparatus, is kept on Queenscliff southern jetty.

The description of the western shore continued later.

The southern shore of Port Phillip from Point Nepean to Observatory Point, 1¼ miles eastward from it, forms a bight ¼ mile deep; but the depth of water is less than 3 fathoms, and there are numerous sunken patches; the 3-fathom edge of this shallow water and foul ground extends from the shore to 200 yards outside the line of the points of the bay.

Observatory Point (lat. 38° 18′, long. 144° 41′) **and Quarantine station.**—There is a flagstaff on Observatory Point, which marks the western boundary of the quarantine station, and from this flagstaff the coast trends 1¼ miles eastward to another flagstaff, the eastern boundary of the station.

The whole of the shore fronting the quarantine station is steepto, there being 3 fathoms at 100 yards and 5 fathoms at 200 hundred yards off.

The quarantine ground extends between Observatory Point and the eastern boundary flagstaff, the anchorage being in 8 and 9 fathoms 1,500 yards from the shore, with quarantine jetty bearing 188°, and the high light at Queenscliff bearing 315°.

Jetty.—There is a jetty 400 yards eastward of Observatory Point and another at the quarantine station 255 feet long, with a face 63 feet long and 17 feet wide, on the outside of which there is a depth of 17 feet at low water.

Lights.—An occulting white light, 55 feet above water, visible 7 miles, is exhibited from an open steel framework beacon, painted white, situated on the beach at Observatory Point.

A light is shown from the end of the quarantine station jetty. This light is shown only when required for quarantine purposes.

Portsea Jetty.—From the eastern flagstaff the coast trends eastward for 1,500 yards to Point Franklin, the eastern point of Weeroona Bay; in the depth of the bay, at Portsea, there is a jetty 166 yards long and 15 feet wide, with a face 235 feet long and 25 feet wide, at the outer side of which there is a depth of 12 feet at low water.

Portsea, a much frequented sanatorium and watering place, is a telegraph station.

Limestone is abundantly prevalent, and there are numerous kilns, the best lime of the State coming from here.

Population about 300, many of whom are lime burners.

Light.—A flashing green light, visible 3 miles, is shown from the outer end of the jetty.

Buoy.—A red conical buoy is moored in 11 feet water about 100 yards off Point Franklin.

The coast (lat. 38° 19', long. 144° 43').—From Point Franklin the coast takes an easterly direction to Point King; eastward of Point Franklin a bank extends 35 yards from the shore and is steep-to.

Beacon.—A white lattice beacon has been erected on Point King.

Buoys.—Two barrel buoys marking the western end of the channel to Sorrento are moored off Point King; the northern, a black buoy, lies in 8½ feet eastward; the southern, a red buoy, in 7½ feet southeastward from the point.

Submarine cable—Prohibited anchorage.—A submarine cable is laid across the south channel between Points Franklin and Arthur and the south channel fort, passing 300 yards northwestward of No. 5 buoy and 300 yards southeastward of No. 2 buoy. Vessels are prohibited from anchoring within 800 yards on either side of a line drawn from the south channel fort to No. 2 buoy, and thence to the southern shore midway between Points Franklin and Arthur.

Directions—Entering Port Phillip by day.—Vessels usually make the high bold land at Cape Otway, which is easily distinguished by its circular white lighthouse. It is desirable to round Cape Otway at a distance of not less than 3 or 4 miles, and when the lighthouse bears 357°, 6 miles distant, the course and distance to Port Phillip Heads will be about 53°, 65 miles, passing 3½ miles outside Henty Reef. All other dangers are cleared by giving the coast a berth of not less than 2 miles.

If Cape Otway should be rounded early in the evening, with a fresh southerly wind, beware of overrunning the distance, as a strong current after a prevalence of southerly gales often sets northeastward along the land. Bearings of Eagles Nest Light give a good check.

After passing Eagle Nest Point, 36 miles to the northeast of Cape Otway, if the weather be at all clear, Arthur's Seat will be seen rising inland over the waters of Port Phillip before the lower and nearer land in that direction becomes visible. Proceeding onward, the land about Cape Schanck will be seen to the eastward, appearing at first like a long low island trending to the southeast. On nearing the entrance, Barwon Head will open out on the port bow. This headland is a good mark for making the port; but in thick, hazy weather care must be taken not to mistake it for Port Phillip Heads, as by so doing vessels have gone ashore.

Since the channel at the entrance to Port Phillip has been deepened to 40 feet for a width of 1,000 feet to the eastward and for 300 feet to the westward of the line of the range lights, and as there is an additional width to the westward of 400 feet, with a least depth of 37 feet, mariners may now navigate the fairway with greater safety than before by keeping eastward or westward of the mid-channel course, according as they are entering or clearing the heads, while still conforming to the rule of the road.

On occasions when two or more vessels are being navigated through the fairway in opposite directions, the practice of keeping near the line of the range lights, whether entering or clearing, is attended with considerable risks in a place where strong tidal currents and swirls prevent vessels being kept on a straight course.

Deep draft vessels are directed, when crossing the Rip, to keep the high light a little westward of midway between the low light and the Obelisk.

When the lighthouses at Queenscliff are made out they must be brought in line, bearing 42°. This line is 100 yards eastward of the western edge of the 40-foot channel. The Obelisk and Queenscliff high light in line, bearing 38°, marks the eastern edge. Between these lines is the 40-foot channel. When the white beacon on Point Nepean is well open to the northward of the Red Rock Beacon, a course may be shaped for the South or West Channel.

Swan Beacon just open of Shortland Bluff, bearing 48°, leads about 100 yards to the eastward of Lonsdale Rock.

Corsair Rock is cleared by keeping the low lighthouse at Queens-cliff in line with the east end of the houses near the high lighthouse 33° until the white beacon on Point Nepean is well open to the north-ward of the red beacon when going in, or well open to the southward when going out.

Vessels drawing less than 14 feet may, in the day time, pass between Lonsdale Reef and Rock by keeping Swan Point in line with Short-land Bluff, bearing 50°.

Entrance is prohibited when deep-draft vessels are leaving.

At night—Point Lonsdale Light.—The white light marks the safe navigable waters clear of Port Phillip Heads. It also enables the mariner to pick up the entrance to Port Phillip. The red sectors are intended to warn mariners of their proximity to the shore, and should such red lights be seen when navigating outside the heads, the vessel's course should be altered to seaward to get within the white arc of light.

After picking up Point Lonsdale white light, the white lights at Queenscliff from the high and low lighthouses will, respectively, come in sight, then steer to get into the white sector shown from the low lighthouse, which will first be seen on a 46° bearing, then bring the low white light in line with the high white light bearing 42°, and keep those range lights on (which leads across the bar of the newly dredged channel in a least depth of 40 feet) until the red sector show-ing inside the heads eastward from Point Lonsdale Lighthouse has been crossed. A course may then be shaped for the South or West Channel as desired. Deep-draft vessels have to take South Channel, which has a depth of 36 feet in the fairway.

Sailing vessels making Port Phillip Heads from the westward and obliged to stand on and off shore between Eagle Nest (Split) Point and the heads should not shoal the water to less than 20 fathoms nor lose sight of the red light on Eagle Nest Point until well within range of Point Lonsdale white light.

Tidal signals.—The state of the tide is signaled from Point Lonsdale day and night.

Deep-water channel between the heads. This channel is some-what across the tide; to avoid using it and to enter on a straight course a direct channel across the bar has been dredged to 40 feet, as before mentioned. The ranges for the deep-water channel are as follows: Two beacons, the outer white with disc as topmark and the inner an inverted triangle painted black with a vertical white central stripe, are situated in Lonsdale Bight. In range bearing 343° they lead in with not less than 42 feet at low-water spring tides until the beacon on Point Nepean is in range with Rock Beacon.

Caution.—Vessels bound eastward and clearing the heads are cau-tioned against leaving the line of range lights and altering course to the southeastward on reaching the line of the Lonsdale Bight Beacons when other vessels are being navigated inward through the Rip.

Deep-draft vessels should not be taken out on the line of the Lonsdale Bight Beacons, but should be steered straight out with the high light showing midway between the low light and obelisk until Point Lonsdale is well abaft the beam.

Fort Port Phillip (lat. 38° 18′, long. 144° 38′) **from the eastward—By day.**—When steering for Port Phillip from the southward and eastward it is usual to make the land about Cape Schanck, 17 miles to the southeastward of Port Phillip Heads, which can not be mistaken, owing to its bold character and white circular lighthouse on its west and highest part. It is recommended to sight Cape Schanck before running far into the bight for Port Phillip, and should the wind blow strong from the southward it is not safe to run without having sighted it. Having passed Cape Schanck, keep a good offing in proceeding toward the heads until Queenscliff Lighthouses open out, the intervening land of Point Nepean preventing their being seen before the high lighthouse bears 2°, the low lighthouse 14°.

Vessels entering the heads from the southeastward should alter course to starboard before reaching the line of range lights, and steer in to the eastward of that line or on the starboard side of the channel, as previously directed.

By night.—Vessels having passed Cape Schanck Light from 3 to 4 miles off, should steer to make the white light on Point Lonsdale, and keep it in sight until the range lights at Queenscliff are brought into line, then steer through the entrance fairway as directed for vessels from the westward.

Caution.—At night a sailing vessel should keep a good offing, and on no account be hove to when waiting for daylight near Port Phillip Heads. Several vessels that have done so have drifted into danger; others have been lost from this cause, combined with inattention to the lead and the tidal currents.

A vessel should not enter at night without a pilot or against a strong current.

Waiting for tide.—By the tidal signals on Point Lonsdale the time and state of the tidal current may be known; it is advisable for vessels waiting for the turn of the current outside the heads to favor Point Lonsdale shore, where the current runs fairer, and in bad weather small vessels incur less danger from the tide rips, besides having much smoother water.

Pilots.—As there is constantly one pilot vessel outside the heads, when it is practicable to keep at sea, no stranger should attempt entering without taking a pilot; but the channels are so carefully lighted and buoyed that it is quite possible to do so. If proceeding from sea to Geelong and requiring a harbor pilot, save time by sending a telegram from the heads, stating draft of water to the harbor master, in order to have a pilot ready to board the vessel off Point Henry.

Signals—Vessels aground.—Vessels grounding in the South, West, or Hopetoun Channels, thereby obstructing the navigation, are to show the undermentioned signals: By day, two balls or shapes placed vertically 6 feet apart; by night, in addition to the ordinary lights, two red lights, placed vertically 6 feet apart, in globular lanterns of not less than 8 inches in diameter and in such a position with respect to the ordinary white light as to indicate as nearly as possible the position and extent of the obstruction. A lookout is to be stationed on board, or in a boat, to give warning to approaching vessels.

Sailing vessels—To enter the heads with the in-going current.—If a pilot has not been taken aboard a sailing vessel outside the heads, and the last quarter ebb current signal be up, or the in-going current be made, steer, when within 8 or 10 miles of the entrance, to bring the high lighthouse on Shortland Bluff in line with the low lighthouse bearing 42°; and with a fresh fair wind and current steer so as to keep the two lighthouses in line, until the red beacon on the rocky islet off Point Nepean is open south of that point, or, if using the deep channel, steer in with the beacons in Lonsdale Bight in line, until the beacons on Point Nepean and rock are in range. It is, however, not recommended without a pilot or local knowledge of the tidal currents and only when the vessel is fully under command.

To pass eastward of Lonsdale Rock keep Swan Island Beacon open of Shortland Bluff 48° until Point Lonsdale signal house, white with a slate roof, opens well to the northward of the tidal signal flagstaff. Vessels drawing less than 14 feet may, in the daytime, pass between Lonsdale Rock and Reef by keeping Swan Point in line with Shortland Bluff 50°.

To pass westward of Corsair Rock keep the low lighthouse at Queenscliff in line with the east end of the houses near the high lighthouse, 33°, until the white beacon on Point Nepean is well open north of the red beacon on the rocky islet off that point.

With a scant or light easterly wind and in-going current Swan Island Beacon must be kept well open of Shortland Bluff, so as to avoid Lonsdale Rock.

To enter the heads against the outgoing current steer, when within 2 miles of the heads, to get the low lighthouse open east of the high one, until near Point Lonsdale, when haul as close round Lonsdale Reef as practicable, taking care, however, if the draft be more than 14 feet, to avoid Lonsdale Rock by not shutting Swan Island Beacon in with Shortland Bluff, and on no account to shut off Swan Point with Shortland Bluff, until clear of Lonsdale Reef, and the red beacon on the rocky islet off Point Nepean is open south of that point, when the rocks and reefs in the entrance are cleared.

Working in between the heads is best done near the time of slack water, when the race is nearly quiet, and the vessel will be much more under command. In standing to the westward, Swan Island Beacon must be kept open of Shortland Bluff until Lonsdale Rock is cleared. Vessels of light draft may stand more inshore, keeping Swan Point a little open of Shortland Bluff, making due allowance for the set of the ingoing current. After clearing Lonsdale Rock and Reef do not bring Queenscliff low lighthouse east of 59° in order to avoid Victory Shoal and the foul ground between Point Lonsdale and Shortland Bluff. Lonsdale Bight should be avoided by all vessels.

In standing to the eastward, do not proceed farther than when the obelisk on Shortland Bluff touches the east side of the high lighthouse bearing 39° to avoid the tide rips near Point Nepean.

At night.—The passage through the heads should not be attempted at night, except with steam or a commanding fair wind; to enter under either of these favorable circumstances, when the high and low lights at Queenscliff are clearly distinguished, bring them in range bearing 42°, which leads through the fairway in 40 feet.

Should the wind become scant and a vessel be compelled to tack when near Lonsdale Reef or the Corsair Rock, these dangers will be avoided by vessels of light draft so long as Queenscliff low white light is kept in sight; but they must go about or haul toward mid-channel before the low light changes from white to red.

Towing hawsers.—The marine board of Victoria is of opinion that the only form of towing gear which may be relied on for safe towage when navigating the channel through Port Phillip Heads consists of a length of 15 to 20 fathoms of wire attached to a full length, 120 fathoms, of rope hawser.

Anchorages.—Having entered and cleared the dangers which lie between the heads, proceed northeastward for the anchorage off Shortland Bluff, toward the West Channel; or if of heavy draft, eastward for the anchorage off the Quarantine Station, in from 8 to 9 fathoms water, in the entrance of the South Channel.

Vessels detained inside Port Phillip Heads may during northerly or westerly winds anchor in 6½ fathoms, with Queenscliff high lighthouse bearing 286° and Swan Island Beacon in line with No. 1 buoy, West Channel; or at night with the light on Observatory Point bearing 194°. Vessels of light draft when anchoring off Shortland Bluff should, in order to keep the fairway to the West Channel clear, bring up as close as possible to the northwest side of the channel.

If it is necessary to anchor off the quarantine station before proceeding through the south channel, after getting well inside the heads, steer to the eastward along the north side of Point Nepean, avoiding the shoals which front the shore by keeping Barwon Head just open of Point Lonsdale; or, at night, by keeping within the limits of the light shown from Point Lonsdale Lighthouse, between the bearings of 280° to 266°; when the high light at Queenscliff bears about 315°, keep it so astern and anchor in 8 or 9 fathoms, abreast of the quarantine station, half or three-quarters of a mile from the shore.

It is not advisable in bad weather to anchor in either the West or South Channel, on account of the current and the loose nature of the bottom; but in southwesterly gales small sailing vessels find good shelter in 16 to 18 feet of water about ¼ mile 87° from the northeast point of Swan Island. Vessels bound up, and caught in the South Channel with a north or northwesterly gale, will find anchorage in Capel Sound, in 5 to 7 fathoms sand, by bringing the White Cliff to bear 233°, and Eastern Lighthouse bearing 98°; but if daylight and the wind permit, it would be better to get back to the anchorage off Shortland Bluff.

No stranger should anchor close to the heads, except to save the vessel from going ashore; although coasters sometimes, to avoid being carried by the current inside the heads in a calm, anchor at about a mile outside, where the bottom is sandy; and sometimes in the bight between Barwon Head and Point Lonsdale.

Caution.—Strangers who through stress of weather bring up here or at the anchorage off Shortland Bluff should not attempt to proceed above these anchorages without a pilot.

Examination anchorage.—The examination anchorage is situated close northwestward of Observatory Point. It is an oblong area, the limits running 1,500 yards 27°, and 2,000 yards 297°, from its southeastern corner, which is situated 300 yards 27° from Observatory Point Lighthouse.

The coast.—The southern shore of Port Phillip from Point King curves southeast about 1¾ miles to the Sisters, a double point with a sand beach between, from the east of which the coast, after trending east-southeastward 2½ miles to White Cliff, a conspicuous cliff of bare sand over 80 feet high, takes an east-northeast direction for 6½ miles to the foot of Arthurs Seat. Between Point King and the Sisters is Sorrento.

Sorrento (lat. 38° 20′, long. 144° 45′) is a watering place. The population is about 900. There is a telegraph station here.

Jetty.—A jetty projects 183 yards from the shore in a northeast direction, with a face 160 yards long and 22 feet wide, having a depth of 10½ feet at low water along the outer side.

Light.—A flashing white light with red and green sectors, visible 2 miles, is shown from a lamp-post on the end of Sorrento Jetty.

Buoys.—A black cask buoy is moored in 18 feet on the northeastern edge of the channel, 400 yards northward of the outer end of the jetty.

A small black cask buoy is moored in 6½ feet on the northern edge of the channel, southeastward from the same part of the jetty.

A red barrel buoy, marking the southern side of the eastern channel to Sorrento, is moored in 17 feet off the Sisters.

Directions.—Vessels entering the channel to Sorrento at its western entrance off Point King should keep between the two buoys off the point, the least depth in the fairway being 9 feet.

To enable vessels up to 7 feet draft to make Sorrento Jetty, narrow sectors of white light will be shown to the northward between the buoys marking the entrance channel off Point King and to the eastward over a safe course from Canterbury Jetty to Sorrento. On either side of the white sectors the light will show green on the starboard and red on the port hand when approaching Sorsento Jetty by the above-mentioned channels.

Life-saving apparatus.—There is a life-saving rocket apparatus kept at Sorrento.

Canterbury Jetty, about 1 mile westward of White Cliff, is 400 yards in length, with 12 feet of water at its extremity.

A red cask warping buoy in 5 fathoms lies in line with the jetty at about 60 yards off its outer end.

A black barrel buoy, moored in 4 fathoms, marks the southern edge of the bank off Canterbury Jetty, which extends northward of the southern side of South Channel.

Aspect (lat. 38° 22′, long. 144° 48′).—The land from Nepean to White Cliff has hills 100 to 225 feet high on it with numerous lime kilns, wells, and some ponds. Between White Clig and Arthurs Seat the country is flat, and at 3 miles to the eastward of the cliff it appears to be swampy, with a creek intersecting the shore midway between White Cliff and Arthurs Seat.

Rye.—About ½ mile eastward of White Cliff is the village of Rye, a favorite summer resort for visitors.

The population is about 100.

Jetty.—A jetty projects 533 yards from the shore in a northerly direction, with 12 feet at low water along its outer end.

A warping buoy is moored in 40 feet of water 390 feet off the northwest corner of the jetty.

Light.—A flashing white light, 25 feet above water, visible 3 miles, is shown from a lamp-post on the outer end of Rye Jetty.

South Sand—Banks.—From Point King to White Cliff the coast is fronted by a bank, mostly of sand with weeds, extending, midway, 2 miles from it. This bank, named the South Sand, has generally 8 to 10 feet water upon it, with some small hollows of deeper water and numerous knolls, on some of which there are only 1 to 6 feet water. Two other banks extend together about 2 miles eastward from the South Sand and terminate near the South Channel Pile Lighthouse.

Sorrento Channel—Buoys.—Between the southeastern extremity of South Sand and White Cliff is the entrance of a channel ¼ to ⅓ mile wide, with 4 to 9 fathoms water, trending along the coast toward Point King, but a bar of from 8 to 10 feet extends from Point King to the shoal bank eastward. This passage is named Sorrento Channel and is buoyed.

The east side of this channel is marked by four black buoys and the west side by three red buoys. The channel has a minimum depth of 9 feet at low water, but there is a small patch of 6 feet about 100 yards west-northwestward of the northern black buoy.

Capel Sound is a clear space 2 miles long, east and west, and 1½ miles wide, bounded to the northward by the banks just described and to the southward by the coast extending eastward from White Cliff. There are regular depths of 6 to 8 fathoms throughout the greater portion of the sound, over a bottom of sand and shells and mud, but shoal water extends from ¼ to ⅓ mile off the south shore.

From Capel Sound to Arthurs Seat the shore continues bordered by a shoal ⅓ mile broad, the depths increasing from 3 fathoms at the edge of the shoal to 7 fathoms at 1½ miles from the shore over a bottom of sand and shells.

Rosebud Jetty, about 1 mile west-southwestward of the Eastern Lighthouse, is 167 yards in length, with a depth of 5 feet at its outer end.

Light.—A light is exhibited from a post at the end of the jetty.

Anchorage.—Vessels entering, and caught in South Channel by a northerly or northwest gale, will find anchorage in 5 to 7 fathoms in Capel Sound, with White Cliff bearing 233°, and the top of Arthurs Seat 98°; but if daylight and the wind permit, it would be better to run back to the anchorage off Queenscliff Lighthouses.

By night the Pile Light shows red over safe anchorage in Capel Sound.

Arthurs Seat is situated nearly 15 miles from Shortland Bluff; it is a conspicuous bluff 975 feet high, sloping down to the southeast. From the southward its northwestern extremity appears precipitous, and being the highest land on the coast, is a remarkable object by which to distinguish the entrance to Port Phillip. On its summit the old lighthouse tower still stands.

Light (lat. 38° 21′, long. 144° 55′).—A flashing white light with red sector, 100 feet above water, visible 13 miles, is shown from a white iron lighthouse on the shore northwestward of Arthur's Seat.

Water.—The land for 3 miles northeastward of Arthur's Seat is low, with good spring water near the shore north of Arthur's Seat.

Coast.—Continued later.

Western shore of Port Phillip.—

Swan Island is low and marshy, with a ridge of sand hummocks along its southeastern shore, and is separated from the northeast point of Queenscliff Peninsula by a shallow opening 100 yards wide, forming the south entrance to Swan Bay. Swan Island is nearly 2 miles long, east-northeast and west-southwest, and one mile across at its broadest part; but it is nearly divided in two by a bight, with a small islet in it, on its north side. There are three islets close to the southwestern extremity, and another close to the north point.

Queenscliff Bight.—Between Shortland Bluff and the southeast part of Swan Island is a bay half a mile deep; but it is fronted by a bank having irregular depths of 3 to 16 feet water on it, the outer edge of which, from 300 yards off Shortland Bluff, trends east-northeastward nearly 2 miles to Swan Spit.

Many shoal patches, with from 3 to 6 feet upon them, have formed in Queenscliff Bight.

Buoy.—A black can buoy, No. 1, marks the outer edge of this shoal ground, at 1,400 yards southward from Swan Island beacon.

Swan Spit.—A bank with irregular depths of 3 to 15 feet extends out from the southeastern shore of Swan Island, a distance of 1,600 to 1,800 yards eastward and southeastward of Swan Island beacon; the edge of the bank is not well defined, and subject to constant changes in depth, with formation of sandy knolls, carrying 14 to 15 feet on them, and with 19 to 20 feet between them. The end of the spit, with 16 feet on it, is 1,800 yards southeast from Swan Island beacon and shows on the chart as a detached shoal.

At night a red ray from Queenscliff low light shows over the end of Swan Spit.

Buoy.—Buoys described later.

Swan Island Beacon (lat. 38° 15′, long. 144° 41′) is a steel framework structure with sloping sides and is painted white with a red top. It is surmounted by a staff and globe.

Submarine cables.—Two submarine cables are laid between Swan Beacon and Observatory Point. Both cables cross Royal George Shoal about 400 yards eastward of No. 1 Royal George Buoy, they then divide, one passing about 100 yards eastward and the other 300 yards westward of No. 2 beacon (Pope's Eye Fort) on Pope's Eye Bank. The cables lie respectively 200 and 400 yards eastward of Pope's Eye Buoy, when they run almost in a direct line to a point on the southern shore about 100 yards westward of the quarantine west boundary flagstaff.

Vessels are cautioned that anchorage is prohibited within 400 yards on either side of a line from Swan Beacon to No. 2 beacon (Pope's Eye Fort), thence to the quarantine west boundary flagstaff.

Submarine mining.—A submarine mining practice ground has been established on the eastern side of Swann Island in the entrance to Port Phillip. It is bounded by the undermentioned imaginary lines:

(a) On the south by a line drawn 313° from Swan Spit Buoy No. 3 to Swan Beacon.

(b) On the east by a line drawn 12° from Swan Spit Buoy No. 3 to a small black buoy moored at a distance 1,000 yards 86° from the northeastern extremity of Swan Island.

(c) On the north by a line drawn approximately 281° from the black buoy mentioned above to the outer pile marking the channel to Swan Island Jetty.

(d) On the west by a line drawn 188° from the above-mentioned pile to the coast, and from that point the east coast of Swan Island as far as Swan Beacon.

Vessels are prohibited from entering this area, the limits of which have been inserted on the charts, during the occasions when mining practice is in operation, of which previous notice will be given.

Swan Bay (lat. 38° 15′, long. 144° 42′) is a large shallow lagoon northward of Queenscliff, 5¼ miles long and 1¼ miles across, with an opening 1¼ miles wide, between Swan Island and a narrow tongue of land projecting nearly 1½ miles from the north-northeastward.

From the north point of Swan Island a mud flat stretches nearly across the opening to Duck Islet, between which and the south end of the tongue of land is a narrow boat channel, having 6 to 13 feet water, marked by white beacons on the northeast side, and by a black beacon on the southwest side of the entrance; but a bank extends from the northeastern extremity of Swan Island to the tongue of land, forming a 3-foot bar across the mouth of the boat channel.

The eastern and southern parts of Swan Bay are mostly occupied by mud flats, leaving only portions of the west side accessible even to boats, there being generally not more than 2 to 5 feet water in the bay.

Coast.—The shore from the northern entrance point of Swan Bay extends 2¼ miles northward to South Red Bluff and thence 1,500 yards farther in the same direction to a point close to the northward of St. Leonards Jetty. A continuation of the bank which stretches northward from Swan Island borders this shore, from which it projects from 200 to 300 yards, with 2 to 6 feet water on it.

Beacons.—There is a white beacon on South Red Bluff and another on the coast at 1¾ miles southward of it.

St. Leonards is a small postal fishing village and watering place, 28 miles southwestward of Melbourne, with a population in 1911 of about 100 persons. Communication by coach to Drysdale, thence by rail; also by steamer to Portarlington.

Light.—A fixed green light, visible 2 miles, is shown from a post on the end of St. Leonards Jetty.

Jetty.—A jetty 500 feet long, with a face 115 feet long and 22 feet wide, is built at St. Leonards, with a depth of 9½ feet over rocky bottom, at outer end. From the southern end of the face a breakwater of close piling extends southwestward for 110 feet, and affords shelter for boats.

Buoy.—A black buoy is moored 200 yards southeastward of St. Leonard Jetty in 9 feet of water.

Coast.—Continued later.

Sand banks and channels within the entrance.—The first 2¼ miles within the heads is free from dangers, but above that distance it is crowded with sand banks, radiating nearly 9 miles from their southern and western extremities. Between these banks there are several channels, three being buoyed, namely, the South, West, and Coles Channels; the others are narrow and intricate.

SAND BANKS.

Great Sand and Middle Ground, together with the shoal spit extending eastward from the latter, form a large triangular shoal, with its southwestern angle situated 2 miles northeastward from Observatory Point, its southern side extending 9 miles to the eastsoutheastward, and its western side 5 miles to the northeastward.

Great Sand is a flat which occupies all the western portion of the above triangle, and is about 4½ miles long, northeast and southwest, and 2¼ miles wide. The depths on it are from 1 to 6 feet at low water, and there are several patches which dry.

Mud Islands are three low wooded islands on a bank of less than 1 mile in width, on the center of the Great Sand.

Middle Ground lies 1½ miles southeastward from Great Sand, the depth between them varying from 3 to 20 feet. It is about 1¾ miles long, east and west, and ¼ mile wide, and has 1 to 5 feet water on it. Other banks with less than 6 feet water lie to the eastward from it, and shallow water—6 to 10 feet—extends 4 miles eastward from it in the form of a spit, which terminates in a point at nearly 2 miles northwestward from Eastern Lighthouse.

Pinnace Channel is a gully with 22 feet water, on the eastern side of the Great Sand and Middle Ground Shoal.

Popes Eye Bank (lat. 38° 17', long. 144° 42'), which forms the northeast side of the south entrance of West Channel, is a bank of sand 1 mile long, east-northeast and west-southwest and about 200 yards broad, with from 3 feet to 3 fathoms of water. Its southwestern extremity is 1¾ miles eastward from Shortland Bluff.

A small detached shoal, with 18 feet water on it, lies 250 yards southwestward from Popes Eye Bank; from it the depth increases to 30 feet at a distance of 1,100 yards southwestward of the bank.

Light.—A flashing white light, 24 feet above water, visible 7 miles, is shown from a red beacon on the western side of the stone annulus 8 feet high, known as Popes Eye Fort.

Buoy.—The southwestern extremity of the shallow water extending from Popes Eye Bank is marked by a spherical buoy, painted in black and white horizontal stripes and surmounted by a cage, moored in about 6 fathoms, situated 1,200 yards southwestward from Popes Eye Beacon.

West Middle Sand.—West Middle Sand, which separates Symonds and Lœlia Channels, is about 5 miles long northeast and southwest. Its width at the northeast end is 2 miles, in the middle about ½ mile, and its southwest end terminates in a spit 800 yards northward from the center of Popes Eye Bank. It is irregular in shape. The depth of water on it varies from 1 to 15 feet. There are several ridges, with less than 6 feet water on them.

Beacon.—A black single-pile beacon stands on the northeast end of the shoal, and indicates the east entrance of the Symonds Channel.

Buoy.—The southwestern end is marked by No. 4 red conical buoy.

William Sand, which forms the northwest side of Lœlia Channel and the southeast side of West Channel, is 4 miles long in a southwest and northeast direction, and ¼ mile to ½ mile broad within its 3 fathoms edges. From ½ mile within its southwestern extremity to about 1,500 yards within its northeast spit William Sand rises to a narrow ridge, with 1 to 7 feet water over it.

Its southwestern end is 1.1 miles eastward from Swan Island Battery.

West Sand.—This shoal, which forms the western side of West Channel and divides it from Coles Channel, is about 3 miles long, northeast and southwest, and at its northern end about 1¼ miles wide. The depths on it vary from 1 to 17 feet. It rises to several narrow ridges with 1 to 5 feet water over them. The longest of these ridges extends from a position 1,300 yards northeastward from Swan Point, 2 miles northward, and is from 50 to 500 yards broad. The northeastern end of another ridge, about 1 mile long, northeast and southwest, lies about ½ mile west-southwestward from West Sand Lighthouse. The northern end of West Sand is ½ mile eastward from the white beacon on South Red Bluff. The northeastern end is marked by West Sand Lighthouse. Between these extremities a bight, 1,340 yards wide, with 4 to 3¼ fathoms water, trends ½ mile southwestward into the shoal.

Light (lat. 38° 12', long. 144° 45').—An occulting white light with red sector, 35 feet above water, visible 11 miles, is shown from a screw-pile lighthouse, in 15 feet water, on the northeast side of West Sand, West Channel.

South Sand.—Has been described.

CHANNELS.

South Channel (lat. 38° 19', long. 144° 45')—**Depth.**—The least water which a vessel must pass over in South Channel is 33 feet at low water.

The South, or Great Ship Channel, is bounded to the southward by the South Sands, and on the north side by the southern edge of Great Sand and Middle Ground.

South Channel is 1 mile wide at its western entrance, between No. 2 red buoy and No. 3 black buoy, and ½ mile wide abreast of No. 6 red buoy; but only ¼ mile wide at its eastern entrance, between the Pile Lighthouse and No. 11 black buoy. The depths in the channel are very irregular, varying from 10 fathoms in the middle of the western entrance to 20 fathoms at 1½ miles farther eastward; thence the depth varies from 11 to 16 fathoms between No. 4 red and No. 5 black buoys, gradually decreasing eastward to 33 feet north of the Pile Lighthouse.

The banks on either side of South Channel are steep-to.

Between Nos. 9 and 11 black buoys and almost in alignment therewith, the channel has been dredged to a uniform depth of 33 feet at low water for a width of 450 feet extending southerly from the north edge of the channel.

Speed.—The speed of steam vessels must not exceed 7 knots when passing through the dredged cutting.

Tidal currents.—The ingoing current sets through the South Channel at a rate of 1 to 1¼ knots and sets strongly over the northern banks, and the outgoing ¾ knot to 2 knots and sets strongly over the southern banks.

Shoals at the entrance.—Southeastward 800 to 1,600 yards from the southwestern extremity of Great Sand Shoals, with 21 to 23 feet water over them, extend 200 yards into the channel beyond the line of the buoys.

A shoal with 24 to 30 feet water on it, ¼ mile long and about 200 yards wide, lies in the fairway, its western end being 1,300 yards southward from the southwestern extremity of Great Sand.

A shoal, with 26 feet water (marked by a light buoy) lies 250 yards southwestward from the southwestern end of the shoal in the fairway.

A shoal, with 28 to 30 feet water, lies 800 yards westward from the 26 feet shoal.

A shoal, with 28 to 30 feet water, lies 800 yards west-southwestward from the 26 feet shoal.

South Channel Light (lat. 38° 20′, long. 144° 51′).—An occulting white light, with red sectors, 27 feet above water, visible 9 miles, is shown from a pile lighthouse, in 21 feet of water, at the eastern end of the channed.

Lightbuoy.—A black lightbuoy, showing a flashing red light every 6 seconds, is moored on the southern side of the 26 feet shoal, at a distance of nearly 1 mile, 11° from Portsea Jetty Light.

Buoys—North side of channel.—No. 1 black can buoy with cage is placed at the west end of Great Sand.

No. 3 black can buoy on the south edge of Great Sand.

Nos. 5, 7, 9, 11, and 13 black can buoys on the south edge of Middle Ground.

No. 15 black lightbuoy at the east end of Middle Ground. This buoy exhibits 10 feet above water a white flashing light every 6 seconds.

All buoys appear white from a distance, on account of birds.

Buoys—South side of channel (lat. 38° 17′, long. 144° 43′).— The western spit of the South Sand is marked by No. 2 red conical buoy; the north edge of the bank is marked by Nos. 4, 6, and 8 red conical buoys.

The south side of the western end of the dredged portion of the channel is marked by a lightbuoy, No. 10, showing a flashing red light every 4 seconds, and the south side of the eastern end of the dredged channel is marked by a lightbuoy, No. 12, showing an occulting green light every second.

Directions—South Channel (lat. 38° 18′, long. 144° 44′).—For the South Channel, after having entered and cleared the dangers between the heads, steer for a position about 600 yards to southward of the Popes Eye black and white buoy, from which the range South Channel pile lighthouse and the eastern lighthouse at the foot of Arthurs Seat are in line bearing 107°. Alter course to this range, 107°, and when about 1 mile westward of the black light buoy south-southeastward of No. 1 black buoy, leave the range and steer to pass about 200 yards to southward of this lightbuoy, avoiding the 25-foot spot 800 yards southwestward of the lightbuoy. The range passes over the fairway shoal in 25 feet of water, but 100 yards to the southward of the lightbuoy the depth is 34 feet, with 37 feet outside this again for

300 yards. When past this lightbuoy haul up again onto the range, passing between the black and red buoys which mark the northern and southern sides of the channel.

When approaching Nos. 8 and 9 buoys, haul to the northward and steer through the dredged channel, leaving the lightbuoys close-to on the starboard hand.

Having passed to the southward of No. 11 black buoy, continue steering about 109°, so as to pass 400 to 600 yards southward of No. 13 black buoy, and then steer 87° and round, on the southeast side, No. 15 black lightbuoy, which marks the eastern spit of the Middle Ground.

Signals, when aground in this channel, have been given previously.

Working through.—When working through the South Channel be guided by the lead, not standing into less than 4 fathoms on either side nor within the line of buoys, bearing in mind the tidal currents which set over the banks. After passing South Channel pile light-house there is plenty of room between the Middle Ground and the shore, which may be approached to nearly ¾ mile in 5 fathoms. When clear of the Middle Ground and to the northward of Martha Point a vessel may stand westward until Station Peak is open of the high-land on Point George, bearing 309°, which will lead northeastward of all the banks.

At night, after getting well inside the heads with the Queenscliff Lights in line and Point Lonsdale white sector of light in sight, bear-ing about 273°, steer about 87°, taking care to keep clear of the shoals which border Point Nepean by keeping in the sector of white light shown from Point Lonsdale Lighthouse. As the low lighthouse at Queenscliff shows a white light up the South Channel, the north banks in the west entrance of the channel are avoided by not shutting in the white light on a 289° bearing, but this line passes very close to the shoals. When south of Popes Eye Shoal, the range through the South Channel (the Eastern light under Arthurs Seat in line with the South Channel pile light) will come on bearing 107°. Steer as directed for daytime; bearings of Popes Eye bank light will indi-cate when to leave the range lights temporarily so as to pass south-ward of the flashing red light on fairway shoal buoy; when the West Sand light turns from white to red alter course to pass close north-ward of the lightbuoys marking the dredged channel.

On passing the lightbuoy at the east end of the cutting, steer 109° for 1 mile (the pile light will open first red, then white), when alter course to 92°, taking care not to change the pile light from white to red till east of No. 15 lightbuoy or till the Eastern light has changed from white to red; the vessel is then clear of the east end of the Middle Ground. Attention must be given to the tidal currents.

Sailing vessels working down, and when to the northward and in the vicinity of the Middle Ground and Great Sand, will know they are getting into danger when either the South Channel pile light or the Eastern light under Arthurs Seat shows red. They will also know their proximity to the eastern shore when the Eastern light ceases to be visible.

Directions for Hobson Bay and Geelong given later.

West Channel.—West Channel, between West Sand and William Sand, is 5 miles long in a north-northeast direction, and is from 200 yards to ½ mile wide. It is the channel most frequently used.

Depth.—The least depth in the fairway is 17 feet, over a bottom of sand and shells.

West Channel is bounded on the west by the bank extending off Queenscliff Bight and by Swan Spit farther on by the shoal water extending eastward from the south end of West Sand, and in the northern part by West Sand itself; on the east by the southwest end of West Middle Sand and by William Sand.

The narrowest and shoalest part of the channel is about 2 miles above the southwest entrance, the width being little over 200 yards and the depth 17 feet.

The north entrance to West Channel is marked by the West Sand Lighthouse.

West Channel may be considered safe for vessels not drawing more than 16 feet. If the tide could be depended upon it would be quite possible and safe for vessels of 17 feet draft to use this channel, but the tides are so influenced by the wind that it is not safe to trust to the calculated time of high and low water for rise and fall. An easterly wind has a similar effect to that which it has on the outer coast, viz,. that of keeping the tide low; a westerly or southerly wind keeps the tide up.

The southwest entrance is divided into two channels by the Royal George Shoal, that to the northward between Royal George Shoal and Swan Spit is 600 yards wide with 22 feet least water; the other channel between the east end of Royal George Shoal and the southwest spit of West Middle Sand is about 450 yards wide with 19 feet least water.

SHOALS IN WEST CHANNEL.

Royal George Shoal (lat. 36° 16′, long. 144° 42′), which lies between the two entrances of the West Channel, and is midway between Popes Eye bank and the bank extending from Swan Island, is ¼ mile in length, east and west, with a depth of 13 feet least water on it.

At night a red ray from the Queenscliff low light shows over this shoal.

Buoys.—The west end of the shoal is marked by No. 1 Royal George buoy, a sperical buoy, painted with black and white horizontal bands, moored in 19 feet water. No. 2 Royal George buoy is a light buoy painted in red and white horizontal bands (on account of birds it generally looks white) which marks the east end of the shoal, and shows a green occulting light, two flashes every six seconds. It is on the alignment of Observatory Point and Popes Eye bank lights.

A shoal, with 13 feet water on it, lies 1 mile north-northeastward from the east end of Royal George Shoal.

A small patch, with 15 feet water on it, lies about midway up the channel and 1.1 miles eastward from the white beacon on the northern entrance point of Swan Bay.

Shoals with 17 feet or more water on them are not here separately mentioned.

Buoys in West Channel—The western side of the channel is marked by:

No. 1 black can buoy, on the edge of the bank off Queenscliff Bight, at 1,400 yards southward from Swan Island beacon.

No. 3 black can buoy, on the end of Swan Spit.

A black can buoy, on the eastern side of the 13-foot shoal noticed above.

A black can buoy nearly 1,600 yards eastward from Swan Point and 300 yards westward of the line of the range of the channel.

No. 5 black can buoy, on the west side of the 15-foot patch noticed above.

Nos. 7 and 9 black can buoys, on the edge of West Sand, in the northern part of the channel.

The eastern side of the channel is marked by No. 4 red conical buoy, on the southwestern extremity of West Middle Sand, and by Nos. 6, 8, 12, 14, and 16, which are red conical buoys, except No. 12, which is a spherical lightbuoy showing a flashing red light every six seconds, on the edge of William Sand.

Directions—West Channel (lat. 36° 16′, long. 144° 42′).—The West Channel may be considered safe for vessels drawing not more than 16 feet of water.

If bound through the West Channel, after entering the heads and clearing the dangers in the entrance, steer about 66° to pass between No. 1 Royal George buoy and Popes Eye beacon. From thence steer to pass 200 yards to the southeastward of No. 2 Royal George lightbuoy, and bring Observatory Point Light beacon in line with the beacon on Popes Eye Bank, 207°; this line will lead through the channel as far as No. 12 lightbuoy, bearing in mind that a patch of 15 feet lies 100 yards eastward from No. 5 buoy. From No. 12 lightbuoy steer to the north-northeastward, passing about 200 yards to the southeastward of the West Sand Lighthouse.

With a scant wind, proceeding up against the outgoing current, do not stand too near the eastern bank, as the current sets upon it, especially at the northern end of the channel.

The West Channel may also be entered between Royal George Shoal and the bank extending from Swan Island by steering about 56° from the entrance between Points Lonsdale and Nepean to pass between Shortland Bluff Buoy and No. 1 Royal George Buoy, then steer 74° and round the Swan Spit Buoy at the distance of 200 yards, after which bring Observatory Point Light Beacon in line with the beacon on Popes Eye Bank, 207°, and proceed as before directed.

Signals when aground in this channel have been given.

At night, enter between the heads with the Queenscliff Lights in line, bearing 42°, and when Point Lonsdale red sector of light is crossed leave the line of range lights and steer about 66° to pass to the northwestward of the light on Popes Eye Beacon; thence steer to pass about 200 yards to the eastward of the Royal George Light Buoy.

After passing Royal George Buoy, bring the light on Observatory Point in range with the light on Popes Eye Bank, bearing 207°, which leads through the channel as far as the No. 12 Lightbuoy, whence steer to pass about 200 yards to the southeastward of the West Sand Lighthouse Light.

To enter between Royal George Shoal and the bank extending from Swan Island, steer with the range lights at Queenscliff in line 42° until Lonsdale Point Light changes from red to white; then alter course to about 56°, and when, after crossing the red ray, the white light is exhibited from Queenscliff low lighthouse comes in sight, steer about 74°, taking care to keep in the sector of white light; then bring the light on Observatory Point in range with the light on Popes Eye Bank, bearing 207°, and proceed as before directed.

The West Sand Light changes in the channel from white to red on a 42° bearing. The above directions lead through the channel in not less than 17 feet at low water.

Attention must be paid to the tidal currents, which do not set fairly through this channel.

It is recommended that vessels from seaward should enter West Channel to the southward of Royal George Shoal, and those outward bound should leave to the northward of it.

Coles Channel, between West Sand and the western shore of Port Phillip, is a 2-fathom passage used by small vessels acquainted with the locality.

The west side of Coles Channel is formed by the shoal extending northward from the east end of Swan Island and along the western shore. The channel is ¼ to ½ mile wide, with 2¼ to 3½ fathoms in its north and south entrances, but only 2 fathoms in its central and widest part; the water shoals suddenly toward West Sand, but gradually toward the shore.

Buoys.—The eastern side of the channel is marked by three red conical buoys on the west edge of West Sand.

Symonds Channel, between Great Sand and Middle Sand, is ¾ mile wide at its southwest entrance, between Great Sand and Popes Eye Bank, whence the channel extends 6 miles in a northeasterly direction, and is 1,340 to 2,000 yards wide, until within 1½ miles of the beacon on the east point of West Middle Sand, where detached banks, with 16 and 17 feet of water on them, so encumber the channel that at ½ a mile southwest of the beacon there is only a width of about 300 yards, with a depth of 17 feet, and there may be other shoals with less water.

There are two knolls with 15 and 16 feet water on them, on the southeast side of Symonds Channel, eastward from Popes Eye Bank; but, with these exceptions, there are generally 9 to 5 fathoms from the southwest entrance to within 1½ miles of the beacon.

Symonds Channel may be made available in northerly or northwest winds, when unable to fetch through West Channel; but it is only recommended for small vessels, not being buoyed.

Lœlia Channel extends from the southwest part of West Channel 4 miles in a northeasterly direction, between West Middle Sand and William Sand; its southwest entrance is 100 yards wide, with 3¼ fathoms water; thence the channel increases 600 to 800 yards in width, with 3½ to 5 fathoms, until within ⅛ mile of its northeast entrance, which is only 100 yards wide, with 3¼ fathoms water. This channel is not buoyed.

WESTERN SHORE OF PORT PHILLIP.

Coast.—Continued.

Point George.—From St. Leonards the shore trends northward 1¼ miles to Indented Head, whence it recedes in a northwest direction 1¼ miles to Point George, close to the northward of which is White Woman Rock. From the point of St. Leonards to Point George a bank, with 2 to 3 feet water on it, borders the shore, from which it extends from 200 to 300 yards.

Buoy.—A black cask buoy, in 9 feet of water, marks the northern edge of a 4-foot shoal situated nearly ½ mile north-northeastward from White Woman Rock.

Governor Reef, about ½ mile southeastward of Indented Head, is a patch with 1 foot water on it, marked by a beacon, consisting of a pole and ball, painted black, and about 16 feet above high water.

Prince George Bank (lat. 38° 06′, long. 144° 45′)—**Lightbuoy**—
Buoys.—From ½ mile off the point of St. Leonards the 3-fathom
edge of the shoal water, which extends from the shore, trends in a
northerly direction 3¼ miles to the northeastern extremity of Prince
George Bank, ¼ mile off which is moored a black lightbuoy in 6
fathoms of water, which shows a flashing white light every three seconds, a little more than 2 miles from Point George.

Two black can buoys eastward of Governor Reef mark the southeastern edge of Prince George Bank.

From the northeastern extremity of Prince George Bank, its
northern edge, extends 3 miles in a westerly direction, to ¼ mile
off the shore to the westward. There are two 4-foot patches on
the northern edge of the bank nearly in line with Prince George
Lightbuoy bearing 78°, one patch being distant 1 mile and the other
1¾ miles from the shore.

Compass adjustment.—To enable vessels navigating Port Phillip entrance to test their compasses, the following true bearings of
various beacons and marks are given:

Outside Port Phillip Heads.

High lighthouse on with low lighthouse	42°	29′
Rock beacon on with—		
Lonsdale bight black and white beacon	326°	56′
High lighthouse	19°	28′
Swan Beacon	34°	20′
Pile, Pope's Eye Fort	57°	55′

West Channel.

West channel pile lighthouse on with—		
Cheviot Hill	210°	38′
Swan Beacon	221°	09′
High Lighthouse	223°	31′
White Beacon, South Red Bluff	283°	11′
Swan Beacon on with high lighthouse	229°	00′
Pile, Popes Eye Fort on with—		
Eastern flagstaff, quarantine station	173°	30′
Cheviot Hill	215°	05′
West Channel pile lighthouse	28°	43′

South Channel.

South Channel pile lighthouse on with—		
High Lighthouse	291°	12′
St. Pauls Hill	252°	36′
White Cliff	221°	17′
Eastern Light	107°	28′
Western flagstaff, quarantine station with Cheviot H.ll	249°	23′
Pile, Popes Eye Fort on with—		
Swan Beacon	349°	21′
West Channel pile lighthouse	331°	17′

The western arm of Port Phillip has at its head the town and harbor of Geelong.

Geelong approaches.—The southern shore of the western arm, after a slight curve for a little more than 1,500 yards in a north-westerly direction from Point George, extends westward 1½ miles, and thence, with a slight bend to the southward, nearly west 1⅜ miles to Point Richards.

For the first 2 miles from Point George shoal spits and detached patches, with from 2 to 6 feet water on them, project upward of ⅛ mile to ¼ mile from the shore. From 1 mile northwestward of Point George to ½ mile eastward of Point Richards the 3-fathoms edge of a continuation of Prince George bank extends 670 yards to ¼ mile from the shore; but from Point Richards it projects northwestward ¼ mile to a spit with 10 feet water on it.

Lightbuoy.—A quarter of a mile to the northward of the spit is a black spherical lightbuoy, moored in 4½ fathoms. From this buoy is exhibited a flashing red light, flash half a second, eclipse one and a half seconds.

Portarlington (lat. 38° 07′, long. 144° 39′) is a small township about a mile eastward of Point Richards. It is a watering place, with telegraphic communication. The population is about 500.

Jetty.—A jetty 15 yards wide projects 345 yards from the shore at Portarlington in a northerly direction into 12½ feet at low water. On the eastern side of the jetty at 83 yards from the outer end a cross breakwater of close piling projects eastward for 67 yards and affords shelter for boats.

Light.—A fixed green light, 22 feet above water, visible 2 miles, is shown from a lamp-post at the outer end of Portarlington jetty.

Jetty.—Form Point Richards the shore trends southwestward 4¾ miles to Drysdale jetty, which projects 167 yards from the land into a depth of 5 feet at low water.

For the first 1½ miles southwestward of Point Richards a bank, with 3 to 4 feet water on it, extends nearly 1,350 yards from the shore. From the outer edge of this bank, close to which there are depths of 3 to 4 fathoms water, the 3-fathoms edge of the shoal water border-ing the shore trends southwestward to ½ mile off Drysdale jetty. Three or four detached patches, with 3 to 6 feet water on them, lie between 1½ and 2¼ miles southwestward of Point Richards. There is only a depth of 6 feet water at about 200 yards off the jetty, and be-tween ½ mile and 1¼ miles to the northeastward of it spits, with 3 to 4½ feet water on them, project about 670 yards from the shore.

Drysdale (lat. 38° 09′, long. 144° 33′) is a postal town, 212 feet above sea level; about 1 mile from the town is the celebrated Clifton Spa, the springs of which are strongly impregnated with iron, magnesia, sulphur, and seltzer. Population of the town is about 600. There is communication by railway and telegraph.

The coast from Drysdale Jetty trends west-southwestward 2¼ miles to Bellarine white beacon on a slight projection of the land forming the southern point of the south entrance of the channel to Geelong Outer Harbor; some rocks lie close to the shore on either side of the beacon, and between one and two thirds of a mile to the westward of it. The 3-fathom edge of the shoal water fronting the shore extends from ½ mile off Drysdale jetty to 800 yards off the southern entrance of the channel.

Geelong Outer Harbor (lat. 38° 09′, long. 144° 30′).—The outer harbor extends 3½ miles north and south between the 3-fathom edges of the banks fronting the north and south shores, and is 2¾ miles wide between the bank which extends southward from Wilson point to Wilson spit, and the bank which incloses the inner harbor.

The main entrance into Geelong Outer Harbor lies between East Bellarine beacon and Wilson spit.

Depth.—The depth in the entrance channel, between the Wilson spit light beacon and the black can buoy, is 25 feet at low water. The channel is about 200 yards wide.

The depths in the outer harbor are remarkably even, 4 to 4¾ fathoms, except on the western side, where there are depths of 5 to 5½ fathoms. The bottom is mostly mud, with some patches of clay.

Ballarine Beacon is a large white pyramidal beacon situated on the southern shore of the outer harbor, 4¾ miles 187° from Wilson point.

East Bellarine Beacon is a single pile beacon surmounted by a staff and cage, standing in 20 feet of water and marking the edge of the bank which extends over ½ mile from the southern shore. It bears 43°, 1,400 yards from the Bellarine beacon. This beacon is useful to vessels which fail to pick up the lightbuoy in thick weather.

Wilson Spit Light Beacon.—A red pile beacon lies in 28 feet of water off the southern end of Wilson Spit in a position 1,840 yards northwestward from East Bellarine Beacon. A light is shown from the beacon.

Buoy and lightbuoy.—A black can buoy, marking the southern side of the entrance channel, lies about 250 yards southward from Wilson Spit Light Beacon, in about 4½ fathoms of water.

Between Wilson Spit and Hopetoun Channel a black spherical lightbuoy is moored in 25½ feet at low water about 1¼ miles east-southeastward from No. 2 beacon of Hopetoun Channel. It exhibits 10 feet above sea level a flashing red light every 3 seconds.

Light.—The above-mentioned buoy is temporarily replaced by a barge with a black structure, from which a flashing red light is shown.

The coast from abreast the south entrance of the channel forms a bay extending 4 miles west-northwest to Point Henry. It is barely 1 mile deep, and is mostly occupied by a bank, the 3-fathom edge of which projects from 800 yards off the south entrance point to ¾ mile eastward of Point Henry.

Aspect.—The land between Points Richards and Henry is mostly low, the hills scattered over it rarely exceeding 120 feet in height, except the summit of Bellarine, situated 2¾ miles within Point Richards, which attains an elevation of 447 feet. Most of this land is under cultivation, and several villages and country residences of the merchants and inhabitants of Geelong are on it.

Point Henry is low, the bluff, about ¼ mile within it, which is its most elevated part, being only 25 feet high.

Jetties.—A jetty projects about 616 yards from the east side of the point into 11 feet water, the inner portion has been removed to low-water level for a distance of 400 yards from the shore. A green barrel buoy, denoting foul ground, is moored between the shore and the standing portion of the jetty. Another jetty projects from the west side, 400 yards into 4 feet water.

Anchorage.—There is good anchorage in the outer harbor in 4 to 5 fathoms, mud, between 1 and 1½ miles eastward of Point Henry.

Coast.—The northern shore (lat. 37° 58′, long. 144° 41′) of the western arm from Werribee River trends 2¾ miles, southwestward, and thence 2¼ miles westward to a low point, on the east side of which is a small stream flowing from the northward, and on the west side of the point is the mouth of Little River, which winds through the low land from the west-northwest. From the projection of the land midway between Werribee and Little Rivers, a spit, with 4 feet water on its extremity, projects 1 mile to the southward.

Buoy.—A red conical buoy is moored in 14 feet water ½ mile south of the spit.

Beacon Point—Beacon.—From Little River the shore extends south-southwestward 1¼ miles to Beacon Point, from which a shoal spit projects 1,340 yards toward a beacon, distant 1 mile southeastward from the point. This beacon, on the east part of the spit off Beacon Point in 6 feet water, consists of a mast and ball, painted red, 11 feet high. From Werribee River to the point half way between the river and Beacon Point the 3-fathom edge of shoal water projects irregularly ½ mile to 1½ miles; and from the halfway point to the beacon it extends 2¼ miles from the shore, the edge of the bank closing ¼ mile of the beacon.

Kirk Point.—From Beacon Point the shore extends south-southwestward 1⅜ miles to Kirk Point, and is fronted by a bank, of which the 3-fathoms edge projects from $\frac{1}{2}$ mile southward of the beacon to about the same distance from Kirk Point.

From Kirk Point the low shore trends west-southwestward 2 miles and thence southward 2¾ miles to Point Wilson, forming a bay, of which the bight for a distance of 1½ miles is filled by a mud flat. For about a mile southwestward from Kirk Point rocky spits project from 200 yards to $\frac{1}{4}$ mile off the shore.

Buoys.—From $\frac{1}{4}$ mile off Kirk Point the 3-fathom edge of an extensive bank, with some shallow patches on it, curves in a southerly direction to a spit named Arthur the Great with 9 feet of water on it, marked by a red conical buoy in 20 feet water. From the end of this spit, the 3-fathom edge of the bank, after turning 1,340 yards to the northwest, extends south-southwestward 2¾ miles and is marked at 1.3 miles south-southeastward from Wilson Point by a red conical buoy, surmounted by staff and ball, called Steamboat Buoy, in 16 feet water; a partion of the spit, to the edge of the 3-fathom curve, extends $\frac{1}{4}$ mile eastward of the red buoy.

Wilson Spit.—From the buoy just described a continuation of the bank on which it is situated extends southward 1¾ miles, and is little more than 300 yards in breadth, with depths of 14 to 17 feet; its extremity, Wilson Spit, forms the north side of the south entrance of the channel into Geelong Outer Harbor.

Wilson Spit Beacon has been described.

Point Wilson (lat. 38° 05′, long. 144° 30′) is low, with a small islet close off it and numerous rocks extending about 200 yards to the southward. Two detached patches, having 5 and 6 feet water on them, lie, respectively, east-southeastward $\frac{1}{4}$ mile and southeastward $\frac{1}{4}$ mile from Point Wilson.

From Point Wilson the coast extends in a west-northwest direction 1.4 miles to the central and longest of some jetties, projecting into 2 or 3 feet water, and thence nearly $\frac{1}{2}$ mile westward to a low point, close off which lies Snake Island. The coast is bordered by mud and sand, with isolated rocks in places, which for upward of 1,500 yards westward from Point Wilson extend for more than $\frac{1}{4}$ mile from the land.

Beacon.—A beacon, surmounted by a basket ball, painted red, and 9 feet above high water, has been erected at the extremity of the foul ground 1,100 yards southwestward of Point Wilson.

Snake Island extends from 100 to 900 yards from the mainland westward of Point Wilson, with which it is connected by a flat, terminating in a rocky spit, projecting 400 yards southward from the island.

From 1¼ miles southward of Point Wilson, the 3-fathom edge of the bank fronting the shore trends northwestward to a 17-foot spit, northwest a little more than 1 miles from Point Wilson. From this spit the 3-fathom edge of the bank trends irregularly, in a west-northwest direction nearly 2 miles, to within 100 yards of the ledge of rocks projecting from the shore midway between Snake Island and Point Lillias. There are patches, with 3 and 4 feet water over them, between the edge of the bank and the rocks extending south from Snake Island.

Buoys.—A red cask buoy, situated 900 yards southward from the south point of Snake Island, marks the outer 4-foot patch. A similar red buoy, in deep water 1,600 yards southwestward from Snake Island, marks the dredgers' spoil ground.

Point Lillias (lat. 38° 06′, long. 144° 27′).—Point Lillias is the end of a tongue of land not more than 400 to 600 yards broad, projecting southwestward from the line of coast.

The space between Point Lillias and Point Henry, nearly 2½ miles, is encumbered with shoals, through which have been dredged the channels to the inner harbor.

Bird Rock—Beacon.—From the western projection of Point Lillias a narrow rocky ledge extends south-southwestward ⅓ mile to Bird Rock, on which is a beacon. The beacon consists of a red mast and cage 14 feet high, and is on the center of the rock. This rock and the ledge connecting it with Point Lillias are inclosed by a rocky shoal, with 2 to 5 feet water on it, extending 200 yards from the east side of the ledge, and southwestward 400 yards from the beacon.

The 3-fathom edge of the bank bordering the shore from 200 yards southward of the shore midway between Snake Island and Point Lillias trends south-southwestward to 800 yards eastward of Bird Rock and then curves round in a southeast direction to a position southeastward 1 mile from Bird Rock; it then turns southwestward ⅓ mile to the artificial cut that forms a channel through the narrowest part of the bank.

Buoy.—The middle of the southeast edge of the bank extending from Bird Rock is marked by a red conical buoy, situated southeastward 1,850 yards from Bird Rock Beacon.

From the south side of the Artificial cut the bank, here only 200 yards broad, stretches south-southwestward 1,400 yards. There are depths of 14 to 17 feet water over this bank, which is detached, owing to the Hopetoun Channel being cut through it.

Channels to inner harbor.—The inner harbor is entered by the Hopetoun Channel (see next page) or by the north channel after passing through the artificial cut.

These channels have been cut through the bank (about 1½ miles wide and extending from Point Henry to Point Lillias) which incloses the inner harbor. A ridge 1,500 yards long near the center of this bank dries at low water.

Artificial cut (lat. 38° 06′, long. 144° 27′):—This passage, which is not buoyed and which bears 157°, 1,800 yards from Bird Rock, was dredged through the bank 200 yards in a southeasterly and north-westerly direction; it is 100 yards wide and available for vessels up to 9 feet draft.

South Channel.—At ¼ mile northwestward from the artificial cut is the eastern entrance of the south channel. It is nearly 1¼ miles long east and west, but has a navigable width of only 47 feet in places; it is navigable only by boats and small crafts.

Beacons.—The north side of the south channel is marked by two large beacons at its outer and inner ends, with three single pile beacons between them.

North Channel.—This channel is available at low water for vessels drawing 9 feet; it lies about ½ mile off the north shore, with its eastern entrance situated between Bird Rock Beacon and the eastern beacon of the south channel. From the red pile beacon, situated about 400 yards southwestward of Bird Rock Beacon, the channel curves in a west-northwesterly direction for 1,600 yards to the east red dolphin beacon, at the east entrance to the channel, with a general width of 600 feet. At the east red dolphin beacon the channel is about 400 feet wide, and thence trends west-southwestward 350 yards to the west red dolphin beacon, and a farther 550 yards to the black single pile beacon at the western end of the channel. The width of the channel between the east and west red dolphin beacons is about 150 feet, increasing gradually to 400 feet at the black single pile beacon. It is only available with local knowledge.

Beacons.—The channel is marked on the north side by two red single pile beacons and two red dolphin beacons, and on the south side by three black single pile beacons.

Hopetoun Channel (lat. 38° 07′, long. 144° 21′).—The east entrance of Hopetoun Channel bears 300°, 2⅜ miles from the light beacon off the southern end of Wilson Spit. The channel extends in a westerly direction, 2 miles, into Geelong inner harbor.

The depth in the channel is 25 feet. The surface width is 300 feet and the navigable width 230 feet. The line of beacons on either side should not be approached nearer than 35 feet.

Tide gauges on Nos. 2 and 8 beacons show the actual depth in the channel.

Beacon Lights.—The north side of the channel is marked by four lights, each elevated 26 feet above water, and exhibited from red pile beacons situated about 35 feet northward of the north side of the navigable channel. The beacons have been numbered from 2 to 8.

These lights extend from the east to the west entrance of the channel, and are at equal distances of 4,040 feet apart.

The south side of the channel is marked by three lights, each elevated 26 feet above the sea, exhibited from black pile beacons, numbered 1, 3, and 5, situated about 35 feet southward of the south side of the navigable channel; these beacons are at equal distances of 4,040 feet apart, placed so as to alternate with those on the north side of the channel.

Beacons.—There are two single pile beacons to assist vessels in keeping the center of the channel during daylight; one a red single pile beacon with staff and disk on the north side of the channel, and in line with the red beacons, and opposite No. 3 black beacon; one a black single pile beacon on the south side of the channel, and in line with the black beacons, and opposite No. 6 red beacon.

Dredging.—When dredging operations are in progress vessels must pass the dredger on the side on which a basket ball is exhibited by day and two red vertical lights by night. The dredger also exhibits an anchor light at the stem and stern.

The pilot or master of every steam vessel must, when at a distance of not less than 200 yards from the dredger, cause the engines of the vessel to be put at dead slow, and before passing over or along the mooring chains thereof or any work in progress, the engines must be stopped and kept stopped whilst passing.

During the night, or when a vessel is about to pass, the dredger will be hove as close as practicable to the side of the channel.

When not working, the dredger will (weather permitting) be moored on the north side of, and clear of, the channel.

In connection with the dredging operations, two outriggers will be placed on the north (or channel) side of the black (or port hand) beacons, and extending 17 feet from them; care should be taken to avoid fouling them.

Caution.—Should the extra lights required on the dredger cause any inconvenience to persons in charge of steamers navigating the channel it is requested that one prolonged blast of the steamer's whistle should be given, when the extra lights will be extinguished and remain so until the vessel has passed the dredger.

Directions for Hopetoun Channel are given later.

Geelong Inner Harbor (lat. 38° 08′, long. 144° 24′), the most spacious and secure anchorage in Port Phillip, extends from Limeburners Point on the southern side, nothward nearly $4\frac{1}{4}$ miles to the entrance of Limeburners Creek on the northern side, and is $2\frac{1}{2}$ miles wide between the western shore and the 6-foot edge of the bank which extends from Point Henry to the northern shore.

Depths.—The depths are remarkably regular, over mud, gradually increasing from the 3-fathom edge of the bank on the eastern side, to 5 and $5\frac{1}{2}$ fathoms within $\frac{1}{4}$ mile of the western shore of the harbor, and to 4 fathoms at 400 yards off the town of Geelong, in Corio Bay, the southwestern corner of the harbor.

Between Point Henry and Limeburners Point is. Stingaree Bay.

The eastern shore of the harbor, from Point Henry, extends southward 2 miles to an elbow of the coast, between which and a low point ¼ mile to the west of it, a shallow inlet nearly ¼ mile wide trends 1,500 · yards into the low flat land in an east-southeast direction, toward Point Henry township; but the inlet is filled by a mud flat, which dries from 1 to 2 feet at low water.

From the western entrance point of the inlet the low shore trends west-northwestward 1¼ miles to Limeburners Point. On the western side of the former point is a bight in the land, ¼ mile in extent, partly inclosing a remarkable pond in the mud flat which projects from the bight.

From Limeburners Point the water frontage of the town of Geelong forms Corio Bay, 1,340 yards deep, extending from the point to Hutton Wharf. There are depths generally of 6 feet water within 150 yards, and 22 to 24 feet within 300 yards of the shore.

Conspicuous chimney.—The chimney of the electric power station, 128 feet high. situated to the southward of Yarra Wharf, is a conspicuous object from the anchorage.

From Hutton Wharf the west shore treads northward a little more than 1 mile to the south point of a cove about ¼ mile in extent, into the head of which Cowies Creek flows through the lowland from the northwest. The shore from the north point of this cove turns round 1 mile to the northeast, and then curves in nearly the same direction 1¾ miles to the entrance of Limeburners Creek.

Limeburners Creek from its entrance trends east-northeastward 1,340 yards, and thence about the same distance in a northerly direction, and is from 400 to 800 yards wide, with 5 to 11 feet water.

Beacons.—Six pile beacons, surmounted by disks, in 4 feet of water, mark the channel into Limeburners Creek; starboard-hand beacons are painted red and port-hand beacons black.

The shore, from Limeburners Creek, trends to the eastward for 1¼ miles, and thence to Point Lilias forms a bight ½ mile in width and depth, which is filled by a bank of mud, sand, and weeds.

Directions—West Channel (lat. 38° 12′, long. 144° 45′) **to Geelong.**—From the West Channel to Geelong, after rounding West Sand pile lighthouse, steer 357° 5¼ miles and pass eastward of the black light buoy off the northeastern extremity of Prince George Bank; when ¼ mile to the northward of it, steer 273° for the black light buoy off Point Richards, and having passed at the distance of 200 yards north of it, steer 244° for the red-light beacon off Wilson Spit; pass to the southward of this light beacon, and midway between it and the black can buoy moored 200 yards to the southward.

If drawing less than 14 feet steer 256° from Point Richards Buoy, pass south of the red conical buoy, with staff and ball, about 1¼ miles south-southeastward from Point Wilson, and cross over the bank extending from it.

Working up.—From the West channel to Geelong, with a contrary wind between the West Sand lighthouse and the northeastern extremity of Prince George Bank, do not stand into less than 5 fathoms, nor bring the lighthouse east of 177° until north of the Prince George lightbuoy, between which and the lightbuoy off Point Richard do not stand into less than 5 fathoms, nor bring that point west of 267°. From Point Richards to Point Henry the south shore should not be approached to less than 4 fathoms, and the north shore to less than 5 fathoms, until west of Wilson Spit.

Clearing mark.—Station Peak, 1,132 feet high, on the northwest side of Port Phillip, open of the high land of Point George, bearing 309°, leads northward of the northeastern extremities of all the entrance banks, at a distance of a mile from the nearest.

At night.—From about 400 yards eastward of West Sand pile light steer 357°, keeping the light white, as it shows red over Prince George Bank, and pass east of Prince George lightbuoy. When ½ mile northward of this buoy steer 273°, and pass north of Point Richards Lightbuoy. Thence steer about 244°, and pass 100 yards southward of the Wilson Spit light beacon, thence for Hopetoun Channel or the anchorage in the outer harbor.

From Point Richards buoy, if drawing less than 14 feet, steer 250°, and when the eastern light of the Hopetoun Channel bears 270°, proceed across the bank into the outer harbor with it on that bearing.

Anchorage.—To anchor in the outer harbor of Geelong (lat. 38° 08′, long. 144° 27′), steer about 301° from Wilson Spit red lightbuoy, and come to in 4½ fathoms, with Point Henry Bluff bearing 256°, at about 1 mile from the shore.

Directions—Hopetoun Channel (lat. 38° 07′, long. 144° 21′).— From Wilson Spit red light beacon steer about 314° for 1.7 miles to the black spherical lightbuoy; after passing this buoy steer for the outer beacon of Hopetoun Channel. Then enter the channel and keep the red beacons marking the north side of the channel on the starboard hand and the black beacons marking the south side on the port hand. The line of beacons on either side should not be approached nearer than 35 feet. As the navigable width of the channel is only 230 feet careful steering is necessary.

Caution.—When using the Hopetoun Channel the limited width must be especially remembered, so that every precaution may be taken to avoid collision with other vessels, or with the beacons. Before proceeding through the channel care must be taken to straighten the vessel on the course of the center of the channel, as accidents have occurred through vessels entering with their heads athwart the channel. It is not advisable for vessels to pass each other from opposite directions in this channel, but a vessel should wait outside either entrance when another vessel is seen entering from an opposite direction until the other vessel has passed through.

The speed of steam vessels navigating the Hopetoun Channel is not to exceed 7 miles an hour, and no sailing vessel is to enter this channel whilst any other vessel is proceeding through in an opposite direction.

At night.—In navigating the Hopetoun Channel at night leave the white lights marking the north side of the channel on the starboard hand entering from the outer harbor and the red lights marking the south side of the channel on the port hand.

Before entering the Hopetoun Channel vessels should be straightened up on the line of the center of the channel, as accidents have occurred through vessels not taking this precaution.

Vessels proceeding outwards from Geelong should not leave the Hopetoun Channel before being well clear of the outer red beacon (No. 2) which marks the eastern end of the channel, and should then steer for the black spherical lightbuoy. After passing this lightbuoy, the course may be laid to pass between Wilson Spit light beacon and the black can buoy to the southward.

Signals to be made by vessels aground in this channel have been given.

North channel.—Having passed No. 2 beacon of Hopetoun Channel, steer for Bird Rock Beacon, passing about 200 yards eastward from the red beacon at the eastern end of the South Channel. Pass about 100 yards southward from the red pile beacon on Bird Rock and steer between the black and red single pile beacons marking the port and starboard sides of the channel until abreast of the east red dolphin beacon, when steer to pass about 80 feet southward from the west red dolphin beacon and 100 feet northward from the black single pile beacon at the western end of the channel, after clearing which a course of about 216° for 3¼ miles may be shaped to Yarra Wharf.

South channel.—Previously referred to.

Anchorage (lat. 38° 08′, long. 144° 22′).—There is convenient anchorage in 4½ fathoms 600 yards northward of the Geelong wharves. The bottom being soft mud mixed with sand and clay, a long scope of chain is necessary in strong winds to prevent the vessel from dragging.

Vessels, however, may anchor anywhere in the inner harbor according to draft.

Tide gauges.—There are tide gauges on Nos. 2 and 8 beacons, Hopetoun Channel and at Yarra wharf, the 25-foot mark of which corresponds to the low-water datum, to which all depths both in Hopetoun Channel and at Geelong wharves are reduced.

Tide.—It is high water, full and change, at Geelong wharves at 3 h. 17 m. The average rise of the tide is 1 foot 9 inches, but during a prevalence of westerly winds the tides may rise 2 feet 10 inches at high water and at low water above the datum mark; during prevailing easterly winds the tide may rise 1 foot 3 inches only, above, at high water and fall 1 foot below the datum mark.

Explosives anchorages.—Geelong outer harbor: Anywhere except within ½ mile of any fairway or channel. Inner harbor, northward of a line within the western beacon of North Channel in line with Bird Rock beacon on a bearing of about 89°, but not within ¼ mile of the shore.

Geelong (lat. 38° 09', long. 144° 22') lies 45 miles southwestward of Melbourne. The town is well laid out on ground sloping to Corio Bay, with broad streets at right angles to each other, and large public buildings. There are several cloth manufactories, also meat-preserving works, and extensive tanneries. The country surrounding Geelong is agricultural.

The population of Geelong and suburbs in 1918 was 32,000.

Pilot.—There is a pilot stationed at Geelong.

Communication.—The Ballarat and Melbourne railways form a junction at about 1½ miles to the northward of Geelong station. From this junction the Geelong and Melbourne Railway curves in a northeasterly direction nearly 32 miles over a low flat country to the Geelong Junction northwest of Williamstown. Steamers run daily between Geelong and Melbourne.

Trade.—The chief articles of import are coal, hardware, machinery, and timber. The chief exports are wool, leather, sheepskins, frozen meat, wheat, oats, barley, cordage, and cement. The principal trade is with the United Kingdom and interstate.

The port of Geelong includes all inlets, rivers, harbors, and navigable waters southward and westward of a line from the right bank of Little River to the east side of Mercer Street, Portarlington.

Water.—Fresh water is laid onto the wharves at Geelong and can be obtained by vessels lying alongside; if at anchor, the water is brought alongside in a steam water boat.

Repairs.—There is a firm at Geelong capable of executing large repairs.

Eastern Pier is about 420 feet long, and built of stone for 260 feet from the shore, the remainder being of wood, and 52 feet wide at the outer end, where there is accommodation for vessels of 9 feet draft.

Yarra Wharf is on the western side of Eastern Pier, nearly 1 mile west of Limeburners Point, is 277 yards long, and 31 yards wide, with sufficient water for a width of 200 feet from both sides of the pier for vessels drawing 23 feet.

Moorabool Wharf, 250 yards westward of Yarra Wharf, is about 100 yards long, 130 feet wide, with accommodation at both sides for vessels of 16 feet draft. The customhouse is situated near the inner end of this wharf.

Railway Wharf.—The railway wharf, about 70 yards westward of Moorabool Wharf, is 44 feet wide and extends 280 yards from the shore into about 25 feet of water. The pier is available for a width of 200 feet on either side for vessels drawing 23 feet.

Hutton Wharf, at the northwest point of Corio Bay, projects about 180 yards from the shore into 8 feet at low water. A new wharf has been built about 50 yards northward from the old wharf, which is in ruins. It is about the same length.

Corio Quay Wharf.—At 2,200 yards northward from Hutton Wharf is the southeastern end of Corio Quay Wharf; thence the wharf extends northwestward 500 feet, and then westward for the same distance. It is used in connection with the freezing works and known as Corio Quay.

Each berth alongside the wharf has been deepened to 26 feet at low water by a dredged channel 400 feet wide and 1,000 feet long. Each berth is fitted with four electric conveyors for loading wheat or frozen meat.

A chimney lies close southward from the freezing works and another to the southward; they are both conspicuous.

Geelong Freezing Works Pier.—At $\frac{1}{4}$ mile northward from Corio Wharf the Geelong Freezing Works Pier extends 462 feet in a southeasterly direction, with a depth of 28 feet at low water alongside. The outer end has a width of 36 feet for 142 feet, thence a width of 15 feet to the shore.

A chimney, 85 feet high, its top being about 100 feet above high water, is built near the base of the pier.

Small piers.—At Limeburners Point a pier 57 yards in length, with a depth of 7 feet alongside at low water, projects in a northerly direction. Eastward of Yarra Wharf and westward of the railway wharf, several bathing establishments extend out into the harbor.

The naval college jetty is situated 700 yards northward from Hutton Wharf.

Lights.—A light is exhibited from a lamp-post at the outer end of Yarra Wharf.

A light is shown from the outer end of Eastern Pier.

A light is exhibited from each corner of the outer end of Moorabool Wharf.

A light is exhibited from the outer end of the railway wharf.

A light is exhibited at the end of the new Hutton Wharf.

A light is shown from the northeastern angle of Corio Quay Wharf.

Buoys—Mooring buoys.—Three sets of moorings for destroyers are laid down to the eastward from the outer end of Yarra Wharf. They buoys are painted red and are in 23 to 20 feet at low water, shoaling to 15 feet at 100 yards inshore.

Two warping buoys are moored off the railway wharf, the outer one in 26 feet of water, lies 200 yards northwestward from the outer end of the wharf; the inner one, in 14 feet of water, lies about 100 yards westward from the center of the wharf.

A warping buoy in 26 feet of water is moored 100 yards northeastward from the center of Corio Quay Wharf.

Beacons.—Two beacons, west-northwestward from the wharf mark the northern side of the dredged channel to Corio Quay Wharf.

Three dolphins lie to the northward of the channel. A pile stands 132 yards west-northwestward from the western dolphin for a mooring when a vessel is berthed alongside the dolphins.

Rifle range beacons.—The rifle butts are situated ½ mile southeast of Limeburners Point; two black beacons, surmounted by a red basket ball, are placed in Stingaree Bay, which denote the western extremity of the danger zone when firing is being carried on at the range.

Caution.—When the red flag is hoisted at the rifle butts boats should keep westward of the line of beacons in Stingaree Bay.

Compass adjusting buoy (lat. 38° 08′, long. 144° 22′).—A buoy for the adjustment of compasses is moored off the end of the railway pier, Geelong.

The true bearings of the undermentioned marks from the buoy are as follows:

Station peak, 13° 57′.

Conspicuous tree, Stingaree Bay, 103° 11′.

Flagstaff, Botanic Gardens, 130° 19′.

Conspicuous tree in hollow, Mount Anakie, 342° 59′.

Chimney, Geelong gas works, 329° 00′.

Directions—From Geelong to Hobson Bay.—Vessels should leave the inner harbor by the Hopetoun Channel, thence steering southeastward, and after passing between Wilson spit lightbuoy and the black can buoy steer 72°, for 2 miles, then alter to 58°, to pass half a mile northward from the light beacon off Point Richards. When 2 miles northeastward of this buoy alter course to 47°, to make the white sector of Point Gellibrand Light.

From Geelong to sea by the west channel.—From the east end of the Hopetoun Channel steer to pass the black spherical lightbuoy

and then for Wilson spit light beacon, and from thence northeastward passing northward of the lightbuoy off Richards Point. Having passed this buoy, steer 92°, to pass northward of the lightbuoy
off Prince George Bank, and when the buoy is passed steer 174°, for
the west sand lighthouse. The light shows red over the Prince
George Bank. Proceed through the channel and to sea, as directed
when going from Hobson Bay.

By the south channel.—From Geelong to sea, proceed as just
directed, to half a mile northward of the black lightbuoy off Prince
George Bank; and from thence proceed 149°, 15 miles for No. 15
lightbuoy, which marks the eastern spit of the Middle Ground; and
having rounded this, follow the directions given later for proceeding
to sea from Hobson Bay by the south channel.

Werribee River (lat. 37° 59′, long. 144° 41′) has a 3-foot bar
across its entrance, within which the first reach trends westward
about 1 mile. It is about 200 yards wide, with from 1 to 2 fathoms
water. Above this reach the river is merely a small stream, flowing
in a winding direction from the north-northwestward. From the
mouth of the river a spit with 16 feet water on its extremity projects
1¼ miles to the southeastward.

Werribee, situated on the river of the same name and 20 miles
southwestward of Melbourne, is a postal township, and has communication by rail and telegraph; population in 1911 was 2,301.

A jetty, situated about ½ mile to the southwestward of the entrance
to Werribee River, is 586 yards long, 18 feet wide, with a depth of
15 feet at low water at its extremity.

Light.—A fixed green light is exhibited from the outer end of
Werribee Jetty.

Aspect.—The country between Melbourne and Geelong is generally low, flat, and partially wooded; it is intersected by several
creeks, already noticed, and there are many small lagoons, most of
which are situated near the shore within about 8 miles of Williamstown.

Station Peak.—The only hills in the neighborhood worthy of
notice appear to be Youngs, the most elevated of which is Station
Peak, rising from the south portion of the group to the height of
1,132 feet. It bears 0°, distant 10¾ miles from the Bluff on Point
Henry.

From Werribee River the low northwestern shore of Port Phillip
trends northeastward for 6 miles to Point Cook. At 2 miles southwestward of Point Crook there is a low projection, whence rocky
shoals, with from 3 to 4 feet water on them, projects ½ mile.
The 3-fathom edge of the shoal water, which borders the shore, extends a mile off Werribee River to about the same distance off the

rocky shoals just mentioned. Thence to Point Cook the 3-fathom
edge of the shoal water generally extends ⅓ mile from the shore.

Point Cook is low and rocky, with a spit, extending 1 mile to the
eastward, having a depth of 10 feet water at its extremity, at halfway between which and the shore there is a rocky patch with from
3 to 6 feet water on it.

Buoy.—A black buoy is moored in 4¾ fathoms water at ¼ mile off
the spit.

Coast.—From Point Cook a low shore, with several small lagoons
close behind it, forms a shallow bay, barely 1 mile deep, extending
northward 4 miles to Altona. At midway between the two points is
the mouth of Skelton Creek, which winds through the low swampy
ground to the northwestward. The 3-fathom edge of the shoal,
which fills this bay, extends beyond the line of the two points, and
forms, midway, a spit projecting southward with its extremity 1¼
miles northeastward from Point Cook.

Truganina Jetty.—This jetty, which is used for the transfer of
explosives, is situated 2½ miles northward of Point Cook, and is 533
yards in length with a depth of 9 feet at low water at its end.

Two red cask warping buoys are moored off Truganina Jetty,
nearly 200 yards southward, and 500 yards eastward from the end
of the jetty.

Lightbuoy—Explosives anchorage.—A lightbuoy, which exhibits a flashing light, is moored in 28 feet at low water, at 1.3 miles
141° from the outer end of Truganina Jetty. This buoy marks the
northeastern boundary of the inner anchorage for vessels carrying
not more than 250 tons of explosives, exclusive of any for Defense
Department.

Jetty.—From Altona, 1¼ miles northeastward from Truganina
Jetty, a jetty 433 yards in length, projects to the southward; there
is a depth of 9 feet at its outer end at low water.

Buoy.—A black can buoy is moored 500 yards to the southeastward of the end of Altona Jetty. When approaching the jetty this
buoy should be kept on the starboard hand.

From the point of Altona the shore trends northeastward 1.7 miles
to Kororoit Creek, and is bordered by a rocky bank, extending 1⅛
miles southeastward off Altona Point, with 7 to 18 feet water and
shoal patches upon it. From the southern point of this bank the
southern extremity of Point Gellibrand bears 70°, distant 2 miles.
Two gullies, having 3½ and 3¼ fathoms water, extend ½ mile into
this bank from the southeastward. The northeastern one, which is
close to the low rocky point before noticed, approaches the mouth of
Kororoit Creek to ¼ mile, with 3 fathoms water.

Beacon (lat. 37° 53', long. 144° 51').—On the outer shoal patch in Altona Bay is a black beacon, consisting of a mast and ball, 10 feet above sea level.

Rifle Practice—Danger beacons.—Two red spar beacons are placed off the rifle ranges at Williamstown about 2,000 yards from the shore; these in line with the black beacon, marking the outer shoal patch in Altona Bay, indicate the line within which vessels are cautioned not to go whilst firing is going on at the ranges.

To Point Gellibrand, the coast from Kororoit Creek trends southward and eastward 2 miles. The 3-fathom edge of the foul rocky ground which borders the shore projects 300 to 600 yards from it.

EASTERN SHORE OF PORT PHILLIP.

Coast.—

Dromana Bay.—From the foot of Arthurs Seat the coast curves northeastward for 4 miles to Martha Point, within which point is Dromana Bay, where there are depths of 3 fathoms water ¼ mile from the shore. Tides described later.

Dromana is a small town, with a population of about 400 persons. There is a telegraph office here, and daily conmmunication by coach and steam vessel with Melbourne.

Jetty.—At 2 miles northeast from the Eastern Lighthouse is Dromana Jetty, 28 feet wide, which projects 500 yards from the shore into a depth of 13 feet at low water. At 160 feet from its outer end, and pointing northeast, is a cross jetty 170 feet long and 25 feet wide, with 7 to 9 feet of water along it on the inside.

Light (lat. 38° 20', long. 144° 58').—A fixed red light, visible 2 miles, is shown from the end of Dromana Jetty.

The shore from Martha Point may be approached to ¼ mile in 3 fathoms, and trends north-northeastward nearly 2 miles to Martha Cliff, which forms the southwest point of Balcolms Bay. The land between the point and cliff rises to a ridge, of which the southwest and highest part is Mount Martha, a hill 527 feet high, distant 4¼ miles northeastward from Arthurs Seat.

Balcolms Bay extends from Martha Cliff nearly 3 miles to Fisherman Point, and is 1,340 yards deep; except within ½ mile of Martha Cliff the shore may be approached to a ¼ mile, where there is a depth of about 3 fathoms, but there are some rocks close along shore, of which Shag Rock lies 1¾ miles northwestward of Martha Cliff; at 1,500 yards northeastward of the cliff is Balcoms Creek.

Fishermans Bay and Mornington.—Fishermans Bay, which is the water frontage at Mornington, is merely a slight indentation of the coast extending from Fisherman Point nearly 1 mile northward to Snapper Point. Shoals project ¼ mile from the southern portion of the bay, but the shore north of these shoals may be approached to the distance of 200 yards in 3 fathoms.

Mornington is a watering place. It is in daily railway communication with Melbourne. There is also telegraphic communication. Population is about 1,500.

Snapper Point is narrow, about 50 feet high, and projects ¼ mile from the adjacent coast. Tides given later.

Jetty—Wharf.—From Snapper Point a jetty 35 feet wide projects 150 yards northeastward into 25 feet of water, shallowing to 16 feet at 50 yards in from the outer end. An L end extension projects west-northwestward 57 feet from the outer end of the main jetty, with depths of 25 to 30 feet alongside.

Firewood Wharf, 67 yards long and 12 feet wide, lies on the southeastern side of the jetty, at its inner end, with depths of 6 feet at its northwest to 13 feet at its southeast end and 12 to 15 feet at a distance of 25 yards.

Lights.—A flashing green light, 50 feet above water, visible 10 miles, is shown from a white wooden house 10 feet high on Snapper Point. A fixed red light 35 feet above water, visible 3 miles, is shown from a mast on the end of the jetty.

Mount Eliza (lat. 38° 12′, long. 145° 06′).—From Snapper Point the coast trends northeastward 4 miles to Davy Point. It is slightly embayed, and intersected by four small creeks flowing northwestward from the hills at the back. The most conspicuous of these hills is Mount Eliza, 527 feet high, distant 11¼ miles northeastward from Arthurs Seat. This coast may be approached to ¼ mile in 3 fathoms, but it is rocky for about 1½ miles southward from Davy Point.

The coast from Davy Point, after receding nearly ½ mile to the eastward, extends northeastward 1¼ miles to the village of Frankston. The country behind is hilly and is intersected by two or three small creeks.

The southeastern shore of Port Phillip, which is mostly wooded, has several townships and numerous houses and other buildings are scattered along it.

Wooleys Reef.—Between Davy Point and Frankston a shoal with 3 to 17 feet water on it extends ⅓ mile from the shore, and at a little more than ½ mile northward from the point a spit projects northwestward ¼ mile farther to Wooleys Reef, with only 4½ feet water on it.

Buoy.—A red conical buoy, in 16 feet water, marks the outer edge of Wooleys Reef.

Frankston (lat. 38° 09′, long. 145° 07′) is the center of a large fishing and firewood trade. The population in the township is about

1,200. It is connected with Melbourne by rail and there is a telegraph station.

Jetty.—The jetty at Frankston is 373 yards long and 13½ feet wide with a face 175 feet long and 20 feet wide, with a depth on the outer side of 14 feet at low water.

Light.—A fixed green light, visible 3 miles, is shown from the outer end of the jetty.

Garrum Swamp.—From Frankston a low uniform coast curves in a north-northeast direction 8½ miles to the point of Mordialloc and is separated by a narrow piece of wooded land from Garrum Swamp. This swamp has been drained and hundreds of acres of splendid land rendered available for cultivation. The coast from Frankston nearly to the point of Mordialloc may be approached to the distance of 600 yards in 3 fathoms; but a ledge of rocks projects ¼ mile south from the point.

From Mordialloc to Ricket Point 2¾ miles to the westward the coast forms a bay 2¼ miles across and 1,500 yards deep. From 4 and 4½ fathoms in the entrance of this bay the water shoals to 3 fathoms at ¼ mile from the shore. Three patches, on the central and smallest of which there are only 6 feet of water and on the others 12 and 15 feet, lie west-northwestward 1,340 yards, 1 mile and 1½ miles, respectively, from the eastern point of the bay.

Mordialloc is a township on the creek of the same name, with a population of about 1,500 persons. Steamers ply to Melbourne, with which place there is also communication by rail and telegraph.

Jetty.—There is a jetty 433 yards long and 14 feet wide, with a depth of 9 feet at low water at its extremity. The shoals northwest of it are referred to above.

Light.—A fixed red light, 30 feet above water, visible 3 miles, is shown from a lamp-post at the outer end of Mordialloc Jetty.

Mentone is at the head of the bay. There is communication by railway and telegraph. The population is about 1,200.

Jetty.—The jetty at Mentone is 294 yards long and 20 feet wide, with a depth of 12 feet at low water at its extremity.

Ricket Point (lat. 38° 00′, long. 145° 02′).—Ricket Point, 800 yards westward of the western end of the bay northward of Mordialloc, is a low point about 30 feet high, but rapidly rising to 60 feet at a few yards inland. Foul ground and shallow water, less than 3 fathoms, extends ¼ mile southward of Ricket Point and the point just eastward of it.

Measured mile—Beacons.—A measured mile has been established off Ricket Point. The front and rear beacons, which come into line on a bearing from seaward of 56°, and mark the extremities of the mile, are each surmounted by a triangle, painted white, those inshore being inverted, and so arranged that their vertices appear to make contact with those in front when running the course, 146°, about 1¼ miles from the shore in 9 fathoms of water. The distance on this course between limits is 6,080 feet.

The front southeastern beacon is situated 226°, 400 yards from Ricket Point; and the front northwestern beacon off Quiet Corner marks the outer limit of foul ground off Black Rock.

Owing to the foul ground in the locality, boatmen should keep outside the line of front beacons, which are in 20 feet of water.

Caution.—Vessels, sailing or steam, should keep clear of vessels on the measured mile course.

Picnic Point.—Picnic Point is a well-defined projection with a grassy top and fringed with rocks. From Ricket Point a mostly rocky coast extends 3¼ miles northwestward to Picnic Point. The coast between Ricket and Picnic Points is bordered by foul ground and sunken patches, some with only 4 and 5 feet water on them, extending off nearly ¼ mile. A spit projects southwestward from Picnic Point with a depth of 3 fathoms at ¼ mile off.

Black Rock, about 1 mile northeastward of Ricket Point, is a detached piece of rock near high-water mark.

Red Cliff, situated about 1¼ miles southward of Picnic Point, is very conspicuous. It is a high reddish-colored cliff, and the only cliff in the vicinity that is bare of vegetation.

Buoy.—A red conical buoy marks the end of the shoal extending northward from Picnic Point.

Vessels bound for the jetty should pass northward of this buoy until in line with the jetty, when if drawing not more than 6 feet, steer straight in, but if drawing up to 8 feet, keep to the east-north-eastward, until the outer end of the jetty is in line with the extremity of Picnic Point, then steer in.

Anonyma Shoal is a rocky patch ⅓ mile long, in a northwest and southeast direction, and 300 yards broad, with one foot water on its shoalest part; there are 4 fathoms at 200 yards from its outer edge, and 3¼ fathoms between it and ¼ mile off the beach. It lies 1,500 yards southward from Picnic Point.

Buoy.—A red conical buoy marks the northwest edge of Anonyma shoal.

Jetty.—From about 200 yards to the northeast of Picnic Point a jetty 12 feet wide extends 273 yards to the northwestward, having a depth of 9 feet at low water at its extremity.

Light.—A fixed green and red light, 32 feet above water, visible 3 miles, is shown from a lamp-post at the outer end of the jetty. The green sector of the light indicates a safe approach to the jetty, and the red sector marks the foul ground and shoal water which lies to the westward of the jetty.

A timber breakwater about 180 yards long, which affords shelter for boats, is constructed on the reef northwestward of the outer end of the jetty.

From Picnic Point to Green Point, 1¼ miles northward from it, the coast forms a slight indentation, bordered by a shoal, of which the 3-fathom edge extends from ½ mile off Picnic Point to 400 yards close to the southward of Green Point.

A rocky patch, with 3¼ fathoms water over it, lies nearly 1¼ miles westward from Picnic Point; there are depths of 5 and 4¼ fathoms between this patch and the shore.

Green Point.—Shoal water with less than 3 fathoms extends west-southwestward ¼ mile from the point.

Brighton Beach Jetty.—About 200 yards to the southeastward of Green Point a jetty 22 feet wide extends 233 yards to the south-westward, having a depth of 15 feet at low water at its extremity.

From Green Point the coast extends north-northwestward 1½ miles to Point Cole, and thence in the same direction 1¼ miles to Point Ormond, the eastern point of Hobson Bay. For about 1 mile north from Green Point the coast is bordered with rocks, and from ¼ mile off the point, the 3-fathoms edge of the shoal water fronting the shore trends irregularly to 1,340 yards off Point Cole and thence northwestward to 1,500 yards off Point Ormond.

Brighton (lat. 37° 55′, long. 144° 59′).—The southern and greater portion of the coast from about Green Point to Point Ormond forms the water frontage of the town of Brighton, a watering place, and a suburb of Melbourne, from which it is distant 8 miles. It is a favorite place of residence, abounds with handsome villas, and there is a long sandy beach. Market gardening is the chief industry of the district. The population is about 11,000.

Park Street Jetty.—From nearly 1 mile northward of Green Point, Park Street Jetty extends 466 yards, with a width of 11 feet, west-northwestward, and thence north-northwestward 97 yards, with a width of 20 feet. There is a depth of 10 feet at the outer end.

Light.—A fixed red light, 30 feet above water, visible 2 miles, is shown from a lamp post on the outer end of the jetty.

Buoy.—A red conical buoy, moored in about 10 feet of water, 500 yards southward from the outer end of the jetty, marks the outer limit of the foul ground between the buoy and the shore.

Point Ormond, the eastern point of Hobson Bay, lies 1¼ miles northward of Point Cole, and is a round, sloping point about 40 feet high.

Point Ormond—Jetty.—A jetty, 220 yards in length, extends in a northwesterly direction from Point Ormond, into a depth of 8 feet.

Beacons.—Two beacons, in 4 feet of water, are placed 120 yards southwestward and 253 yards north-northwestward from the outer end of the jetty, to mark the outer limits of the foul ground on either side of the approach to the jetty. In approaching from the southwestward, westward, or northwestward vessels should keep within the waters between the beacons.

Hobson Bay.—At the north end of Port Phillip Bay the waters contract, forming the portion known as Hobson Bay, which between Points Gellibrand and Ormond is near 3½ miles across, and into which at the northwestern corner the river Yarra empties itself.

The bay consists of all inlets, rivers, bays, etc., within a line drawn from a point at the termination of northerwest side of Fitzroy Street, St. Kilda, eastern shore Hobson Bay, to the Time Ball tower in Williamstown.

Depths.—The western portion of the bay is the deeper water side, the depths ranging from 18 to 26 feet at low water; the eastern side is mostly occupied by a flat bank with 12 to 15 feet on it at low water, the 2-fathom curve being from 400 yards to 1,400 yards from the shore.

Landmarks.—When approaching Hobson Bay the most easily identified building is that of the sugar works, situated close to the shore immediately westward of the Lagoon Piers at Port Melbourne. It consists of a very tall, red-brick block, and is very much higher than any of the surrounding buildings.

Government house, standing on rising ground, is also conspicuous and easily identified, being built of gray stone and having a tall, square tower with a balcony.

The gasometer of the gas works, westward of Pickles Street, is also an unmistakable mark. The time-ball tower on Point Gellibrand is not very conspicuous, owing to some white buildings near it.

Coast.—From Point Ormond (lat. 37° 53′, long. 144° 58′) the low shore of Hobson Bay trends north-northwestward a little more than a mile to a pier at the west point of the town of St. Kilda. For about 800 yards northward of Point Ormond there are numerous rocks scattered over the shoal extending 400 yards from the shore.

The shore from about ¼ mile northward of St. Kilda Pier extends in a straight line west-northwestward for a distance of about 2 miles to Port Melbourne Town Pier. There is a depth of 9 feet water within 200 yards of the shore from St. Kilda to Port Melbourne Piers.

The shore from the Melbourne & Hobson Bay Railway Pier, which is nearly 600 yards to the west of Port Melbourne Town Pier, trends westward 400 yards to the new railway pier and then 1 mile in the same direction to the entrance of the Yarra River. The 1-fathom curve of the shoal which borders the shore extends from it about 300 yards at the new railway pier to 300 yards from the northern side of the entrance of the river.

St. Kilda, about 3½ miles southward of Melbourne, is a fashionable watering place, with an esplanade along the sea beach. The population is about 25,000.

Pier.—St. Kilda Pier is 25 feet wide and projects westward from the shore for 733 yards to a face 200 feet long and 25 feet wide, with a depth of 10½ feet at low water on the outside. At 500 feet in from the outer end there is an arm 360 feet long and 25 feet wide, with a depth of 6 feet at low water on the outside.

Light.—A fixed red light, elevated 19 feet above high water, visible 2 miles, is exhibited from the outer end of St. Kilda Pier.

Port Melbourne (lat. 37° 51′, long. 144° 56′), formerly called Sandridge, is situated on the north side of Hobson Bay, 3 miles southwest of Melbourne, and has accommodation for vessels of 35 to 37 feet draft of water.

Town Pier.—The pier projects from Port Melbourne, in a southwesterly direction, for a length of 720 yards, into 30 feet water, and has berthing accommodations on both sides for vessels of 27 to 30 feet draft, with berths dredged to 29 feet at the pier for a distance of 1,300 feet on its northern side and 1,200 feet on its southern side.

On this pier is a hand crane to lift to 5 tons.

Railway Pier.—At about 400 yards westward from Port Melbourne Town Pier, the Railway Pier extends from the shore in a southerly direction 716 yards, into 30 feet water, and has berthing accommodation for vessels drawing 30 feet, and all facilities for loading and unloading. The space between the two piers has a depth of 30 feet at low water over its greater part, and between Railway and New Piers the depth is 37 feet at low water.

New Pier.—The New Pier, about 400 yards westward of Railway Pier, extends about 620 yards from the shore, the outer 440 yards of it being in 37 feet of water.

Channel.—The dredged channel leading to Port Melbourne Railway Pier, the Town Pier, and New Pier, is 600 feet wide, with a depth of 37 feet. The channel is being extended to the southward, is dredged to 34 feet, and will be deepened to 37 feet.

Range marks.—The front mark is the mast on the head of the Railway Pier. The rear mark is a white mast, with red diamond-shaped topmark, situated about 900 yards, 2°, from the front mark.

By day.—These marks in line, bearing 2°, lead through the center of the dredged channel.

By night.—A light on the front mark in line with a light near the rear mark indicates the same line.

Beacon light.—The entrance to Port Melbourne dredged channel is marked on the eastern side by St. Kilda piled beacon, placed 40 feet eastward of the eastern side of the channel and about 1,150 yards eastward from the outer end of Williamstown Breakwater Pier. The piled structure is red with a deck level 10 feet high, surmounted by a white skeleton superstructure about 20 feet above the deck level; from it a light is shown at an elevation of 22 feet.

Lightbuoy.—A red cylindrical buoy, exhibiting an occulting white light, is moored on the eastern side of the dredged channel, at the inner end, about 700 yards, 203°, from the end of the Town Pier.

Lights.—A fixed red light, visible 3 miles, is shown at the outer end of Port Melbourne Town Pier.

A fixed green light, visible 3 miles, is shown at the outer end of the Railway Pier.

Near the inner end of the Railway Pier on each side is a small light.

A light is shown from each outer corner of the New Pier.

Prohibited anchorage.—Anchorage is prohibited eastward of the line indicated by 2 tripod beacons in line bearing 2°. The front beacon is situated at the inner end of the New Pier, and is surmounted by a black triangular topmark. The rear beacon is about 200 yards distant, 2°, from the front beacon, and is surmounted by a black spherical topmark.

Dredged area.—The western limit of the dredged area near the piers is indicated by beacons.

A compass-adjusting beacon, a white tripod with a black globe over a white triagle as topmark, is situated 304°, distant 1,425 yards from the light on the end of the Railway Pier. There is a buoy for compass adjustment nearly ½ mile southward of the Town Pier, in 23 feet water, and 5 others for light-draft vessels, westward of the Railway Pier, in 15 feet water.

There is also a buoy for compass adjustment in the center of Victoria Dock.

Point Gellibrand (lat. 37° 52′, long. 144° 55′).—From the southern extremity of Point Gellibrand a low rocky shore trends northeastward 1,340 yards to the eastern extremity of the point, on which stands the Time Ball Tower, and is bordered by ledges of rocks, with spits of foul ground, which, midway, extend ¼ mile from the shore toward the pile lighthouse. These spits are inclosed by a shoal bank, the 3-fathom edge of which, from 500 yards south of the southern extremity of Point Gellibrand, trends eastward 700 yards and northeast 600 yards to the southeast elbow of the bank and from thence extends north ½ mile to within 200 yards of the outer end of Williamstown Breakwater Pier.

Light.—A flashing white light with red sectors, 50 feet above water, visible 12 miles, is shown from a pile lighthouse, the iron tower of which is painted white, $\frac{1}{2}$ mile southeastward of Gellibrand Point.

Fogsignals.—Fogsignals are made from the lighthouse.

Storm signals are exhibited from the yardarm of the flagstaff on the lighthouse.

Williamstown (lat. 37° 52′, long. 144° 54′), on the southwest side of Hobson Bay and 8 miles from Melbourne, with which it is connected by a railway, has a population of about 12,000 persons; their business is principally with shipping. There is accommodation alongside the piers for vessels of various sizes; there is also provision for the repairs of vessels.

Piers.—From the eastern extremity of Point Gellibrand the breakwater pier extends 56° 500 yards from the shore; there is berthing accommodation on the northwest side for vessels of 29 feet draft.

From about 50 yards to the northwestward of the Time Ball Tower on the eastern extremity of Point Gellibrand, the Railway Pier extends north-northeastward 1,750 feet, and has accommodation in 1,150 feet of its length on the east side for vessels of 29 feet draft and 1,000 feet on the west side for vessels of 28 feet draft. A narrow 3-fathom shoal extends out nearly 400 yards midway between the breakwater and the railway piers.

Dock Pier extends northeastward 350 feet from the shore, with a depth of 28 feet at low water along the outer 200 feet of its length.

New Railway Pier, 100 yards westward of the Dock Pier, extends northeastward 600 feet from the shore, with accommodation for vessels drawing 29 feet.

Harbor Trust (Reid Street) Pier, about 260 feet westward of the New Railway Pier, projects from the shore north-northeastward 500 feet into a depth of 24 feet.

Ann's Wharf (Harbor Trust Pier), 800 feet westward of the New Railway Pier, projects from the shore north-northeastward about 500 feet, with depths at low water from 17 feet at inner to 22 feet at outer end. From about 400 feet within the end of this wharf the dockyard wharf extends to the inner part of the patent slip jetties, and incloses the dockyard reserve.

Gem Pier (lat. 37° 52′, long. 144° 54′).—From the inner patent slip, the shore trends westward a little more than 200 yards to the Gem Pier, which projects northward about 500 feet from the shore into $8\frac{1}{2}$ feet at low water.

Lights.—A fixed green light, visible 3 miles, is shown on the end of the Breakwater Pier.

A fixed green light, visible 3 miles, is shown from the outer end of Ann's wharf.

A fixed red light is shown from the outer end of Gem Pier.

Channel to the piers.—The channel, about 500 feet wide, leading to the various piers at Williamstown, with a depth of 28 feet at low water, has been dredged, and is being continued to a depth of 34 feet.

Docks and slips.—See Appendix I.

Repairs.—There are several foundries in Williamstown, and also in Melbourne, capable of undertaking marine repairs of every description.

The Victorian Government has a complete set of workshops and factories in connection with the Alfred Dock.

Caution.—A square red flag is exhibited near the entrance to the Alfred Dock when docking operations, requiring undisturbed waters at the entrance to the dock, are in progress. When the signal is shown, mariners are cautioned to proceed at the vessel's slowest speed when passing the entrance to the dock, otherwise serious damage may be caused.

Water.—Good fresh water can be obtained from the mains at the Alfred Dock and Dock Pier, also at all piers and wharves.

The western shore (lat. 37° 51′, long. 144° 54′).—From about 150 yards westward of Gem Pier the shore trends north-northwest-ward 1,000 yards; it then curves to the northward for nearly a mile to a point near which are the directing walls of the River Yarra, and where is the Williamstown steam ferry. The water frontage of North Williamstown extends to the northwest about 1 mile from the Gem Pier. About 600 yards north-northwestward of the Gem Pier is the Stevedore Pier, 550 feet long, with a depth of 7 feet at its outer end. There are also several boat jetties and sheds along this frontage.

The northwestern bight of Hobson Bay (lat. 37° 52′, long. 144° 55′) is occupid by a bank extending nearly across it. The 3-fathom edge of this bank from off the end of Ann's Wharf trends to the eastward to about 300 yards northward of the Williamstown railway pier; it then extends northeastward and northward 1,500 yards, and from that curves to the northeastward toward the dredged area of the New Pier. The entrance of the River Yarra has been dredged through this bank.

Lightbuoy.—A lightbuoy is placed in 24 feet at low water, with the end of Williamstown breakwater pier bearing 206°, distant 350 yards. From it is exhibited a flashing red light. It marks the southwestern edge of the dredged channel to the entrance of the Yarra River.

Tides.—It is high water, full and change, at Williamstown at 2 h. 53 m.; springs rise 2¼ feet.

Anchorage.—The bay is open to southerly gales, which send in sufficient sea to interrupt traffic; but small vessels can at all times find shelter off Williamstown if they can not get within the shelter of the piers. The depth of water is 18 to 23 feet. Vessels must moor. Eastward of the submarine mining ground are depths of 26 to 31 feet, with mud bottom.

Explosives anchorage.—The anchorage for vessels with explosives is: Westward of an imaginary line bearing 236° from Point Gellibrand pile lighthouse, and exceeding a distance of 1,500 yards from the shore. The lightbuoy, moored about 3¾ miles 245° from Point Gellibrand pile lighthouse indicates a suitable anchorage.

The submarine mining ground lies south of Williamstown breakwater pier, between lines drawn from Gellibrand pile lighthouse to the outer end of the Breakwater Pier, and from the pile lighthouse to the Time ball tower. Mariners are cautioned not to anchor in the vicinity.

Time signal.—A time ball is dropped daily, Sundays excepted, from a staff on the old lighthouse at Point Gellibrand, at 1 h. 00 m. 00 s. p. m. standard time of Victoria, equivalent to 15 h. 00 m. 00 s. Greenwich mean time. The signal when hoisted is 72 feet above high water, and the drop is 11 feet. The ball is hoisted halfway up as a preparatory signal at 0 h. 55 m. 00 s. p. m. and dropped by electricity from Melbourne observatory. An error of one-third of a second is notified in the next day's newspapers. Should the signal fail, a red flag is hoisted and the ball is dropped a second time at 2 h. 00 m. 00 s. p. m. The old lighthouse is a square castellated tower. The upper half of the ball is painted black and the lower half red.

Yarra River is a narrow winding stream, taking its rise about 65 miles to the eastward in the Dividing Range; from Hobson Bay to the Queen's Bridge at Melbourne its length is about 5½ miles, the average breadth of the stream being about 300 feet, with a depth of 26 feet at low water in the fairway.

Entrance channel.—A channel 400 feet wide and 27 feet deep has been dredged from seaward to the entrance; the southwestern edge of the dredged part is marked by the lightbuoy, already mentioned.

The course of the river is northwestward for 1¼ miles from the entrance, thence northward for a mile, whence it curves northeast and eastward to Melbourne.

At about 2 miles from the entrance is Stony Creek, on the west side; and 400 yards farther, on the east side, are timber jetties. Twelve hundred yards from Stony Creek the old course of the river turns northward for 1,500 yards, where it is joined by the Saltwater River. The Yarra then curves to the eastward on the south side of Coode Island, through New Channel, and thence to Queen's Bridge.

Within about 1½ miles from the Queen's Bridge there are wharves on both sides of the river; 1,500 yards from the bridge is a swinging basin, 550 feet across.

Works in progress.—The work of widening the swinging basin and setting back the south wharf is in progress. The swinging basin, now 550 feet wide, is to be enlarged to 650 feet. The depth is to be increased from 26 feet to 30 feet throughout. A clean cut will be made from near the Coode Canal to the outer end of the swinging basin. Opposite the entrance of the Victoria Dock the width will be increased to 800 feet.

Regulations—New Channel.—The speed of vessels proceeding to Melbourne must not exceed 5 knots. Only masters exempt from pilotage for the port are permitted to navigate their vessels in the river without the services of a pilot.

No inward-bound vessel shall enter the New Channel (abreast Coode Island) until any outward-bound vessel which may be in or approaching the channel and will in all probability be in it before the inward-bound vessel could have passed through the channel shall have passed clear of the lower or western end of the channel, the intention being that no two vessels shall be navigated in the channel at the same time. This regulation applies only to an inward-bound vessel whose gross tonnage, with that of the outward-bound vessel, shall exceed 8,000 tons. The inward-bound vessel shall remain at a safe distance below the lower or western entrance of the channel until every outward-bound vessel shall have passed clear of that entrance. Fishing without permission is prohibited in the River Yarra between Queen's Bridge and New Channel.

Lights.—The undermentioned lights are exhibited in the River Yarra:

A fixed red light is shown from a beacon erected on the north (starboard) side of the entrance of the river and a light from a beacon on the south (port) side of the entrance.

In addition to the above, 4 lights are exhibited on the starboard bank of the river (entering), and 13 on the port bank. · There are also two light buoys showing one occulting red and one white. The positions of these lights and buoys will be better understood by the chart than by a written description.

Tide gauges.—For guidance while navigating the river tide gauges showing the depth of water are erected at the following places:

1. At red beacon, mouth of river, on starboard side.
2. At short road ferry, on starboard side.
3. At junction of river and Stony Creek, on port side.
4. At west end of Coode Canal, on port side.
5. At east end of Coode Canal, on port side.
6. At Victoria dock entrance.
7. At Johnston Street Ferry, on starboard side.
8. At Spencer Street Dock, on port side.
9. At Queen's Wharf, opposite customhouse, on port side.

Tides.—It is high water, full and change, in Yarra River at Queen's Wharf, Melbourne, at 2 h. 48 m.; springs rise 2 feet 8 inches.

The waters of the River Yarra are almost continually running outward, there being practically no flow in during flood tide, the rise of the sea level merely banking up the waters of the river and checking the outflow. Under the influence of strong west to southerly winds, however, an upstream current is caused, but it only extends to a few feet below the surface, and exerts practically little or no effect on other than very light-draft vessels. The normal rate of outflow is from 1 to 2 knots; this is slightly accelerated during heavy rains.

Floods are not infrequent, overflowing the banks and causing great destruction of property. That of December, 1863, rose 7 feet above the mean level of the river below the falls. Above these falls, the rise was stated on that occasion to have been 30 feet, and in the gorges above Melbourne 50 feet.

Dry docks.—There are dry docks on the south bank of the river river about 200 yards below Spencer Street Ferry. See Appendix I.

Wet docks.—Four miles up the river, on the north bank, and adjacent to the city wharves and Spencer Street railway station, Victoria Dock, a wet dock, has been excavated. On the south side are 6 closed sheds, each 300 feet long, between each of which are similar lengths of open sheds. On the east side are two closed sheds, each 300 feet long, with similar lengths of open sheds between. On the north and east sides of the dock are lines of rails in connection with the Victorian railroads. See Appendix II.

Spencer Street Dock is for the use of small craft.

Cranes.—A 70-ton steam crane on the south side of the river opposite the customhouse; 35 and 15 ton hydraulic cranes on the north side of the river near the middle swinging basin; also a number of hand cranes ranging from 1 to 10 tons.

Repairs.—There are several foundries in Melbourne, and also in Williamstown, capable of undertaking marine repairs of every description.

Coal, coke, liquid fuel, and lubricating oil.—About 780,000 tons of coal are imported annually and from 30,000 to 31,000 tons are kept in stock in Melbourne; 10,000 tons of coke are imported annually, and 1,000 tons are kept in stock; from 5 to 6 million gallons of liquid fuel are imported annually and about three-quarters of a million gallons are kept in stock; about three-quarters of a million gallons of lubricating oil are imported annually, 200,000 gallons being kept in stock; 12,000 gallons of kerosene are imported annually, and 6,000 gallons are kept in stock; 300,000 gallons of coal tar are usually in stock. There are four coaling wharves 160 to 400 feet long, with depths alongside of from 22 to 23 feet.

Water.—Facilities for watering vessels exist at all wharves and piers. Water is supplied to vessels alongside at 3 shillings per 1,000 gallons. Vessels lying in the bay are supplied by private water boats, the charge being a matter of arrangement, generally 5 shillings per ton.

Signals.—For signals in use at the ports of Victoria, see Chapter I.

Wharf regulation.—The propelling engines of any vessel are not to be worked whilst moored alongside any wharf in the ports of Victoria without permission of the port or wharf authority.

Salvage tugs, etc.—The powerful steam tug *J. A. Boyd*, Melbourne Harbor Trust, is fitted with salvage and fire pumps of the latest type. The steam tug *James Paterson*, slightly smaller, and the Melbourne Harbor Trust launch *Commissioner* are also equipped with salvage and fire pumps.

Wireless telegraph.—A wireless telegraph station has been established at Melbourne. It is open to the public at all times. Weather reports are dispatched between 7.30 and 9 p. m.

Wireless time signals, in accordance with the International Time Convention, have been established at Melbourne. The signals are made at 00 h. 00 m. 00 s. and 12 h. 00 m. 00 s. Standard time, corresponding to 14 h. 00 m. 00 s. and 2 h. 00 m. 00 s. G. M. T., respectively.

Adjustment of compasses.—The fall of Mount Macedon is a good distant mark to use, when swinging in Hobson Bay, to ascertain the deviation of the compass.

Port regulations—Quarantine.—Masters of vessels arriving report to the pilot the places at which they loaded and touched, and answer all questions respecting the health of the crew and passengers, under penalty of £100.

The pilot is to give notice to the master if the vessel is liable to quarantine, whereupon he shall hoist a yellow flag, under penalty on the master of £100.

All vessels from other than Australian ports must undergo an examination at the heads by the health officer.

Melbourne (lat. 37° 50′, long. 144° 58′).—The area of greater Melbourne was 162,600 acres, and the population in 1917 was about 708,000.

It is the capital of the State of Victoria, and stands on seven hills, rising gradually from the Yarra; it is laid out in broad straight streets generally at right angles to each other, and has many handsome public buildings. Its secure port and central position, with the network of railways and rivers connecting Melbourne with a large portion of Australia, command for it the chief export and import trade of an immense pastoral and agricultural district, independently of the demands of the gold fields.

The principal articles of import are manufactured goods of all kinds, provisions, machinery, railway materials, coal, timber, wine, spirits, etc. The principal exports are gold, live stock, hides, wine, timber, and wool.

DIRECTIONS.

South channel (lat. 38° 19′, long. 144° 54) **to Hobson Bay.**— Having rounded No. 15 black light buoy, which marks the eastern spit of Middle Ground, steer 2° 27 miles, which is the course and distance thence to Hobson Bay; enter eastward of Point Gellibrand pile lighthouse and anchor or moor, as previously directed.

For Geelong.—From No. 15 buoy a 334° course leads about 1 mile eastward of Prince George bank lightbuoy.

Working up.—As the eastern shore of Port Phillip is free from outlying dangers, it may be approached within 1 mile from Arthur's seat all the way up to Red Cliff. Between Anonyma Shoal and Hobson Bay shoal water extends farther from the shore, which should therefore be given a berth according to the draft of the vessel.

West channel to Hobson Bay.—From 200 yards eastward of west channel pile lighthouse, the course is 22° and the distance 20½ miles to Point Gellibrand pile lighthouse. There are no dangers in the way, and the depths are regular, gradually increasing from 9 fathoms a mile northward of west channel lighthouse to 12 fathoms midway, and thence decreasing to 6 fathoms 1 mile southward of Point Gellibrand pile lighthouse; and the bottom being soft mud and shells, a vessel may anchor anywhere along this route.

Working up from the west channel lighthouse to Hobson Bay, do not stand into less than 5 fathoms on either side, nor approach the western shore nearer than 3 miles, until Station Peak is in range with Point Cook, 264°, when, in standing to the westward, Point Gellibrand pile lighthouse must not be brought eastward of 53°, nor must Point Gellibrand be approached within ½ mile until north of it. The bottom, at the distance of 1 mile offshore, from Point Gellibrand to Point Wilson, is rocky, with shoal patches.

From Hobson Bay (lat. 37° 52′, long. 144° 56′) **to sea by the west channel.**—Sailing vessels bound to sea from Hobson Bay by the west channel will generally clear the heads the same day by leaving Hobson Bay two or three hours before daylight, when there is frequently a moderate land or northerly wind. A 202° course for 20 miles from a fair berth off the Gellibrand Pile Lighthouse leads to the West channel lighthouse, where, if prevented by southerly gales from proceeding through the channel, there is good anchorage with the lighthouse bearing 199° and Station Peak just in range with Indented Head.

From the lighthouse at the northern entrance proceed through the west channel, following inversely the directions already given for going northward.

At night.—Vessels bound for sea through the west channel at night should pass a good 200 yards eastward from the pile lighthouse at the northeastern entrance to the channel, then bring the light on No. 12 lightbuoy about half a point on the port bow and steer 223°. Continue this course until nearing the lightbuoy, then alter course to port gradually, and bring the light on No. 2 beacon, Popes Eye Bank, in line with the light on Observatory Point, bearing about 206°. Keep these lights in line until 1 mile past No. 12 Lightbuoy, then bring them slightly open to the eastward, keeping them so until entering the occulting white sector of Queenscliff low light, then steer for that light, keeping in the white sector until Observatory Point Light is obscured, when a course 239° may be shaped for the fairway between Port Phillip Heads.

NOTE.—Although the west channel is available for vessels of 17 feet draft, yet on account of its banks and knolls being subject to constant change those in charge of vessels proceeding through it should obtain the latest chart of the resurvey of the channel issued by the Department of Ports and Harbors, Melbourne, and kept corrected to date.

From Hobson Bay to sea by the South Channel steer from Point Gellibrand Pile Lighthouse about 180°, 27 miles, for No. 15 black light buoy, which marks the east spit of the Middle Ground, after rounding which change to the westward to pass 400 yards southward of No. 13 black buoy, then steer to pass close northward of the light buoys marking the southern edge of the dredged channel, and about 400 yards southward of No. 9 buoy, then gradually bring the South Channel Pile Lighthouse in range with the Eastern Lighthouse; these kept in range astern, bearing 107°, lead down in midchannel; the red conical buoys are left on the port, and the black can buoys on the starboard hand.

When approaching No. 3 black can buoy, alter course to pass about 200 yards to southward of the black lightbuoy which lies southward of the fairway shoal. The range passes over the shoal in 25 feet of water, but 100 yards to the southward of this lightbuoy the depth is 34 feet, with 37 feet outside this again for 300 yards. On passing the light buoy change again at once toward the range, so as to give a wide berth to the 25-foot patch which lies 800 yards southwestward of the lightbuoy.

From 800 yards south of No. 1 buoy steer 272° leaving the Pope's Eye black and white buoy 1,200 yards on the starboard hand, taking care to keep Barwon Head just open of Point Lonsdale until Queenscliff lighthouses are in line 42°, with which range astern, proceed out between the heads to sea.

The beacons in Lonsdale Bight in range bearing 344° lead out in deeper water, but since the direct channel was deepened to 40 feet there is no necessity, except perhaps for a sailing vessel, to use it. Due attention must be paid to the tidal currents.

At night.—From Hobson Bay at night steer about 180°; the eastern white light will then be made nearly ahead, and the red light when nearing No. 15 buoy. When the flashing white light of No. 15 buoy is seen, steer to pass to the eastward of it. Having passed the buoy at a distance of about 400 yards, continue to the southward till the South Channel Pile Light has fully changed from red to white, then alter course toward it, bearing in mind that the northern limit of the white light passes less than 200 yards south of the Middle Ground. Pass close northward of the lightbuoys marking the dredged channel; then gradually bright the Pile and Eastern Lights in range astern bearing 107°, which leads through the South Channel.

A bearing of Portsea Light will indicate when, on approaching the fairway shoal, to change to the southward of the lightbuoy moored southward of it, and to follow carefully the day directions given above for this detour.

When Point Lonsdale Light bears about 267°, steer for it, until Queenscliff low light changes from red to white. Then bring the Queenscliff lights in line astern, bearing 42°, which lead between the heads to sea. Attention must be given to the tidal currents.

Working.—Vessels leaving Hobson Bay against strong southerly winds, especially during the summer months, when these winds prevail, will get to sea sooner by working down the eastern side of Port Phillip, and going through the South Channel, where, having smooth water, they will be enabled to pass through the South Channel, down to the entrance between the heads; by bearing down the middle of Port Phillip, and taking the West Channel, more swell will be experienced, and a large vessel probably have to anchor off the West Channel Lighthouse, and wait for a shift of wind.

To work out between the heads, the tidal currents must be attended to, and it is better to do so near slack water, when the " Race " or " Rip " is nearly quiescent and greater command of the vessel obtained. With an outgoing tidal current and light winds, be careful not to be drawn into the bight between Shortland Bluff and Point Lonsdale, the current setting from thence directly over Lonsdale Reef. A vessel within its influence, nearly becalmed, and having only the sails to trust to, has no resource but that of dropping her anchor, which she is nearly certain to lose, from the rocky nature of the ground.

Anchorage.—Vessels having good ground tackle may, if necessary, anchor in any part of Port Phillip above the entrance banks, there being nowhere a greater depth than 15 fathoms, and good holding ground everywhere.

Tides.—Between the heads of the entrance to Port Phillip the tides are most irregular, the narrow entrance to the large basin within checking the fair course of the tidal wave; hence after southerly gales it may be high water all day, and the contrary with northerly gales.

It is high water, full and change, at the under-mentioned places in Port Phillip as follows:

	Time.	Springs rise.	Neaps rise.
	H. m.	*Ft. in.*	*Ft. in.*
Entrance to Port Phillip	[1] 2 00	7 0	5 6
Queenscliff	2 00	3 0	2 0
Dromana Bay	2 19	3 0	2 6
Snapper Point	2 14	2 8	2 0
Point Henry, Geelong	2 39	3 0	2 6
Williamstown, Hobson Bay	2 53	2 3	.. ..
Queens Wharf, Melbourne	2 48	2 8	2 2

[1] Approximately.

The mean water, or half-tide level, varies as much as the rise and fall of the tide, it being influenced by the strength and direction of the wind outside the heads. Southerly gales cause an elevation of both high and low water, and northerly gales have a contrary effect; the latter sometimes keep back the flood tide for an hour or even 1½ hours later than the time by calculation.

On an average it is high water at Port Phillip Heads about 3½ hours before that at Williamstown, but owing to weather causes this interval may vary from 3 to 4 hours. As the tidal currents in the fairway run for about 3 hours after the time of high and low water by the shore at Point Lonsdale, it will, therefore, be approximately slack water flood at the heads when it is high water at Williamstown.

Tidal currents.—The current turns at about high water and at about 2 hours after low water by the shore. In the middle of the entrance between Point Lonsdale and Point Nepean the period of slack water is very limited. Tidal signals have been described.

Set of the in-going current.—The in-going current comes from the southward and eastward, increases in strength as it nears the heads, sets right into the entrance, across and through the reefs, with great force, and spreads toward Shortland Bluff and Point King. The current decreases in strength as it enters the channels, setting toward Swan Point and through the West Channel in an oblique direction, tending toward Coles Channel and Indented Head; and above the West Channel Lighthouse, northwestward across Prince George Bank; spreading from thence toward Geelong, Point Cook, and Hobson Bay. In the South Channel the current sets to the east-northeast, across the Middle Ground, through Pinnace Channel, and spreads along the eastern shore toward Hobson Bay.

Set of the outgoing current.—The outgoing current sets out of Hobson Bay southeastward for a few miles, when it takes a more southerly direction toward Prince George Bank; it then passes obliquely through the various channels, the current from Symonds Channel joining and turning that of the West Channel below the Royal George Shoal, setting toward the bight between Shortland Bluff and Point Lonsdale, and from thence out through the heads at a great rate; the body of the current setting athwart the entrance toward Point Nepean, and away southeastward along the land and in to the shore between Point Nepean and Cape Schanck.

Between the heads the current runs from 5 to 7 knots; in the West and South Channels between 2 and 3 knots; and about 1¼ knots in the broad space above the channels. In Hobson Bay, during the winter months, there is always a surface current running out, owing to the freshets from the River Yarra; this current frequently sets along both sides of the bay at the rate of 2 knots. The current is weak in Geelong Bay, but in the North Channel it sets 2¼ knots across the bar and becomes weaker as it spreads over the Inner Harbor of Geelong.

CHAPTER IV.

AUSTRALIA—SOUTH COAST, PORT PHILLIP TO CAPE HOWE—BASS STRAIT.

The coast from Point Nepean trends southeastward 16 miles to Cape Schanck, and may be approached to 1 mile in depths of 8 to 16 fathoms. The highest hill along it is 433 feet above the sea, 2 miles north of the cape.

It is desirable to sight Cape Schanck before running far into the great bight for Port Phillip; and if the wind blows strong from the southward it is unsafe to run without having seen it. The cape is also an excellent mark for Port Western, the western and principal entrance of which lies between 7 and 10 miles eastward of the cape.

Cape Schanck (lat. 38° 30′, long. 144° 53′), the southern extremity of the peninsula which separates Port Phillip from Port Western, is a narrow cliffy headland, 278 feet high close off which is the remarkable Pulpit Rock, with a smaller rock lying southward, nearly ¼ mile from the cape.

Caution.—The reef to the southward of Pulpit Rock lies about 1,400 yards southward from the lighthouse; in passing the light it must therefore be given a wide berth.

Lloyd's signal station.—There is a Lloyd's signal station at Cape Schanck Lighthouse and vessels can communicate by the International Code, and at night by Morse lamp. It is connected by telephone to Dromana, which is connected with the telegraph system.

Life-saving apparatus.—A life-saving rocket apparatus is kept at Cape Schanck.

Light.—A fixed and flashing white light, 328 feet above water, visible 24 miles is shown from a white circular stone lighthouse on the highest part of Cape Schanck, at about ½ mile northwestward from Pulpit Rock. During bush fires, which occur during the summer months, the smoke from them has been observed to give this light a red appearance.

Fog signal.—A fog signal is made from the lighthouse.

On the east side of Cape Schanck a rocky bight extends east-northeastward 1½ miles to Barker Point; and at ½ mile westward of the point is a small stream of fresh water. From Barker Point the coast, which is closely bordered with rocks, trends east-northeastward 5¼ miles to West Head.

West Head (lat. 38° 29', long. 145° 02'), distant 7 miles eastward from Cape Schanck, is a narrow cliffy projection 85 feet high, and inclosed by reefs, which at 1,500 yards westward of the head, extend from the shore for 1,300 yards. This head forms the western point of Port Western; a rock with 1¾ fathoms lies southeastward 1,200 yards from it.

Port Western is an extensive bay, protected from the sea by Phillip Island, between the western point of which and West Head is the west entrance to the port; the east entrance being a narrow channel separating the eastern end of the island from the mainland to the eastward.

The northern shore of Port Western, from west head, curves north-northeastward 7¼ miles, and thence eastward 5 miles to Sandy Point, forming a bight, the northeastern and greatest portion of which is occupied by shoal water, thickly strewed with knolls, extending from the shore to Middle Bank, which trends west-southwestward 4½ miles, and south-southwestward 2 miles from Sandy Point; and is reported to have extended half a mile further to the southwestward. The high land (275 feet) near Barker Point open of West Head bearing 263° until Tortoise Head bears 62° clears the southwest edge of the bank. Between the southwestern spit of this bank and West Head there are depths of 4 to 10 fathoms.

Depths.—**The west entrance of** Port Western, which lies between West Head and Grant Point, is 3½ miles wide between Black Rock and the 10-foot rock off West Head, with 15 fathoms in mid-channel and 5 to 6 fathoms close to the rocks on either side; and being open and free from any other hidden danger it is easy of access, and affords sufficient room for a vessel of any size to work in or out. From 3½ miles within the entrance a clear channel 1 to 1½ miles wide, with 6 to 17 fathoms water, extends northeastward, between Phillip Island and Middle Bank, to abreast of Sandy Point, where the port divides into two arms, one trending northward and the other eastward.

Limits of port.—The port of Port Western includes all inlets, rivers, bays, harbors, and navigable waters north of and within a line bearing 117° from West Head to Grant Point, and also north of a line bearing 53° from Cape Wollamai to the opposite shore.

Flinders is a postal town situated just inside West Head; population in 1911 was 390. It is a signal station and connected by telegraph with all parts; a coach runs daily from Bittern.

Jetty.—At Flinders there is a jetty 356 yards in length and 13 feet in width, with a depth of 8½ feet at low water at its outer end.

Light.—A light is exhibited from a lamp-post 24 feet high on the outer end of Flinders jetty.

Anchorage—Buoy.—There is a red warping buoy anchored 150 yards off the end of the jetty in 14 feet of water; to avoid fouling the telegraph cable, anchor off the jetty with the light bearing westward of 273°.

Prohibited anchorage.—Vessels are not to anchor with the light bearing between 273° and 325°, or showing white at night.

Grant Point (lat. 38° 31′, long. 145° 07′), situated 4½ miles eastsoutheastward from West Head, is a craggy projection, forming the western extremity of Phillip Island; a reef extends westward, ¼ mile from it to Round Islet, which is 98 feet high. At 1,500 yards southwestward from this islet is Black Rock, which is 34 feet above high water and fringed by a reef, on which the sea breaks heavily with a southerly swell; between the reef and Round Islet is a passage with 5½ to 7 fathoms water, available for small craft on an emergency.

The southern shore of Port Western, or the northwestern coast of Phillip Island, from Grant Point trends north-northeast one mile to the rocky southwestward point of Cat Bay; reefs project a quarter of a mile from this shore and from the south side of the bay. From the bight of this little bay the coast sweeps round north-northeastward 2½ miles to McHaffie Reef; then northeastward 2 miles to Red Rock, and eastward 5 miles to Observation Point, the northeastern end of the island; for the first 4½ miles it is bordered by reefs, and thence to the point it is steep and sandy; none of these reefs project beyond ¼ mile from the shore.

Buoys.—A red conical buoy, in about 5 fathoms water, lies off the northern end of the McHaffie Reef. A red cask buoy marks the edge of the shoal water westward of Red Rock.

Cowes is a post and telegraph town on the northern side of Phillip Island. Communication is by steamer to Stony Point, thence by rail. It is a watering place, and there is excellent shooting and fishing. Population is about 300.

Jetty.—There is a jetty at Cowes 400 feet in length and 25 feet in width, with a depth of 13 feet at low water at its extremity.

Light.—A light is exhibited from the end of Cowes Jetty.

Buoy.—A red warping buoy is moored in 7 fathoms off the eastern end of the jetty.

Mussel Rock lies ¼ mile eastward from Cowes Jetty. Tides are described later.

Anchorage.—There is anchorage in 7 fathoms from ½ to 1 mile eastward of Cowes, about ½ mile from the shore.

Observation Point (lat. 38° 27′, long. 145° 18′), which is low, broken, and swampy, is separated from the higher land south of it by a shallow creek, 400 yards wide at its mouth, whence it branches to the southwestward.

A sand spit extends eastward from the point.

Beacon.—A red beacon, about 10 feet high, marks the end of the sand spit.

Anchorage.—There is good anchorage at ½ mile off Observation Point in 8 fathoms water.

Tortoise Head, east-northeastward 1¾ miles from Sandy Point, is the southern end of a table-topped isle nearly 1 mile long, with a low point projecting from its west side; reefs extend ¼ mile from this and the head, and a spit, with 13 feet water on it, projects southwestward 1,500 yards from the latter.

Directions.—The tidal currents always raise a sufficient ripple to break on the banks which form the northwest side of the main channel, giving timely notice of the shoal water on that side, but if this can not be trusted steer for Tortoise Head, bearing 61°, and well open of the northwest point of Phillip Island, to avoid McHaffie Reef, which projects from it. To clear the southwest edge of Middle Bank keep the high land (275 feet) near Barker Point open of West Head, bearing 263°.

The red buoy off McHaffie Reef, and the red buoy on the edge of the shoal near Red Rock, must be left on the starboard hand when entering.

The north coast of Phillip Island, although bold, should not be approached within ½ mile, as the tidal currents set along it at a great rate, and may, in light winds, sweep a vessel too near the shore. Having arrived abreast of Sandy Point, proceed northward or eastward; if proceeding eastward be careful to avoid the shallow spit 1 mile northward of Observation Point.

Anchorage.—There is anchorage at ¼ mile eastward of Sandy Point in 7 fathoms water.

French Island, on the north side of the eastern part of Port Western, is 11 miles long, east and west, and 7½ miles broad at its western end, between Tortoise Head and Scrub Point, whence it narrows to 4 miles toward Spit Point, the eastern extremity of the island. The southern and greater portion of the island is hilly, the highest part being Mount Wellington, 314 feet high, but the northwestern part and northern coast are low and marshy. From the southwestern extremity of French Island its southern coast trends east-southeastward 3¾ miles to Long Point, its southern extremity; between this and Finger Point, at 2 miles northeastward from it, is a shoal bight, in the entrance of which is Elizabeth Isle, 60 feet high. Tides at Spit Point described later.

The bights formed between Tortoise Head and Long Point, and that from thence to Finger Point, are filled with mud flats, having 1 to 5 feet water on them at high water, with navigable creeks reaching in to the shore. From a 15-foot spit southwestward 1¼ miles from

Tortoise Head, the 5-fathom edge of the bank, which mostly consists of mud flats, trends eastward 4 miles, and then sweeps round close outside Elizabeth Isle to Finger Point.

Buoy.—A red cask buoy, surmounted by a diamond, and moored in 6 fathoms, lies off the end of 15-feet spit 1¼ miles southwestward from Tortoise Head.

Long Point—Jetty.—The jetty at Long Point projects 400 yards south-southeastward from the points.

Rock.—A rocky patch, with a depth of 4 feet, lies ¼ mile west-southwestward from the outer end of the jetty. To avoid this danger vessels should steer in with the end of the jetty bearing 53°; when leaving the jetty the course should be 233°.

The North Arm of Port Western is 1 mile wide between Sandy Point and the spit 1¼ miles southwestward from Tortoise Head; and from its entrance trends northward 10 miles to Watson Inlet, the width increasing to 3 miles between Scrub Point and the western shore.

Depths.—The channel at the North Arm is on the west side, between the mud flat which projects 200 to 800 yards from the shore of the mainland, and the banks, mostly dry at low water, which extend ¼ mile to 2 miles from the west side of French Island. This channel is 1,500 yards to 1 mile wide with 11 to 6 fathoms water, from the entrance to abreast Scrub Point. The banks on the east side of the arm are separated from French Island by a passage about ¼ mile wide, with 3½ to 8 fathoms water in it, but it is encumbered with shoal patches. The southern portion of these banks is called Middle Spit.

Beacon.—The southern end of Middle Spit is marked by a red single pile beacon in 3 feet of water.

At 1½ miles northward from Sandy Point is a shallow inlet; and between Crib Point and Long Islet, 6 miles from Sandy Point, the low western shore forms a bight, in the southern part of which is Sandstone Isle. This bight is filled by a flat, intersected by creeks, the largest of which, from ½ mile northeastward of Sandstone Isle, winds northwestward about 1½ miles to the town of Hastings; this creek is marked by single-pile beacons, black on port and red on starboard side, and has irregular depths of 1 to 5 fathoms.

Shoals.—Off Stony Point at 2½ miles, and at 4¼ miles northward of Sandy Point, 4-fathom shoals project ½ mile from the bank which borders the western shore.

Stony Point (lat. 38° 22′, long. 145° 13′), situated 2½ miles northward from Sandy Point, is a railway terminus and is provided with a jetty 305 yards long and 10 feet wide, having a depth of 12½ feet at low water at its outer end.

Light.—A light, 33 feet above high water, is exhibited from a lamp-post on the end of Stony Point jetty.

Crib Point, 1¼ miles northward of Stony Point, has telegraph and railway stations.

Tea-tree Point, on the east side of the North Arm and 1¼ miles northward of Tortoise Head, is provided with a jetty 245 yards in length and 12 feet wide with a depth of 2 feet at its outer end. It is locally known as Tankerton Jetty.

Fairhaven Jetty, 3 miles northward from Tea-tree Point, is 400 feet long. The jetty is available for boats only. At high water there is a depth of 4 to 5 feet at the outer end, but at low water the bank dries 300 yards from this end. It is approached by a narrow creek, the entrance to which is 1½ miles southward from the jetty.

Beacons.—The eastern side of the entrance to the creek is marked by a red single pile beacon. The western side is marked by a black single pile beacon, standing nearly ½ mile northward from the red beacon and about ¼ mile northward of the end of the spit of the bank westward of the creek.

Shoals—Beacons.—There are several shoal patches northwest of Scrub Point, that on the northern side of the fairway being Eagle Rock, which is awash at low water and marked by a black beacon, consisting of a staff and ball 12 feet high, northwestward 1¼ miles from the point.

A white beacon northward a little more than 1,500 yards from Scrub Point is on Crawfish Rock on the south side of Bagge Harbor. The beacon is 15 feet high, with a 5-foot square top.

Bagge Harbor.—Between Scrub Point and Quail Island, 2 miles northward of it, is Bagge Harbor, about 1½ miles long, and 1,500 yards wide between the drying banks. The depths are 8 to 13 fathoms.

At Bagge Harbor the North Arm turns to the eastward into a sheet of water extending 10 miles east and west, and 4½ miles across between the north side of French Island and the low mainland to the northward and eastward. The north shore is intersected by numerous creeks and inlets. This sheet of water is occupied by a mud flat, which dries at low water and carries 6 to 8 feet on it at high water. ordinary springs. It is intersected by numerous channels branching into it from the North Arm. There are depths of 13 to 4 fathoms for about 4 miles eastward from Eagle Rock. A few of the smaller branches af this navigable water carry 6 feet of water to within ½ mile of the low woody shores.

Hastings, in the bight northward of Sandstone Isle, is a post town and telegraph station, with a population of about 500; it is an extensive fishing station and supplies the Melbourne market to a considerable extent, with which place it is connected by rail. The neighborhood is also noted for its orchards.

Jetty.—The jetty at Hastings is 127 yards long, 10 feet wide, with a face 26 yards long, on the outer side of which there is a depth of 6 feet at low water.

Light.—A light is shown from the end of Hastings Jetty.

Tooradin (lat. 38° 13′, long. 145° 22′) is a postal township 6 miles northeastward of Scrub Point. It has railway and telegraphic communication. Population is about 250.

Jetty.—There is a jetty at Tooradin 25 yards long, 12 feet wide, with a face 21 yards long, with a depth on its outer side of 5 feet at low water, but the channel to it is shallow and tortuous, and only suitable for boats or flat-bottomed craft.

Light.—A light is shown, at an elevation of 30 feet, about 200 yards southwestward of the jetty on the mainland on the eastern side of the channel.

The East Arm of Port Western, between the northern side of Phillip Island and the bank which extends from the southern side of French Island, is 1¼ miles wide, with regular depths of 7 to 9 fathoms. At 1 mile northward of Observation Point is the western point of a narrow spit, with 12 to 6 feet water on it, projecting westward for 2¾ miles from the shoal flat which nearly fills the eastern part of Port Western. Between this spit and the northeastern extremity of Phillip Island there is a bight in the shoal flat, extending 2 miles east and west, and 1 mile across, where vessels may anchor in 6 to 8 fathoms, sand and shells. From the western point of the spit the northern branch of the East Arm sweeps round eastward and northeastward past Elizabeth Island and Finger Point, and is 1,340 to 1,000 yards wide, with 4½ to 11 fathoms water between the banks which border the south coast of French Island, and the shoal flat on its south side.

Settlement Point (lat. 38° 25′, long. 145° 25′), situated 1¼ miles eastward of Finger Point, is a rocky projection of the mainland, between which and the southeastern extremity of French Island are Pelican Islet and Schnapper Rock. The islet lies ½ mile westward from Settlement Point, with which it is connected by a reef. Schnapper Rock, which lies between Pelican Islet and French Island, divides the East Arm into two narrow channels with only 3¼ to 4 fathoms water.

Seen from the southward, the land in the vicinity of Settlement Point presents the appearance of cliffs of a reddish color, about 50 feet high with a rocky foreshore extending out at low water about 200 yards to the southwestward and 300 yards to the northwestward. Foul ground exists to the northwestward of the point for a farther distance of 200 yards from low-water mark to the 1-fathom curve.

Jetty.—Six hundred yards eastward from Settlement Point a jetty projects 127 yards northward from the shore at Corinella, with a depth alongside of 10 feet.

Beacons.—A black beacon, surmounted by a ball, is placed 400 yards northwestward of Settlement Point, and marks the north side of the deepest water leading into Corinella Channel.

A red beacon, 16 feet high, with a square top, marks the Schnapper Rock.

Buoy.—A red buoy is moored off the northern extremity of the reef at Settlement Point, distant 230 yards southward from the black beacon.

Corinella Channel.—From its entrance, 400 yards northwestward of Settlement Point, this channel or creek, extends in an easterly direction along the shore at an increasing distance therefrom for 1½ miles, and gradually decreasing in depth and width until it terminates in the mudflats of the bay. At half-tide the channel has a width of about 400 feet for a distance of ½ mile from the entrance, with a mean depth in the center of 16 feet. At the entrance to the channel is a bar, 200 yards long, with a depth of 4 feet over it; inside the bar, depths of 10 to 12 feet are carried to the jetty.

Pelican Islet, 4 feet high, is 200 yards long by 100 yards wide and covered with a few bushes. The reef forming the islet extends all around about 100 yards. Between the islet and Settlement Point the bottom is rocky and uneven, with a narrow channel of 10 to 12 feet of water lying about 300 yards from the islet.

Schnapper Rock is a half-tide reef about 400 yards long and 200 yards wide. The main channel of the East Arm runs between the rock and French Island, where it is about 400 yards wide, with depths of 7 to 8 fathoms, decreasing to 4 to 5 fathoms at 600 yards northward from the rock. Between the rock and Pelican Islet is a channel about 300 yards, with a least depth of 14 feet in the middle. It is marked by a beacon.

From Settlement Point the coast trends eastward 3 miles to Tenby Point, Queensferry, between which and Passage Point, at 1 mile northeast of Spit Point, is a bay 2 miles deep, forming between it and the island a sheet of water 2½ to 3½ miles wide, the southeastern and greater portion being filled by mudflats, having from 6 to 9 feet on them at high-water springs. The East Arm branches into this space and round Spit Point much as the North Arm does into the mudflat north of the island.

Corinella is a small postal village situated on Settlement Point. Population is less than 100.

Tenby Channel, from 400 to 100 feet wide, with a least depth of 6 feet at low water, extends southerly and easterly from the main channel toward Tenby Point, near Queensferry; there is a jetty at Tenby Point; it is 190 yards long with a depth of 6 feet at its end at low water. The jetty is not used.

Queensferry, 1 mile east-southeastward from Tenby Point, is a small township having postal and telegraphic communication with Melbourne. It is the shipping port for the Great Victoria Colliery Co. Population about 80.

Jetty.—The jetty at Queensferry is 221 yards in length, with a depth of 6 feet at high water at the outer end. The jetty is in an unsafe condition and should not be used.

Grantville (lat. 38° 25′, long. 145° 31′), nearly 1½ miles eastward of Queensferry, has postal and telegraphic communication with Melbourne. A coach runs daily, to Lang-Lang, thence by rail. Population is about 150.

Jetty.—Grantville Jetty is 333 yards in length, with a depth of 6 feet at high water at its outer end. At low water the bank dries for about 1 mile outside the outer end of the jetty.

Light.—A light is exhibited from the outer end of the jetty.

Lang Lang Jetty, 6 miles northward from Grantville, is 400 feet long. It is dry at low water at the outer end, but at high water there is a depth of about 6 feet.

Phillip Island is 12 miles long and 4¾ miles across at its western and broadest part; the eastern end of the island being a peninsula, connected with the western part by an isthmus ½ mile broad, at 8½ miles eastward of Grant Point.

Quoin Hill, situated 3¼ miles east-northeastward from Grant Point, is 218 feet high.

South coast.—From Grant Point the irregular and rocky south coast of the island curves eastward 5 miles to a point, close-off which is the high needle-shaped Pyramid Rock. Between Pyramid Rock and Cape Wollamai, 7 miles eastward from it, the coast forms a bay 2 miles deep, affording anchorage at 1 mile northeastward of Pyramid Rock, sheltered from northwest and northerly winds. The northern shore of the bay consists of a range of low sandhills covered with scrub, and is bordered by reefs, none of which appear to extend beyond ⅛ mile offshore.

Cape Wollamai (lat. 38° 34′, long. 145° 22′), the southeastern extremity of Phillip Island, is a remarkable helmet-shaped granite headland, of a reddish color, rising abruptly from the sea to a height of 332 feet, whence it slopes toward the northwest, forming a peninsula nearly 1½ miles long, northwest and southeast, and 1,500 yards broad. This head is the more conspicuous from its being the highest land on Phillip Island, all the remaining portion of it being low hills, clothed in an almost impervious scrub. The cape is fringed with dry and covered rocks; but none extend beyond ¼ mile from the shore.

Sandy Peak, 161 feet in height, is the highest and most southerly of the sandhills on the east end of Phillip Island. The peak is steep, well defined, and is a good mark for the anchorage.

Eastern entrance to Port Western.—The eastern entrance to Port Western is narrow and tortuous. The tidal currents run with great force through the Narrows—the northern end of the channel, attaining at times a velocity of 5 to 6 knots.

Depths.—The least depth of water in the channel is 7 feet. It is available at high water for vessels of 14 feet draft as far as San Remo, and of 10 feet draft through the inner passage of Port Western.

Western shore of entrance.—**Red Point,** situated 1 mile northward of the southeastern point of Cape Wollamai, is a mass of red granite boulders, 50 feet high.

Beacon.—A black beacon with square top, 15 feet high, stands about 50 feet from the water's edge.

Coast.—Between Red Point and Woody Point, at 2 miles northward from it, the east end of Phillip Island forms a bay nearly 1¼ miles deep, with rocky points and sandy beaches, bordered by a bank, of which the 3-fathom edge projects 200 yards to ½ mile from the shore.

Woody Point, on the western side of the Narrows, opposite Davis Point, is low and forms the northeastern corner of Phillip Island; a dry reef borders the point to the southward, and to the eastward of it are two detached rocks dry at low water. Tides are described later.

Newhaven.—The postal township of Newhaven is situated on Woody Point. There is communication by telephone with Melbourne. Communication by steamer to Stony Point, thence by rail.

Jetty.—The jetty at Woody Point is 80 yards long, with a head 21 yards long, on the outside of which is a depth of 10 feet at low water.

Life-saving apparatus.—A life-saving rocket apparatus is kept at Newhaven (lat. 38° 31′, long. 145° 22′).

Submarine cable—Prohibited anchorage.—A submarine cable is laid between Newhaven and San Remo, crossing the channel about 400 to 500 feet to the northward of San Remo Jetty. To avoid fouling this cable anchorage is prohibited within a distance of 300 yards north-northwestward from San Remo Jetty.

The northeast coast of Phillip Island, between Woody Point and Flagstaff Hill, close to the southeastward of Observation Point forms a bay 2 miles deep, divided by a broad projection into two bights, that to the southward being Swan Corner. Between 1 and 2 miles northwestward of Woody Point is Churchill Isle.

This bay is filled by a mud flat, mostly dry at low water, the outer edge of which, from Woody Point, trends north-northeastward 1¼ miles to a spit, and from thence northwestward 4½ miles to Observation Point. At the north end of this mud flat, at about half a mile eastward of Flagstaff Hill, are the entrances to three small channels, the eastern of which winds 2¼ miles through the flat, carrying depths of 3 to 6 and 2 fathoms to within half a mile from the shore at Swan Corner, and the western channel round Flagstaff Hill to the jetty at Rhyl.

Flagstaff Hill, lying about ½ mile southeastward from Observation Point, with shallow flats between, is the rounded point forming the northeastern extremity of Phillip Island. The higher ground of this point and a clump of pine trees on the north side make the hill a conspicuous object when seen from the north and east arms of Port Western.

Rhyl, nearly 1 mile southward of Observation Point, is a small postal village in a farming district.

Jetty.—Rhyl Jetty is 100 yards in length, with a depth of 9 feet at low water on the outside.

Eastern shore of entrance—Griffith Point (lat. 38° 32′, long. 145° 22′), 1 mile northeastward from Red Point, is a bold sandstone bluff 70 feet high, bare of trees for some distance inland, and fringed by a reef. From Griffith Point the coast trends northwestward nearly 1¼ miles to Davis Point, which is low, sandy, and wooded to the water's edge. A mud flat, which dries at low water, extends 600 yards from the shore eastward of Davis Point. Between this and Woody Point, ¼ mile to the northward of it, are the Narrows.

Rhyl Channel—Beacons.—The channel to Rhyl Jetty carries about 7 feet of water and is marked on the starboard side by five red beacons, and on the port side by two black beacons, one at the entrance to the channel and one opposite Rhyl Jetty.

An extensive bank, with dry patches on it and only 4 feet water over most parts of it, projects about 1 mile from the coast between Griffith and Davis Points.

In bad weather, especially during the ebb, the sea breaks over the edge of the bank.

San Remo, just inside Davis Point, is a watering place with postal and telegraphic communication with Melbourne. Communication by steamer to Stony Point, thence by rail. San Remo is remarkable for its freedom from hot winds and the coolness of its climate during the summer months, when the thermometer ranges from 10° to 15° below that of Melbourne.

Jetty.—There is a jetty at San Remo, 63 yards, with a face 46 feet long, on the outside of which is a depth of 9 feet.

Light.—A light is shown from San Remo Jetty.

Water.—Excellent water can be obtained at all times at the fisherman's hut just within Red Point.

The East Entrance Channel, which lies between the bank just described and that which borders the western shore, is 200 to 600 yards wide, with 5 to 3 fathoms for $\frac{1}{3}$ mile above Red Point. From 400 yards below the first black beacon to the Narrows the channel varies from 50 to 60 yards near the beacon to 250 yards in width, with depths of $3\frac{1}{4}$ to 5 fathoms, and 6 fathoms in the Narrows.

From Red Point to San Remo the channel is $2\frac{1}{4}$ miles long, with depths of 2 to 6 fathoms, the least depth being opposite Black Reef Beacon and the greatest depth off San Remo Jetty.

From the Narrows into Port Western the channel, which has only 7 to 12 feet water in it, leads from the east side of the black beacon off Woody Point to the east side of the black buoy off the point, then curving to the northwestward it takes a north-northeasterly direction between the black and red beacons.

Beacons and buoys.—Entering from seaward, the following beacons and buoys, marking the eastern entrance channel and inner passage should be passed in the following order. Port side: Red Point Beacon, Black Reef Beacon, black with spherical top, on the northern edge of Black Reef, 400 yards from the western shore and $\frac{1}{2}$ mile northwestward from Red Point. Starboard side: A red conical buoy on Middle Sand, on eastern side of channel, 950 yards north-northwestward from Black Reef. A red cask buoy on eastern edge of channel 870 yards northward from Middle Sand buoy. In the narrows the western side of the channel is marked by a black beacon with spherical top about 300 yards westward from San Remo Jetty, and a black beacon on the outer detached rock off Newhaven Jetty. Starboard side: A red pile beacon on the eastern edge of the channel, about 800 yards northeastward from San Remo Jetty. Port side: A black cask buoy on the eastern edge of the shoal ground, and marking the crossing from the main channel to the inner passage. Thence the inner passage is marked on the eastern side by two red pile beacons, on the western side by three black pile beacons in a line, and the Fairway Buoy, a black cask buoy, moored in 8 feet, which marks the inner end of the passage.

Directions.—After rounding Cape Wollamai (lat. 38° 34′, long. 145° 22′), haul in for Red Point, passing it within 200 yards, until it bears 234°. If desired a vessel may anchor in $3\frac{1}{2}$ fathoms between it and the first black beacon. From this outer anchorage pass northward of the black beacon, where the channel is only 50 or 60 yards wide, and then steer to keep the red conical buoy on the starboard bow; give it a berth of 100 yards; and if not intending to anchor in the channel, follow its course to the northward and northeastward, leaving the red cask buoy on the starboard hand. From the Narrows, after passing east of the black beacon southward of Woody Point and between the first black and red beacons north of it, steer to pass eastward of the next black buoy, and turn sharply to the north-westward toward the southern of the 3 black beacons. The chan-nel here is only about 70 yards wide between banks having only 1 to 3 feet water on them. Then proceed, leaving the black beacons and Fairway buoy on the port, and the red beacons on the starboard hand.

Caution.—If drawing 12 feet water, bound into Port Western by the east entrance, wait in the inner anchorage (to the northward of the first red buoy) till nearly slack water, as the current runs at a great rate through the Narrows.

The chart and lead are the best guides for this entrance.

Depths—Anchorage.—From a depth of 8 fathoms northeast-ward of Cape Wollamai the soundings decrease gradually toward the East entrance, close within which is the outer anchorage, in 3 to 4 fathoms, sand, between Red Point and the black beacon to the north-westward. Vessels drawing 15 feet, seeking shelter, and unable to fetch the West Entrance, need not lose ground by running back east-ward, round Wilson Promontory, but may find anchorage within the entrance, sheltered from all winds except southeast gales, inside Red Point, between that point and Black Reef beacon; the available an-chorage space is about ¼ mile long and 300 yards broad, the northwest limit of the anchorage being about 400 yards from Black Reef beacon.

Vessels of 12 feet draft may bring up in the inner anchorage, be-tween the black beacon and the Narrows, in 15 to 20 feet water, sand, and mud, between the two red buoys 1,500 yards inside Black Reef beacon, the available anchorage space being about ½ mile long and 400 yards wide. The most convenient anchorage is between the red cask buoy and Davis Point, where the channel, being widest, affords more room for getting underway. As the currents run through the channel at a great rate, it is advisable for vessels at anchor to lay out a kedge to keep them from fouling their anchors. At the above anchorages the tide usually runs from 2 to 3 knots.

Explosives anchorage.—At a distance exceeding 500 yards from any jetty.

Eastern shore of Port Western (lat. 38° 28′, long. 145° 24′).—From Davis Point the shore forms a bay extending 3¾ miles north-northeastward to Reef Islet, which is surrounded with rocks that con-nect it with the low northeast point of the bay. This bay is nearly

2 miles deep; but it is shallow throughout, there being only 6 to 12 feet water across its entrance from point to point.

Between Reef Islet and Settlement Point, at north 3¾ miles from it, the eastern shore of Port Western is divided into two small bays of nearly equal extent by Cobb Bluff, from which a reef projects about ¼ mile.

Bass River is a small stream winding through the low marshy land into the bay at 2 miles eastward from Maggie Shoal.

Beacon.—On a rock about ⅓ mile to the westward of the south-west corner of Reef Islet is a red beacon, consisting of a pole and ball, 16 feet high.

Maggie Shoal—Buoy.—The bank, which mostly fills the bay, projects as a shallow spit, close off which is Maggie Shoal, with a red buoy on it, lying 2 miles northeastward from Davis Point.

Lœlia Shoal, situated 1¼ miles west-southwestward from Reef Islet consists of two patches with a depth of 3 feet.

Buoy.—The southwest end of Lœlia Shoal is marked by a black and white striped cask buoy.

Sand flat.—With the exception of the northern branch of the East Arm and the bight between Observation Point and the spit northward of it, the whole of the eastern part of Port Western is filled by a flat of sand and mud, with rarely more than 3 fathoms water over any part of it.

Aspect.—The mainland about the eastern end of Port Western is moderately elevated and thinly wooded with short trees; the soil is rich, especially near the banks of Bass River, and is clothed with coarse grass to the water's edge. From the hilly promontory forming the east side of the East entrance, a range of wooded hills stretches away in an east and northeast direction; River Hill, one of the summits, is 816 feet high, and situated northeastward, distant about 10 miles from Cape Wollamai.

Tides.—It is high water, full and change, at Spit Point, French Island, at 1 h. 00 m.; springs rise 10 feet, and neaps 8 feet, the latter ranging 6½ feet; at Bourchier Channel, north of French Island, at 1 h. 13 m.; springs rise 10¾ feet, and neaps 8¾ feet, the latter ranging 7½ feet; at Mussel Rock, north side of Phillip Island, at 0 h. 12 m.; springs rise 8½ feet, and neaps 6½ feet; at Woody Point at 0 h. 50 m.; springs rise about 8 feet, and neaps 5 feet.

Tidal current.—The current in the main channel, between Phillip Island and Middle Bank, runs 3 knots, and in the East arm 1 to 2 knots. The currents run at a great rate in the east channel.

The coast from the eastern entrance to Port Western forms a slight curve trending eastward 4½ miles to Black Head, and thence southeasterly 3 miles to Powlett River, continuing on in the same direction for a further distance of 5 miles to Coal Point.

Coal Point has numerous sunken rocks off it at the distance of 1 mile southward of the point; one rock uncovers at low water spring tides. The heavy break shows the point to be dangerous of approach.

From Coal Point the land takes an east-southeasterly direction to Cape Patterson, from which it is distant 2¼ miles. The whole coast southeastward of Black Head is little more than a succession of sandy hillocks, from 100 to 140 feet high, covered in most places with dwarfed ti trees, but occasionally bare.

Cape Patterson (lat. 38° 41′, long. 145° 37′) is an ill-defined point, rounded and low, and the least conspicuous point along the whole coast; the highest land within a mile of the point is 127 feet above the sea, and this elevation scarcely increases until it joins a range of hills of over 900 feet high to the east and northeastward of River Hill, at a distance of 11 miles from the cape.

There is nothing to point out Cape Patterson. A conspicuous rock, 59 feet high, about 3 miles to the eastward of the cape, known as the Eagle's Nest, lying 100 yards off the coast at its turn towards Anderson Inlet, serves to distinguish it. East and westward of this rock the coast has a cliffy appearance.

Reef.—A reef, dry at low water, extends 600 yards in a southeasterly direction from Cape Patterson, and there are depths of 3 fathoms at 800 yards southward of the cape.

The coast from Cape Patterson trends 2 miles in an easterly direction, and thence northeastward 4 miles to the mouth of Anderson Inlet.

Petril Rock, only 2 feet above high water, with 3 fathoms close-to, lies nearly half a mile from the shore, about midway between Eagle's Nest and the mouth of Anderson Inlet.

Anderson Inlet, which by its two streams, Tarwin River and Screw Creek, drains about 300 square miles of country, is not navigable except for small steamers and auxiliary craft up to 5 or 6 feet draft according to the condition of bar and tide. There is depth of water in patches sufficient to allow the small craft which enter to anchor.

The entrance to the inlet (lat. 38° 39′, long. 145° 44′) lies between Point Norman and Point Smythe, and can be distinguished by the sand hummocks, which, commencing at Point Smythe, form the eastern shore of Venus Bay. From its mouth at Point Norman the inlet extends northeast and eastward 3 miles to Screw Creek, and thence east and southeastward to its head at the Tarwin River.

The inlet, which contains an area of about 5,000 acres, consists mostly of mud flats uncovering at half tide, and intersected by narrow shallow channels, only navigable by boats and small craft. A large sand bank, dry at low water, almost entirely occupies the mouth of the inlet, leaving only a narrow entrance channel between it and Point Norman.

The channel, at half a mile seaward of Point Norman, is fronted by a bar nearly 400 yards wide, on which the depth may vary from 4 to 6 feet at low water. From Point Norman a sandstone reef projects in a southerly direction for 800 yards, the deepest water will be found along its eastern edge; off Point Norman the bottom is rocky, with the reef showing along the edge of the channel. About 1,500 yards northeastward of Point Norman is Point Hughes, from which a sandstone reef projects, and from Point Hughes a sandy beach, with the reef showing at intervals, extends northeast to the jetty at the town of Inverloch, which forms the present terminus of ocean traffic.

The land in the vicinity of Point Norman is low and backed by thick ti-tree and honeysuckle bush, but rises at ¼ mile back to the higher ridge of ground 80 to 100 feet high, which fronts the town of Inverloch, and extends almost to Point Hughes.

Inside the bar the channel, which lies between the outer edges of the reefs projecting from the northwestern shore and the extensive sandbank to the eastward, is about 133 yards wide, with depths varying from 10 to 15 feet, until reaching Point Hughes, where the channel contracts to 100 yards, with depths of over 4 fathoms at about 180 yards from the shore. After rounding Point Hughes the channel follows close to the northwestern shore, at about 80 yards from the beach, with widths of 75 to 100 yards, and a least depth of 9 feet.

Beacons.—Two single staff range beacons, about 250 feet apart, are on the foreshore northeast of Point Norman; the front one is surmounted by a triangular topmark and the rear one by an inverted triangular topmark, both painted white. The beacons are in line on a 3° bearing.

Buoy.—A small red buoy is moored in about 8 feet inside Point Hughes, to mark the western end of the bank, which occupies the middle of the inlet between Point Hughes and Inverloch Jetty.

Inverloch Jetty is 75 yards long and 21 feet wide, with a depth of 9 feet at its outer end.

Light.—A light is exhibited at the outer end of the jetty.

Life-saving rocket apparatus is kept at Inverloch, on the northern shore of the inlet.

Directions.—The entrance to Anderson Inlet lies about 1 mile east-northeastward from Petril Rock, and can be distinguished by the houses on the higher ground fronting Inverloch. Generally, the line of range beacons marks the best crossing over the bar, but, owing to the changes that may occur in the shape and position of the western edge of the sandbank forming the eastern side of the entrance channel, the line of beacons may not always clear the outer spit of the bank, so that the beacons may have to be kept slightly open to the westward.

By bringing the entrance to bear about 8° when approaching the bar the range beacons should be picked up; on bright sunny days they are hard to see, in which case caution should be used and a good lookout kept, for the break on the eastern bank.

Proceeding inward the best time to cross the bar is on the rising tide, from half flood to high water, and about high water when bound outward. ·After crossing· the bar the beacons may be kept in range, or slightly open to the westward, until almost abreast the dry reef off Point Norman, then bring the beacons in range, which will lead about 80 yards eastward from the foul ground off Point Norman. After passing that point keep about 100 to 135 yards ·off the eastern bank and round Point Hughes sharply at about 50 yards off, keeping the small red buoy on the starboard hand; thence keep about 85 yards from the northwestern shore to the jetty.

Venus Bay.—The bight formed about the mouth of Anderson Inlet is known as Venus Bay, but it does not afford good anchorage.

Tides.—It is high water, full and change, at Venus Bay at 11 h. 56 m.; springs rise about 7 feet. Neaps are reported to rise 4 to 5 feet.

The coast.—From Point Smythe, the east entrance point of Anderson Inlet, the coast trends southeastward 13 miles with a slight curve to Watercress Creek; all this coast is a succession of sandhills 110 to 160 feet in height, which for the last 5 miles are almost destitute of verdure.

Watercress Creek is at the foot of the table-land of Cape Liptrap. On the coast, ¾ of a mile northwestward of the mouth of Watercress Creek, is a small rock of sandstone 15.feet above high water, and from this to a distance of 800 yards seaward are several sunken rocks. The coast line here is composed of low sandstone cliffs.

For 1 mile southward from Watercress Creek is a very rugged coast of overhanging sandstone, with jagged and pointed rocks strewn along it.

Arch Rock.—At 1 mile southward of Watercress Creek and off this rugged coast, at the distance of 400 yards, lies Arch Rock, 82 feet high, having a natural arch on its eastern side. There is a rock awash at half-tide, lying 200 yards to the westward of it. The same character of coast continues for a mile beyond Arch Rock in a south-easterly direction, having innumerable pinnacle rocks of various heights strewn along the whole distance, with other outlying sunken and half-tide rocks, in some places nearly ½ mile from the shore. Off this coast craw-fish abound.

Hence the land trends in a south-southeasterly direction 3 miles to a conspicuous islet, 63 feet high, off the western part of Cape Liptrap; half this distance being a straight piece of sandy coast, with the tableland of Cape Liptrap getting nearer as the coast runs southward. There are outlying sunken rocks about 600 yards off this coast. From the islet the cost forms three small bays to the cape.

Cape Liptrap (lat. 38° 55′, long. 145° 55′), which is nearly perpendicular, and 297 feet high, forms the southwest extremity of a table-topped promontory 551 feet high, joining the base of the Hoddle range of hills, which are 968 feet above the sea, at 16 miles northeastward from the cape; these again join the Fatigue mountain range, the highest part of which is above 2,000 feet high.

Light.—A group flashing light, 300 feet above water, visible 23 miles, is shown from a white skeleton steel tower 27 feet high, about 400 yards from the extremity of the cape.

The coast.—From Cape Liptrap the land forms a bight to Grinder Point, which is 2 miles east-northeastward from the cape. This bight is fringed with low water and sunken rocks, which in some places extend 600 yards from the shore.

From Grinder Point the land takes a northeasterly direction for a further distance of 2¼ miles to Bell Point, and a similar description of coast to the last is found, with the exception that the land is somewhat lower, and a number of rocks from 10 to 30 feet high are found at short distances off it.

On the whole of the coast from Cape Liptrap to Bell Point the sea generally breaks heavily for nearly ½ mile off.

Bell Point (lat. 38° 53′, long. 146° 01′) may be known by a large broad-topped rock about 40 feet high, and 200 yards from the shore. From this point the land takes an abrupt turn into Waratah Bay, trending about north-northwestward for 2 miles. At 600 yards from Bell Point is a small islet 60 feet high, and about the distance of 1 mile are Bird Rocks, three in number, and from 40 to 60 feet high, the outer rock being 400 yards from the shore. These rocks are guides to mariners using the bay, enabling them to ascertain their position.

Waratah Bay, from Bell Point to Shallow Inlet, is 8 miles wide and 4 miles deep in the middle, and affords good anchorage, except during south and southeasterly gales. In the depth of the bay, at 4 miles from Bell Point, the coast falls to a height of only 100 feet, when the ordinary feature of sand hills, generally covered with ti tree, is again met with.

Depths.—The depths in the bay vary from 10 to 5 fathoms, the latter depth being found about ½ mile from any part of the shore.

Light.—A fixed red light, visible 3 miles, 120 feet above water, is exhibited from a lamp-post on the shore in Waratah Bay, 400 yards westward from Bird Rock.

Jetty.—In the recess immediately to the northward of Bird Rocks there is a jetty which projects from the shore into 10 to 12 feet of water.

Life-saving apparatus.—There is a life-saving rocket apparatus at Waratah Bay.

Anchorage.—In Waratah Bay there is anchorage in 6 fathoms water, with the light bearing 234° distant 1,400 yards.

Or in good holding ground more than 1 mile from the shore during southwesterly gales, and with plenty of room for working out in the event of the wind chopping round to the eastward.

Tides.—It is high water, full and change; in Waratah Bay at 0 h. 00 m., springs rise 8 feet.

Directions.—Vessels entering Waratah Bay should, after passing Bell Point, steer outside the outer Bird Rock, which may be passed at a distance of 200 yards, then steer for the jetty or anchorage as desired.

Caution.—Mariners are warned when approaching Waratah Bay from the westward to keep at least 1 mile from the coast between Cape Liptrap and Bell Point, as foul ground is reported to exist in this locality.

Several outlying rocks, varying in height from 5 to 30 feet, partially fringe the coast south and east of Cape Liptrap, but none extend more than 400 yards from the shore.

Shallow Inlet.—From the head of Waratah Bay the coast trends east-southeastward until within 2 miles of the entrance to Shallow Inlet, when it becomes low, bare, and sandy, scarcely above high water. The east entrance point of this shallow inlet or lagoon is much higher, but is likewise of bare sand. The surveyors found it was not possible to sound the entrance of the inlet on account of the heavy break; but the depth varies with the prevailing winds and freshets, being occasionally dry at low water, and at other times having sufficient water for a large boat to enter.

From the mouth of this inlet the land trends with a slight curve in a southeasterly direction for nearly 6 miles to Black Rock, which is about 30 feet high, and 200 yards from the coast.

From a position 4 miles northwestward of Shallow Inlet to about 1 mile from Black Rock, shoal water with a sandy bottom extends about ½ mile from the shore, and off the inlet this shoal water extends nearly 1 mile. From about 1½ miles northward to 2½ miles southward of Black Rock, the sandy bottom is interspersed with rocks, some of which uncover at low water.

Shellback Island (lat. 38° 58′, long. 146° 14′).—About 1½ miles southwestward of Black Rock lies Shellback Island, 357 feet high; it is the northernmost of the islands on the west coast of Wilson Promontory.

Tongue Point, 167 feet high, lies 2½ miles southward of Black Rock, the coast between forming a deep bight, in the depth of which are a few low red cliffs, but they are not conspicuous. Tongue Point has a remarkable conical white rock, 30 feet high, close-off it to sea-ward. Abreast of the point at the distance of 1 mile the coast is high, and rises at a distance of 4 miles to parts of the promontory range, which are here about 2,000 feet above the sea. Mount Vereker, the northwest mountain of the promontory, bearing 65°, 6 miles from Tongue Point, is 2,092 feet high, and has a spur 1,654 feet high, running northwestward about 2 miles from it; this spur gradually falls in a westerly direction, and forms the northwestern termination of the high land of the promontory.

From Tongue Point the coast trends southeastward, forming a bight to Leonard Point, southward of which and on the same bearing are Pillar and Norman Points, forming the southern sides of Leonard and Norman Bays.

Norman Island lies nearly 1¼ miles southward of Tongue Point, and may be known by its two peaks, the higher and northern of which is 316 feet high.

Anchorage.—At 200 and 400 yards off this island on its eastern side are depths of 9 to 11 fathoms, where, in the course of the survey, it was often found convenient to drop the anchor during a prevalence of southwesterly winds. Coasting steam vessels of little power bound to the westward, having rounded the promontory and being met by a southwesterly gale, might find it convenient to anchor here in preference to running back again and anchoring in Waterloo Bay, to the eastward of the promontory.

Oberon Bay (lat. 39° 04′, long. 146° 20′), lying southward of Norman Point, is the largest of the three bays on this coast and affords the best anchorage. The bay is 1 mile deep, 1½ miles across, and has a broad, sandy beach, upon which the sea breaks heavily. Landing can in general only be effected in the southeastern corner. From the prevalence of southwesterly winds, none of these bays can be recommended as anchorages for other than steam vessels.

Good shelter has been found in this bay with Oberon Point bearing 234°, about 600 yards, during a heavy gale from the eastward.

Caution.—From experience in this locality easterly gales appear to die away at east or northeast, but a southwesterly gale may spring up with scarcely any notice of its approach, when sailing vessels would find themselves on a lee shore with a swell setting them dead to leeward.

The coast from Oberon Point trends for 1 mile in a southerly direction and then gradually takes a more easterly turn to a moderately deep bight, whence it again trends southward to Southwest Point. All this coast is bold and cliffy, the cliffs in some places being several hundred feet in height, and rising again at the back toward the mountain land of the promontory.

Glennie Group consists of four islands, which lie about 4 miles west-southwestward from Oberon Point, the nearest land of Wilson Promontory.

Great Glennie Island, the largest, is 455 feet high, nearly 2 miles long in a northwest and southeast direction, saddle-shaped, and strewn over with blocks of granite (of which it is composed), which give it a castellated appearance. A rock 3 feet high, over which the sea generally breaks heavily, lies about 400 yards northward of its northern end, and another somewhat larger, 15 feet high, lies about 200 yards off its northeastern end. Three smaller islands lie off the southern point of Great Glennie Island; the southernmost, 367 feet high, is named Citadel Island, from its resemblance to an ancient fortress.

Light.—A flashing white light, 380 feet above water, visible 15 miles, is shown from a red skeleton structure 23 feet high, on the summit of Citadel Island.

Anchorage.—There is anchorage eastward of Citadel Island in 10 fathoms water, with the monolith on the island bearing 257°, and the eastern point of Great Glennie Island just eastward of the middle island bearing 0°. The landing place is on the northeastern side of the island.

Anser Group, of three islands, apparently takes its name from the numerous geese frequenting them. Anser Island is the highest, and rises to a nipple point, 498 feet above the level of the sea; it is cliffy in all directions, but least so to the northward, where landing may often be effected. This island lies about 1 mile southwestward from Southwest Point.

Cleft Island (lat. 39° 09′, long. 146° 18′), the most remarkable of this group, lies nearly 1½ miles southwestward from Anser Island. It is 371 feet high, of a round form, and may be known from having a large slice out of its northwest side, which gives it a cavern-like appearance; it is also perpendicular, and white on all sides.

Two small islets, between 40 and 50 feet high, lie between it and the next or middle island of this group, equally dividing the distance between them. A third islet not quite so high lies 200 yards off the northwest point of the middle island, which is 312 feet high.

Carpentaria Rock.—This rock has a depth of 6 feet over it and lies 1,600 yards 1° from the summit of Cleft Island.

There is deep water between the Anser and Glennie groups and also between them and the mainland. The whole of the coast adjacent to the islands is bold, and all dangers, except the Carpentaria Rock, are visible.

Caution.—Although there is deep water between Glennie and Anser groups, mariners are cautioned against navigating the passages between Citadel and Cleft Islands, and between the northern and southern Anser Islands.

Seals inhabit these rocks and islands during the breeding season.

Tides.—It is high water, full and change, at the Glennie Islands at 11 h. 44 m.; springs rise about 9 feet.

The coast.—From Southwest Point the land trends east-southeastward 1½ miles to a projecting low and stony point, named South Point, the southernmost point of Australia, off which, at a distance of 200 yards westward, is a rock 15 feet high, and in the same direction at a farther distance of 400 yards is Wattle Island, 270 feet high, which from being so close to the shore appears connected, but between the island and 15-foot rock just spoken of is a deep channel through which a strong tidal stream constantly sets. A rock awash lies nearly 200 yards southwestward of the west point of Wattle Island, but with this exception the coast is bold.

Half a mile from the Southwest Point a freshwater creek flows into the sea, and at a short distance inland, eastward of this, is a remarkable stone, 741 feet high, near the summit of the coast range, which closely resembles a tower.

From South Point the coast trends in a general east-northeasterly direction 2½ miles to Southeast Point, where is the promontory lighthouse, but in this distance there are two deep bights, the westernmost running more than ½ mile into the land and forming at its termination a natural basin where there is a running stream.

From Southwest Point to Southeast Point the land rises suddenly from the water's edge to nearly 1,000 feet.

From Southeast Point the land trends northward 2½ miles to Waterloo Point, the southwestern point of Waterloo Bay. At 1 mile in that direction is a point with a small islet off it, but almost connected with the shore by large bowlders; the islet does not extend beyond the line of the coast, and immediately northward of it is an indentation ½ mile deep.

Depths.—South of Cape Liptrap the depths are 28 and 30 fathoms at 4 miles, and 38 and 46 at 14 miles distance. For 13 miles southward of Citadel Island the depths are from 37 to 43 fathoms.

Wilson Promontory (lat. 39° 08′, long. 146° 25′) is a lofty peninsula, 22 miles long, north and south, and 8 miles broad at the center. It is connected with the mainland to the northwestward by a low sandy neck, 10 miles long and 3 to 5 miles broad, which

separates Waratah Bay from Corner Basin. This promontory
rises to rugged mountains, some of which are above 2,000 feet in
height, thickly wooded on their upper and less exposed parts; but
toward the coast they are nearly destitute of vegetation, and descend abruptly to the sea. The soil is shallow arid generally barren;
though the brushwood, dwarf gum trees, and some smaller vegetation, which mostly cover the granite rocks,. give the country a deceitful appearance to a distant observer. See view A, on H. O. Chart
No. 3442.

Light.—A group flashing light, 381 feet above water, visible 26
miles, is shown from a white circular stone tower, 70 feet high, on
Southeast Point,. Wilson Promontory.

Lloyd's signal station.—There is a Lloyd's signal station at
Wilson Promontory Lighthouse, and communication can be made
by the International Code of Signals, and at night by Morse lamp.
It is connected by telephone with Foster, which is connected by
telegraph.

Masters of vessels sheltering in the bays or coves eastward of
Wilson Promontory during heavy southwesterly and westerly
weather and desirous of being reported should either take their
vessel or send a boat within signaling distance of the station during daylight, to signal the necessary message. The rough and inaccessible nature of the coast in the vicinity does not allow of mes-
sengers being sent from the station to such places or vice versa.

Life-saving apparatus.—There is a rocket life-saving apparatus
at the signal station.

Landing can be effected on either side of Southeast Point, according to the direction of the wind.

Forty-foot Rocks lie 4½ miles southward from the lighthouse, and
consist of three separate and distinct islets of granite, of which
the largest and westernmost is 165 feet long, with a breadth of 50
feet at the broadest part near its center; this islet is 20 feet high,
and on its southern extremity there is a granite bowlder 20 feet
in height and 40 feet above high-water mark, which, when the sea
is breaking over these rocks, is probably the only part of them
visible. These rocks are steep-to in all directions.

Rodondo Island (lat. 39° 14', long. 146° 23'), situated 6 miles
southward from Wilson Promontory Lighthouse, is a conspicuous
conical mass of granite, 1,500 yards across, rising to a distinct peak
1,150 feet above the sea, and visible in clear weather from a distance
of 30 miles. It has high cliffs on all sides, the surface above being covered with a dense dwarf scrub, and is steep-to in all directions. The Forty-foot Rocks lie 39°, 2 miles from Rodondo, and
between there is a clear channel with a depth of 36 to 39 fathoms.

As the tidal currents about Redondo Island are reported to run with considerable velocity, sometimes 4 to 5 knots, the neighborhood should be avoided.

Moncœur Islands.—East and West Moncœur Islands, 1½ miles apart, 331 and 318 feet high, lie nearly in line eastward from Rodondo Island, at 5 miles and 6½ miles, respectively, from it. The west island is nearly ½ mile long north and south, and about 200 yards wide, with a small islet 100 yards southward of it. The east island is ⅓ mile long in a north and south direction and rather more than 200 yards wide. These islands, which are almost bare, are bold-to and apparently free from danger.

Tidal currents.—Off Wilson Promontory the tidal currents as a general rule set east-northeast and west-southwest, the west-going current running during the rising tide. The currents turn at nearly high and low water on the shore, but the direction of the currents is much influenced by the winds. Near the promontory, after an easterly gale, the ebb or east-going current which has been checked during the gale continues to run to the eastward when the flood should have made, and at the strength of the flood the current sets to the northward, except inshore, where the tidal currents follow their general law; a southwesterly gale has an opposite effect.

The velocity of the currents off the promontory, where they are strongest does not exceed 2½ knots. Along shore, from the promontory to Cape Wollamai, and to a distance in the offing of 7 or 8 miles, the currents are scarcely felt, but run with their greatest strength off the several points.

Current.—During and after heavy easterly and westerly weather a current sets in the direction of the wind which is blowing or has just ceased to blow.

Offshore depths.—Off Wilson Promontory the soundings afford little guide, but in the bight between Tongue Point and Cape Liptrap. or off Cape Liptrap, a depth of 30 fathoms insures a vessel being 3 miles offshore. This depth likewise insures this distance from the land all along the coast to near Cape Wollamai.

South and southwestward of Cape Patterson the soundings shoal much more gradually than on any other part of this coast, and there are 30 fathoms or less 6 miles offshore. From thence westward the 30-fathom curve again nears the coast until off Cape Wollamai, where it is distant only 1½ miles.

Waterloo Bay, extending from Waterloo Point 2¼ miles northeastward to Cape Wellington, is 1⅓ miles deep, with 14 fathoms, sand, in the center, whence the depth of water decreases gradually to 6 fathoms at 400 yards from the shore, but increases toward the outer points.

175078°—20——12

The western shore of Waterloo Bay forms the eastern end of a low valley 3 miles long, which stretches across the promontory to Oberon Bay. The valley makes a conspicuous break in the high land and divides the Bowlder Range from the Wilson Range. At the east end of this valley a fresh-water stream empties into Waterloo Bay.

Anchorage.—The best anchorage is about 800 yards from the southwest shore in 9 to 12 fathoms water. Steam vessels bound westward met by a southwesterly gale may anchor close in to the land in a small cove under Waterloo Point. The holding ground is good.

Waterloo Bay being so immediately under the high land of Wilson Promontory, and exposed to the swell from both sides of Bass Strait, is not recommended as an anchorage for sailing vessels.

Cape Wellington (lat. 39° 04′, long. 146° 29′), a headland, 255 feet above the sea, rising to 442 feet ¼ mile inland, and forming the northeastern point of Waterloo Bay, projects 1½ miles southeastward from the line of the coast; Kersop Peak, its most elevated summit, rises to a height of 729 feet, at 1 mile northwestward of the cape.

The bold eastern face of Cape Wellington extends northward ¼ mile from its southern extremity, thence the land trends north-north-westward 1 mile to Brown Head, with a cove midway extending ¼ mile to the southwestward.

Mount Wilson, on the northern side of the valley and 3½ miles westward of Cape Wellington, rises abruptly from the southward until its wooded summit reaches the height of 2,320 feet. On the south side of the valley opposite Mount Wilson is a mountainous range known as Bowlder Range, which at its highest part rises to an elevation of 1,725 feet. The whole mountainous range on Wilson Promontory is of granite, with immense bowlders generally visible, but more particularly on the part known as Bowlder Range.

Mount Latrobe, which reaches an elevation of 2,434 feet, lies north-northwestward distant 3½ miles from Mount Wilson; nearly midway between them is Mount Ramsay, 2,313 feet high.

Refuge Cove (lat. 39° 02′, long. 146° 28′), ¼ mile northwestward from Brown Head, and the only anchorage on this side of Wilson Promontory sheltered from the eastward (unless Corner Basin and Bentley Harbor are considered exceptions), is the central of three small deep-water bights between Brown Head and Horn Point, which latter lies 1 mile northward from the head. Hobbs Head, ½ mile southward of Horn Point, forms the north side of the entrance to Refuge Cove, which is only 300 yards wide.

The cove may be easily recognized from being distant midway between Kersop Peak and Horn Point and from its having the first sandy beach which opens north of Cape Wellington. It is about ¼ mile in extent, with 8 fathoms in the entrance, from which the depth gradually decreases to 3 and 4 fathoms, in most places close-to, but near the sandy beaches at 100 yards from the shore.

The anchorage in the cove is in the south part. Refuge Cove is not much used, owing to the difficulty of getting out, the high hills around almost completely screening it from any winds off the land.

The cove between Brown Head and Refuge Cove trends ⅓ mile to the southward, with 9 to 4 fathoms; that between Hobbs Head and Horn Point has the same depth of water, but is open to the eastward.

Tides and tidal streams.—It is high water, full and change, in Refuge Cove, at 0 h. 05 m.; springs rise, 8 feet.

Off Cape Wellington the tidal flood stream divides and runs in opposite directions, one portion of the current which comes from the northeastward turning and running along the shore to the northward, the outer portion of the same currents continuing its course round the promontory to the westward. The ebb currents meet and act in an opposite manner.

Sealers Cove (lat. 39° 01′, long. 146° 27′).—From Horn Point, 200 yards from which there is a rock with 9 feet of water on it, the coast trends west-northwestward for 1,500 yards, and thence westward for another 1,500 yards to the southern point of Sealers Cove, which is 1,200 yards wide north-northwest and south-southeast at its entrance, and about 1,500 yards in extent within. There are depths of 4 and 5 fathoms at the entrance, within which the water shoals gradually, the 3-fathom curve being only 400 yards inside. A heavy swell often rolls into Sealers Cove.

Water.—Fresh water may be obtained either in Refuge or Sealers Cove, but in the latter it might be necessary to go some distance up the creek in the southeast corner. At Refuge Cove water may be obtained with greater facility, principally in the southwest corner of the northern sandy beach.

Five-mile Beach.—From the northern side of Sealers Cove the coast trends northerly and northwesterly 1¼ miles to the south end of Five-mile Beach, and thence extends with a slight curve northward 4¼ miles, being intersected at each end by a stream of fresh water. The beach may be approached to 1,340 yards in 5 to 6 fathoms. At the back of this beach is flat swampy ground, which extends for 1½ and 2½ miles until met by the slopes of Mount Vereker.

At the north end of Five-mile Beach the higher part of the promontory again approaches the coast, forming a small point from which in a northeasterly direction about ⅓ mile is another point abreast of Rabbit Island. Off this point in a southerly direction is a rock which, from its resemblance to the island of the same name, has been named Rabbit Rock; this rock is 50 feet high, and has a small detached rock close-to on its west side.

Rabbit Island, so named from the number of rabbits upon it, lies eastward distant 1,500 yards from the eastern point of Wilson Promontory. It is nearly ½ mile long northeast and southwest, and being 194 feet high, is an excellent mark when proceeding northward to Corner Inlet.

Anchorage.—There is good anchorage in all but southeasterly or easterly gales in 4½ to 5 fathoms 1 mile northeastward of Rabbit Island. Traders bound westward anchor here during southwesterly gales. Small craft bound westward during westerly gales often anchor near Rabbit Island, but in a seaworthy vessel such a course should not be adopted, unless the gale is of unusual violence.

Wood and water may be obtained at this anchorage; the water will be found in the little sandy valley on Rabbit Island by sinking a cask, and the wood may be obtained on the adjacent mainland, or both may be obtained on the mainland.

Tides.—It is high water, full and change, at Rabbit Island at 0 h. 14 m.; springs rise 8 feet.

The coast.—From the point abreast Rabbit Island the coast trends northward 1 mile to a point behind which are two good freshwater streams. At a farther distance of ½ mile in the same direction is another small point, whence the land trends in a westerly direction for ½ mile, and thence northward 2 miles to a point at the southeastern base of Mount Hunter, and thence again in a northerly direction to a smaller point eastward of the same mount, whence the coast, which then becomes low and sandy, trends northward about 2¼ miles to Entrance Point, at the entrance to Corner Basin. Between the several points here spoken of are sandy beaches.

Abreast of Rabbit Island and to the northward of Five-mile Beach and the hill over the coast rises to an elevation of 778 feet; and north-northwestward 1½ miles from this, and nearly a mile from the coast, is Mount Roundback, 1,050 feet high. At 3½ miles northward of Mount Roundback and 1 mile from the coast is Mount Hunter, 1,175 feet high, which is conspicuous, being of a pyramidal shape and the northern high hill of the promontory.

Between Mounts Roundback and Hunter the range falls considerably, but about midway is a wedge-shaped hill 716 feet high. At 2½ miles north-northwestward of Mount Hunter, and on the northernmost point of the promontory, is Mount Singapore, 480 feet high; this hill forms a useful leading mark into Bentley Harbor. One mile eastward of Mount Singapore is Entrance Point.

Seal or Direction Islands (lat. 38° 55′, long. 146° 39′).—About 7 miles eastward of Rabbit Island is Seal Island, the largest of the Seal Islands. The group consists of four small islands and three rocks, which latter extend in a northwesterly direction from Seal

Island. The largest of the rocks, White Rock, is 33 feet high, and distant from Seal Island 1¼ miles. The two other rocks, with rocks awash to the northward of each, lie northwestward 200 yards and 800 yards, respectively, from Seal Island, the one farthest from the island being 8 feet high. A rock with 3 fathoms water on it lies 200 yards southward from White Rock.

Seal Island is 154 feet high, about a mile round, and covered with tufts of coarse grass, among which are the burrows of penguins and mutton-birds.

Notch Island, the second largest, is 123 feet high and lies nearly 1 mile southeastward from Seal Island; it has two hills upon it, and the valley between giving it a notched appearance caused it to be named Notch Island.

Nearly 1,500 yards south-southeastward from Notch Island is Rag Island, 94 feet high; and 1¼ miles eastward is Cliffy Island, 144 feet high.

With the exception of rocks awash 200 yards off the west side of Rag Island, a rock awash 200 yards off the northeast part of Cliffy Island, and the rocks mentioned as lying off Seal Island to the north-westward, the islands are all steep-to.

Cliffy Island (lat. 38° 57′, long. 146° 42′)—**Light.**—A flashing white light, 180 feet above water, visible 19 miles, is shown from a stone lighthouse 25 feet high on the southern side of Cliffy Island.

Fogsignal.—A fogsignal is made at Cliffy Island Lighthouse.

Life-saving apparatus.—A life-saving rocket apparatus is kept at Cliffy Island.

Signals.—In favorable weather signals are exchanged between Cliffy Island and Wilson Promontory stations.

Corner Inlet, lying between Entrance Point and La Trobe Island, is the entrance to Corner Basin, an extensive sheet of water between Wilson Promontory and the land to the northward, with deep channels leading between its numerous mud flats.

Depths.—The inlet is fronted by a bar (which is 2 miles across), the deepest water over which at low water is 21 feet. Within the bar 4 miles east-southeastward from Entrance Point there is a depth of 5½ fathoms of water. Here the channel is more than a mile wide, decreasing to ½ mile between Entrance Point and La Trobe Island; having crossed the bar the depths in the channel gradually increase until nearly abreast of Entrance Point, where the depth is 18 and 19 fathoms.

The port of Corner Inlet and Port Albert includes all inlets, rivers, bays, harbors, and navigable waters north of and within a line bearing 43° from the south end of Rabbit Island to the entrance buoy at the eastern entrance to Port Albert.

Buoys.—There are two conical black buoys moored on the southern side of Corner Inlet, the outer lies in 5 fathoms southeastward, 3½ miles, and the other in 3½ fathoms, southeastward, 1 mile from Entrance Point.

No good mark can be given for entering the inlet. The chart and lead are the best guides; no stranger, however, should attempt to enter Corner Inlet without a pilot.

Corner Inlet sand banks.—The coast of the promontory northward of Rabbit Island is fronted by an extensive shoal sand bank. The southeastern extremity of the shoal with 17 feet water lies northeastward 2½ miles from Rabbit Island, and its inner part trends thence westward to the shore, having depths of 3½ to 4 fathoms between it and Rabbit Island, as well as in a gully between it and the land.

The end of the bank on the southern side of the entrance to Corner Inlet is northeastward 5 miles from Rabbit Island. The 3-fathom curve forming the edge of the southern bank turns from this point westward for 2 miles, when it trends northwestward for 5 miles to Entrance Point.

The northern banks of Corner Inlet extend from the point of La Trobe Island, opposite Entrance Point, in a southeasterly direction for 5½ miles, then curving to the westward for more than 1 mile, in about 12 feet, at low water. The deepest channel to Corner Inlet lies between these banks and is about ½ mile wide, with 21 feet at low water.

The outer part of the northern banks just described trend to the northwest for about 2 miles, and then in a westerly direction for 1 mile to Townsend Point, where they approach the land to ⅓ mile and from the southern side of the channel into Bentley Harbor.

Clearing mark.—Mount La Trobe, open south of Rabbit Island, 231°, leads ¼ mile southeastward of the 3-fathom edge of the banks.

Corner Basin extends 4 miles north and south and 14 miles from Entrance Point to the northwest corner of the basin, into which flows Tarwin Rivulet. The northern and southeastern shores are fronted by swampy mangrove islands, and the basin is mostly filled by mud banks.

Franklyn Channel is the main channel in Corner Basin. It extends about 5 miles in a westerly direction from between Entrance Point and La Trobe Island, where it branches off into three smaller channels. It has a width of from 600 yards to nearly 1 mile, with depths of 5 to 18 fathoms.

Buoys.—A conical black buoy is moored on its south side at about 1 mile northwestward of Mount Singapore, and at the end of a spit where the channel divides.

A narrow sand bank on the northern side of the channel is marked by a black and white striped buoy at its east end.

Toora Channel.—Immediately to the northward of the entrance to Franklyn Channel, and between it and Lewis Channel, is the channel which runs westward and northwestward for about 6 miles to the Toora Jetty. The depth in the main portion of the channel gradually decreases to 2 fathoms at its terminus from whence a subsidiary channel has been dredged to a depth of 4 feet, with the width of 60 feet.

Beacons.—Toora Channel is marked by pile beacons, red on the starboard and black on the port side.

The three branches of the Franklyn Channel at its western end are as follows: The northern branch, trending north and westward to Franklyn River, is marked by pile beacons, red on starboard and black on port hand; the middle, called Stockyard Channel, trends west-northwestward to Stockyard Creek; the south branch extends westward to Golden Creek.

Franklyn River.—The channel leading to Franklyn River has general depths of $1\frac{1}{2}$ to $2\frac{1}{2}$ fathoms for a distance of about $2\frac{1}{2}$ miles, where it shoals to about 6 feet. The remaining portion to Bowen Wharf is shallow, being obstructed by shoals formed by deposits of silt brought down the river by freshets.

Bowen Wharf is on the western bank of the Franklyn River, $\frac{3}{4}$ mile above its mouth; it has a face 70 feet long, with an approach wharf from the bank about 30 feet long and 24 feet wide. There is a depth of about 5 feet on the outer side.

Light.—A light is shown from Bowen Wharf.

Light beacon.—A light is shown from a beacon on the south bank at the entrance to Franklyn River. The beacon is near the outer edge of the mangroves, $1\frac{1}{2}$ miles below Bowen Wharf.

Toora is a post and telegraph town; it is an agricultural and tin-mining district. Population is about 800.

Toora Jetty is connected to Toora railroad station and township by a good road about $1\frac{1}{2}$ miles long.

Lewis Channel—Beacons.—The entrance to Lewis Channel is marked by a pile beacon, on the eastern side. The channel then curves to the north and northeast 4 miles to the township of Welshpool and is marked by beacons on both sides, those on the starboard side, with depths of 2 to 3 fathoms, being painted red and those on the port side black.

Light beacon.—A white light is shown from a beacon erected on the outer end of the spit on the western side of the entrance to Lewis Channel.

Welshpool (lat. 38° 42′, long. 146° 28′), a railway and telegraph town on the north side of Lewis Channel, is in an agricultural and pastoral district. Population of the district is about 500.

Jetty.—Welshpool Jetty projects about 277 yards from the shore and has a face, 33 yards long, with a depth of 12 feet alongside. The jetty is connected by a railway about 3 miles long to Welshpool Station on the Great Southern main trunk line.

Light.—A light is exhibited from the outer end of the jetty.

Benison Channel extends to the southwestward from the western side of Mount Singapore for about 9 miles, and Middle Channel to the southwestward from the northern side of Mount Singapore for about 9 miles.

Bentley Harbor.—During the continuance of strong easterly or southeasterly gales, vessels may anchor in Bentley Harbor, which, as already described, lies between La Trobe Island and the northern banks of the inlet, and is from 1 to 400 yards wide.

Depths.—It has depths of from 8 fathoms off Townsend Point at its east end to 3 fathoms ½ mile from its west end; vessels are there protected from all winds, and there is a strong ebb current to assist them in getting out again. Bentley Harbor is protected to the southward by a sand bank which dries at low water. The pilot stationed at Port Albert often takes vessels through Bentley Harbor into Corner Basin.

Buoy.—The west end of the passage is marked by a black and white striped buoy, on either side of which there is only a depth of 9 feet at low water.

Range.—A good range in is Mount Singapore open of Townsend Point, the southern point of Latrobe Island 276° true. When abreast of Townsend Point keep along the shore and anchor as convenient.

Pilots are always on the lookout at Port Albert, and will come off, if possible, at any time.

Caution.—Mariners are recommended to avail themselves of their services, and not to attempt to enter either Corner Inlet or Port Albert without local knowledge.

Latrobe Island, which lies between Corner Basin and Port Albert entrance, extends from the point abreast Entrance Point, 2¼ miles eastward to Townsend Point, and thence with a curve inward northeastward 4 miles to its most eastern point, at the outer end of Snake Channel entrance to Port Albert.

Latrobe Island is low, but the trees on it give it an apparent elevation of 40 to 60 feet.

Snake Island (lat. 38° 45′, long. 146° 38′) is a narrow island about 1¼ miles long at the east end of Latrobe Island, with which it is almost connected.

Port Albert.—The entrance to Port 'Albert is over 1 mile wide between the east point of Snake Island and the beach on the opposite side. But this entrance is divided by a large bank of sand, which extends from midway between these two points for 2 miles in a southerly direction.

On this large bank of sand there are two parts which form islands, from 2 to 3 feet above high water. From the north 'point this sand bank also extends in southwesterly direction for 1 mile, with a small detached sand bank off its southwestern extremity.

The population of Port Albert (lat. 38° 40', long. 146° 41'), a township at the mouth of the Tarra River, is about 300; of Alberton, on the east bank of the River Albert, 4 miles distant, about 600. Port Albert is one of the principal fishing grounds in the State. Its exports are wattle bark, leather, raw hides, and grain. There is communication with Melbourne by rail; there is also telegraphic communication, and regular steam communication with Melbourne. Port Albert has monopolized the trade of the district.

Port Albert Wharf is about 57 yards long with 9 feet at low water, but vessels drawing 10 feet can lie alongside, as the bottom is all soft mud.

Light.—A light is shown from Port Albert Wharf.

Life-saving apparatus.—A lifeboat is kept at Port Albert Wharf and a rocket life-saving apparatus is kept at Sunday Island pilot station.

Snake Channel.—The northwest side of the bank described forms the southeast side of a very narrow channel running close along the shore at Latrobe Island, only navigable for fishing boats. This is known as Snake or Western Channel. This channel silted up in 1914, and the main channel is again the entrance to Port Albert. Owing to constant changes in the banks and channels leading to Port Albert, the chart must not be relied on.

Main channel.—The main channel lies between the large sand bank occupying the middle of the entrance and the sand spit projecting from that portion of the beach to the eastward, once known as Clonmel Island, and from the tails of these banks breakers extend in a southeasterly and southerly direction, those to the eastward extending nearly a mile and those to the westward half a mile.

Depth.—The bar of Port Albert is, strictly speaking, only navigable for vessels of 9-feet draft. Vessels of greater draft run the risk of being detained inside either for high tides or smooth water.

Though the general feature of the bar of Port Albert remains the same, yet in so far as the navigation is concerned it is continually shifting.

A body of sand appears to be perpetually driving from the east to the west side of the channel, but more particularly during strong

breezes from the eastward or southeast. The tidal current out of
the port is considered, however, to keep a channel across the bar of
6 feet at low water, though it may be doubted whether at all times
even this depth is maintained.

Buoys.—The channel has red buoys on starboard and black buoys
on port side, and vessels proceeding to Port Albert should endeavor
to pass close to these buoys until reaching the eastern channel to
Port Albert.

A fairway buoy painted black with two red bands lies about 2
miles southward of Snake Island. The water eastward of this buoy
is reported to have shoaled.

This buoy is moored in 5 fathoms, and lies about 2 miles southwestward of the entrance to main channel; it does not in any way
mark the entrance but only indicates the navigator's proximity to it.

Owing to the changes in the sand banks in the main channel, no
definite positions can be given for the buoys, as they have frequently
to be shifted to suit the trend of the deep water. The chart must
not be relied on.

Caution.—Sailing vessels navigating the main channel should
take every care, as the tide sets athwart the bar, and on no account
should the bar be attempted without a commanding wind.

Within the entrance there are two channels, the deeper one sweeping round westward, northward, and northeastward for 7 miles; and
the other northward and westward for 5 miles, when they reunite
at a point 3 miles northwestward from Sunday Island Light. The
western branch, northeastward of Sunday Island, is called the Midge
Channel. The space inclosed by these channels is occupied by Sunday and a few smaller islands separated at low water by mud banks.
Sunday Island is mostly covered with ti trees, the highest being
about 41 feet above high water. It is surrounded by mud flats.

Sunday Island—Light.—A light is shown from a pile beacon
erected on the edge of the bank off the southeastern end of the
island, and about 1 mile southward of the pilot station. The light
is only for the guidance of fishermen and others within Port Albert.

Caution.—As the shoals at the entrance at Port Albert extend for
over 2 miles from the coast, vessels sighting the light should immediately haul out to seaward.

Signal and pilot station.—The signal and pilot station is on
Sunday Island, and tidal signals are made from the flagstaff. Communication can be made by the International Code signals.

Inside waters.—From the inner end of the main channel the
direct channel to Port Albert trends northerly for 4 miles to Port
Albert Wharf. As, however, this channel may become blocked
across by sand shoals of 3 and 4 feet in the vicinity of the light

beacon, the deeper and circuitous channel round Sunday Island may have to be taken, an extra distance of 8 miles.

The latter channel carries 4 to 5 fathoms until joining the Midge Channel.

Midge Channel, with a depth of 2 fathoms, runs off the northern side of Sunday Island for a distance of $2\frac{1}{2}$ miles to the northeastern end of the island, where it joins the direct channel to Port Albert. Thence the channel is 50 to 67 yards wide, with a least depth of 7 feet, to Port Albert.

The channel round Sunday Island is marked by stakes on the starboard hand, and the direct channel to Port Albert by red pile beacons on the starboard and black pile beacons on the port side.

The direct channel runs between numerous low islands, generally mangrove, with other smaller channels between them, mostly dry at low water.

This channel to Port Albert is at present disused, the western and Midge Channels are used.

The continuation of the eastern channel is known as the River Tarra upon which, at 2 miles from the town of Port Albert in a straight line, is the township of Tarraville.

Tarraville is a small township on the left bank of the Tarra, near its mouth. Small vessels can get up to it at high water. It is a telegraph station. There is communication by coach with Port Albert, and thence with Melbourne by rail. The population is about 200.

The western channel (lat. 38° 44′, long. 146° 39′).—Inside the narrow western channel is a red buoy moored close to the sand bank which extends to the bar; it lies 800 yards northeast of the east point of Snake Island, and nearly 400 yards north of this buoy is a black buoy on the south edge of a sand bank. From the northeast end of Snake Island the channel trends west $3\frac{1}{2}$ miles, north 1 mile, and northeasterly for $2\frac{1}{2}$ miles, to the Midge Channel.

Depth.—The least depth to Port Albert by the western channel is 7 feet with 7 feet on the bar.

A branch, which takes the name of the Albert River, continues to trend in a northeasterly direction for a further distance of 2 miles, whence it becomes more winding, and at a distance of 4 miles is Alberton. Coal of good quality is found to the westward of the township, but the country is densely wooded.

On both sides of this channel, which for 3 miles is 600 yards broad, there are mud banks, and when clear of Latrobe Island, at $3\frac{1}{2}$ miles from the entrance, there is a channel leading to Welshpool, fit only for boats and small craft. Midway the meeting of the tidal currents has formed mud banks, which at low water spring tides completely block the channel.

Tides.—It is high water, full and change, at the entrance of the main channel at 0 h. 00 m.; springs rise 8 feet.

The tidal currents run strongly in the western channel and way should be maintained on the vessel in narrow places to avoid being set aground on the banks.

The tides generally are greatly influenced by the winds, and no reliance can be placed on calculated times of high water during unsettled weather. Strong west to southwest winds cause the flood to run from 1 to 1½ hours after, and easterly winds 40 minutes sooner than the expected time of high water.

The signal master at the port, who has great opportunities of watching the tides, reports that for six months of the year, ending with the month of February, the highest tides occur in the morning. The p. m. tides begin to be the highest in March. Winds from west-southwest cause the highest tides. When the wind is eastward of south a low tide follows.

Tidal signals denoting the tidal current and the depth of water over the bar in the main channel, leading to Port Albert, are exhibited daily between sunrise and sunset from the flagstaff on Sunday Island, as follows:

Cone at eastern yard-arm—Ebb current.

Cone at western yard-arm—Flood current.

Cone at half-mast—Under 6 feet depth of water.

One ball suspended from yard—7 feet depth of water.

Two balls suspended from yard—8 feet depth of water.

Three balls suspended from yard—9 feet depth of water.

Four balls suspended from yard—10 feet depth of water.

Five balls suspended from yard—11 feet depth of water.

Six balls suspended from yard—12 feet depth of water.

Seven balls suspended from yard—13 feet depth of water and over.

The average depth of water over the bar in the main channel is 6 feet at low-water springs.

Directions.—To Port Albert (lat. 38° 41′, long. 146° 42′) from the westward, after rounding Wilson Promontory, steer for Cape Wellington, after passing which keep Rodondo Island just open of it, bearing 206° until Mount Singapore is in line with Townsend Point, 273°, which will insure a distance of a mile from the bar; or, in the event of Rodondo being obscured, Mount Latrobe kept just open south of Rabbit Island, 231°, will lead rather more than a mile to the southeastward.

When a mile off the bar, the break and fairway buoy will be observed; the pilot will at once board the vessel. It must be borne in mind that, as the coast of La Trobe Island is low, Townsend Point is not visible off the bar, except from a height of above 15 feet; and the lead should be attended to.

From the eastward vessels may approach the shore to a distance of 3 miles, and if at that distance, and coasting to the southward, they will observe the break on the bar; or they may bring Mount Latrobe just open south of Rabbit Island 231°, and make out the bar from that line.

Caution.—Mariners are cautioned against attempting Port Albert entrance without a pilot. It is not recommended to approach the port by night, but to keep a good offing until daylight, and attend to the lead. Vessels bound for the port should not approach the shore to the northward of the line of Mount Singapore in line with Townsend Point until they have picked up the outer buoys of the entrance channel, or have the pilot aboard. As the bar and entrance channel is liable to changes in position and depths and also the buoys, owing to their exposed positions frequently break adrift, the greatest care should be used in vessels which are regular traders to the port.

Explosives anchorage.—The explosives anchorage at Port Albert is at a distance exceeding ¼ mile to seaward of every wharf or jetty.

The coast.—About 1½ miles eastward of Sunday Island is a low sandy island, which extends in an east-northeast direction for 1¾ miles, where there is an entrance between it and the island forming the western side of Shallow Inlet. This latter island is nearly 4 miles long with a few hummocks upon it, the highest being those near its eastern extremity, about 42 feet high. The island, except where the hummocks rise, or a few scattered bushes grow, is scarcely above high water and composed of sand. All is covered with a dwarf vegetation.

Shallow Inlet is about 800 yards wide from shore to shore. From either side of the entrance, sandpits extend in a southerly direction, the eastern for 1,500 yards, the western for ½ mile, leaving a channel between them with not less than 3 feet at low water. The inlet is never used by shipping.

The coast.—From the eastern point of Shallow Inlet the coast trends northeastward about 25 miles in a nearly straight line to Merriman Creek entrance.

As the coast from Shallow Inlet ceases to have any more openings for a considerable distance, by which even boats may enter, this distance may be spoken of as the first portion of the Ninety-Mile Beach, which may be said to end at Conran Point, though the Red Bluff at the old entrance to the lakes is a break to its uniformity.

From Shallow Inlet to Merriman Creek the coast line is nearly separated from the land at the back, which is somewhat higher and thickly timbered, by fresh and salt lagoons or ti-tree swamps, generally salt. At a distance of 17 miles from the inlet, a slightly elevated piece of country, thickly timbered, about 150 feet in height, nears the coast to half a mile, and just to the eastward of this is situated Lake Denison, whose waters discharge into and near the mouth of Merriman Creek.

This district is all low, having an elevation from 50 feet to the westward to only 25 feet to the eastward. Here and there the hummocks in places are considerably lower, and much of the coast is scarcely above high water, while in heavy rains the water of the lagoons breaks through the coast line.

Ninety-mile Beach.—The line of coast between Shallow Inlet (Lat. 38° 41′, long. 146° 50′) and the Red Bluff is locally known as the Ninety-mile Beach. Landing may be effected on it, but such a measure is extremely dangerous, as the beach is treacherous, being what is commonly known as a double beach.

When only a few miles from the land on the western part of the Ninety-mile Beach nothing can be seen but the back ranges of mountains. These extend in a southwesterly direction for 27 miles, from Tom's Cap, 1,196 feet high, lying 19 miles westward of Merriman Creek, to Mount Fatigue, which is 2,050 feet above the sea. The range between rises to summits of even greater elevation than Mount Fatigue, the highest being 2,453 feet. A range of hills, the highest of which is Mount Albert, 1,050 feet high, lies southeast and eastward of Mount Fatigue at a distance of 6 to 12 miles.

From Corner Inlet, northeastward, to the Red Bluff, eastward of the old entrance to the Gipps Land Lakes, the coast is a continuous sandy beach, much broken by inlets and small streams, the latter breaking through the narrow strip of sand after a heavy rainfall. Although a sandy beach is again found northeastward of the Red Bluff for a distance of 30 miles, yet this is not a part of the well known, and hitherto dreaded, Ninety-mile Beach.

From Merriman Creek, which lies northeastward 24½ miles from Shallow Inlet, the coast stretches, with a slight curve inwards, 47 miles northeastward to the new entrance of the Gipps Land Lakes. All this coast is low, from 40 to 85 feet in height, in some places densely covered with ti trees, in others sparsely timbered with honeysuckle, the whole of so uniform and monotonous appearance that, with one execption, no objects easy of identification to the mariner present themselves. The exception is a group of houses immediately at the back of the entrance to Merriman Creek, known as Buckley's Station.

Lakes or lagoons extend close inside the sand-hummocks the whole distance; inside these lakes the land is low and densely timbered, and the same low country interspersed with lakes and marshes extends for miles inland, much of it being subject to floods.

At distances from Merriman creek of 18, 24, and 28 miles respectively are three hummocks, the easternmost, 85 feet high, is named Stockyard Hill. The middle hummock is covered with ti trees, and easily identified by coasters.

At a distance of 7 miles westward of the entrance of the Gipps Land Lakes, and 3 miles from the outer line of coast, is Tambo Bluff, about 250 feet high, from which comparatively high land continues to Mount Barkly, at the entrance to the Gipps Land Lakes, then follows to Red Bluff, skirts the arms and streams of Lake Tyres beyond it, and following the line of the shore at about 2 miles inland, is not again lost; and giving, as it does, a higher appearance to the coast line, clearly marks the difference between the land to the eastward and that to the westward of the entrance to the Gipps Land Lakes.

Mount Barkly, the most conspicuous portion of the land just described, lies 4½ miles westward from Red Bluff, and ½ mile from the outer line of coast. It is 233 feet high, partly cleared of timber, and forms a useful mark for the entrance of the Gipps Land Lakes. The artificial (New) entrance supersedes the old natural entrance, which has been filled up by drifting sands.

Gipps Land Lakes.—The port of Gipps Land Lakes includes Lakes Wellington, Victoria, King, Reeves, Bunga, and all inlets, rivers, bays, harbors, and navigable waters northward of and within a line bearing 71° and 251°, across the outer end of the eastern pier forming the entrance to the Gipps Land Lakes.

The new entrance to the Gipps Land Lakes (lat. 37° 53′, long. 147° 58′) is situated 4½ miles westward of Red Bluff. The entrance works consist of two timber piers 466 yards long, built of close piling; from high-water mark the western pier extends seaward about 500 feet and the eastern pier about 380 feet, with a clear width between them of 250 feet at the outer end. Beyond the existing heads of these piers stumps of piles, remaining from the destruction of piers by the sea, extend seaward for 200 feet from the western and 70 feet from the eastern pier.

The entrance is easily recognized by Mount Barkly and by the flagstaff at the pilot station erected on the sand hummocks immediately to the eastward of the piers. The piers are of a dark color, and surmounted by a footway with white hand railing, which renders them fairly conspicuous from seaward.

The bar.—A sand bar has formed and extends across the entrance in a semicircular shape. Under normal conditions the bar is about 300 feet across on the line of the fairway, with its inner edge about 300 yards from the outer ends of the piers, with a depth of 10 feet.

To the southwestward and southeastward of the entrance the bar is about 300 yards wide, with depths of 9 to 13 feet to the southwestward, and 7 to 9 feet to the southeastward. At about 300 to 400 yards east-southeastward from the outer end of the eastern pier is the southern extremity of the shallow ridge of the bar which extends north-northeastward toward the shore eastward of the entrance, with depths of 4½ to 5½ feet.

During the prevalence of southwest and westerly winds the bar protrudes further to seaward and eastward of its general position, especially so if the lakes be in flood; but with long continued. easterly winds the bank approaches closer to the entrance, with a general decrease of depth from 1 to 2 feet. During southerly and southeasterly gales the changes of the bar may be marked and sudden, a reduction of depth of 5 feet being known to have occurred.

Save in the finest weather, the sea always breaks over the shoal ground eastward of the fairway. The swell generally rolls in across the fairway without breaking, except in heavy southerly to westerly weather, with the ebb tide running strong, when the bar and entrance becomes dangerous and unfit for navigation. During easterly gales the force of the sea is broken by the eastern portion of the bar, and is thus rendered harmless to navigation, besides affording protection to the pierheads.

The navigable part of the bar is usually within the triangle formed by the line of the signal flagstaff and eastern pierhead, and the line of the flagstaff and the red beacon, 300 feet inward from the outer end of the eastern pier. Between these lines the fairway over the bar is 200 feet wide. Within the bar and navigable lines the soundings increase gradually until reaching the pierheads, between which there is a depth of about 50 feet. At 117 yards inside the pierheads the depth decreases to 20 feet, which is maintained to the inner ends of the piers.

Inside the piers the soundings decrease for a length of about 600 yards, forming an inner bar until reaching Jemmy's Point at the foot of Mount Barkly. The best water is on the eastern or Bullock Island side, where a channel 100 yards wide, with depths of 9 to 10 feet, is usually available, but after dry seasons, when the volume of fresh water coming down through the lakes is reduced to a minimum, the eastern end of the bank may so extend toward Bullock Island as to reduce the width of the channel to 200 feet, and the depth to 8 feet. Under these conditions the southern end of the bank, generally 100 yards westward of the entrance to Cunningham Arm, may project as an 8-foot spit across the channel into the arm.

Steam vessels trade regularly to the lakes. The worst months for navigation are March, April, May, and June.

Lights (lat. 37° 53′, long. 147° 58′).—A light, elevated 50 feet above sea level, is exhibited from the flagstaff at the pilot station to indicate the position of the entrance, but it does not in any way mark the fairway.

A white light is shown from the extremity of the platform on east pier when the entrance is safely navigable, and a red light is shown when the entrance is dangerous.

Tides.—It is high water, full and change, at the entrance of the lakes at 8 h. 30 m.; springs rise 3 feet. The rise depends largely upon the wind, rising highest with southwesterly winds. In calms, or with other than southwesterly winds, or even with southwesterly winds if light, the rise was almost nothing, upon one occasion during a calm only giving a range of 1 inch.

The flood makes from the eastward, and the ebb runs to the eastward, but a mile or two offshore the current is barely perceptible. Within the entrance the tides are irregular, owing to its narrowness and the ever-varying volume of water in the various lakes. Between the piers the direction of the current is not dependent on or coincident with the time of high or low water, but is influenced by the state of the weather and the volume of water in the lakes. The flood and ebb current under ordinary circumstances average from $2\frac{1}{2}$ to 3 knots an hour; but under exceptional circumstances, as in time of floods, the current may obtain a velocity of 7 knots.

Tidal signals (lat. 37° 53′, long. 147° 58′).—Semaphores and wicker balls, painted white, denoting the state of the tidal stream, depth of water, and condition of the entrance for navigation, are exhibited daily between sunrise and sunset, from the pilot flagstaff erected on the sand hummocks at the eastern side of the entrance, as follows:

Outgoing stream running_____Ball at eastern yardarm.
Slack at low water_____ Two balls at eastern yardarm.
Ingoing stream running_____Ball at western yardarm.
Slack at high water_____Two balls at western yardarm.
Wait for tide_____One ball half-mast.
Entrance dangerous_____Two balls half-mast.
Depth of water—
    9 feet_____One semaphore.
    9 feet 6 inches_____Two semaphores.
    10 feet_____Three semaphores.
    10 feet 6 inches_____Four semaphores.
    11 feet_____Five semaphores.
    11 feet 6 inches_____ Six semaphores.
    12 feet and more_____ Ball at masthead.

All other signals are made by the International Code. The station is connected by telegraph.

During strong flood tides a dangerous set exists across the channel inside piers, therefore the "wait for tide" signal should be attended to.

Life-saving apparatus.—There is a life-saving rocket apparatus at the pilot station.

Pilot.—A pilot is stationed at the Gipps Land Lakes entrance.

Gipps Land.—The climate and soil of Gipps Land are well fitted for the growth of oranges, limes, hops, tobacco, and opium, and the rivers abound in fish. The most important exports of Gipps Land are gold, wattle bark, cattle, wool, leather, grain, hops, dairy produce, sheep and kangaroo skins, and a large fish trade.

Sale is the principal town in the Gipps Land district and is situated on the Thomson River, 3 miles above its junction with the Latrobe, which falls into Lake Wellington, and about 45 miles from the entrance to the lakes. It is connected to Melbourne by rail and telegraph, has a population of about 4,000. Steamers ply daily between Sale and the Gipps Land Lakes entrance. The district is pastoral, agricultural, and mining; it is also a good sporting center, good fly-fishing and shooting are to be obtained.

Bairnsdale (lat. 37° 50', long. 147° 38'), 20 miles from the entrance to the lakes, lies 8 miles up the River Mitchell, which falls into Lake King. There is a telegraph station here, and it is in connection with Melbourne by railway. Steamers trade regularly with Melbourne. Hops are largely grown in the vicinity, and the country is both pastoral and mineral. The population is about 4,000.

Lights.—Lights are exhibited at several places in the Gipps Land Lakes, for which see Light List.

Adjustment of compasses.—A swinging station consisting of one central beacon and four surrounding beacons has been established in Eagle Point Bay, Lake King, at which vessels may be swung for the adjustment of compasses. The true bearings of several conspicuous landmarks from the central beacon are as follows:

Pile No. 1	9°	28'
Pile No. 2	99°	16'
Pile No. 3	189°	51'
Pile No. 4	279°	34'
Tambo River, light beacon	66°	12'
Raymond Island, northeast end	85°	41'
Cowl, James, hop kiln	300°	39'
Mount Look-out Gap	318°	13'
Mount Taylor Gap	326°	41'

Directions.—Vessels having made out Mount Barkly and the pilot flagstaff at the entrance, can stand in until the tidal signals are made out, which indicate the navigable condition of the bar and channel.

It is not advisable to enter without local knowledge, as probably the entrance is subject to change.

The navigation of the entrance under present conditions requires great care, and the entrance to be navigated only under favorable weather conditions.

The best fairway over the bar is usually found by keeping the signal flagstaff midway between the outer end of the eastern pier and the red beacon situated 100 yards from the same end of that pier; this range leads over the bar in a depth of 10 feet. After crossing the bar haul slightly to the eastward until the flagstaff and the outer end of the eastern pier are in range, and when 100 yards from the eastern pier steer to enter midway between the pierheads. In order to avoid the broken water of the bar and the submerged obstructions which project seaward from the western pier vessels must be kept within the navigable lines mentioned, and not to approach on the line of the eastern pier closer than 150 feet from its outer end.

The safest time to take the bar is at slack water, and then only if a sailing ship when in tow or with a good commanding breeze from any point between southwest and southeast. Mariners are warned against either entering or leaving unless with a good fair wind, as light winds are baffling between the high structures of the piers, and on account of the wreckage at the outer ends of the piers. If bound for Cunningham Arm, it is better to wait for slack tide, and then only under steam or in tow.

At night.—Vessels under steam may enter at night provided that the light on the outer end of the east pier shows white, and the night is sufficiently clear to distinguish the piers, but there is considerable risk. The exhibition of a red in lieu of a white light indicates that the bar is dangerous.

Caution.—Vessels are warned against a strong set to the westward across the entrance during the prevalence of easterly winds, especially on the flood; also that they incur danger by taking the entrance against the signals shown from the pilot station, and by navigating the bar when it is shallow, except under fairly smooth conditions of the sea.

Coast.—Red Bluff (lat. 37° 52′, long. 148° 04′) is situated 4½ miles eastward of the entrance to the lakes; it is over 100 feet high, and conspicuous from its color. It rises gradually to a height of 160 feet, and like the land about it is thickly timbered, though not so much near the coast as inland; the bluff has a few rocks off it, which do not extend far to seaward.

From Red Bluff the coast, which is similar to the land about the bluff, trends east-northeastward 1½ miles to the entrance to Lake Tyers.

Lake Tyers.—The entrance to Lake Tyers is generally barred across during the dry seasons by a sand bank, but after heavy rains the lake breaks through the bank, forming one or two channels to the sea. This entrance is unfit for navigation. A settlement for the education and religious instruction of the aborigines is situated on the northern shore of the lake.

Mount Taylor.—The first hill of importance is Mount Taylor, 1,630 feet high; this hill lies west-northwestward 22½ miles from the entrance to the lakes; as the trees on the summit have been cut down, it presents a tablelike appearance.

A hill of greater extent, but not so high, lies 2 miles westward of Mount Taylor.

Little Dick, situated 24 miles northward from the entrance of the lakes, is 3,154 feet high, and shows generally with three round summits; being a high and large range, it is a conspicuous landmark.

Mount Willie.—Southeast of Little Dick, and 9 miles from the coast at Lake Tyers, is a prominent hill known as Mount Willie. It is 1,182 feet high, has a flattish top, and is conspicuous as being nearer the coast than any hill from Wilson Promontory to Lake Tyers.

Mount Tara, situated 9 miles north-northeast from Mount Willie, has two conspicuous summits with other smaller summits of lesser importance; the principal summit is flat-topped, and 1,993 feet high. Much timber has been cut down here, but a solitary large tree was left standing; the large gap on the summit of the hill and the solitary tree if it still exists would enable strangers to recognize Mount Tara.

From Mount Tara eastward and northeastward the country is mountainous, some of the ranges approaching within a few miles of the coast.

The coast from the entrance to Lake Tyers trends with a curve east-northeastward 21 miles to the Snowy River entrance, and is similar to that westward of the entrance to the lakes, though the sand-hummocks are higher, especially toward Snowy River, near which they attain a height of 176 feet. Immediately at the back of the coast, extending the whole distance, is a fresh-water morass, and generally ½ mile from its margin is the higher back country, which along this part of the coast is about 200 feet high and densely timbered. The hummocky coast is faced with sand cliffs or patches, but of so uniform an appearance that only one patch close to the Snowy River is worthy of notice. This patch, lying ½ mile to the westward of the entrance, is a good guide to it. These patches have probably undergone considerable change since the survey some 60 years ago.

Shoal.—At 4 miles south-southwestward of the Snowy River entrance is a patch of uneven rock bottom, upon which the least depth found was 8 fathoms; close to this foul ground is 20 fathoms, sand.

Snowy River (lat. 37° 48′, long. 148° 32′).—The entrance to the Snowy River lies about 4 miles west of Ricardo Point, and is the outlet of the Snowy and Brodribb Rivers. There is a sandbar at the entrance, the depth of water on which varies from 5 feet at high water in dry to 6 to 8 feet in wet seasons. After passing through the entrance there is usually a good channel for 1⅓ miles to Marlo, with depths of 6 to 9 feet at low water. After passing Marlo the Snowy River is usually navigable at high water for about 4 miles for vessels drawing not more than 5 feet. At 1¼ miles past Marlo is the junction of the Snowy and Brodribb Rivers, the latter winding northeastward 4½ miles to the wharf at Richardson's sawmill; it is 120 to 250 feet wide, with depths of 4¼ to 10 feet in the middle, muddy bottom.

There is a jetty at the foot of the hill at Marlo, with usually a depth of 4 to 5 feet alongside.

Caution.—Owing to the shifting nature of the bar at the entrance the above information can only be given as generally reliable, and mariners without local knowledge are cautioned against navigating the place. There is a considerable settlement on the banks of the Snowy River, the cultivation of maize and the rearing of stock being carried on to an important extent.

Light.—From a post erected about 1 mile westward of the entrance to Snowy River a light is exhibited at 100 feet above high water. This light does not in any way indicate the entrance; it is only to show vessels they are nearing the Snowy River, the bar of which is always shifting.

Ricardo Point lies 4½ miles eastward of the Snowy River entrance. Very many years ago this entrance was close to the westward of the point, but a low sandy shore now occupies the space between, at the back of which is a salt-water lake, with an occasional opening into the sea, near the mouth of the Snowy River. The hummocks on Ricardo Point are about 100 feet high, the point itself is rocky, sunken rocks extending more than 400 yards from the shore in all directions.

Mount Raymond, situated 6 miles northward from Ricardo Point, is a conspicuous hill, 992 feet high at its north elevation; mountain spurs extend in a southerly and southeasterly direction.

Conran Point (lat. 37° 49′, long. 148° 43′) lies 5 miles eastward of Ricardo Point; it projects nearly 1 mile from the general line of coast; its highest part is 192 feet high, but is not easily distinguished; the land about the point is flat and covered with a dense dwarf scrub. The coast between Conran and Ricardo Points forms a sandy bight, skirted with grassy hummocks over 100 feet high. In

the center is one conspicuous hummock 163 feet high, with a sand
patch near its summit, over which is a grove of ti trees.

Under the eastern part of Conran Point, extending ⅛ of a
mile offshore, are numerous sunken rocks upon which the sea breaks
heavily at times.

In fine weather there is landing to the westward of the point;
landing will also be found to the eastward, but it is not good. At 1½
miles inland from Conran Point, and extending at that distance from
the coast to the Snowy River, the higher ground is densely timbered,
with an average height of about 300 feet.

Beware Reef lies about 2¾ miles eastward from Conran Point;
the reef is 8 feet above high water, and has sunken rocks (upon which
the depth is uncertain) lying eastward and southeastward of it to a
distance of ⅓ mile.

Pearl Point is situated 7½ miles east-northeastward from Conran
Point; the low sandy coast between forms a bight, and the bank
range rises to a height of about 300 feet. In the bight are two small
fresh-water streams; off the mouth of the westernmost are two
patches of sunken rocks more than ¼ mile from the shore.

To the eastward of Pearl Point are two conspicuous conical sand
cliffs, which render the point easy to identify. Scattered rocks lie off
Pearl Point to the southward for a distance of 400 yards, and 1 mile
to the eastward sunken rocks extend from the shore ⅓ mile. The
ridge of densely timbered broken country referred to as lying at the
back of Conran Point is also found at the back of Pearl Point,
whence it extends more inland toward Mount Cann at the back of
Sydenham Inlet.

The aspect of the land at the back of Pearl Point is marked in
character, as from the Snowy River eastward to Cape Howe there is
no part of the coast that is not defined by some conspicuous mountain or hill. At the back of Ricardo Point, 6 miles inland, is Mount
Raymond, which has been already described.

North-northeastward of Mount Raymond, distant 12 miles, is the
conspicuous long range of Diana, over 3,000 feet high; and in the
same direction at a distance of 21 miles from Mount Raymond is
Mount Ellery, a double-peaked mountain 4,300 feet high.

Sydenham Inlet.—From Pearl Point the coast runs in a nearly
straight line eastward for 6½ miles to Sydenham Inlet, the entrance
to Lake Bemm; over this district are grass-covered hummocks about
100 feet in height, northward of which the land is low for some
distance. At the back of Lake Bemm the land is densely timbered,
with an elevation of about 300 feet. Sydenham Inlet is small and not
worthy of particular notice. North of Lake Bemm, at a distance of
8½ miles from the coast, is Mount Cann, a peaked hill 1,885 feet
high, the summit of which was used as a surveying station.

From Sydenham Inlet the coast trends eastward 5½ miles to the Tamboon River, and thence about 2 miles in the same direction to a rocky stretch of coast, off which are several sunken and dry rocks, the highest of which, named Cloke Rock, is 25 feet above high water. One and a half miles inland is Point Hicks Hill, 924 feet high.

Cape Everard (lat. 37° 48′, long. 149° 16′), situated 6½ miles eastward from the Tamboon River, is easily recognized by a sandy peak 538 feet high, lying about 1 mile northward of the cape. This summit has (or had) a gradual fall to the westward of bare sand and is more remarkable when viewed from that direction.

Cape Everard has four points, the southernmost of which projects nearly 1½ miles from the line of coast. The cape is composed of granite with bowlders strewed over the whole face. There is landing in fine weather in the western bight. A deep but exposed bight lies to the southeastward of the cape.

Rocks.—A rock above high water, and several awash or sunken, are scattered 400 yards off the southern points of Cape Everard.

A reef of rocks, nearly awash at high water, lies nearly 400 yards eastward of the eastern part of the cape. There is deep water between these rocks, the outer of which has 18 fathoms close-to to seaward.

Landing can be made in fine weather in the bight to the westward of the cape, where the sandy beach commences.

In addition to the scattered rocks there is a rock 3 feet above low water and therefore barely covered at high water, lying 119° 1,500 yards from the southern part of the cape; and another the same distance, 94°, from the same point, with only 7 feet on it at low water. These detached dangers, occupying a position close to the fairway of steam vessels between Melbourne and Sydney, require due caution to avoid them.

Light.—A group flashing white light with red sectors, visible 20 and 10 miles respectively, 185 feet above water is shown from a white concrete tower 120 feet high, on the south extremity of the cape.

Red sectors of the light give warning of a too close proximity to the shore or to outlying dangers, and when seen, course should be altered to seaward until they are out of sight. In thick weather mariners should not rely upon seeing these red lights, but should keep a good offing.

Life-saving apparatus.—A life-saving rocket apparatus is kept at the lighthouse.

Everard Hill, 5 miles north of Cape Everard, is densely timbered and 1,200 feet high. From Cape Everard a sandy beach trends in a northeasterly direction 2 miles to the mouth of the little River Toolaway, and thence eastward for 5½ miles to Island Point.

Island Point, named from a rock, 30 feet high, which lies close off it to the southward, at about 7 miles eastward of Cape Everard, is 233 feet high and bordered by half-tide and sunken rocks for more than 200 yards off. At 400 yards southward of the point is a small rock only 1 foot above high water. The coast between Cape Everard and Island Point consists of sandy beaches with rocky points having reefs lying off them for 400 yards. About midway, and close to the coast, is a group of conspicuous bare sand hummocks, and to the eastward of this group are several sand patches.

From Island Point the land trends with a curve eastward 2½ miles to Rame Head. Between Cape Everard and Rame Head the coast rises gradually inland until at 2 or 3 miles from the shore it attains an elevation of about 600 feet. The country is densely timbered and undulating.

Rame Head (lat. 37° 47′, long. 149° 28′), of granite formation, rises to 378 feet on its eastern side; another summit of the same elevation rises close to the southwestward. To the northward the land falls, but again rises gradually, until at 4 miles distance it attains an elevation of 896 feet. The western part of the head is fringed with rocks, and a rock awash lies close to the southeastward of the extreme point. Densely timbered ranges occupy the district between Rame Head and Genoa Peak; the latter being 16 miles to the north-northeastward.

From Rame head the coast trends in a northerly direction for 1½ miles as far as a sandy beach; thence in a northeasterly direction for nearly 1 mile to Wingan Inlet. Over the sandy beach, and near its western part, is a sandcliff 204 feet high

Wingan Point, which forms the eastern side of the entrance into Wingan Inlet, lies 2 miles northeastward of Rame Head. Wingan Inlet is difficult of access. The best time for entering in a boat is after westerly winds. Landing may sometimes be effected outside the inlet to the westward of Wingan point. Oysters are found in this inlet.

The Skerries.—Off and southward of Wingan Point are the three Skerries rocks, the highest and central of which is 42 feet above high water. Close to the Skerries there are several detached rocks, some above high water; the outer of these, which is covered at high water, is ½ mile from Wingan Point. From Wingan Point the coast trends in a northeasterly direction 4 miles to Sand-Patch Point.

Sand-patch Point is well named; a large body of drift sand near the point making it conspicuous. The only part of the coast which at all resembles Sand-Patch Point is Cape Everard, but there the drift sand is not so conspicuous when seen from the eastward; the sand also at Cape Everard is higher than at Sand-Patch Point.

Danger off Sand-patch Point.—Nearly $\frac{1}{2}$ mile southward from Sand-Patch Point is a pinnacle rock with $1\frac{1}{4}$ fathoms on it at low water, known as the Long Reef. It is a dangerous rock, on which the sea breaks occasionally.

A rock awash lies close to the southeastward of Sand-Patch Point.

Little Rame Head (lat. 37° 41′, long. 149° 41′) lies $4\frac{1}{4}$ miles northeastward from Sand-patch Point; the coast between is about 300 feet high, and forms a rocky bight with a few sandy beaches. Upon a hill 240 feet high, immediately over the head, a survey station was erected. Eastward of the head, at a distance of 400 yards, is a rock 10 feet above the sea.

Race.—Off Little Rame Head, at a distance of 4 miles in a south-southeasterly direction, there is a depth of 19 fathoms, rock, near 35 fathoms, sand; and at a distance of 7 miles in the same direction 27 fathoms, rock, near 50 fathoms. This uneven bottom extends over a distance of 3 miles, causing a confused sea; in heavy weather small craft should avoid the place.

The coast.—From Little Rame Head the coast trends 8 miles north-northeastward to Bastion Point, the coast between being about 300 feet high. As a continuous heavy swell rolls on this coast, it should not be approached nearer than 1 mile; it is also fringed with sunken rocks. One mile to the southwest of Bastion Point is a conspicuous sand patch.

Bastion Point is comparatively low, being only 75 feet high; the land behind the point, and between it and Little Rame Head, is densely timbered, and rises to the height of about 300 feet.

South and southeastward of Bastion Point are numerous rocks on which the sea breaks, one, with $1\frac{1}{2}$ fathoms upon it, lies 600 yards to the southeastward, and another, 3 feet above high water, lies 600 yards to the southwestward. The land of Bastion Point forms the western side of Mallagoota Inlet. There is a landing place in fine weather on the northern side of the point.

Mallagoota (Mallacoota) Inlet.—From Bastion Point the rocky coast trends 800 yards northwestward to the sandy beach which fronts Mallagoota Inlet, and forms the shore to the eastward. From the junction of the coast and the beach the rocky shore continues north-northeastward 600 yards to Captains Point, inside the inlet.

The entrance channel through the coastal beach fronting the inlet is a shifting one, varying from a position close to the rocky shore inside Bastion Point to a point 600 yards eastward along the beach. The position alters according to the varying volume of water coming down through the inland rivers and lakes.

During heavy floods the waters usually break through close to the rocks about 800 yards inside Bastion Point, with a channel 7 to 9 feet deep. This channel may be maintained for several months by recurrent freshets, but, when these discontinue, the entrance gradually works 200 to 300 yards to the eastward, the depth on the bar decreasing to 4 feet. During long spells of dry weather, the channel reaches its eastern limit, south-southeastward to southeastward from Captains Point, and may shoal on the bar to 1½ to 2 feet.

The uncertainty of the position of the entrance and depth on the bar renders navigation of the entrance hazardous, and requires good local knowledge and great caution.

The inland waters of the inlet consist of the lower and upper Mallagoota Lakes, connected by a passage. The greater portion of the lower lake is filled by drying banks or with a depth of about 1 foot, but a channel, with depths of 13 to 20 feet, leads from the lake to the passage. The northwestern shore of the lake consists of rocky, timbered spurs, forming indentations with deeper water close to the points. The passage connecting the lakes is about a mile long, 200 to 300 yards wide, with depths decreasing from 5 fathoms at its southeastern to 3 fathoms at its northwestern end, and bold, high-timbered land on either side.

The upper lake is of smaller area, with depths of 8 to 12 feet, muddy bottom, along the navigable track to its northwestern corner, into which the Genoa River falls.

Genoa River, from the upper lake, winds along a tortuous course northwestward for 3½ miles to Gipsy Point, the water being very deep but shoaling to 7 to 8 feet in the vicinity of the point; thence the river continues westward for 7½ miles to Genoa settlement, and consists of sharp bends and shallow reaches of 3 to 4 feet, decreasing to 2 feet at the settlement.

There is a landing at Gipsy Point, which is the inland water terminus of good traffic through Mallagoota Inlet. Small motor boats ply above the point to convey excursionists from Genoa to Mallacoota.

Wallagaraugh River joins Genoa River about ½ mile above Gipsy Point, and extends 10 miles northward into New South Wales. The river varies in width from 100 to 400 feet, with a general depth of 6 feet, shoaling in places to 2 or 3 feet.

Tides.—Off the entrance to and on the bar of Mallagoota Inlet springs rise about 5½ feet, neaps rise 4½ feet. Off Captains Point 1½ feet is a good rise of tide, and 1 foot within the lakes. From this point the land forms a sandy bight to Telegraph Point, which lies 6½ miles east-northeastward from it.

At 4 miles from the inlet the coast projects to the southward, toward Tullaburga Island.

Tullaburga Island, 28 feet high, lying 4 miles east-northeastward from Bastion Point, is a rock with little soil, and a few bushes on the northeast part.

At 1¼ miles southwestward of Tullaburga there is uneven bottom, but not less than 7 fathoms water was found.

Gabo Island (lat. 37° 34', long. 149° 55') lies 3 miles eastward from Tullaburga Island. It is nearly 1½ miles long, in a north and south direction, and about half a mile broad near the center and southern end; toward the northern end the island tapers gradually to a point, which latter consists of low granite boulders, separated by a channel about 200 yards wide, from Telegraph Point on the mainland. Gabo Island, composed of red granite, is steep-to in all directions, except to the northward. To the westward the slopes of the island are covered with grass and dwarf bushes. Near the center are a few sandhills whose bare sides face the southeast, and only show as sandhills in that direction; the highest of these hills has an elevation of 171 feet.

On the northwestern side of the island is a small sandy bay, with 5 fathoms in the central part, where there is good anchorage for one vessel, except in southwesterly gales. There is a small jetty in this bay used for landing stores for the lighthouse. Several moderate southwesterly gales have been ridden out in this bay. It is probable that a gale of some continuance would have to blow direct in before the swell would make the anchorage unsafe. Though there is often a heavy swell outside, scarcely any was felt at this anchorage. Sailing vessels using the anchorage should get under way directly the swell reaches into the bay. Small craft, using the boat harbor in the southern corner of the bay, should approach with the outer end of the jetty bearing about 132°, to avoid the sunken and visible wreckage lying about 50 yards northwestward from the jetty.

A small but constant stream of fresh water runs out on the small sandy beach at the anchorage.

In the narrow channel between Gabo Island and the mainland the deepest part has about 6 feet water; boats occasionally use this channel, but a confused sea, caused by the meeting of the swell from opposite sides of the island, renders it unsafe.

Light.—A group flashing white light with red sectors, visible 19 and 10 miles, respectively, 179 feet above water, is shown from a red circular granite lighthouse, 156 feet high, on the southeastern extremity of Gabo Island.

Fogsignal (lat. 37° 34', long. 149° 55').—A fogsignal is made at the lighthouse.

Lloyd's signal station.—There is a Lloyd's signal station at the lighthouse and communication may be made by the International code of signals, and at night by Morse lamp. It is connected by telephone with Green Cape, thence via Twofold Bay to Sydney.

Life-saving apparatus.—A rocket life-saving apparatus is kept at Gabo Island Lighthouse.

Tides.—It is high water, full and change, at Gabo Island at 8 h. 50m.; springs rise 6 feet.

Tidal currents.—At the western part of the Ninety-mile Beach tidal currents exist, which are gradually lost in proceeding northeastward, and near the entrance of the lakes are not observable. The flood runs to the southwest, and the ebb to the northeast, with 20 miles eastward of Shallow Inlet a rate of one knot at springs.

Currents.—A current, averaging from ½ to 1½ knots an hour, generally sets eastward through Bass Strait with westerly winds, and westward with easterly winds, continuing for one or two days after the respective winds have ceased.

During the progress of the survey, no current was felt inshore between Wilson Promontory and Gabo Island.

Weather.—The experience of the weather on this coast, which was obtained during the survey, does not point to any great hazard in approaching the Ninety-mile Beach. In westerly gales comparatively smooth water is obtained by working up inshore when to the westward of the entrance of Gipps Land Lakes. Easterly gales are not without warning signs, therefore a sailing-vessel inshore when an easterly gale threatens should at once get an offing.

As the western part of the Ninety-mile Beach is approached, easterly gales are not so generally felt; Wilson Promontory appears to be the dividing point. In the months of May to August, as westerly gales usually shift to the southward, it is more advisable to stand toward the Tasmanian coast, and so be ready to take advantage of the shift of wind.

In the months of September, October, and November, the same course can not be recommended; the wind during these months does not change for a continuance, but constantly shifts to the west-northwestward.

Soundings.—Southward and southeastward from Wilson Promontory the depths are 43 to 32 fathoms. Soundings about here give no indication of the vicinity of the promontory nor of the islands lying south and southeastward from it.

Toward the Seal Islands the depths decrease to 20 fathoms, the 20 fathoms' line running from Cape Wellington to Cliffy Island, and thence northeastward parallel to the land at a distance of 10 miles.

From the neighborhood of Port Albert to Merriman Cove there are depths of 6 to 9 fathoms at 1 mile from the shore, gradually increasing to 13 and 14 fathoms at a distance of 6 miles.

From Hogan Group, 29 miles eastward of Wilson Promontory, to 7 miles south of Gabo Island, there are 28 to 41 fathoms for the first 60 miles. On this line no soundings have been taken after the first 60 miles until within 17 miles of Cape Everard, where there are 65 fathoms. Few sounding have been taken beyond 10 miles from the coast from Shallow Inlet to Gabo Island. Twenty-five miles eastward of the inlet there are 21 fathoms; at the same distance southward of the entrance to the lakes are 30 fathoms, and at 27 miles southeastward of the entrance are 29 fathoms, whence the depth increases to 67 fathoms at 7 miles to the eastward. South-southeast of Conran Point at 10 miles there are 29 to 34 fathoms, at 40 miles south 150 fathoms, and at 35 miles southeast 140 fathoms. At 5 miles off Cape Everard there are 60 fathoms, and this depth is found between Little Rame Head and Gabo Island at about 7 miles off the land.

Coast.—Cape Howe lies 5 miles northeastward from Gabo Island. The description of the coast from Gabo Island to Port Jackson will be found in Chapter VIII.

EASTERN ENTRANCE OF BASS STRAIT.

The eastern entrance of Bass Strait is the space included between Wilson Promontory and the northeast part of Tasmania. Between these two headlands are numerous islands, occupying an extent of about 120 miles, which, from their formation of granite, and the manner in which they lie, as a connecting chain, would appear to have been the upper part of a range of hills which once joined the two lands before natural causes produced the opening which bears the name of Bass Strait.

That such was the original formation of this part, or at least its disposition a comparatively few years ago, appears extremely probable on inspecting the chart; as also that Wilson Promontory and Cape Liptrap were formerly insulated.

The depths in Bass Strait are fairly regular, ranging from 30 to 48 fathoms, with generally 5 fathoms within 1 mile of its shores. The bottom mostly consists of sand and shells in the northwestern and greater portion, and more of mud, marl, and ooze in the southeastern part of the strait.

Eastward of Bass Strait between 35 fathoms at 20 miles eastward of the southern extremity of Barren Island, and 38 fathoms at about 40 miles northeastward of the north point of Flinders Island, the depths range from 20 to 42 fathoms, and thence the depths increase rapidly to more than 200 fathoms in the direction of Rame Head and Cape Howe.

Tidal current.—At the eastern part of the fairway of Bass Strait the flood current sets to the southwest and the ebb to the northeast.

Vessels proceeding eastward from the neighborhood of King Island usually find the tidal current against them of longer duration than the current in their favor.

ISLANDS AND DANGERS IN THE FAIRWAY.

Curtis Group (lat. 39° 28′, long. 146° 39′).—**Curtis Island** is 1⅓ miles in length in a north-northeast and south-southwest direction. It rises in two peaks; the southern is square-topped and 1,100 feet high; the northern, 736 feet high, has a bare granite summit. The coast is precipitous all around, but especially so at the south end of the island. Landing may be effected close to the north point, but only in very fine weather. See view C on H. O. Chart No. 3442.

Cone Island, 1¼ miles southeastward of the southern point of Curtis Island, is a rocky islet 368 feet high; on its northern side there are two small rocks; the outer, Passage Rock, is distant 600 yards and uncovers 3 feet at spring tides; the other is 8 feet above high water; on its south side, and close-to, there is a remarkable leaning pinnacle 82 feet high.

Sugarloaf, so named from its appearance, the southernmost of the group, is 308 feet high, and lies 2½ miles southward of the southern point of Curtis Island; on its north side and close-to there is a small rock 8 feet high.

Clarendon Rock, lying 1,600 yards eastward of the northern point of Curtis Island, has 4 feet of water on it at low-water springs. This rock is of very small extent, and is surrounded by deep water; it breaks in heavy weather, but is generally difficult to distinguish.

Crocodile Rock lies 9¾ miles northwestward from the northern point of Curtis Island and almost in the center of the channel between Dodondo and the Moncœur Islands and Curtis Island. It is a large, smooth bowlder of granite, about 2 feet above high-water springs. The sea almost continuously sweeps over it, so that the rock itself is seldom visible; as a general rule, however, its position is well marked by the breakers.

Cutter Rock, 1½ miles northeastward from Crocodile Rock, has 4 fathoms water on it at low-water springs; it is of small extent, surrounded by deep water, and has not been observed to break.

Devil's Tower is a rugged islet, 363 feet high, lying 6½ miles north-northeastward from the north point of Curtis Island; there are a few detached rocks quite close-to on the southwestern side. otherwise it is free from danger or fringing reefs. When viewed from the northwest or southeastward, the Devil's Tower shows a double summit, the northeastern being the higher of the two.

Hogan Group.—Hogan Island, the largest of the group of the same name, the summit of which lies 14½ miles northeastward from

Devil's Tower, is 1¼ miles long, north and south, by 1,500 yards wide; it is 428 feet high, and precipitous on the south and west sides. See view B on H. O. Chart No. 3442.

To the northward and eastward of Hogan Island there are several small islets and rocks; the outermost, Seal Rock, is 15 feet above high water, and lies about 2½ miles northeastward of the summit of Hogan Island. There are no hidden dangers in the neighborhood of the group.

Anchorage.—Excellent shelter in heavy southerly winds may be obtained on the northeast side of Hogan Island, in the bay formed by the rocks projecting to the northward, and Long Island. The water is rather deep (about 20 fathoms), but the holding ground is good. With westerly gales, the bay on the eastern side of Hogan Island and south of Long Island is a better anchorage than the above bay. The depth is 10 to 15 fathoms, with a bottom of rocks, stones, and sand, affording good holding ground.

Water.—Fresh water in small quantities may be obtained on Hogan Island, either in the bay on its eastern side, or in the small cove on its northwest side.

The Hogan Group is infested with black snakes.

Soundings.—There are depths of 40 to 30 fathoms from Curtis Isle to Hogan Group, and between them and Wilson Promontory there are similar depths of water, the bottom being generally sand, shells, and coral.

Directions.—No other sunken dangers are known to exist between Curtis Isle and Wilson Promontory; in the night, or during thick weather, it is prudent for a stranger who is desirous of clearing the strait to obtain a sight of Curtis Isle, and pass on its south side, to the southward of the Sugarloaf, between the latter and Judgment Rocks, as its high summit, Cone Islet and Sugarloaf Rock to the southward of it, are remarkable objects, by which its identity can not be mistaken unless passing between the promontory and Rodondo).

Kent Group (lat. 39° 29′, long. 147° 19′), situated 18 to 23 miles southeastward from Hogan Group, consists of Deal, Dover, and Erith Islands, and N. E. Isle, which latter lies east-northeast 1¼ miles from Garden Point, the northern extremity of Deal Island. The channel between Kent and Hogan Groups has depths of 30 to 35 fathoms.

Deal Island, the largest of the group, is 3¼ miles long and 2½ miles broad. It rises to conical granite hills, some of which are clothed to their summits with an impervious scrub. The highest of these hills, near which is the lighthouse, rises from the south point of the island to the height of 949 feet. The coast is generally precipitous, especially on the south side, and is indented with numerous bays. View E on H. O. Chart No. 3442.

Light (lat. 39° 29', long. 147° 19').—A revolving white light, 1,000 feet above water, visible 36 miles, is shown from a white circular tower on Deal Island, situated near the summit, at the southwest side, and 1,500 yards northward of the south point of the island. The light, from its great elevation, is frequently obscured by fogs.

The lighthouse is built 400 yards southeastward of the highest part of the island, but is sufficiently high for the light to show over it.

Lloyd's signal station.—There is a Lloyd's signal station at Deal Island Lighthouse, and communication can be made by the International Code of signals.

Murray Pass, the channel which separates Deal from Erith Island, is nearly ½ mile wide, with depths of 25 to 33 fathoms in mid-channel.

Erith and Dover Islands are connected at low water, but at high water there is a boat passage between them, and together are 3¼ miles long north-northeast and south-southwest and about 1¼ miles across, at the broadest part.

Erith, the northern island, is irregular in shape with a deeply indented coast line; a grass valley, in which is a great number of rabbits, trends in a northwesterly direction through it.

Dover Island rises abruptly to a height of 744 feet; the coast, especially on the south side, is precipitous.

North Rock, a small black rock 27 feet high, lies 800 yards west-northwestward of the northern bluff of Erith Island; there are two smaller rocks about 100 yards southward of it.

Northeast Isle is small and 345 feet high. It lies about 1 mile northeastward of Deal Island. At 400 yards northwestward of the north point of the isle is a remarkable rock, 40 feet high, named from its appearance the Anvil Rock.

Anchorages.—Of the numerous bays with which the coast line of this group is indented, there are only two where it is at all safe to anchor, East Cove and West Cove.

East Cove, on the west of Deal Island, northward of the lighthouse in Murray Pass, affords good shelter in easterly and southerly winds, but it is a dangerous place in a southwesterly gale; the bottom is uneven, the holding ground not good, and in the strength of the tidal current the swirls and eddies come well into the cove. At the bottom of this cove there is a sandy beach, and a boathouse; it is here the stores for the lighthouse are landed. There is a fairly good road from the beach to the lighthouse.

West Cove (lat. 39° 29', long. 147° 19') also in Murray Pass, east side of Erith Island, gives protection from all but northeasterly gales, and as an anchorage generally is to be preferred to East Cove.

Tides.—It is high water, full and change, in East Cove at 10 h. 15 m. The greatest range of tide observed in January, February, and March was 8 feet. The tides at neaps are very irregular.

Tidal currents.—The flood current comes from the northeastward, the ebb from the southwestward. In fine weather it is slack water at the time of high and low water; but this is not always to be depended on, as the duration of the current appears to be greatly influenced by the wind.

Both flood and ebb run with considerable velocity through Murray Pass, frequently causing a very turbulent sea; in bad weather there are heavy tide rips off all the salient points of this group, and especially at the exposed entrance to the pass.

Supplies.—The light keepers and their families are the only inhabitants of this group; they all reside on Deal Island, and keep a number of sheep; a few may be purchased at a reasonable price; excellent water cress grows in profusion in the stream which runs into Garden Cove, and can easily be obtained.

Water.—Fresh water is abundant in the northern part of Deal Island.

Communication.—The Tasmanian Government communicates with Deal Island periodically for the purpose of landing stores and provisions.

Steamers plying between Sydney and Launceston pass close to this group, and communicate if signaled from the lighthouse flagstaff to do so.

Islands and rocks.—**Southwest Isle** lies 9¼ miles westward from the south bluff of Deal Island; it is a small rocky islet ½ mile long, north and south, and 323 feet high.

Judgment Rocks are a group lying to the northward of Southwest Isle. The northern and largest of these rocks is 105 feet high, and distant 1,600 yards from the north point of Southwest Island. See view E on H. O. Chart No. 3442.

The passage between these rocks and Southwest Isle should not be used.

The Pyramid, situated 19¼ miles southward from the south bluff of Deal Island, is a bare, square-topped mass of granite, 243 feet high; there is deep water fairly close to its western side, and from 11 to 20 fathoms for about ½ mile from its eastern side; it is usually surrounded by tide rips, except when there is no wind. This rock has frequently been mistaken for a sail.

Wakitipu Rock.—This rock, which consists of several pinnacle rocks, with a least depth over it of 1¼ fathoms at low water, is about

175078°—20——14

¼ mile in extent with deep water around it, and is situated at a distance of 9 miles, 63°, from Pyramid Rock.

Wright Rock (lat. 39° 36', long. 147° 32'), 124 feet high, is situated southeastward distant 11¼ miles from the south bluff of Deal Island. This rock should be given a berth of 1 mile; there are irregular depths within that distance around it, and there are several small rocks close-to, both on its northern and southern sides; it is also generally surrounded by tide rips. View on H. O. chart No. 3442.

The channel between Kent Group and Wright Rock, which is 10 miles wide, has tolerably regular depths (except within 1½ miles of the rock, where there are depths of 11 to 13 fathoms) in 24 to 29 fathoms, sand and shells; there are depths of 29 fathoms gravel and small stones, at 2 miles to the northwest of Wright Rock, and the same depth, on a coarse sandy bottom, 7 miles to the northward. Depths of 25 to 40 fathoms continue for 20 miles farther in a northerly direction.

Endeavor Reef lies 2¾ miles southeastward from Wright Rock; it is about ½ mile long, in a north and south direction; the reef does not uncover, and is usually marked by a heavy breaker.

Beagle Rock, situated 5½ miles eastward from Wright Rock, is of small extent, 5 feet above high water, and is steep-to.

Rock.—A rock on which a steamer struck is reported to exist about 7 miles eastward of Beagle Rock, but its position is doubtful.

Craggy Isle, situated 4 miles southward from Beagle Rock, is bare and rocky, about 1,340 yards in length, and rises at its western end to a height of 371 feet. Rocks extend about 400 yards from the western extremity of the isle; and east-northeastward of it for rather more than 800 yards from the eastern extremity. View on H. O. Chart No. 3442.

Dangers.—An extensive rocky patch, with a least depth of 2½ fathoms, lies upward of 1 mile eastward of the east extremity of Craggy Isle; and irregular soundings extend south and southwestward from that isle for a distance of about 1½ miles.

Craggy Rock, 152°, 2¼ miles from the summit of Craggy Isle, has a least depth of 4 fathoms. The soundings between this rock and the patch eastward of Craggy Isle are irregular. The dangers in this locality break heavily in bad weather, and both they and Craggy Isle are usually surrounded by tide rips. View on H. O. Chart No. 3442.

Tidal currents.—The flood and ebb currents run through these channels at a rate of about 2 knots at springs. In strong breezes there are generally heavy tide rips in the vicinity of the reefs.

Furneaux Group, the southeasternmost of the chain of islands between Wilson Promontory and the northeast extremity of Tasmania, consists of Flinders and Barren Islands, the largest of the group; and numerous smaller islands, rocks, and shoals. This group extends from the Sisters in a south-southeastward direction nearly 60 miles to Moriarty Banks, and is 32 miles across.

These islands are inhabited by a small number of people, who procure a living by seal-fishing and preserving mutton birds; many of them are half-casts, the offspring of marriages between the sealers and aboriginal women.

The Sisters are two high islands, from 1 to 5 miles off north point of Flinders Island, and are visible in clear weather at the distance of 30 miles. The Sisters have rather uneven surfaces with not much vegetation, but they harbor numerous sea birds.

East Sister (lat. 39° 39', long. 148° 0') is 2¼ miles long, 615 feet high, and lies 3½ miles northward of Flinders Island. Southward of East Sister, and separated from it by a narrow channel, is a ledge of rock, part of which is 10 feet high; the south point of this ledge is 1,600 yards from the shore.

West Sister is 2½ miles long, almost divided in the middle by a deep valley. The western part is 526 feet high, and the eastern 636. The east point of this island lies 1 mile northward of the north point of Flinders Island and 2 miles southward of the west point of East Sister.

The soundings around these islands and the northern part of Flinders Island from North Point to Bligh Point, are very irregular and are apparently the result of the action of the tidal currents.

When passing between the Sisters, take care to avoid a spit which extends 1,500 yards in a northwesterly direction from the northeast point of West Sister Island.

Tidal currents.—In the channel between the Sister Islands the current runs at the rate of 1½ knots at springs. In the channel between West Sister Island and Flinders Island the current runs at a much greater rate, causing tide rips and overfalls, and has scoured out the narrow channel to a depth of over 70 fathoms.

Flinders Island, the largest of the Furneaux Group, is 36 miles long, northwest and southeast, 20 miles broad at the center, and has an area of 513,000 acres; from Bligh Point, the northwestern point of the island, a reef projects a short distance. View D. on H. O. Chart No. 3442.

The principal ridges on the island take a general south-southeasterly direction from its northwestern point to its southern extremity and are barren and mountainous, presenting a bold abrupt front to the westward and sloping to low land on the eastern side, which is bordered by a sandy beach. These ridges are separated at about the middle of the island by Healthy Valley, which stretches across it. The western side of Flinders Island is fronted by several small islands, under the eastern side of which vessels may find shelter from westerly winds.

The coast of Flinders Island from North Point to Cape Frankland is rocky and broken; between Bligh Point and Sentinel Island it falls back considerably, forming a bay on the southwest side of Mount Killiecrankie, in the middle of which there is a small rocky islet 22 feet high.

This bay is entirely exposed to the northwestward, but affords protection from southwest gales, the violence of the sea being broken by Sentinel Island and the surrounding rocks; a moderate swell only comes into the bay. Although the bottom is sand, the holding ground is good.

There are two places in this bay where boats may find shelter, one amongst the rocks in the south part of the bay, and the other at the northeast point under Mount Killiecrankie. When entering and leaving this bay it is advisable to pass east of the islet, so as to avoid a sunken rock which lies 600 yards southwestward of it.

Radio.—A radio station has been established on Flinders Island. It is open to the public from 9 a. m. to 6 p. m. of meridian 150° E. Call letters VIL. The station is closed on Sundays. Weather forecasts are received and transmitted on request.

Mount Killiecrankie rises in the northwest part of the island to the height of 1,035 feet.

Cape Frankland (lat. 39° 52′, long. 147° 45′), situated south-southwestward 9 miles from Bligh Point, is the western and central extremity of a hilly peninsula extending 5 miles in a north and south direction, and 4 miles from the western coast line of Flinders Island.

A reef projects a short distance from the cape, and an islet lies ½ mile and Sentinel Islet 2½ miles northward of it; rocks extend for some distance from these islets.

A reef extends ⅓ mile from the southern point of the peninsula; and in the bight to the eastward of it is the cluster of Flat Rocks, between which and the shore to the northward there is a boat harbor with 3 fathoms water in it.

Marshall Bay.—Between the southern point of the peninsula of Cape Frankland and Settlement Point 6 miles southeastward from it, the west coast of Flinders Island forms Marshall Bay, 3½ miles deep, with 8 to 10 fathoms across its entrance, and 9½ to 4 fathoms along its southern shore; but it is exposed to the westward.

Roden and Pasco Islets, mostly connected by reefs, extend from the shore at 2 miles southeastward of the cape, 4 miles in a southerly direction. North Pasco Islet is 242 feet high and South Pasco 36 feet. They are situated in the northern approach to Marshall Bay.

Neither Marshall Bay nor Pasco Islets give any protection during the presence of westerly winds.

Tides.—At Roden Island it is high water, full and change, at 10 h. 07 m.; springs rise 10 feet, neaps 7½ feet, neaps range about 5 feet.

Frankland Rock, situated 4¼ miles westward from Cape Frankland, is a double rock awash at half tide, with 18 to 23 fathoms close about it, except on its east side, where a bank of not less than 7 fathoms extends ½ mile.

Settlement Point (lat. 40° 01′, long. 147° 52′) is hilly, having a small reef on its south side, and a cluster of islets and rocks on which is Rabbit Island, extending from ¼ to 1¼ miles southwestward from it, with depths of 11 to 4 fathoms between Rabbit Island and the point.

The west coast of Flinders Island, from Settlement Point, curves east-southeastward 2 miles to a projection, between which and Long Point, 4½ miles southeastward from it, is a bay 1½ miles deep, with mountainous land behind it; the highest summit being the Sugarloaf, 1,472 feet high, 6½ miles to the eastward from Settlement Point.

Hummock Island, the northern point of which lies 3¾ miles westward from Settlement Point, is 5¼ miles long, in a north and south direction, and is 1 mile broad at either end, between which it is only ½ mile wide. Its two highest hills are on its northern and southern ends, the former being 483 and the latter 570 feet high. From the northern point a reef stretches 1,500 yards to the northward, and this reef continues to the eastern point of the island, the 5-fathoms line being about 800 yards offshore. Shoal water extends 1,500 yards southward from the eastern point, and there is a rock dry at low water, 600 yards southward from the point; along the island from the eastern point to the Koh-i-noor Rock, there is deep water to ¼ mile offshore.

Koh-i-noor Rock, nearly awash at low water, lies about ¼ mile offshore and 2½ miles south-southwestward from the most easterly point of Hummock Island.

Anchorage.—There is good anchorage in 6 fathoms water on the east side of Hummock Island, at about 1,500 yards to the northward of the Koh-i-noor Rock, just noticed, and ½ mile from the shore. It may be approached by passing round either the north or south end of the island.

Entering the anchorage from the southward the island must not be hugged too closely, to avoid the Koh-i-noor Rock.

Tides.—It is high water, full and change, at the anchorage, Hummock Island, at 10 h. 30 m.; springs rise 10 feet.

Passage Islets are three in number, lying from 1 to 1½ miles southward of the south point of Hummock Island, the southernmost and largest islet being 600 yards in extent.

The reef on which these islets are situated extends nearly $\frac{1}{4}$ mile
northward from the northeastern islet, between which and the south
point of Hummock Island is a safe passage $\frac{1}{2}$ mile wide.

Myrmidon Rock (lat. 40° 02′, long. 147° 44′), on which there is
9 feet at low-water springs, lies 1$\frac{3}{4}$ miles westward from the north
point of Hummock Island; it is of small extent, and there is deep
water between it and the island.

Swires Patch, on which there are 5 fathoms at low-water springs,
lies 2$\frac{3}{4}$ miles northwestward from the south point of Hummock
Island, and irregular depths extend 1$\frac{1}{4}$ miles to the northward of it.

Rock.—A rock which dries about 6 feet at low water is reported
to exist about 9 miles westward from the southern end of Hummock
Island. The position is approximate.

Long Point, which has a rock awash close off it, is a peninsula
stretching about 1$\frac{1}{4}$ miles southward from the line of coast, from
which it is nearly separated by a shallow inlet having a narrow entrance, with a small islet close to its eastern point, and another on
the west side of the inlet. Between Long Point and another projection 2$\frac{1}{4}$ miles to the eastward of it, the bight is full of shoal patches,
which prevent a near approach to the fresh-water stream situated
close to the shore, at 1,500 yards to the northward of the east point
of the bight.

From the eastern point of the bight, just noticed, the west coast of
Flinders Island takes a south-southeast direction 6$\frac{1}{2}$ miles to the
northern point of a hilly promontory, extending 1$\frac{1}{4}$ miles north-
northwest and opposite direction and 1 mile within the line of coast.
Between 2 and 3 miles northward of the point is a slight projection
of the coast, close behind which is a fresh-water swamp.

The southwestern point of Flinders Island lies 1$\frac{1}{4}$ miles southeastward from the southern extremity of the hilly promontory just
described, and there is a small bight on either side of it.

Strzelecki Peaks.—At 3 miles northward from the southwestern
point of Flinders Island, Strzelecki Peaks, the highest mountains on
the island, rise to the height of 2,550 feet.

Reef Isles (lat. 40° 06′, long. 147° 54′) are four in number, with
several rocks above water, connected by reefs extending from 3 miles
westward to 4$\frac{3}{4}$ miles southwestward from Long Point. Chalky ·
Island, 79 feet high, the northern, is 1,500 yards long, north and
south, and is inclosed by rocky shoals; the island should not be approached on its west side within a mile in 11 fathoms water. The
other three, which are small islets, lie 2 to 2$\frac{1}{2}$ miles southward and
southeastward from the northern island, and are connected by a
narrow continuous reef extending westward and southwestward
from the eastern to the western and southern islets.

Another small island, Isabella, lies 1¾ miles southward from Long Point, between which and Reef Isles there are depths of 4 to 5 fathoms water; the depth decreases northward to 2 fathoms abreast of Long Point. Between this island and the coast eastward of it there are depths of only 6 to 9 feet water at 1,000 and 1,500 yards from the shore.

Channels.—There is a clear channel nearly 2 miles wide, with 8 to 10 fathoms water, between the north end of Hummock Island and the islets off Settlement Point; and another 5 miles wide between the island and Reef Islets; the depths gradually increasing to the southward to about 20 fathoms between Passage Islets and the southern Reef Islet.

Tidal currents.—The tidal currents in this channel follow nearly the trend of Hummock Island, the flood setting to the southward, three-quarters of a knot, and the ebb to the northward, half a knot.

Kangaroo Island, which lies 5 miles southward from Chalky Island, is of a crescent form, with its points to the southward and southeastward, each having a reef, projecting a short distance from it. This island is 1½ miles long northeast and southwest, and ⅓ mile broad at the center. Reefs extend 400 and 600 yards from its northwest side and northeast point, and for a mile to the northward.

Anchorage.—There is anchorage in 7 fathoms water off the northeast end of this island.

Green Island, 3 miles eastward from the northeast point of Kangaroo Island, is 1¼ miles long north and south, and ⅓ mile broad at the center, where it rises to a hill. A cluster of islets extends nearly 1 mile northward from the north end of the island. There is deep water close round the rocks which skirt the southern half of Green Island, but there are only 2½ to 1½ fathoms along the southern edge of a shoal which connects the island with the mainland.

Anchorage.—There is anchorage in 4 fathoms water at ⅓ mile off the southeast side of the island.

Soundings.—There are depths of 7 to 13 fathoms water between Kangaroo and Green Islands, from whence the depth gradually decreases northward to 4 and 5 fathoms between the eastern Reef Islet and Isabella Island.

Chappell Islands are three in number, with numerous islets and rocks, lying from 5 to 12 miles westward of the southwest point of Flinders Island.

Goose Isle (lat. 40° 18′, long. 147° 48′), the westernmost of the group, lies 12 miles westward from the southwest point of Flinders Island; it is 1½ miles long north-northwest and south-southeast, and ¼ mile broad, with an islet about ¼ mile in extent surrounded by a reef close off its northwest extremity. The island is 54 feet high at its highest and northern part and consists of granite bowlders with a shallow covering of soil in the hollows where a few sheep are grazed. There are depths of 6 fathoms water close to the south point, and from 7 to 9 fathoms near the eastern side of Goose Isle.

Light.—A fixed white light, 100 feet above water, visible 16 miles, is shown from a white, circular, stone lighthouse 74 feet high on Goose Isle, ¼ mile from its south point.

Anchorage may be obtained in from 9 to 11 fathoms, sand, to the eastward of Goose Island, with the northernmost of the lighthouse-keepers' dwellings in line with the flagstaff, and Hummock Island open of the northeast point of Goose Island. This anchorage is good during westerly winds, but being so close to the shore, is unsafe for sailing vessels should the wind shift to the southeast.

Tides and tidal currents.—It is high water, full and change, at Goose Island at 10 h. 48 m.; springs rise 9 feet. The flood current sets to the northwest, the ebb to the southeast, at an average rate of 1 knot.

Badger Island (lat. 40° 19′, long. 147° 52′), the central and largest of the Chappell Islands, is flat, about 4 miles in extent, 109 feet high at its northeastern point, and sparsely covered with timber. A rocky spit, with 12 feet upon it, extends nearly 1½ miles to the northward of its northwestern extremity. From the northwestern point the west side of Badger Island trends southward 1,500 yards and thence, forming a bay 1¼ miles in length and ½ mile in depth, to Unicorn Point, the southwestern point of the island, from whence the south and east coasts of the island sweep round 4 miles to its northeastern point. Over Unicorn Point there is a conspicuous granite bowlder.

The northern side of the island forms a bay extending westward from the northeastern point 1½ miles, and having from 9 to 3 fathoms across its entrance. The western point of this bay has a reef projecting nearly 1,340 yards northward from it, between which and the northwestern point of the island is a projecting point fringed by a reef.

At ¼ mile eastward of the eastern point of Badger Island is Little Badger Island, 17 feet high, and at 1¼ miles southeastward of the southeastern point of Badger Island is Beagle Island.

Anchorage.—Eastward of Badger Island good anchorage may be obtained in about 7 fathoms, sandy bottom, with Little Badger Island in line with the summit of Mount Chappell Island.

The channel between Goose and Badger Islands is 1¾ miles wide, and has from 7 to 19 fathoms water in it.

Boxen Island and Double Rock.—Boxen Island, 22 feet high, lies 6 miles southeastward from Goose Island lighthouse, having rocks and foul ground all round it, but principally on its west and north sides, where they extend ¼ mile. Double Rock, 34 feet high, lies 1 mile northward from Boxen Island. Six hundred yards from Double Rock in a northwesterly direction is a rock, dry at low-water springs. Foul ground extends ¼ mile to the southeast of Double Rock, and there is only a very narrow channel with not more than 4 fathoms between it and Boxen Island.

Beagle Island, 2¼ miles northeastward from Double Rock, is 21 feet high, with a reef extending about ¼ mile off its western side. A 3½-fathom bank extends off its north side for ¼ mile, on which are several rocks above water; and there is a sunken rock ¼ mile to the southward of it.

Rochfort Rock.—Nearly midway between Double Rock and Beagle Island lies Rochfort or Lucy Rock with 6 feet water over it, on which the sea breaks in westerly winds toward low water.

Mount Chappell Island (lat. 40° 16′, long. 147° 56′), 1¼ miles in length in a north and south direction, and 1,500 yards in breadth, lies 1 mile northeastward of Badger Island, and rises to a rounded summit 653 feet high. To the westward and extending northward for 2¼ miles and northeastward for ½ mile of this island, there are several groups of rocks 15 to 25 feet high, the highest being 1¼ miles north-northwestward from the north end of the island.

A shoal patch with 4¼ fathoms on it, which breaks in heavy westerly weather, lies 1½ miles 66° from the northeast end of the island; and in the center of the channel between Mount Chappell and Badger Islands is a rock which dries 3 feet at low water.

Mount Chappell Island is a favorite breeding place of the mutton bird or sooty petrel, and during the season above 200 men, women, and children are employed in salting the young birds (which form an article of diet amongst the poorer population in Tasmania) and collecting their oil, which is used for softening leather, lubricating machinery, and other purposes. The birds deposit their eggs about the last week in November. The season for obtaining the young birds and oil is from January to March.

South coast of Flinders Island.—From the southwestern point of Flinders Island the coast trends 2¼ miles, eastward 1½ miles, and east-northeastward 3½ miles to Badger Corner, and is formed by numerous small sandy bays and sharp projecting points, the spurs from Strzelecki Peaks generally extending to the points.

Dangers.—The dangers near this part of Flinders Island are Entrance Rock, with 2 feet water, lying 204° 1.3 miles from the southern point of the hilly projection; a rock with 12 feet, nearer the point; a bank with 12 feet, lying 1 mile off the southwestern point; and a shoal of 9 feet, extending to a distance of over ½ mile from the shore.

Badger Corner is a small cove in the western corner of a bay which extends from the southeastern point of the cove 4 miles east-northeastward and is 1¾ miles deep. This bay is mostly occupied by small islands and shoals, with generally very shallow water between them, and is fronted by the two Dog Isles.

Between the Little Dog Isle and Badger Corner there is a group of rocks covered at high water, to the westward of which is a narrow channel, with 12 feet water, leading to the anchorage for small vessels in Badger Corner; the channel to the eastward of these rocks is blind.

Dog Isles are situated between Badger Corner and the southeast point of Flinders Island. Little Dog Isle, about ½ mile across and 118 feet high, lies 1,200 yards eastward of the southeastern point of Badger Corner. Shoal water extends from it in a southwesterly direction 1½ miles.

A rock 3 feet high lies 400 yards from its southern extremity, and another 300 yards from its northwestern extremity. Other rocks, with numerous sand banks, lie to the northward, and a sandy flat, nearly dry at low water, extends ½ mile to the northeastward.

Dog Isle (lat. 40° 15′, long. 148° 15′), nearly 1 mile from Little Dog Isle, is about 1¼ miles in extent, having a flat-topped, conspicuous hill 254 feet high at its northwest point; there is a shoal sandy bay eastward of the southwest point of the isle, dry at low water, except near the center, where there is a depth of 3 feet; at the head of this bay there is a house off which small vessels anchor.

At 1,200 southeastward of the southwest point is a rock 2 feet above high water, from which a sand bank extends nearly ½ mile to the eastward, parallel to and at a distance of ¼ mile from the south coast of the island. Fifteen hundred yards northward of Dog Isle is Little Green Island, and 1 mile northeastward of the northeast point of the isle is a sandy point of Flinders Island, from which the coast trends northeastward for 3 miles to the southeastern extremity of Flinders Island, thence in a northerly direction to the mouth of a lagoon extending above 2 miles to the northwestward, and nearly 2 miles to the southwestward from its mouth. From the northeast point of Dog Isle shoal water extends almost to the nearest point of Flinders Island.

In the depth of the bay north of Dog Isles there is the small Samphire River. Midway between Dog and Vansittart Islands is a rocky islet, with shoal water extending more than ¼ mile to the northward; the remainder of the space between Dog, Vansittart, and

Flinders Islands is clear of danger, with from 6 to 25 fathoms of water.

The east coast of Flinders Island is low and sandy; from North Point it trends in a southeasterly direction for a distance of 18 miles, and then curves to the eastward for about 5 miles, there forming a junction with Babel Island in a low sandy spit. South of Babel Island the coast trends southward for a distance of 15 miles to the northern side of Franklin Sound.

Aspect.—Quoin Hill, situated 3¼ miles southward of North Point, rises to the height of 810 feet; but there are no conspicuous objects along this coast between the hill and the Patriarchs, three remarkable peaks, rising from the low sandy land behind the east point of the island and separated from the mountainous ranges to the westward by a low sandy plain. The northeast and highest of the Patriarchs, situated 19½ miles southeastward from North Point, is 772 feet high, and has a very sharp, conspicuous appearance when seen from the southeast.

Beagle Spit (lat. 39° 45′, long. 148° 03′).—At about 3¼ miles to the southeastward of North Point, a dangerous, sandy spit stretches out from the shore 5¼ miles in an east-northeast direction, with only 9 feet water on a patch at 1¾ miles within its extremity, where the depth is 18 feet; the east extremity of this spit lies 103°, distant 7¾ miles from North Point.

Soundings.—There are depths of 10 to 18 fathoms between the Sisters and the end of the spit, off which there are from 10 to 18 fathoms; and from 7 fathoms at 1,340 yards southeast of it there are regular depths of 12 to 13 fathoms to Babel Isles, off the east point of the island.

At 3 miles eastward of East Sister Island there are depths of 18 fathoms, thence decreasing to 12 fathoms and increasing again to 22 fathoms at 2 miles northeastward of Babel Isles.

Babel Isles were so named by Capt. Flinders from the discordant and various notes of the innumerable birds on them. The principal isle, situated 22½ miles southeastward from North Point, Flinders Island, is about 1½ miles in extent east and west, and the same distance north and south. The summit of this isle is a flat-topped wooded peak, 656 feet high; and near its western end there is a remarkable pyramidal hill 446 feet high. Two islets lie close to the eastern side of the principal isle. The northern and larger islet is 105 feet high and the southern islet 55 feet high.

Dangers.—The bay southward of Babel Isles is free from dangers, in that to the northward are several small rocks, but they all show above water. The outermost rock lies 1½ miles northwestward from the northwestern point of the principal Babel isle and is 4 feet high. Another rock, awash at high water springs, lies ½ mile northwestward from that point. Three miles westward of the principal Babel isle and about 1 mile from the beach are two small groups of rocks.

Anchorage.—With westerly winds there is anchorage on a sandy bottom, either northward or southward of the sandy spit connecting the principal Babel isle with Flinders Island. The depths are regular, decreasing gradually to 3 fathoms close to the beach. With winds eastward of north or south, no vessel should anchor in this neighborhood.

Tides.—It is high water, full and change, at the Babel Isles at 10 h. 05 m.; springs rise 7 feet.

Tidal currents.—North of Babel Isles the flood sets to the northward and ebb to the southward, parallel to the coast, and generally with regularity, especially near the shore. South of Babel Isles the tidal currents are weak and irregular.

Minnie Carmichael Shoal (lat. 40° 01′, long. 148° 33′).—This shoal, off the east coast of Flinders Island, has an estimated depth of 3½ fathoms and lies 11 miles 114° from Babel Island summit.

Close southward of this position no bottom could be obtained at 21 fathoms.

This shoal was searched for by H. M. surveying vessel *Flying Fish* in 1887, under favorable circumstances with regard to weather, but no indication of shallow water could be found in its assigned position.

Rock.—A rock, reported by the master of the New South Wales pilot steam vessel *Captain Cook*, also by the master of the bark *Woollohra,* and on which the bark *Lawrence* was wrecked in 1869, is situated 7°, distant 16 to 18 miles from Cape Barren; it is charted about 8 miles southward of the Minnie Carmichael. This rock was searched for unsuccessfully by H. M. surveying vessel *Dart* in 1888, but a thorough examination could not be made of the locality. It is marked E. D. (existence doubtful) on the chart.

Caution.—Seeing the doubtful nature of the above reported dangers, as also that several vessels are reported to have been wrecked on detached dangers off the east coast of Flinders Island, when seeking shelter from westerly gales, mariners are cautioned accordingly.

Franklin Sound, 4 miles wide, between Flinders and Barren Islands, is fronted to the westward by the Chappell Islands; the only navigable channel into this sound from the westward is between Flinders Island and the Oyster Rocks.

Depths.—A bar with 3½ to 5 fathoms extends across its entrance; the summit of Vansittart Island in line with the south point of Flinders Island, bearing 89°, leads over the bar in 4½ fathoms water.

A quarter of a mile southward of this range on the bar there is 14 feet water, at ½ mile south of it, 10 feet; and on the north side at ¼ mile distance, 12 feet. After crossing the bar the only danger to be avoided is a sand bank with patches 8 and 11 feet, parallel to and ½ mile northward of East Anderson Isle.

At this part of the sound the channel is divided into three arms, the northern running northeastward as far as Little Dog Isle and Badger Corner; the middle, eastward to the south end of Dog Isle, narrowing there to a bar of 3½ fathoms, which separates it from the main channel of the sound; and the southern or main channel which turns to the southeastward round the rocks off East Anderson Isle toward Barren Island, thence to the northeastward, maintaining a broad channel up to the passage between Dog and Vansittart Islands.

Anchorage.—South of East Anderson Isle a branch of the main arm extends westward, and good anchorage may be obtained in 4½ or 5 fathoms water, near the south side of the isle.

The southwest entrance to Franklin Sound is blocked by numerous rocks and banks, and can not be navigated without considerable risk.

Oyster Rocks (lat. 40° 17′, long. 148° 04′) are two rocks ¼ mile apart, northeast and southwest, from each other; the northern rock, 35 feet high, lies west-northwestward 2 miles from the summit of West Anderson Isle, and its northern side is bold; from these rocks a sand bank, dry in some parts, extends 1½ miles in a westerly and southerly direction; thence easterly to the west coast of West Anderson Isle.

West Anderson Isle, 211 feet high, lies 1.4 miles southward from the south point of Flinders Island; this isle was originally named Woody Island, but every particle of timber has been cut down or burnt, nearly its only vegetation being a coarse grass which supports a few cattle and sheep. The island is about 1 mile long north-northeast and south-southwest and 1,340 yards broad. The south side of West Anderson Isle is foul to a distance of ¼ mile; thence a narrow shoal, with from 2 to 18 feet water, extends westward to a distance of 3 miles.

Two hundred yards north of West Anderson Isle is an islet the north side of which is bold; this islet is connected with West Anderson Isle at low water by a sand bank having on it a smaller islet.

East Anderson Isle, known locally as Tin Kettle, is 1½ miles long in an east and west direction, and from ¼ to ½ mile broad. Off its east point four rocks, only a few feet above high water, extends ¼ mile in a northerly direction, and 300 yards northeastward of the outer rock is a rock which dries 2 feet at low water. About these rocks and extending 1,500 yards to the eastward of the east point is a shoal bank.

Sand bank.—At 1¼ miles northeastward of the eastern point is the western extremity of a sand bank, with 15 feet water; this bank extends eastward toward the south end of Dog Isle and has from 2 to 18 feet water on it.

A flat, in some places dry, connects East and West Anderson Isles, which are distant from each other 1,500 yards.

Long Island is situated at the southwest entrance of Franklin Sound and 4¾ miles southeastward from Badger Island; it is 2½ miles long, in a northeast and southwest direction, from ¼ mile to 1,500 yards broad, and is bordered with rocks; on its northern part, 165 feet high, there is a conspicuous granite bowlder, and near its western extremity, 98 feet high, there is another granite bowlder. It is connected with Barren Island at low water.

Doughboy Island (lat. 40° 20', long. 148° 03') is small, and lies 1¼ miles northeastward from Long Island; the depths between this island and Ned Point are very irregular.

Barren Island, the second in size of the Furneaux Group, contains about 110,000 acres, and between Cape Sir John, its west point, and Cape Barren, its east point, is 22 miles in length; it is 12 miles broad between its north and south points. The island is high, rocky, and irregular with some rounded hills near its northwest coast. There is also a remarkable peak on the southeastern part of the island.

Northwest coast of Barren Island.—Ned Point, on the south side of Franklin Sound, projects nearly ½ mile from the coast line, having to the eastward a small bay named Munro Bay, used by small vessels. Several rocks extend from the point in a north-north-westerly direction, and at a distance of ½ mile there is a rocky islet. The coast from Ned Point trends southwesterly 4 miles to Franklin village, the population of which is about 250 people. Their principal occupation is salting the mutton bird and procuring the oil.

North coast of Barren Island.—From Ned Point the coast trends in an easterly direction 5 miles to Lee River, thence in a northerly direction 2 miles to Apple Orchard Point; from which three small islets extend ½ mile in a northwesterly direction, affording good shelter to the eastward of them, but the tidal currents run very strong in different directions, and an anchor invariably comes up foul. Eastward of Ned Point for 3 miles, as well as northeastward along the south coast of East Anderson Island, there is deep water, but at 2 miles east-northeastward from Ned Point is the west tail of a large flat which extends the whole distance to Apple Orchard Point, and inshore to Lee River.

Mount Munro, 2,348 feet high, situated 2¼ miles southeastward of Ned Point, has a rounded summit densely timbered; between it and the coast are ranges, sterile in appearance, composed chiefly of granite.

Puncheon Point.—From Apple Orchard Point, the coast trends eastward 1½ miles to Dover Point; thence 1¼ miles to Dover River, and northwestward 4 miles to Puncheon Point, the north point of Barren Island; several islands and rocks surrounded with sandy flats lie westward and southwestward of this point over a space of more than 1 mile.

Vansittart Island, locally known as Gun-carriage, is 2¼ miles long, north-northwest and south-southeast, and 1¾ miles broad at its southern end, and almost connected with Puncheon Point at low water; it rises at its center to a broad summit 552 feet high, from which a spur extends to its southwest point. On the west side there is a sandy bay, where the owner of the island resides.

To the southward of a line from Apple Orchard Point to the southwest point of Vansittart Island there are numerous shoals dry at low water.

The eastern entrance to Franklin Sound lies between sand banks extending from the southeast point of Flinders Island and others extending from Vansittart Island; these banks have been named Vansittart Shoals. The channels between them are said to shift with every strong wind.

Depth.—The best and probably the most permanent channel was found to have a depth of 4 fathoms. The shifting channel along the coast of Flinders Island is very narrow. Northerly and northeasterly winds are said to have the greatest effect in shifting the Vansittart Shoals. The sea breaks heavily in easterly weather.

Tidal currents.—At the eastern entrance of Franklin Sound the flood currents meet, one coming from the north-northeast, and the other from southeast. The flood current sets to the westward through Franklin Sound, and from thence about west-northwest on the north side, and west-southwest on the south side of Chappell Islands; and the ebb in the contrary direction. In the north channel the currents run 2 to 2½ knots.

West coast of Barren Island.—The west coast is rock bound and trends 2 miles in a southerly direction from a point 1 mile to the southward of the western extremity of Long Island to Cape Sir John. This coast is foul to the distance in places of about ½ mile from it. At 1 mile from the coast and 2 miles northwestward from Cape Sir John is a rock 7 feet high, from which foul ground extends nearly ½ mile in northwesterly and southwesterly directions.

Cape Sir John (lat. 40° 25′, long. 148° 00′), the southwest extreme of Barren Island, has three off-lying rocks from 6 to 8 feet above water, situated respectively southwestward 1,500 yards, south-eastward 1,500 yards, and eastward ½ mile from the cape. One mile to the northward of Cape Sir John is a conspicuous round-topped hill 531 feet high.

The coast from Cape Sir John recedes to the northeastward, forming Thunder and Lightning Bay, nearly a mile deep and 1¼ miles wide. An islet 58 feet high, nearly connected with the shore, lies ¼ mile southeastward of the east entrance point of Thunder and Lightning Bay. East of this islet is another small bay, the south point of which lies 1,500 yards eastward of the islet, and has numerous bowlders lying off it to a distance of ¼ mile; thence the coast trends southwesterly ½ mile and easterly 2 miles to a point 600 yards to the southward of which lies Malms Rock, dry at low water. The latter part has numerous rocks off it to a distance of ¼ mile.

Wombat Point (lat. 40° 27′, long. 148° 08′), locally known as Rocky Head, lies 2½ miles eastward of the point north of Malms Rock; near the extremity of Wombat Point is a small granite island 74 feet high, 300 yards eastward of which there is a rock uncovering at one-third ebb, and to the southwestward at 1,500 yards is a sand bank with 2½ fathoms water.

Sloping Point lies 4 miles eastward of Wombat Point; between these points is a shoal bay having near its center an island about 20 feet high, named Battery Island. A rock awash at low water lies 200 yards southward of Sloping Point, and another above water lies nearer the shore.

Kent Bay extends from Sloping Point 5½ miles eastward to Passage Point, the south point of Barren Island, and is 3 miles deep; it is encumbered with shoals, and the only anchorage is in 11 fathoms water, 600 yards from the shore, 1¼ miles east-northeastward from Sloping Point. Abreast this anchorage is the most convenient place for watering.

From Sloping Point the northwestern shore trends northeast 3 miles to a small point, at 1,500 yards from which a projection of the northern shore divides the head of the bay into two bights.

Shoal.—A narrow shoal upward of 2 miles long in an east and west direction, with from 3 to 18 feet water on it, lies across the mouth of the bay, its west end eastward about 1¾ miles from Sloping Point. There is a clear channel ½ mile wide between the south edge of this shoal and the 3-fathom edge of the bank to the southward.

Sloop Rock, about 2¼ miles northeastward of Sloping Point, is situated on a reef of rocks. It is 12 feet high and lies about 800 yards from the west shore of Kent Bay.

Anchorage.—Small vessels may anchor between Sloop Rock and the shore in 3 to 6 fathoms. Several vessels may here lie at anchor in 4 or 5 fathoms, sheltered from all winds. From this anchorage the depths decrease to 1 fathom at 1,500 yards northward of Sloop Rock; and between the rock and the head of the bay there are irregular depths of 4 to 2 fathoms.

The eastern bight of Kent Bay is filled by a shallow flat, which extends in patches along the eastern shore, nearly to Passage Point.

East coast of Barren Island.—From Puncheon Point the coast trends southeastward 7½ miles, with several rocky points and intermediate sandy bays, to Harley Point, off which several rocks above and below water extend southeastward nearly ½ mile, thence at a distance of nearly 1 mile southeastward is Flat Rock 3 feet high, with a few sunken rocks extending to the northwestward. Harley Point is nearly separated from the land behind by a lagoon. There are lagoons behind the beach, at intervals for 3½ miles to the southeastward of Puncheon Point, and at 2 miles to the northwestward of Harley Point is an inlet barred across at its entrance.

Cape Barren (lat. 40° 26′, long. 148° 29′), the east point of Barren Island, southeastward 3½ miles from Harley Point, is a rounded rocky point having numerous hillocks over it. At nearly 1,500 yards eastward of the cape lies a rocky islet 40 feet high, named Gull Islet; and at 1 mile east is Gull Rock 12 feet high, with numerous other rocks above and below water extending nearly ½ mile in a northerly direction.

Midway between Cape Barren and Gull Islet is a sunken rocky patch with 14 feet over it. There are depths of 17 fathoms water at ½ mile to the eastward of Gull Islet; but there are strong tide ripples near the reef. At 1½ miles southeastward from Cape Barren there is a patch with 6½ to 8 fathoms water on it.

Between Harley Point and Cape Barren the coast forms a double bay.

Cone Point, so named from two conspicuous cone-shaped granite rocks, lies 5 miles southwestward of Cape Barren, with two bays between, having sandy beaches in their depths. The point dividing the bays has two conspicuous sand patches over it.

The northeastern bay extends 3 miles from Cape Barren, and is 1 mile deep. The southwestern bay is 1,300 yards deep, with a lagoon extending 1¼ miles along the back of the beach, at about ½ mile from it.

Soundings.—From 7 fathoms at 2 miles eastward of Puncheon Point the depths increase to 14 fathoms at 1½ miles off Harley Point, and 17 fathoms at 1½ miles off Cape Barren; thence to about 3 miles eastward of Cone Point there are from 13 to 17 fathoms.

Passage Point (lat. 40° 29′, long. 148° 21′), the southern extremity of Barren Island, lies 2 miles westward of Cone Point, having a bay with a steep sandy beach lying between; the bay is nearly 1 mile deep, with 7 to 9 fathoms water in it. Behind the sandy beach on the west side of this bay is a lagoon, the water in which is of a red color and a little brackish. On the west side of Passage Point there are several smaller ponds which contain good water.

Mount Kerford, 1,644 feet high, is the highest peak of the range which rises from the various points on the southeast coast of Barren Island. The whole range is conspicuous from its barren and whitish appearance.

Armstrong Channel, between Barren and Clarke Islands, is seldom used, owing to the numerous banks and strong tidal currents in it. The passage of Armstrong Channel between Barren Island and Preservation Isle is neither so wide nor so straight as that along the coast of Clarke Island.

At the west entrance the passage between Barren Island and Preservation Isle is about ¼ mile wide.

The passage between Preservation Isle and Clarke Island is nearly 2 miles wide; the only dangers in the entrance of this passage are Eclipse Rock, off the west point of Clarke Island, a rock awash ¼ mile off the same point, and the 2-fathom rock to the southeastward of Rum Islet; the better side of the channel is along the coast of Clarke Island; care, however, is necessary to give the coast northward of Clarke Hill a berth of ½ mile, and again to approach the next point, the western point of Kangaroo Bay, to avoid the Middle Bank, with 3 feet upon it, which lies in mid-channel.

From the western point of Kangaroo Bay, Armstrong Channel trends in the direction of Sloping Point, between which and Seal Point there is a depth of 41 fathoms in mid-channel.

From ¼ mile northeastward of Seal Rocks, on the northwest side of Seal Point, several sand banks, dry at low water, and shoals extend to Forsyth Isle. Armstrong Channel continues northward of these shoals, a narrow sand bank occupying a central position, and terminates in a passage 800 yards wide between Forsyth and Passage Isles. A smaller passage exists between Passage Isle and Passage Point, but it should only be used in case of necessity, as the tidal currents run at the rate of 5 or 6 knots, and its center is occupied by a half-tide rock. It should be entered in a sailing craft with the flood current, keeping near Passage Point.

Although there are many sand banks in, as well as on each side of, the wider parts of Armstrong channel, a passage of sufficient width and depth is swept out by the tidal currents for vessels to go through. The bottom is either rocky or sandy; rocky in the deep and narrow parts, and sandy in the bights and shoaler places. A careful study of the chart, with a good look out and attention to the lead, are the safest and best guides for this channel.

Depth.—A least depth of 5¼ fathoms can apparently be carried through.

Night Islet (lat. 40° 29', long. 148° 01'), 46 feet high, lies 4 miles south-southeastward from Cape Sir John. Dry and sunken rocks, nearly 100 yards in extent, lie ½ mile from it in a northeasterly direction; and Little Night Islet lies ¼ mile in a southerly direction.

Preservation Isle, 84 feet high, in the west entrance of Armstrong Channel, is of granite formation, 1¾ miles long in a northwest and southeast direction, and ½ mile across its broadest part. Many small islets and rocks extend ¼ mile from its northwestern extremity, and from its east side a shoal spit, partially dry at low-water spring tides, extend to a distance of 1½ miles.

Tides.—It is high water, full and change, at Preservation Isle at 10 h. 36 m.; springs rise from 5 to 7 feet.

Rum Islet, close to the south point of Preservation Isle, is about ¼ mile in extent; it has a reef projecting a short distance from its southern end, and is joined to Preservation Isle by a reef of dry and covered rocks, which, together with the islet, protects Hamilton Road from the southwestward. A rock with 2 fathoms water on it, 98° 800 yards from the south point of Rum Islet, lies in the track of vessels bound to Hamilton Road.

Hamilton Road, eastward of the southeast point of Preservation Isle, affords anchorage in 4 fathoms water, 600 yards from the point; in this vicinity there are patches of 3 fathoms. After a continuance of heavy gales from the westward a long swell rolls round the south point of Rum Islet, and the swell does not gradually increase but sets in suddenly and may compel a vessel to get underway.

Middle Bank is a long, narrow shoal, the east end of which, in 3 fathoms, lies ½ mile southwestward from Sloping Point; whence it extends west-southwestward 2 miles, with as little as 3 feet on the shoalest part.

Kangaroo Bay.—Westward of Seal Point, the northeastern extremity of Clarke Island, lies Kangaroo Bay, 1 mile deep, having a narrow channel leading into it along the western shore, and anchorage may be obtained in the center of the bay in 3 or 3½ fathoms; the remaining portion of the bay is shoal.

Seal Rocks are a cluster of dry (8 feet high) and sunken rocks on a shoal projecting to the northwestward about ⅓ mile from Seal Point, the northeast point of Clarke Island, leaving a channel 1,340 yards wide between them and Sloping Point.

Forsyth Isle, 3 miles southeastward of Seal Point, is 1¼ miles long, north and south, and ⅓ mile broad. The bank which forms the southern side of Armstrong Channel extends 1¼ miles northward of Forsyth Isle, thence in a westerly direction toward Sloping Point; and numerous banks and channels exist between Forsyth Isle and Clarke Island. A sand bank extends 600 yards from the east side of Forsyth Isle, and a shoal of 3½ fathoms extends nearly 1 mile to the eastward of the southeast point, and almost unites with another shoal extending southwest from the south point of Passage Isle. At 1 mile southwestward from the southwest point of Forsyth Isle lies a rock awash at half tide. In the vicinity of the shoal water the sea breaks heavily.

Passage Isle (lat. 40° 30′, long. 148° 21′), 177 feet high, lies 600 yards from Passage Point; it is 1¾ miles long in a northwest and southeast direction and about ½ mile broad. At 200 yards from its northeast point, and nearly in mid-channel, is a rock dry at low water, and 400 yards northward of its north point are several rocks above water.

A few rocks above water extend 200 yards off the south point, and rocks having less than 6 feet water extend ¼ mile southeastward from the southeastern extremity. Off the western side a sand bank extends 800 yards from the shore, except off the northwest point, which is steep-to.

Water.—Good fresh water may be collected at certain seasons in small pools near the southeastern end of Preservation Isle; but that which drains from the rocks appears to possess some pernicious qualities. Small pools or runs of water are to be found almost everywhere under the high parts of Barren Island, and it is probable there may be some on Clarke Island.

Birds.—Preservation Isle and the adjacent rocky islets are visited by numerous sea birds, including the Cape Barren goose, only found in Australia, a few black swans, and great numbers of the sooty petrel, which latter burrow in the ground like rabbits, and when skinned and smoked are passable food.

Clarke Island, the southern of the Furneaux Group, 8 miles long in a northeast and southwest direction and 6 miles across its southwestern part, rises near its northwest coast to a peaked hill 676 feet high; and to a broad-topped hill 525 feet high near its southwestern extremity.

Eclipse Rock.—At ¼ mile westward from the west point of Clarke Island is the Eclipse rock with 11 feet over it, about which there is a confused sea.

Clearing range.—Sloping Point open north of Clarke Island bearing 63° leads to the northward, and a conspicuous bowlder on a hill 783 feet high, near the west end of Barren Island, open west of Rum Islet 347°, leads to the westward of Eclipse rock.

Southwest coast.—The west point of Clarke Island is rocky, and to the southward of it is a deep bay, which was formerly much used by coasting vessels, but can not be recommended.

A few scattered rocks lie off Lookout Head (the southwest point of Clarke Island) to the southeastward and eastward; the outer is distant 600 yards from the shore, and is 2 feet above high water. Between Lookout Head and the south point of Clarke Island, which lies 2¾ miles east-southeastward from the head, the southwest coast of the island forms an exposed bay 1 mile deep, with reefs extending about ¼ mile from the northwest and eastern shores of the bay.

The south point of Clarke Island is inclosed by a reef of sunken rocks, between which and Moriarty Point, the southeast point of the island, is a small exposed bight.

Moriarty Point lies 1,500 yards east-northeastward from the south point, thence the east coast of Clarke Island trends in a northerly direction for 7 miles to Seal Point, its northeastern extremity.

Moriarty Bay (lat. 40° 35′, long. 148° 12′), lying to the northward of Moriarty Point, has bad holding ground, on a broken rocky bottom; there is good landing in a corner of the bay, but the whole of this vicinity should be avoided.

Lookout Rock, 60 feet high, lies 1,500 yards northwestward from Lookout Head, between which and Lookout Rock is the Napper Rock, awash at high water.

A rock lies about ¼ mile northward from Lookout Rock.

Moriarty Rocks.—At 3¾ miles eastward form Moriarty Point lies the outer of two rocks, named Moriarty Rocks from their proximity to the Moriarty Bank; the outer or southeast rock is 20 feet high, and the northwest rock is 25 feet high. At 1¾ miles northwestward from Moriarty Rocks are two rocks 20 and 15 feet high, with several smaller rocks above and below water near them.

Moriarty Banks.—The western of these banks extends in a west-southwesterly direction nearly 2½ miles from the highest Moriarty Rock; the least water upon it is 1 foot, but the general depth is about 9 feet. The eastern tail of the east bank lies east-southeastward 2¼ miles from the same rock; the least water found on it was 16 feet. To the northwestward of the east bank is another bank, the least water upon which is 15 feet. A fourth bank extends in a northeast and southwest direction from the two rocks which lie northwestward 1¾ miles from the Moriarty Rocks; this bank, which has a sunken rock at its northeasterly termination, is nearly 2½ miles long. Nearly midway between it and the shore is a half-tide rock.

Clearing marks.—Mount William, 714 feet high, near the northeast coast of Tasmania, bearing 203° leads eastward of Moriarty Banks; and to pass to the southward of them, the south point of Clarke Island must not be brought westward of 290°.

Caution.—The whole of the space included between the east coast of Clark Island, the Moriarty Rocks, and Passage and Forsyth Isles is either foul ground, or the strong tidal currents cause such a race and heavy break as to make the place dangerous.

CHAPTER V.

Tasmania.—The north coast of Tasmania forms the south side of Bass Strait. It extends for about 165 miles between Eddystone Point and Cape Grim, its northeastern and northwestern points. Near the bottom of the bight, which it forms by curving to the southward, are Port Dalrymple and Devonport, the former being the mouth of the River Tamar. The whole of this coast lies generally in very smooth water, the prevailing winds being off the land, and the long southwesterly swell outside being interrupted by the islands at the western entrance of the strait. Its navigation is represented to be free from dangers to within 1 mile of the coast and of the islands which lie off it, except for Hebe Reef in the neighborhood of Port Dalrymple.

The northeastern extremity of Tasmania is low with a coast range of sandhills; from this level part rise Mounts Cameron and William, the loftiest of a group of peaks cresting a ridge; the latter is used as a guide for vessels working through Banks Strait.

Eddystone Point (lat. 41° 00′, long. 148° 21′), 81 feet high, the northeastern extremity of Tasmania, forms the north point of the Bay of Fires. At 400 yards southeastward from the point is the Eddystone rock, 19 feet high, the southern of the Victoria rocks, having between it and the land two half-tide rocks. The eastern of the Victoria rocks is named Norgate, and the northern, Greyhound rock; the latter lies nearly two-thirds of a mile 80° from Eddystone Point, and has 5 feet water over it; Norgate rock lies 100° about 1 mile from the point, with 14 feet water on it. A small red light shows over the Victoria rocks. See below.

On the north side of Eddystone Point there is good landing.

Lights.—An alternating group flashing white and red light, 139 feet above water, visible 18 miles, is shown from a white, circular, granite lighthouse, 400 yards northwestward of Eddystone Point.

There is also a small fixed red subsidiary light shown from a red circular iron structure, 20 yards eastward of the lighthouse, covering the Victoria rocks.

Lloyd's signal station.—There is a Lloyd's signal station at the Eddystone Point Lighthouse; communication can be made by the

International code of signals. It is a telegraph station, is connected
with the telephonic system of the colony, and is easily accessible by
boat.

Tides and tidal currents.—It is high water, full and change, at
Eddystone Point at 8 h. 10 m.; springs rise 7 feet. The flood
current sets to the northward, the ebb to the southward, but neither
current has any strength to the southward of Banks Strait.

The coast from Eddystone Point trends northwestward 10½ miles
to Cape Naturaliste, with several small points and bays between;
off the points are numerous granite bowlders, some of which are 20
feet high.

Cape Naturaliste (lat. 40° 51′, long. 148° 14′), 71 feet high, is
faced by sand cliffs, but the coast about it is lower than that of Eddystone Point.

Mounts Cameron, William, and Pearson are the only remarkable hills near this vicinity. The highest peak of Mount Cameron,
1,825 feet, lies 18¼ miles westward from Eddystone Point. It is one
of several peaks, and the summit is of a haycock form; the ridge,
of which Mount Cameron is the highest part, is over 3 miles in
length in a northeast and southwest direction. Mount William, 9
miles northwestward from Eddystone Point, rises gradually to a
rounded summit 714 feet high. Mount Pearson, 5 miles westward
from Eddystone Point, is a broad-topped hill 623 feet high.

George Rocks (lat. 40° 55′, long. 148° 20′), a group of granite
bowlders, the highest of which is 64 feet high, lie 4 miles north-
northwestward from Eddystone Point; these rocks occupy a space
of 1½ miles and, with the exception of the largest two, are quite
barren; scattered through the group are a few half-tide rocks, but
most of the rocks are from 10 to 30 feet high.

At 2¼ miles 336° from Eddystone Point and 1 mile from the coast
is a rock awash at low water, and at 1,500 yards 319° is a rock
awash at high water.

In addition to the above there are several patches of less than
4 fathoms, so that the passage between George Rocks and the main
is not recommended.

Eucalyptus Rock, with 10 feet water, lies 3 miles from the shore,
about 8 miles 348° from Eddystone Point; the sea rarely breaks on
it, but the position may be known by kelp in the vicinity. Eddystone Point open of George Rocks bearing 170° leads between
Eucalyptus and Salamander Rocks.

Salamander Rock.—This danger, with 10 feet water over it, lies
3¼ miles 68° from Eucalyptus Rock. No distinct break has been
seen on this rock. The south round hill of Mount Cameron Range
open southward of Mount William bearing 246° leads to the southward, and the same hill well open northward of Mount William

242° leads to the northward of Salamander Rock. The rock lies in the direct track of vessels plying between Melbourne and Hobart.

Black Reef, 9 feet high, lies 1¼ miles northeastward from Cape Naturaliste; rocks above water extend to a distance of ¼ mile north-westward of it, and at 1,500 yards east-southeastward of the reef is a rock, above water, with sunken rocks around it.

Mussel Rock, with feet of water on it, on which the sea some-times breaks, lies 1½ miles 322° from Black Reef. Mount Pearson over Cape Naturaliste bearing 177° leads to the westward of Mussel Rock.

Soundings.—There are depths of 41 fathoms at 5 miles and 42 fathoms at 13 miles off Eddystone Point, and about 35 fathoms from the latter to the same distance off Cape Barren, the bottom being rock at about midway and sand to the northward. There is a depth of 23 fathoms at 1 mile eastward of George Rocks, and from Black Reef to 8 miles east-northeastward the depths range from 14 to 23 fathoms.

Mussel Roe Point (lat. 40° 49′, long. 148° 11′) lies 2½ miles west-northwestward from Cape Naturaliste; to the east and northeast of this point there are many sunken rocks, and others awash at low water, the most outlying of which is situated eastward nearly 1,500 yards from the point.

From Mussel Roe Point the coast trends southward for nearly ½ mile to the mouth of Mussel Roe River, which is small and only navigable for boats. From the mouth of this river the coast trends from southwest to northwest for about 6 miles, forming the bay known as Mussel Roe Bay.

Mussel Roe Bay—Anchorage.—Mussel Roe Bay affords good anchorage in 6 or 7 fathoms, sandy bottom, in the northwestern portion of the bay, with Swan Isle Lighthouse bearing about 0°. Near the anchorage is a sand bank named Cockle Bank, with 19 feet water, which bears 300° from Mussel Roe Point, distant 2 miles.

At 1½ miles southward of Swan Isle is Tree Point, the northwest point of Mussel Roe Bay; thence the coast trends in a westerly direc-tion for 2 miles to Little Mussel Roe River, the mouth of which may be entered by boats at half tide; from Little Mussel Roe River the coast trends 1,500 yards in a northwesterly direction to a point made conspicuous by two bare sand hills.

Little Mussel Roe Bay, between Tree Point and the last-men-tioned point, is used by the smaller coasting vessels on account of the shallow anchorage; it is, however, much more exposed than Mussel Roe Bay.

The coast from the point northwest of Little Mussel Roe River trends westerly 4 miles to Cape Portland; this part of the coast is irregular, and has numerous shoals extending from it. Near Cape

Portland, between it and Foster Islets, is a low islet; there are also other islets nearer the shore connected by ledges; the whole locality is very dangerous for boats on account of the confused sea.

There is a good navigable channel between Swan Isle and the coast just described, but in consequence of the strength of the tidal currents it is well to give a wide berth to the foul ground extending from Swan Isles and Foster Islets. The range for mid-channel is Mount William in line with rocky part of coast north of Mussel Roe Bay bearing 151°.

Swan Isle (lat. 40° 44′, long. 148° 07′), 109 feet high, lying 1½ miles off the northeast coast of Tasmania, is of gray granite, but sand hills covering the granite give it the appearance of being nearly all sand. At ⅓ mile westward of the west point of the island is a sunken rock. A group of rocks, some of which uncover at low-water springs, lies nearly 1¼ miles 306° from the lighthouse.

A rocky patch with 4 fathoms water over it is 1,200 yards 142° from Swan Isle Lighthouse, and another with a depth of 5 fathoms north-northwestward lies 1,600 yards 187° from that lighthouse.

Little Swan Isle lies north-northwestward nearly 1 mile from the west end of Swan Isle, and at ⅓ mile beyond it in the same direction lies Cygnet Isle. Between these isles are numerous rocks, some of which are above water; and shoal water extends 1,500 yards to the westward of the west point of Little Swan Isle.

Harry Rock, lying northward 1 mile from Cygnet Islet and 306°, nearly 3 miles from Swan Isle lighthouse, has 16 feet over it at low water.

At 1 to 1¾ miles northward of Swan Isle Lighthouse there is a race or overfall, but no shoaler water than 7 fathoms could be found.

Light.—A fixed and flashing white light, 100 feet above water, visible 15 miles, is shown from a white circular stone lighthouse on the northeast point of Swan Isle close to the water's edge.

Signal station.—There is a signal station at Swan Isle Lighthouse, and communication can be made by the International Code. It is not connected by telegraph.

Anchorage may be obtained off a small sandy bay on the southeast side of Swan Isle; the bottom is rock, or sand over rock; the holding ground is therefore bad. Sailing vessels working through Banks Strait from the eastward often get as far as Swan Isle with the flood current, and anchor during the strength of the ebb; but it is not advisable to anchor at Swan Isle if westerly gales are expected, as it is not uncommon for the wind to shift to the southeastward.

Tides and tidal currents.—It is high water, full and change, at Swan Isle at 9 h. 16 m.; springs rise 7 feet. The flood current sets to the northwestward, the ebb to the southeastward, at the rate of 3 knots at spring tides, influenced, however, by the wind.

Cape Portland (lat. 40° 44', long. 147° 56'), situated 8¼ miles westward of Swan Isle Lighthouse, is low and rocky; south-south-eastward, 1,500 yards from the cape, is a summit with a pile of stones, 119 feet above the sea; the coast from the cape trends in a south-southeasterly direction, to a point which lies 2¼ miles from the cape.

Islets.—At ⅓ mile northwestward of this point is an islet about 20 feet high, with another islet ½ mile northward of it, from which a reef projects ¼ mile to the northward. These islets are too small to afford protection to the bay immediately east of them, and, although used by small vessels it is only during a continuance of easterly winds; the best channel into the bay is between Cape Portland and the northern islet; between the southern islet and the shore there is also a channel of 2½ fathoms.

Foster Islets, 47 feet high, situated 7 miles westward from Swan Isle Lighthouse, are two islets, connected at low water. A shoal with 3 feet water extends 1¾ miles to the eastward, and a rock awash at low water lies 1 mile northward from the islets; the sea breaks heavily upon this rock, and there is a heavy tide rip in the vicinity. There are depths of 9 fathoms water at 1½ miles northeastward of the outer Foster Islet.

Banks Strait, which separates the Furneaux Group from Tasmania, may be said to extend between Goose Island to the northwest and Eddystone Point to the southeast.

In the strait with a head wind, it is generally impossible for sailing vessels to work to the westward during the ebb current, and the custom is to anchor either under Swan Isle, or in Mussel Roe Bay, the latter anchorage being preferable, as the holding ground is not good at Swan Isle anchorage. Occasionally small vessels anchor off Little Mussel Roe River, the water there being shoaler.

The survey of 1874–1877 proved the existence of many dangerous rocks, and the navigation of Banks Strait consequently requires care. The bottom consists of sand and in some parts rock; the depths in the fairway are from 15 to 30 fathoms.

Directions.—In Banks Strait the chief dangers to be avoided on the southern shore are the reef and rocks off Swan Isles and the foul ground and rocks northward of Foster Islets. It may, however, be noticed that a vessel from the southeastward can close the shore when Mount Pearson is over Cape Naturaliste bearing 177°, as she will then be westward of Mussel Rock, Black Reef, and the rocks that lie off the coast.

Working to the westward in Banks Strait.—In the summer months, when westerly gales are of short duration, it is advisable to stand toward the Tasmanian coast to take advantage of the shift of wind.

Winds.—During the survey of Banks Strait the heaviest and most frequent gales (generally from the westward) were experienced in the months of September, October, and November.

On the termination of a westerly gale the wind in the vicinity of Banks Strait sometimes shifts to the southeastward; the barometer standing a little above 29.60 inches. The wind seldom blows home with much strength, but sufficiently so, the swell rolling in simultaneously, to necessitate leaving the anchorages, which are open to the southeast. All anchorages in Banks Straits which are exposed to the southeast require great caution in their use, owing to the uncertain nature of the winds.

Tidal currents.—The flood current is the west-going current and the ebb the east-going; the currents are each of 6¼ hours' duration at springs; but during neaps the flood runs 7 hours and the ebb 5¼ hours. The interval of slack water never exceeds a quarter of an hour; the west-going current begins 30 minutes after low water at springs and 50 minutes after it at neaps; the east-going begins 40 minutes after high water at springs and 10 minutes before it at neaps.

In the narrowest part of the strait (8¼ miles wide) between Swan Isles and Clarke Island the tidal currents run at the velocity of 3 knots at springs; westerly winds accelerate the east-going current, which occasionally attains a velocity of 5 or 6 knots.

The velocity of the currents, the strongest being the east-going, at springs in the middle of Banks Strait causes, when opposed to the wind, a high topping sea, somewhat dangerous for small craft.

Ringarooma Bay (lat. 40° 44′, long. 147° 56′) extends from Cape Portland to Waterhouse Point 13 miles southwestward, and is 6½ miles deep.

From the point 2¼ miles, southward from Cape Portland, a rocky coast trends south-southwestward nearly 1 mile; the beach which forms the southeastern shore of Ringarooma Bay then curves south-southwestward 5¼ miles to Ringarooma River, the coast thence continues southwestward for 2 miles with the same sandy beach, and westward 4 miles with a rocky formation to the mouth of the Tomahawk River, whence it trends ½ mile northward to Tomahawk Point, close off which, and connected with it at low water, is Tomahawk Islet. Between this islet and Waterhouse Point the coast consists of a sandy bight and rocky points. At 1,500 yards westward of Tomahawk Point is a remarkable bare sandhill. There are depths of 14 to 9 fathoms water at 1½ miles from the southeastern shore, and 9 to 5 fathoms within a mile of the southwestern shore of the bay.

Ringarooma River is used by small vessels occupied in the export of tin ore, quantities of which are found in the vicinity of Mount Cameron, where the Ringarooma takes its rise. The mouth of the river is small, and at low water there is very little more than 1 foot on the bar, but as the rise of tide is 7 to 9 feet, there is generally sufficient water for the small vessels using the river; the principal danger is the surf during strong northerly and westerly winds, at which time vessels do not venture.

Lower Ringarooma or Boobyalla, at the mouth of the river, is the shipping place; there is regular steam communication with Launceston. Population is about 600.

Tomahawk River is not navigable, but small vessels are beached in fine weather at high water, and floated again at the next high tide. The bight about the mouth of Tomahawk River is shoal.

Both rivers have fresh water within 3 miles of their mouths, and there is a fresh water lagoon at 2 miles northeastward of Ringarooma River.

Waterhouse Point, 4½ miles north-northwestward from Tomahawk Point, is the rocky termination of a range of hills descending from Hardwick Hill, 385 feet high; the point has a reef of rocks (pinnacles of which dry at low water) projecting ¼ mile to the northward. Between Tomahawk and Waterhouse Points is a bay in the depth of which rocks awash and sunken extend to a distance of ½ mile from the shore; and at ½ mile eastward of Waterhouse Point there is a reef of rocks, some of which dry at low water.

The coast from Waterhouse Point trends southwestward 1½ miles to a small point abreast Little Waterhouse Island; this part of the coast has a sandy beach, with numerous dry and sunken rocks extending 600 yards from the shore. Off this point, on a spit running in the direction of Little Waterhouse Island, there is a rock dry at low water lying 700 yards from the shore, and at 1,200 yards from the point there is a depth of 14 feet, between which and the shoal extending from Little Waterhouse Island there is a navigable channel 500 yards wide, with 4 to 5 fathoms water.

Waterhouse Island, the north point of which lies nearly 3 miles northwestward from Waterhouse Point, is 2¼ miles long north-northeast and south-southwest, and ½ mile broad.

Waterhouse Island is 144 feet high; it has an even summit, and falls gradually at its north end; in some parts there are a few trees, but the island is nearly cleared of timber.

Off this sandy beach a spit, dry at low-water spring tides, extends in a northeasterly direction about 800 yards; the outer edge of the spit lies 400 yards from the shore, with shoal water 200 yards to the eastward of it. Along the east coast of the island northward of the sandspit a shoal flat extends to a distance of 800 yards until within ½ mile of the north point; thence there are no dangers beyond 200 yards, along the north and west coasts of the island.

Little Waterhouse Island, 38 feet high, lies about ½ mile south-east of the south point of Waterhouse Island, and is about 200 yards across; rocks above and below water, extend nearly 400 yards from it in the direction of the south end of Waterhouse Island, narrowing the principal passage to a width of 500 yards. A spit extends 400 yards in an easterly direction from the east point of the island, the tail of which has 16 feet water on it.

Dangers.—Northeastward of Little Waterhouse Island, in the direction of the channel and occupying a central position, is a sandbank with patches of rock upon it; the southwest end of this sandbank is distant ½ mile from Little Waterhouse Island, and the northeast end 1½ miles. The shoalest part of this bank has 7 feet water, and lies north-eastward 1 mile from Little Waterhouse Island. Several detached sandy knolls of 14, 16, and 18 feet, respectively, extend to the eastward, the outer of which lies 2 miles northeastward from Little Waterhouse Island.

At a distance of about 1,500 yards west-southwestward from Little Waterhouse Island is a rock dry at low water; and in the same direction (254°), distant 1.6 miles, is Barrett Rock with 11 feet water. Between the rocks and Little Waterhouse Island there are depths of 9 to 10 fathoms water.

Anchorage (lat. 40° 49′, long. 147° 38′).—The anchorage between Waterhouse Island and Waterhouse Point, on the east side of the island, is a safe and useful anchorage, affording shelter from easterly or westerly gales. Anchor as convenient in the channel, either near the southwest end in 5 fathoms, about 600 yards off the only sandy beach on the southeast side of the island, or in the northern part of the channel, over a sandy bottom, in about the same depth, near a patch of 3 fathoms, with Croppies Point over the west end of Little Waterhouse Island, bearing 224°.

The northern entrance to the anchorage is encumbered by shoals of 3 to 3½ fathoms water, and by the bank extending ½ mile eastward from Waterhouse Island.

Tides and tidal currents.—It is high water, full and change, at Waterhouse Island at 10h. 16m.; springs rise 8 feet. The flood is the west-going current, and its rate about 2 knots at the anchorage.

Croppies Point, situated 4 miles southwest from Waterhouse Point, is free from off-lying dangers to the eastward as far as the point abreast Little Waterhouse Island; the coast line is rocky, and the few sunken rocks which exist are not more than 200 yards from it.

Papanui Rock.—A rocky patch, about 800 yards in extent, is situated 325° true, 3.8 miles, from the northern extreme of Waterhouse Island. It has a depth of 11 feet over its shoalest part, with 4 to 5 fathoms close around it.

South Croppies Point lies nearly 1 mile from Croppies Point, with two small points and exposed sandy bays between.

Croppie Rock, with 12 feet water over it, lies ½ mile 307° from South Croppies Point.

At 2¼ miles to the southwest of South Croppies Point is a rock 9 feet above high water, and at 2¾ miles in the same direction a rock 2 feet above high water; these rocks are about 1 mile off shore.

Anderson Bay.—From South Croppies Point the coast, a sandy beach backed by sandhills which attain an elevation of 140 feet, trends in a southwest direction for about 11½ miles, to the head of· Anderson Bay, where the rivers Great Forester and Brid discharge themselves by one mouth into the sea. The coast then trends in a northwest direction for about 5 miles to East Sandy Cape, being of a rocky and broken nature. On the east side of the bay the soundings are regular, with a depth of 5 fathoms, generally about ½ mile from the beach; on the west side the depths are variable.

Forester Rivers.—The mouth of Great Forester River is blocked by a sandy bar, which dries at low water springs. The outermost rocks in this vicinity are always above water.

Nearly 1 mile off the mouth of the Little Forester River is Forester Rock, 2 feet above high water.

Rock.—A rock, of small extent, with 6 feet of water on it, lies 2.8 miles 65° from Forester Rock—that is to say, nearly in the center of the bay.

Bridport (lat. 41° 01′, long. 147° 24′).—The town of Bridport, on the left bank of the Great Forester River, ½ mile from the mouth, has a post office and telegraph communication; a small steam vessel calls at Bridport occasionally from Port Dalrymple, and there is a railway from Launceston to Scottsdale, about 13 miles to the southward.

Bridport is a favorite resort for fishing, shooting, and sea bathing. Fruit grows abundantly, and there is good agricultural and grazing country. The population is less than 100.

Anchorage.—Anderson Bay generally affords shelter only with southerly winds, but fair shelter has been obtained in westerly gales, ½ mile southeast of East Sandy Cape, in 5 fathoms, sand, and good holding ground. In approaching this anchorage care must be taken to avoid a rocky ledge projecting from a point situated about 1 mile southeastward of East Sandy Cape, and the rock near the center of the bay, mentioned above.

East Sandy Cape (lat. 40° 57′, long. 147° 20′) is formed by a long low ridge extending in a northerly direction from the high ground inland; the cape terminates in a conspicuous sand hill, 125 feet high, which shows a bare face to seaward except in a westerly direction.

Rocks.—A ledge of rocks, which covers and uncovers, with no outlying dangers, stretches 400 yards northward from the cape.

West Sandy Cape, about 3 miles westward of East Sandy Cape, is formed by a series of low sand hills, fronted by shelving rocks.

Dangers.—A rocky path which breaks heavily in bad weather, and has a depth of less than 6 feet on it at low water, lies ¾ of a mile northward of West Sandy Cape. two miles westward of that cape Flat Rocks, detached ledges which mostly cover at about ¾ flood, extend 1 mile from the shore. The soundings off this coast are irregular, and it should not be approached nearer than 1½ miles.

Ninth Island, situated 6½ miles to the northward of West Sandy Cape, is 108 feet high, flat-topped, devoid of trees, and nearly covered with grass. The depths round the island are irregular, especially northward and eastward of it, and as the examination of that locality was only partial and less water may exist than shown on the chart, the northeast side of the island should be given a berth of at least 1 mile.

Noland Bay.—From West Sandy Cape the coast trends southwestward for about 7 miles, thence westward about 7 miles to Stony Head. Noland Bay is the eastern part of this indentation; its shore is sandy, with sand hills 30 to 70 feet high, and fronted by ledges of rocks which cover and uncover. Nearer Stony Head the land becomes more elevated, and is faced by cliffs 40 to 100 feet in height. The soundings in Noland Bay are tolerably regular, and there are no outlying dangers known.

Great Piper River, which discharges itself into the southwest part of Noland Bay, is blocked at its mouth by a bar of sand which dries at low water springs.

Stony Head is a conspicuous headland, 295 feet high, with cliffs and broken ground, 120 feet high, seaward of its summit. This headland is the extremity of a range of hills sloping down from the inland mountains, the most conspicuous of which is Round Hill, 770 feet high, situated about 2½ miles southward of the head.

Tenth Island, situated 2¾ miles northwestward from Stony Head, is a rock 30 feet high, which may be passed on any side at a distance of ¼ mile.

Tides.—It is high water, full and change, in Tam O'Shanter Bay 2½ miles east of Stony Head, at 11 h. 0 m.; springs rise about 10 feet.

Tidal currents.—The flood is the west-going current and sets parallel to the shore, the ebb is the east-going current. In the channel between Tenth Island and Stony Head, and near salient points such as the Sandy Capes, the currents attain a rate of about one knot; as the distance from the shore increases, the tidal currents become weaker and much affected by prevailing winds.

Five Mile Bluff is about ¾ mile west-southwestward from Stony Head, the coast between forming a bay. One mile westward of Stony Head there is a slight projection fronted by shelving rocks and shallow water, which should be given a berth of 1 mile.

Two Miles Reef.—Nearly midway between Five Mile Bluff and Low Head, a reef extends from the shore in a northwesterly direction for 1,500 yards, and shoal water extends for ½ mile further in the same direction. This reef covers at half tide and is not beaconed.

Vessels are liable to be set by the ebb current into the bay between the reef and Low Head.

Between Five Mile Bluff and Low Head a vessel should keep an offing of at least 2 miles, as inside this limit the sea breaks heavily with on-shore gales.

River Tamar approach—Low Head (lat. 41° 03′, long. 146° 49′).—From Five-mile Bluff the coast trends southwestward 4½ miles to a bight formed on the southwest side by a narrow promontory extending northwestward 1 mile to Low Head, the eastern entrance point of Port Dalrymple and Tamar River.

Lights.—A group flashing white light, 142 feet above water, visible 18 miles, is shown from a white circular lighthouse on Low Head, which stands 400 yards within its extremity.

An additional red light is shown at Low Head Lighthouse 20 feet below the main light which shows a fixed red sector of light over Hebe Reef to as far south as West Head.

Lloyd's signal station.—There is a Lloyd's signal station at Low Head, and communication can be made by the International Code. It is connected by telegraph.

Tidal signals.—The following day signals are shown from the flagstaff of Low Head Lighthouse:

    Blue flag, west yardarm: First quarter flood.

    Blue flag, masthead: Second quarter flood.

    Red flag, west yardarm: Third quarter flood.

    Red flag, masthead: Last quarter flood.

The ebb tide signals are the same, with the addition of a black ball under the flag.

Buoy.—A bell buoy, painted red and brown, with black staff and ball, is moored in 8½ fathoms, 353°, about 2 miles from Low Head Lighthouse. Vessels anchoring outside the River Tamar should not do so to the eastward of a line joining the buoy and Low Head Lighthouse.

Range lights.—Two range lights are exhibited from two towers 30 feet high, painted white, on Shea Oak Point, Port Dalrymple, east side of Tamar River entrance. The lights are 400 yards apart and kept in line bearing 128° lead through the Middle Channel,

Tamar River entrance. The upper light is 55 feet above high water. The lower light is 38 feet above high water. For details see Light List.

A light is shown from a small iron tower at the pilot station. A light is shown from a small iron tower about 400 yards 349° from the preceding light. These lights in line astern lead up Sea Reach.

Hebe Reef (lat. 41° 03′, long. 146° 45′), the outermost danger off the entrance of Port Dalrymple, is about $\frac{1}{2}$ mile in extent, mostly in an east and west direction. The small portion of its center, which covers at half tide, lies westward distant about 2.3 miles below Low Head Lighthouse. A bank with 4 to 4½ fathoms water on it extends $\frac{1}{4}$ mile eastward from the reef; but there are 6 and 7 fathoms at less than $\frac{1}{4}$ mile north and south of the reef. In fine weather the sea runs over the reef without breaking.

Buoy.—A conical buoy painted with horizontal red and white stripes, surmounted by staff and ball, is moored in 11 fathoms water 200 yards northward of the reef.

Port Dalrymple and Tamar River—Depths. —Vessels of all draughts can proceed from sea to Georgetown, but it would be advisable to take a pilot on a first visit. -Vessels of 21 feet draught can reach Launceston at high water 38 miles from the entrance, if in charge of a pilot or by those locally acquainted with the navigation.

The least water appears to be 9½ to 10½ feet at low water and 22 to 23 feet at high water springs, at 25 miles above the entrance of the river and thence to Launceston.

Port Dalrymple, the principal harbor on the north coast of Tasmania, constitutes the entrance of Tamar River, which river formed, by the confluence of the North and South Esk Rivers at Launceston, flows through a valley between two irregular chains of hills that shoot out northwestward from the great body of inland mountains. In some places these hills stand wide apart, and the river then widens to a considerable extent; in others they nearly meet and contract it to narrow limits. Of the two chains of hills which bound the valley, the eastern one terminates at Low Head; the other descends to Badger Head, 6½ miles west-southwestward from Low Head.

The ends of these chains, when seen from directly off the entrance, appear as two clusters of hills having some resemblance to each other; and in fine weather the distant blue heads of the back mountains are seen over the tops of both clusters. These appearances, together with the position of the vessel, are the best distant marks for finding Port Dalrymple.

From the eastward, Ninth Island, and afterwards Stony Head, with Tenth Island lying off it, show the vicinity of the port; and

Low Head, with the conspicuous lighthouse tower on it, will be perceived in the bight to the south-southwestward. At about 10 miles southwestward of the port the back land is high, rising to 1,700 feet, and the top of the ridge is rugged, forming unusual shapes. These mountains, with direction of the coast and the most remarkable of the clusters of hills just noticed, serve as marks for Port Dalrymple, from the westward.

The entrance of Port Dalrymple, between Low Head and Friend Point, situated about 1.7 miles south-southwest from the head, is difficult of access in a sailing craft, on account of the numerous reefs and banks in it, extending a considerable distance from the western side of the entrance; it should, therefore, be avoided and entrance made by Middle Channel.

The greater part of the shoals at the entrance are covered at half tide, so that at half flood, or even a little before, is the best time to enter Port Dalrymple, as almost the whole of the dangers are then visible.

The eastern shore of Port Dalrymple, from Low Head, trends 1¾ miles in a southeast direction to Shea oak Point, the south point of Port Dalrymple, and consists of alternate points and small bights, bordered by a shoal, the 3-fathom edge of which projects 200 to 400 yards from the low-water line; the shoal extends as a spit 300 yards northwestward from Low Head; off Dotterell Point, at 650 yards southward of the lighthouse, it projects 700 yards westward, nearly to the Middle Bank.

Port Dalrymple or Middle Channel, the main entrance, lies between the Middle Bank, on the northeastern, and Yellow Rock, on the southwestern side; it is nearly 400 yards wide, with depths of 10 to 20 fathoms.

Dangers.—Middle Bank (lat. 41° 03′, long. 146° 48′), the most dangerous shoal in the entrance of Port Dalrymple, is a rocky patch, 650 yards long and 400 yards wide, situated ½ mile southwestward from Low Head Lighthouse, with depths of 2 to 3 fathoms at low-water springs. The flagstaff on Low Head, open to the northward of the lighthouse, leads northward of the bank; and the high rock on Black Reef, in line with the first saddle in the hills south of Flinders Point, leads southward of the bank.

The sea breaks heavily over this bank in bad weather; and there are always heavy tide ripples on ebb and flood.

Buoy.—The southwest or channel side of the bank is marked by a black can buoy.

Barrel Rock—Beacon.—Barrel Rock, which uncovers at half ebb, is marked by a black beacon with ball and vane, at 1,340 yards southward from Low Head Lighthouse; a rocky ledge with less than 2 fathoms over it at low water extends from the beacon in a southerly direction, the southern extremity of which, with 2¼ fathoms of water, lies 500 yards southward from the beacon.

Clearing range.—Two red beacons on Cordell Point, 1 mile southeastward from Low Head, in range lead southward of the ledge. The white range towers on Shea oak Point in range lead southward of the ledge in 24 feet of water, but very close to it.

Yellow Rock and West Reef.—On the south side of the middle channel and 1,400 yards southwestward from Low Head is Yellow Rock, an extensive patch of kelp, with a double-headed rock, on which the least depth of water is 9 feet. This rock forms the east extremity of West Reef, the northern edge of which extends from it nearly 1,500 yards in a westerly direction. This reef is about ¼ mile broad, but the only part of it uncovered at high water is Black Reef, which is 2 feet high, near the center, at 1.1 miles southwestward from Low Head.

Buoy.—Yellow Rock is marked on its northern side by a red conical buoy.

Shear Beacon, ¼ mile southeastward from Yellow Rock, is red, and stands on the highest part of Shear Reef, which dries 7 feet; it is connected with West Reef by shoal water where the greatest depth does not exceed 15 feet; a spit, with 1½ fathoms on its extremity, projects 250 yards northeastward from Shear Beacon. Shear Rock, with 2 feet water lies just within the end of the spit.

Buoy.—A red conical buoy lies to the northward of the rock and spit.

Anchorage.—There is anchorage in 4 to 8 fathoms in Port Dalrymple, between the Barrel Spit and Shea Oak Point, avoiding the 4 fathoms spit extending off the shore between Cordell Point and Shea Oak Point. The anchorage is slightly to the westward of the Sea Reach Range.

The old range towers are two circular black stone beacons, built on the southern end of the lagoon beach, east-southeastward, distant about 1,500 yards from Barrel Rock Beacon.

Eastern Channel, lying between Middle Bank and the shoal which borders the west side of Low Head, is only available by small craft with local knowledge. It is 400 yards wide in the outer part, with 4 to 6 fathoms water; but the inner part is only about 100 yards in width, with 3¼ fathoms apparently on a ridge, extending from the southeast extremity of the Middle Bank to the shore. Browne's House open west of Shear Beacon 182° leads through. The flood tide sets across the north part of the channel, and in bad weather the sea breaks from the Middle ground to the eastern shore. Directions are given later.

The southwest shore of Port Dalrymple from Friend Point, the southwest entrance point of the port, trends southeastward about

1,500 yards to Browne's House, the first within Friend Point. From Browne's House the shore trends in the same direction for 1½ miles to the north point of Kelso Bay.

This shore is fronted by a bank, which extends about 1 mile northward and northeastward to West and Shear Reefs, with a narrow inlet—about midway between Friend Point and the outer edge of the reefs—running into the bank from the westward, and carrying from 1¾ to 1¼ fathoms water. Between this inlet and the shore there are numerous patches of reef, dry at low water.

From the spit which projects northward from Shear Beacon, the 3-fathom edge of the bank trends southeastward 1¼ miles, and thence about south-southeast nearly 1½ miles to the north point of Kelso Bay.

Sea Reach—Buoys and beacons.—The western side of Sea Reach Channel is marked by 4 red conical buoys; the spit or elbow buoy has a staff and drum. The eastern side, by a black beacon with ball topmark on Simmons' Mistake Reef, and a black can buoy 1 mile to the southward from it.

Anchorage.—There is good anchorage in 4 or 5 fathoms at about 200 yards above or to the southeastward of the northwest buoy. There is also anchorage in 5 to 7 fathoms about 400 yards northward from Honduras Bank buoy and 50 yards westward of Sea Reach Lights' leading line.

Honduras Bank—Buoy.—Honduras Bank, 1 mile southward from Shea oak Point, with 15 feet water on its outer edge, extending about ½ mile off from the western shore, is marked by a red conical buoy.

The Eastern shore of the River Tamar from Shea oak Point takes a general south-southeast direction for 2 miles to Windmill Point, the southwest point of Georgetown. It is fringed with reefs and a flat which is steep-to, projecting 400 to 900 yards from the shore.

Simmon's Mistake—Beacon.—Simmon's Mistake is a reef lying ½ mile southward from Shea oak Point. The western edge is marked by a black beacon surmounted by a ball.

Beacons and buoy.—At 1 mile southward of Simmon's Mistake beacon is Long Tom beacon, black with square top. This beacon and a black can buoy mark the edge of the flat westward of Georgetown. A black beacon, with triangular top, marks the edge of the flat between Long Tom beacon and Windmill Point.

There are several patches of reef on the flat between the beacons and the town.

Bombay Rock, situated 300 yards northwestward from Long Tom beacon, has 4 feet water on it and is marked on its northwest side by a can buoy painted in black and white horizontal bands;

there is a narrow 4-fathom channel on the east side, but the main channel is on the west side.

Georgetown is situated on the eastern shore, at nearly $3\frac{1}{2}$ miles within Low Head; it is built upon a flat, forming the northwestern side of York or Georgetown Cove, at the western foot of a group of conical hills. Mining is carried on in the neighborhood, but the district is chiefly pastoral. There is steam communication with Launceston and Melbourne. It is also a post and telegraph station. Population of district about 1,100.

Windmill Point—Light.—A light is exhibited from a beacon on Windmill Point.

York Cove, eastward of Windmill Point, is about $\frac{1}{2}$ mile in length and extends along the southeastern side of the town; it is 300 yards wide, with from 8 to 10 fathoms in the entrance, decreasing to 2 fathoms toward its head.

Garrow Rock, which lies in the entrance of the cove, is a cluster of rocks 200 yards long, with a light beacon on it, distant 300 yards south-southeastward from Windmill Point. There is a depth of 9 fathoms between the shoal and the town and from 13 to 16 fathoms between the shoal and the southeast entrance point of the cove.

Light.—A light is exhibited from the beacon situated on the southwest side of Garrow Rock.

Kelso Bay and Arthur Head.—Kelso Bay, on the western side of the port, is about 1,500 yards wide and $\frac{1}{8}$ of a mile deep; the bay, except a small inlet close-to its north point, is filled by a shoal flat, which extends about halfway across toward Georgetown.

Spit—Buoy.—The northern extremity of this flat forms a spit, with 3 feet water on it, marked by a red conical buoy.

Range lights.—Two light beacons, erected $\frac{1}{2}$ mile westward from Arthur Head, give the line of channel past Honduras Buoy and Bombay Rock Buoy.

Garden Isle.—From the red buoy on the northern spit of the flat its northeastern edge trends east-southeastward 1,340 yards to the north point of Garden Isle, which is 400 yards long northeast and southwest, with a small hillock 30 feet high on its northeast end, close off which there is a depth of 14 to 18 fathoms.

Beacons.—A red beacon marks the north edge of the reef 600 yards northwest of Garden Isle, and a similar beacon the eastern edge of the reef 200 yards beyond it. Between Arthur Head and Garden Isle there is a depth of 7 fathoms.

Port Dalrymple Channel from the Middle Bank to Georgetown is $\frac{1}{4}$ mile to 200 yards wide between the flats which front the shores, the narrowest part being abreast of Simmon's Mistake Beacon.

Beacons and buoys.—The shoals on either side within the entrance of Tamar River are marked with beacons and buoys in accordance with the uniform system of buoyage. On entering, all red buoys and beacons must be left on the starboard hand, all black buoys and beacons on the port hand, and parti-colored buoys and beacons passed on either side as directed.

Pilots—Tug.—Pilotage is compulsory to merchant vessels, and it is advisable for strangers to secure the services of a pilot as soon as practicable, especially at night. Vessels making the usual signal should wait in the offing with the range lights in line until they are answered by a rocket from the pilot station when they may run down and pick up the pilot.

A powerful tug is available if requested by signal.

Should the weather be too bad to permit the pilot to proceed outside, the boat will lie in mid-channel and direct the vessel, if necessary, by the pilot flag.

The pilot boat is painted white and carries a red and white flag.

Directions—Port Dalrymple or Middle Channel.—It is not advisable to enter this port on a first visit without the assistance of a pilot. These directions are subject to amendment.

For Port Dalrymple or Tamar River, use the Middle or Port Dalrymple Channel, being the safer. When from the northward or eastward keep the lead going, and, having Low Head Lighthouse or light bearing 180°, it may be approached to 3 miles, when a southwesterly course can be steered until the white range light towers or lights on Shea-oak Point are sighted on a southeasterly bearing, then steer southward and bring them in range bearing 128°, which is the range for entering passing the eastern end of Hebe Reef and buoy at a distance of nearly 600 yards and between the red conical buoy on Yellow Rock and the black can buoy on Middle Bank.

The red conical buoy on Shear Rock will be passed about 50 yards off in about 4 fathoms of water and the west end of the spit near Barrel Rock in about the same depth.

The rear light should be opened northward of the front one when midway between Yellow Rock Buoy and Shear Rock Buoy, and southward of the front one when near the Barrel Spit in a vessel of deep draft.

Entering from the westward, care must be taken not to bring Low Head Light eastward of 110° by day or to enter the arc of fixed red light shown from Low Head Lighthouse over the Hebe Reef at night, before the range lights are brought into line.

When Shear Rock is passed (at night the light at the pilot station comes in sight when past the rock) bring the range lights on Shea-oak Point about 1½ points on the port bow, taking care not to open the high or black tower or high light northward of low light, to avoid the spit extending southward from Barrel Rock, and, having the hand lead going quickly, steer to bring the light tower at the pilot station in range with the light tower to the northward of it, bearing 349° astern, which course will lead up Sea Reach, leaving the red buoys on the starboard hand, and Simmons Mistake black beacon on the port. After passing Simmons Mistake beacon the range lights westward of Arthur Head will be picked up nearly ahead. These lights in range lead through the main channel past Honduras Buoy and Bombay Rock Buoy. The sectors of the lights divide on the range, being red to westward and white to eastward. The green sector of the light on Windmill Point turning red, or the light on Garrow Rock coming in sight will indicate that the Bombay Rock has been passed. Steer for Garrow Rock Lighthouse, leaving the black buoys and beacons on the port hand and the red buoys and beacons on the starboard hand. Garden Isle can be closely rounded. The light on Windmill Point kept in sight leads westward of Garrow Rock, near Georgetown.

Great attention must also be paid to the tidal currents, as they set obliquely across this part of the river; the ebb, for instance, crosses from Kelso Bay to the beacon on the west point of the east flats, and with such strength as to form whirlpools. Both flood and ebb attain a velocity of from 3 to 5 knots an hour at times.

Eastern Channel.—To enter Port Dalrymple by Eastern Channel, which it is not advisable for a stranger to do—and should never be attempted at night—Sheer Beacon being clearly distinguished, bring Browne's house open west of the beacon, bearing 182°, which leads through Eastern Channel. Continue this course until the range beacons on Shea-oak Point are in range, when steer toward them, and as before directed.

Anchorages—Port Dalrymple.—The anchorage in Port Dalrymple is a little to the westward of the line of Sea Reach lights (the high light to westward open of the low) in 8 fathoms water.

The light at Low Head pilot station having become visible or the red beacons on Cordell Point being in line, to anchor at Port Dalrymple haul to the eastward and bring the Sea Reach range lights nearly in line.

Georgetown or York Cove.—Having entered the cove, anchor opposite the wharf, in 5 fathoms, and moor either with 100 yards each way or with a kedge on the shore, or, perhaps, with a hawser to the trees.

Kelso Bay.—To anchor in Kelso Bay, pass on the west side of Bombay Rock and keep near the western shore in order to avoid the northern spit of Kelso Bay Flat, marked by a red buoy. After entering the bay a vessel may moor to the trees.

Proceeding to sea.—From close eastward of Garden Isle, leave the red beacons and buoys on the port hand and the black beacons and buoys on the starboard hand. To clear the shoal westward of Cordell Point, do not bring the red beacons on that point in range until Low Head Lighthouse is open west of Barrel Rock Beacon. Then proceed to sea with the range light beacons in line.

At night, from Garden Isle keep in the white sector of light on Garrow Rock bearing 125° to 129° astern until the lights on Arthur Head are nearly in line. Alter course sharply to the northward and bring them into range astern; the sectors divide on the range, being red to westward and white to eastward. When the Sea Reach range lights come in range steer for them until the white range light on She-oak Point is just open south of the red range light; keep these lights so until the pilot station light becomes white, then bring them in range and keep them so until outside. When Low Head Light bears 82° the vessel is outside the entrance of Middle Channel.

Above Georgetown.—The eastern shore of Tamar River, between the southeast entrance point of Georgetown Cove and Saltpan Point, which lies 1,340 yards southward of it, forms an irregular sandy bay, between which and Garden Isle the river is ⅓ mile wide, with depths of 9 to 25 fathoms, affording room for several vessels to anchor, but the bottom is uneven and the currents are rapid and irregular.

Porpoise Rock—Buoys.—Purpoise Rock lies 200 yards off Saltpan Point and has 5 feet on it at low water; it is marked by a black conical buoy on its northwest end, moored in 20 feet at low water, and a red conical buoy on its southeast end in 15 feet at low water, and the water is deep close round it.

Channels.—There is a channel on both sides of Porpoise Rock. The western channel is the widest; the eastern channel, with 14 fathoms of water, is about 400 feet wide, and though narrow, possesses the advantage of being in a straight line from Garden Isle.

Range beacons, painted white, in Bryants Bay, or a red light beacon on the foreshore east of Windmill Point in Georgetown Cove in line with Garrow Rock Lighthouse, or the white lights on them at night, lead through the eastern channel. Should the black buoy be out of position, or away, two red beacons on Garden Isle kept in range will lead clear westward of the rock, and two red beacons on Roundabout Point in line apparently point to the rock.

Deceitful Cove.—From Saltpan Point the shore trends southeastward 1,500 yards to North Head, between which and Point Effingham, 600 yards to the southward of it, is the entrance of Deceitful Cove, a shoal creek trending to the northward.

Point Effingham—Light beacon.—A light is shown from a beacon on the extremity of the point.

The western shore from Arthur Head curves south-southeastward ⅓ mile to the northwestern extremity of Bryants Bay; thence the bay extends southeastward nearly 1 mile to Anchor Point and is ¼ mile deep, with depths of 4 to 5 fathoms close to the shore. There are from 26 to 7 fathoms between Saltpan and Anchor Points, with anchorage in 4 to 8 fathoms in Bryants Bay, at ¼ mile from the shore.

Shag Rock—Beacons.—Shag Rock, nearly 400 yards southeastward from Anchor Point, is just covered at high water; there is deep water close round the rock, and 19 fathoms between it and the shore; this rock is marked by two red beacons surmounted by triangles.

West Arm—Yorktown.—The entrance of West Arm lies between Anchor Point and Inspection Head at Ilfracombe. The arm trends west and southwestward 2¼ miles to the ruins of Yorktown. It is a shallow inlet ¼ mile wide halfway in, above which it expands to 1,340 yards in width, and has a small fresh water stream flowing into its western corner.

Beacon.—A red beacon marks the shoal ground off Inspection Head.

Ilfracombe is situated on Inspection Head, which separates West Arm from Middle Arm.

Middle Arm is about 1,500 yards wide in a northwest and opposite direction between Inspection Head and Shag Point, whence it trends 2¼ miles to the southward. There are depths of 10 to 13 fathoms in the entrance and 3 fathoms at about 1,500 yards within it, above which the arm is mostly filled by a shallow flat, branching to the southward and southeastward.

Beauty Point, situated on the west shore of Middle Arm about ½ mile within the entrance, is the shipping port for Beaconsfield, and has a pier where vessels of any draft may berth and lie afloat at all stages of the tide. There is 40 feet at low water and ample shed accommodation. There is a deep channel between it and the main river.

Light.—A light is exhibited from the pier at Beauty Point.

Beacons and buoys.—The entrance of the channel to Beauty Point Harbor is marked on the western side by a red beacon and on the eastern side by a black conical buoy. The eastern side of the channel is also marked by a light beacon and a black conical buoy, which lies about 200 yards northeastward from Beauty Point Pier.

Beaconsfield, the fourth town of importance in Tasmania, is situated in a mining district 1½ miles inland from the left bank of the river about 5 miles southward of Georgetown. Its population is about 5,000. Steamers from and to Georgetown and Launceston call daily to embark and discharge passengers and cargo at Beauty Point Pier and Bowens Jetty, connected with Beaconsfield by steam tramway; there is also coach communication with Launceston.

Middle Isle—Middle Bank.—Between Shag Point, off which there is a red beacon, and Middle Point, 2 miles northeast of it, the south shore of the river forms a bay 1,500 yards deep; but it is filled by a shallow flat, Middle Bank, the edge of which from Shag Point extends northward 1 mile to a spit, marked by a black conical buoy, whence it curves round eastward and northward to Middle Isle, which lies ½ mile westward of Middle Point.

The light on Beauty Point shows red over Middle Bank.

Anchorage.—There is a good anchorage in 5 to 7 fathoms ¼ mile northeastward of Middle Isle with soft regular bottom, and out of the strength of the current, where a vessel not having a pilot is recommended to anchor before proceeding farther up the river. A jetty projects in a northwesterly direction from the north end of Middle Isle, with a depth alongside of 9 feet at low water.

The quarantine ground is the bay westward of Middle Isle, which affords anchorage in depths of 4½ to 7 fathoms, with 4 fathoms close to the edge of the flat.

The eastern and northern shore from Point Effingham trends eastward 1½ miles to Sawyers Point; there are depths of 6 fathoms close to this shore, and from 11 to 4 fathoms in the fairway between it and the quarantine ground.

Long Reach—Wolverine Shoals—Beacons.—From Middle Point, the southwestern shore of Long Reach trends east-southeast-ward 2½ miles to Point Rapid. The northern half is fronted by a dry bank to the distance of 200 yards, with Wolverine Shoals, with depths of less than 3 fathoms, extending ¼ mile into the channel. A red light beacon marks the northern extremity of the bank and a red beacon the edge of the drying bank 400 yards southwestward of it. A 17-foot channel between these beacons is indicated by white range beacons on Sawyers Point. Little Dragon Reef, 1.4 miles southeastward from Middle Point, is also marked by a red beacon; a rocky spit (Dragon Reef), which extends 400 yards from the shore and is 1,500 yards northwestward of Point Rapid, is also marked by a red beacon.

Shoals with 13 to 17 feet water on them lie about 400 to 700 yards northward from Point Rapid, nearer to the southern than the north-ern shore. The main channel is to the eastward of the shoals. Clear-ing beacons for leading either side of them are placed on the shore to the southeastward.

The depths of water in Long Reach are irregular, varying from 15 to 6¾ fathoms in the fairway; the deepest water being on the northeast side of the reach except opposite the Little Dragon Beacon. The northeast shore of the river from abreast Middle Isle curves northeast, thence southeastward for 2 miles to Big Bay, where there is a fresh-water stream, close off which there is anchorage in 4 fathoms. Thence the shore extends 2 miles southeastward to the entrance of East Arm; it is intersected by small creeks, and rises to a range of stony but well-timbered hills.

East Arm is 800 yards wide at its entrance, whence it trends east ½ mile, and southeast 1 mile, its eastern corner terminating in Fourteen Mile Creek. From 9 fathoms in the entrance the depths decrease to 1¼ fathoms about 1 mile within it. There is ironstone along the southwest shore of East Arm.

Moriarty Reach.—From Point Rapid the western shore of Moriarty Reach trends southwestward 1 mile to Ruffins Bay, off the south point of which there are some rocks marked by a red beacon, and thence sweeps round in a south-southwest direction 1¼ miles to a point close off which is Drumstick Islet, with sunken rocks along its southeast side marked by two red beacons. This shore is indented by a shoal bight, and may be generally approached within 200 yards in 6 to 8 fathoms. There is anchorage in 6 or 7 fathoms, close off Ruffins Bay; and in 8 fathoms off a similar bight at ⅓ mile northeastward of Drumstick Islet.

Jetty.—There is a jetty in Iron Pot Bay, southwestward from Point Rapid.

Sidmouth is a postal township on the left bank of the River Tamar at the entrance to Whirlpool Reach. There is a daily steamer communication with Launceston and Georgetown.

Lights.—Range lights are established at Sidmouth, arranged in a triangle, one front, two rear, to lead from Point Rapid to Sidmouth, and from Sidmouth to Swan Point.

Between Drumstick Islet and a projecting part of Sidmouth, ⅓ mile southwest from it, is a bay having from 3 to 5 fathoms water, in which there appears to be anchorage, out of the current.

The eastern shore of Moriarty Reach from East Arm trends southwestward 1 mile to Sheeps Tail Point, thence south ½ mile to another point, between which and Rocky Point, 1,500 yards south-southwestward from it, is Redwood Bay, ½ mile deep, with a creek in its bight; the bay is bordered by a flat which extends ¼ mile from the shore.

At the entrance of this bay is Reids Rock, with 1½ fathoms on it, 400 yards long, north-northeast and south-southwest, with a black beacon on it and a black can buoy at the south end. The channel between Reids Rock and the western shore, with Sidmouth range lights in line, is 300 yards wide, with 8 fathoms water in it.

Redwood Islet lies 300 yards southwestward of the southwest point of the bay just described, from which point the shore trends south-southwest, ¼ mile to the east point of the northwestern entrance of Whirlpool Reach; the channel between Redwood Islet and Drumstick Islet is 200 yards wide, with 12 fathoms water in it.

Whirlpool Reach.—Whirlpool Reach, from its northwest entrance between Sidmouth and the opposite point, trends southeastward nearly 1 mile, and is less than 200 yards wide, with irregular depth of 20 to 7 fathoms.

Whirlpool Rock is just within the northwest entrance of Whirlpool Reach; it is composed of numerous pinnacles of blue stone, intermingled with thick clay, and is about 120 feet by 90 feet in extent at low water springs. It lies nearly in the center of the channel and 490 feet from the eastern shore, and has 12 feet over it. This rock has 10 fathoms water on its southwest side, but only 3 fathoms on its northeast side.

Two white beacons on the eastern shore when in line show when a vessel is abreast of the rock, and two white beacons on the western shore in line give a safe course to the westward of it. There is a channel on both sides of the rock, but the western one is the best.

Light beacon.—A light beacon is near the latter pair of white beacons.

Southwestern shore.—From Great Mary Ann Creek, on the southwest side of the southeastern entrance of Whirlpool Reach, the southwestern shore trends southeastward nearly 1½ miles, and thence south-southeastward 1¼ miles to Supply River. About ½ mile southeastward of the creek some sunken rocks lie about 200 yards from the shore. Deviot Jetty is on this shore, about 1 mile northeastward from the creek.

Spring Bay is a bight 1½ miles deep, situated between Barrett's Point, marked by a black beacon, and Mowbray Point 1¼ miles east-southeastward from it. At ¼ mile within the entrance is Middle Bank, with 5 to 9 feet water on it, and marked by a black beacon on its southeastern end and a dolphin near the center. There is a channel, with 6 to 11 fathoms water, all round the bank. There is anchorage in 5 to 8 fathoms, sand and shells, between Barretts Point and Middle Bank.

Supply Bay, between Deviot Jetty and the bank off Swan Point, is filled by a shallow flat except near its western shore, where there is a channel 400 yards wide, with 6 to 3 fathoms water, reaching half way to Supply River.

Exeter—Swan Point.—From 400 yards southeastward of Supply River the river frontage of the postal township of Exeter trends northeasward 1 mile to the west point of a shallow bight, which extends 800 yards eastward to Swan Point. A bank, which dries at low water, extends northward 400 yards and northwestward 800 yards from Swan Point; two red beacons mark its edge. A shoal flat also extends northeastward and eastward from the point; the edge is marked by a red conical buoy at 900 yards eastward of Swan Point.

There is a jetty on the eastern side of the point.

Mount Direction or Macquarie—Signal station.—Mount Direction, nearly 3 miles east-northeastward from Swan Point, rises from Upway to the height of 1,212 feet and has a signal station on its summit.

Dorchester.—Between Mowbray Point and the point near Egg Islet three shallow indentations of the northern shore form the river frontage of Dorchester.

Depth in river.—From Whirlpool Reach to Mowbray Point the depths are 17 to 10 fathoms in the fairway. Midway between Deviot jetty and the Swan spit bank a shoals pit, with $3\frac{3}{4}$ to 4 fathoms water, extends toward Barrett's buoy from the flats in Supply Bay, reaching nearly across the river. The deepest water appears to be $5\frac{1}{4}$ fathoms close to Barrett's buoy. A shoal with $4\frac{1}{4}$ fathoms extends 200 yards southwestward from Egg Islet flat.

There is anchorage in $6\frac{1}{4}$ fathoms at 800 yards southwestward from Egg Islet.

Egg Islet lies northwestward 1,500 yards from Swan Point and 200 yards from the most prominent point of Dorchester; it is 300 yards in length northeast and southwest, and has a spit extending 800 yards to the westward, where it is marked by a black conical buoy, Barrett's buoy. This islet and spit are separated from the shoal which borders the shore by a channel 200 yards wide, having 9 to 15 feet water.

Northeastern shore.—From the point $\frac{1}{4}$ mile eastward of Egg Islet the northeastern shore curves $1\frac{1}{2}$ miles in an east-southeast direction to a fresh-water inlet, and thence southeast 1,500 yards to Faheys Creek, the south point of the mouth of which has a ledge of sunken rocks projecting $\frac{1}{4}$ mile from it. From this point the bay (Swan Bay) extends south-southwestward 1 mile and is $\frac{1}{2}$ mile deep, with streams flowing into its bight, but it is inaccessible on account of the shoal flat which fills the bay, the edge of which is marked by a black conical buoy. From the southern extremity of this bay the shore extends southwestward nearly $1\frac{1}{4}$ miles to Native Point, and is bordered by a bank 200 to 400 yards broad. From Native Point the shore curves round to the southeastward for about $1\frac{1}{4}$ miles to a point (Windermere) opposite Rosevears.

The western shore from Swan Point to Little Swan Point 1 mile south-southeast from it, forms a bay $\frac{1}{4}$ mile deep, but it is filled by a shoal flat. From Little Swan Point the shore trends southward about $1\frac{1}{2}$ miles to Stony Creek.

The shore from Swan Point to Stony Creek is confronted by a bank, the northern portion of which, for about 1,500 yards southeastward of Swan Point, extends nearly 1 mile from the shore, but the outer edge from thence gradually closes southward to 200 yards off Stony Creek. There are general depths of 3 to 12 feet on this bank, the edge of which is marked by the red conical buoy at 900 yards eastward of Swan Point already mentioned, and a light beacon at the bend of the bank 1,800 yards east-southeastward from Swan Point.

.The shore is also bordered by a flat about 300 to 600 yards broad, dry at low water, on the outer edge of which is a beacon, 600 yards east-northeastward from Little Swan Point.

River—Depths.—From Egg Islet the channel trends east-southeastward 1½ miles, and is 500 yards wide, with depths of 14 to 8 fathoms, between the northeastern shore and the shoals which extend from Swan Point. The river then increases to 1 mile in width, but the channel is only 400 yards wide or less, with depths varying from 8 to 5¾ fathoms. Southward from Stony Creek the river contracts to 500 yards in width, the channel here being only 200 yards wide.

Gem Rock at the entrance to Stony Creek is dry at low water and marked by a red buoy.

Blackwall Light Beacon is situated on the west bank southward of Stony Creek.

From Stony Creek the river sweeps round eastward 2¾ miles to Cimitere Point (Rosevears), and is generally about ¼ mile across, from shore to shore; the channel being 150 to 200 yards wide, with from 9 to 4¼ fathoms water in the fairway.

Rosevears is situated at Cimitere Point, where there is a pier and post and telegraph offices. The population is about 200.

Ships of moderate draught can proceed up the river as far as this in charge of a pilot; further up the river becomes shallow and the channel narrow.

It is advisable to moor, when lying off Rosevears, on account of the limited space when swinging.

From Rosevears to Launceston the channel is indicated by 24 light beacons and numerous dolphins and beacons, but it is unnecessary to describe it as a pilot or local knowledge is absolutely necessary.

Launceston approach—Depths.—Vessels drawing 21 feet can proceed as far as Launceston at high water. There is from 12 to 16 feet water in the berths at low water over a soft bottom. There is a basin off Town Point (immediately opposite Launceston), 400 feet long and 50 feet broad, with 20 feet at low water, and head and stern moorings have been laid down. From Swan Bay to Launceston channels and cuttings are being dredged with a view of obtaining a minimum depth of 14 feet at low water up to the wharves. A new channel, cutting off Stephenson's Bend, is under construction.

Caution.—The compass should not be relied on, as there is much local attraction at different parts of the river, more especially where ironstone is marked on the chart. There is, however, not much use for a compass in the upper parts of the river.

Directions—Upper river from Georgetown.—As before stated, this navigation can only be undertaken by a pilot, the employment of whom is compulsory. From Georgetown, haul close round Garden Islet, to avoid Garrow Rock, and having passed the island, steer to leave the black buoy on the west side of Porpoise Rock on the port hand, or at night pass eastward of the Porpoise Rock, with Garrow Light in range with the light on the beacon at Georgetown Cove, then proceed so as to pass midway between Anchor and Effingham Points (there is a light on Effingham Point); and after clearing Shag Rock—if not required to anchor in the Quarantine Ground—steer for the north point of Middle Isle, thence northeastward into Long Reach; and having fully opened its southeastern trend, steer through it, keeping nearer the northeastern shore than otherwise, to avoid the shoals and spit which project from the southwestern shore. The pilots prefer going up on the flood and coming down on the ebb, as the tide then suits better for rounding the Porpoise and Whirlpool Rocks. Going up on the flood they pass to the westward of the Porpoise Rock, but on coming down with the ebb they, as a rule, pass between it and Saltpan Point, which is steep-to.

Launceston, the second city in Tasmania, is situated at the head of Tamar River, which, following the winding course of the river, is 38 miles from the sea. It lies in a valley inclosed with hills, and the lofty Mount Barrow, 4,644 feet high, is 12 miles to the eastward. The city has wide streets, excellent public buildings, an extensive public library, large public gardens, and is lighted with electricity. Launceston is connected with the whole railway system of the island, and there is a telegraph station connected with all parts. Steam vessels run to Melbourne three times a week during the summer and twice during the winter seasons; to the northwest ports and west coast two lines each, weekly; and to Sydney fortnightly. The population in 1917 was about 25,000.

Trade.—The principal imports are manufactured goods, tea, sugar, wine, etc., and the exports are wool, oats, fodder, gold, silver, tin, lead, coal, timber, potatoes, fruit, and bark.

Wharves.—Vessels of 4,000 tons register, 350 feet in length, and about 20 feet draught, can berth in the basin. There is from 12 to 16 feet in the berths at the wharves at low water over a soft mud bottom and 12 feet at the cattle jetties.

The railways are connected with the wharves.

Berthing.—Owing to the narrowness of the channel, passing vessels cause a considerable scend rendering it necessary for those in charge of vessels lying alongside the wharves to keep a careful lookout on their hawsers.

A steam tug, maintained by the Marine Board, is available for towing vessels, as moderate rates; the signal for the tug is a checkered flag, hoisted where best seen. When this signal is made by a vessel in the offing, entering the port, the tug, if at Launceston, will be telegraphed for on that vessel's account.

Water.—Fresh water may be obtained by applying to the Marine Board; it can be sent off in their tug. It is laid on to the wharves.

Dock.—There is a floating dock at Launceston. (See Appendix I.) In York Cove, Georgetown, and other places on the banks of the river, small craft may be safely beached to be cleaned or examined.

Pilotage is reasonable. Vessels anchoring below Georgetown are charged one-third pilotage; at or above Georgetown, and below Whirlpool reach, one-half pilotage. Vessels arriving and sailing in ballast or putting in to seek freight, or from stress of weather, and not breaking bulk, are exempt from all port charges, except only those of pilotage in cases where the services of a pilot have been actually required and received.

Tides and tidal currents.—It is high water, full and change, at the pilot station, Port Dalrymple, at 11 h. 10 m.; springs rise 10 feet. The rise is irregular, the greatest observed being 10 and the least 4 feet. The highest tide noticed was during the neaps, caused by a strong northwest gale. The flood current runs 5 h. 50 m., and the ebb 6 h. 25 m., at a velocity varying from 2 to 5 knots, according as the river is confined or open. The ebb current setting round Low Head into the bay to the eastward drifts vessels in that direction. At 3 miles in the offing the flood stream runs west-northwest 1 to 2 knots.

At Launceton it is high water, full and change, at 1 h. 00 m.; springs rise 12½ feet. During winter, after rains, the stream sets down for days together, at the rate of 1 to 3 knots.

The north coast of Tasmania from West Head or Flinders Point trends westward 32 miles to Round Hill Point, and there are depths of 10 to 15 fathoms at 2 miles off it. There are not many projecting points, but this coast is intersected by no fewer than six rivers and one creek, all of which, except the creek, are accessible to vessels of 200 tons. One of the rivers, the Mersey, is available for vessels 18 feet draft. These streams flow through a hilly country, which is tolerably wooded, to the back mountains. Upon this elevated range are many variously shaped peaks, among which are Mount Roland, 4,047 feet high, southeastward, distant 27 miles, and Black Bluff, 4,381 feet high, southward 24 miles from Round Hill Point.

But the most worthy of notice of these mountains appears to be Valentine Peak, 21 miles south-southwest from the point. This peak is a bare mass of granite 4,100 feet high, and as it glistens in the first beams of the morning sun like an immense spire, it becomes the most remarkable hill feature on the north coast of Tasmania.

From Friend Point, south side of entrance to Tamar River, Flinders Point lies to the westward distant nearly 3 miles, and the coast between forms Boobyalla Bay, consisting of three bights, behind the southeastern of which, Shoal Bay, is a lagoon of fresh water. Boobyalla Bay is fronted by a continuation of the shoal flat which projects from Friend Point, with its 3-fathom edge extending 1,340 yards from the shore. Shoal Bay affords anchorage for small craft in 2 to 3 fathoms, and farther off shore in 6 fathoms, sheltered from all winds between west by south, round by south, to northeast by east, with the extremity of Flinders Point bearing 280°, distant 1.2 miles. A black conical buoy lies off the edge of the reef and flat at the head of Shoal Bay. There is a detached patch between this anchorage and Flinders Point, with 3 feet of water on it, lying eastward 1,340 yards from the northern extremity of Flinders Point. There is a channel 1,500 yards wide, with 6 to 8 fathoms water, between this bay and Hebe Reef.

West Head or Flinders Point (lat. 41° 04', long. 146° 42) projects 1,340 yards from the line of coast, and is nearly ½ mile broad; rocks which dry 2 feet extend 600 yards northward from the point.

Badger Head and Asbestos Hills.—Badger Head, situated 3½ miles southwestward from Flinders Point, and another projection 1½ miles southward of Badger Head are rocky and form the northwestern termination of the Asbestos Hills, in which the mineral of that name is found; the hills are from 1,240 to 1,350 feet high, and in clear weather are conspicuous from seaward. From the rocky projection southward of Badger Head a low coast curves in a southwest direction 4 miles to a spit forming the southeast side of the entrance of Port Sorell.

Port Sorell is only available for small craft and boats. The northwestern entrance head of the port which lies 4¼ miles west-southwestward from Badger Head, projects above 1 mile from the line of coast, and is fringed by a reef of rocks. At about 1 mile southeastward of the head toward the port is Carbuncle Islet, which is connected with the shore by the reef, and forms the west point of the entrance over the bar, where there are 6 to 7 feet water; the bar does not shift, and the only unseen danger is a rock on the east side, with an iron beacon upon it. From the bar the channel trends between the shoals, 1½ miles in a south-southeast direction, with 2 to 5 fathoms, close up to the southeastern entrance point; above this a very narrow channel turns about 1¼ miles southward and eastward into the port, where there are 2 to 4 fathoms water, between the Sisters Islet on the west, and a broad, but shallow creek on the east side, trending northeastward nearly 2 miles; the east point of the Sisters Islet had a temporary beacon on it.

Burges.—This township, which is situated on the west side of Port Sorell, about 2½ miles within the entrance, has an extensive jetty, with tramway and trucks, for the purpose of loading small craft.

Directions.—In approaching Port Sorell it is usual for the small craft that frequent it to make the land a little to the westward of the port, as the wind during nearly nine months of the year prevails from northwest, west and southwest, and there is almost a constant current setting to the eastward.

To proceed for the fairway, avoid the beaconed rock on the east side of the entrance, pass near Carbuncle Islet, leaving it on the starboard hand; keep the houses of Burges right ahead, and run between the east point of the Sisters Islet and a black buoy which lies off it; after which anchor, in 4 fathoms, off the jetty.

Rubicon River, which flows into Port Sorell, is navigable for small craft and boats for a distance of 7 miles from the entrance; but its narrow winding channel requires the aid of an experienced person as a pilot, who may be obtained on the spot.

Heidelberg is a postal township situated near Green Creek, a shipping place about 8 miles up the river. There is daily mail communication with Latrobe, 10 miles distant.

Trade.—The exports of Port Sorell consist of posts, rails and paling, fruit, farm and dairy produce, some of which is shipped at Heidelberg. There is a shipbuilding yard in Port Sorell, where small craft are built. The timber at this port is of excellent quality, and such craft as can enter may be repaired at the current rates.

Tides.—It is high water, full and change, in Port Sorell at 11 h. 35 m.; springs rise, 8 to 9 feet.

The coast.—From the northwestern head of Port Sorell the coast trends southwestward 7½ miles to the entrance of the Mersey River, and may be approached within a mile in from 4 to 7 fathoms, except at about 4½ miles westward of Port Sorell, where the Horseshoe Reef extends 1½ miles from the shore.

Egg and Wright Islets are two rocks, one on the northern and the other on the southwestern part of Horseshoe Reef, which consists of detached dry and sunken rocks extending nearly 2 miles offshore about 5 miles southwestward of Port Sorell entrance.

The light on Mersey Bluff shows red over Horseshoe Reef.

Devonport and Mersey River (lat. 41° 09', long. 146° 28').—Devonport has, next to Launceston, the best and most secure anchorage on the north coast of Tasmania.

The entrance to Devonport may be easily known by its western head, Mersey Bluff, being high land covered with foliage, except the extreme point, or bluff, upon which is the lighthouse.

A reef, discernible from the broken water on it, projects 250 yards from the shore between the bluff and the river entrance.

Bar.—There is a bar across the entrance, consisting of sand, which does not alter, except during a northwesterly gaile, when it may silt up from 12 to 24 inches; this is cleared by the dredger when required.

A least depth of 16 feet at low water is maintained on the bar and in the channel, which is 150 feet wide.

Mersey River, which flows into Devonport, is navigable for vessels with a maximum draft of 9 feet, at high water only, as far as Latrobe, a distance of 6 miles, to which point the tide reaches.

A training wall from the east shore of the Mersey River is 967 yards long, with 79 yards of its outer end submerged, and is awash at high water in two places. A pier or wall, 40 yards long, has been built from the west shore near the range lights.

Lights.—A fixed white 'light with red sectors, 122 feet above water, visible 16 miles, is shown from a white circular brick tower on Mersey Bluff, 42 feet high.

There are two range lights on the west bank of the River Mersey, which in line 205° lead over the bar; the front light is exhibited from a square white wooden tower, and the rear light from a circular white brick tower.

There is also a small light on the west pier.

Lloyd's signal station.—There is a Lloyd's signal station at Mersey Bluff, and communication can be made by the International code. It is connected by telegraph with the harbor office.

Pilots.—Pilots are always to be had, and it is advisable to take one; communication can be made with the lighthouse by the International code; the pilot boards all vessels requiring his services, outside the red buoy; therefore, anchor outside the bar when the weather permits, or stand off and on, keeping the pilot-jack flying at the foremast head until boarded. When the weather does not admit the pilot to come outside, his boat remains in mid-channel with the pilot flag flying.

The charge for pilotage is reasonable.

Anchorage.—There is good anchorage in 7 fathoms outside the bar; but vessels should bring up well under the West head, so as to have sea-room when getting under way, and to avoid the reef which projects from the east side of the entrance.

Tides.—It is high water, full and change, at Devonport at 11 h. 35 m.; springs rise 11 feet, neaps 8 feet.

Tidal currents.—Both flood and ebb currents attain a velocity of from 1 to 2 knots inside the river, but after heavy rains the ebb has been known to run 4 knots.

On both flood and ebb there is always an eddy setting on to the West Devonport wharf. This ceases for an interval of about 10 minutes at slack water. On the flood, this eddy, which runs parallel to the wharf, is only felt at a distance of about 100 feet.

Outside the port the flood is the west-going current, and is not felt beyond 5 miles from the coast.

Tide signals (lat. 41° 09′, long. 146° 23′).—The following signals are shown from the flagstaff at the harbor office indicating the depth of water on the bar:

One ball at west yardarm	14 feet.
Two balls at west yardarm	15 feet.
Three balls at west yardarm	16 feet.
One ball at east yardarm	17 feet.
Two balls at east yardarm	18 feet.
Three balls at east yardarm	19 feet.
One ball at each yardarm	20 feet.
Two balls at west, one ball at east yardarm	21 feet.
One ball at west, two balls at east yardarm	22 feet.

A ball at the masthead denotes flood tide.

Directions.—In approaching Devonport from the eastward, along the coast, keep a good lookout for Egg and Wright islets, on the dangerous Horseshoe Reef; the lighthouse bearing 226° or at night the white light showing leads clear of the reef. But, in a sailing vessel as a general rule, the land should be made a little to the westward of the port, in consequence of the prevailing westerly winds and easterly current.

Having made the land just to the westward of Devonport, and passed Don Bluff—which is a cleared piece of land, with dead trees upon it, about 2 miles westward of Devonport—round Mersey Bluff at the entrance of the port, and steer for the opening of Mersey River. The range light towers will then be seen ahead; keep the towers in line bearing 205°, and proceed inward, leaving the red buoy and beacons on the starboard, and the black on the port hand. Anchor abreast the railway station.

At night.—The following directions are given for entering the river at night, but local knowledge is necessary in addition. Keep Mersey Bluff Light white until the two range lights on the west shore of the river are in range. Proceed in with these lights in range until abreast of Police Point, then keep in mid-channel until abreast the railway station, when anchor. On entering, care should be taken not to open the range lights to the westward, as the western edge of the channel is only a few feet off Aitkenhead Spit.

Devonport (lat. 41° 11′, long. 146° 24′), East and West, is a flourishing seaport town near the mouth and on both sides of the Mersey. Steamers run weekly to Melbourne, regularly to Sydney and other ports of Tasmania, and it is a favorite resort of tourists. It is connected by railway and telegraph. Population is about 5,000.

Wharves.—Devonport has 400 yards of wharfage on the west side of the Mersey River above the railway station, and 400 feet on the east side, with depths of 14 to 18 feet alongside, which will be dredged to 20 feet. The extent of wharfage is being considerably increased.

Warping buoys are laid down in the harbor.

Coal—Water—Provisions.—About 500 tons of coal are kept in stock and can be procured at either wharf. Fresh water is laid on at both wharves, and provisions are easily obtained at moderate prices.

Patent slip.—See Appendix I.

Repairs.—Devonport possesses many natural facilities for repairing vessels, as they may be laid upon the hard shingle without the least danger, and take advantage of the sawmills, where every kind of timber of the best description may be purchased at a moderate cost, and resident shipwrights may be procured.

Small repairs can be executed at Findlayson's foundry. Iron castings up to 15 hundredweight, also gun-metal and phosphor-bronze castings, can be made. The largest lathe will turn up to 14 inches diameter and 20 feet in length.

Trade.—The principal exports are potatoes and grain, chiefly oats. The largest vessel which has cleared at Devonport was 3,684 tons register, drawing 22 feet of water.

Time ball.—A time ball is dropped by electricity from Hobart at 1 h. 00 m. 00 s. p. m. standard time of Tasmania, equivalent to 15 h. 00 m. 00 s. Greenwich mean time. It can not be depended upon.

Latrobe.—The township of Latrobe is situated at the head of the navigable part of the river. The wharves at Latrobe have tramways and trucks for unloading vessels; and good commodious buildings have been erected for stowing grain and other produce. There is railway communication by the Western Line, and it is a telegraph station. Population is about 2,000.

Don River (lat. 41° 10′, long. 146° 22′), 1½ miles westward of Devonport, is narrow, although quite safe for vessels of 100 to 200 tons, which keep up a trade with this port. Don Bluff is higher than Mersey Bluff, and has cultivated land and dead trees upon it. A reef, which projects ¼ mile from Don Bluff, serves to break the sea from the immediate entrance.

Although there is no bar at the entrance of the Don River, there are only 4 to 5 feet at low water; but at high water springs there are 12 to 14 feet. A buoy is moored with a heavy anchor and chain to the northwestward of the mouth of the river and another buoy in mid-channel 200 yards from the immediate entrance. These buoys may be passed on either side and are fitted with shackles to enable sailing vessels to warp in or out.

Pilot.—Assistance may be obtained from the heads of Don River by sending a boat on shore, or a pilot may be procured at Devonport by standing off that port with the pilot jack flying.

Directions.—After making the entrance of Don River (lat. 41° 10', long. 146° 22') at a little to the westward of it, as directed for the neighboring ports, stand in for Don Bluff, and having passed the buoys on either side, proceed in, leaving a beacon at the end of the western reef, on the starboard hand, and with the prevailing north-westerly winds, luff up to the jetty, or run the vessel aground upon the bank, which may be done with perfect safety.

Don is a township near the mouth of the River Don and 2 miles westward of Devonport. The population is about 500, chiefly employed by the Don Trading Co.

Repairs.—Several vessels belong to the proprietors of the coal mines in the vicinity, and there is every facility for repairing small vessels in Don River, there being a gridiron, by means of which the bottoms of vessels of 300 tons may be repaired. There is a steam sawmill in constant work, with excellent timber of all kinds fit for shipbuilding, and resident shipwrights may be engaged on reasonable terms.

Exports.—Piles of the largest dimensions may be procured, and there is a constant export of timber, both sawn and split; also coal and farm produce.

Tides.—It is high water, full and change, in Don River at 11 h. 35 m.; springs rise 8 to 9 feet.

Forth River, the mouth of which forms Port Fenton, lies 4 miles westward of Don River, and has a bar at the entrance, which some 50 years ago was fordable on foot at low water. A reef projects from each head, and the entrance is difficult of access, on account of the changing nature of the channel. At 1 mile northward from the mouth of the river, a bank is said to have been formed, upon which the sea breaks at low water. Forth River is deep within the bar, and vessels of about 100 tons load afloat alongside the stores erected on the bank of the river.

Rock.—A rock with a depth of 2¼ fathoms over it lies 5¾ miles 286° from River Mersey lighthouse. The rock is about 700 yards in extent. The red sector of the River Mersey light shows over the rock.

The estuary of the Forth offers ample and excellent harborage
for moderate-sized craft, vessels drawing 13 feet lying there at any
state of the tide without touching, and the improvement of the
harbor by scouring the bar has further increased shipping facilities.

Pilot.—Regular traders are assisted from time to time by marks
or beacons placed on the land to show the channels; but a pilot may
be obtained by hoisting the signal, and the river should not be attempted without local knowledge, especially in rough weather, when
the sea breaks across the bar.

Directions.—In making the heads of Forth River (lat. 41° 09′,
long. 146° 17′), keep well to the westward to counteract the easterly
set, and steer for the entrance; bring Mount Roland, a precipitous
mountain, 4,047 feet high, which is situated 19 miles southward from
the entrance, to bear 181°, thence proceed as guided by the marks
and beacons, in charge of a pilot.

Leith, a post and telegraph township, is situated on the eastern
side of the river, near the entrance or heads; there is a station of the
Western Railway here.   Population is about 150.

Hamilton-on-Forth is situated on both sides of the river, about
2 miles from Leith, a bridge spanning the river.   It is a post and
telegraph station and coaches run to Leith railway station, meeting
every train.   Population is about 450.

Exports.—The exports of Forth River consists of posts, rails,
paling, and farm produce.

Tides and tidal currents.—It is high water, full and change,
in Forth River, at 11h. 30m.; springs rise 10 feet.   In the Forth,
like the other rivers on this coast, the tidal currents are rapid, and
the ebb is accelerated in winter by the river freshets.   This, together
with the seldom-failing night calms and early morning land breeze,
enables vessels to make a good offing before meeting the sea breeze.

Leven River (lat. 41° 09′, long. 146° 13′).—The entrance of this
river, which is open to the northeast, lies between masses of irregular
and pinnacle-shaped rocks and ledges, with a bar across.   The
mouth, 3½ miles westward of Forth River, is wide and well sheltered
from the prevailing westerly winds by the reefs of rocks extending
1,500 yards from Dial Point, the western entrance head, under the
lee of which there is good temporary anchorage outside the bar
in moderate weather for vessels awaiting the tide.   The coarse sandy
bar at the entrance of Leven River, which seldom alters, is fordable
on foot at low water, but at high water generally has 9½ feet water
on it.   Efforts are being made to get the bar deepened.

Bar or **Channel Rock** lies ¼ mile in a northeast direction from
Black Jack Rock, which has a white beacon on it, and this bar rock
is the principal danger in Leven entrance.   Half-tide Rock lies
about ½ mile eastward of Bar Rock; this rock is sufficiently distant

from the entrance not to form a danger if its position is known; its name signifies its depth.

Bar or Channel Rock dries 1 foot at low water, and in respect of its hidden character at nearly all times of tide, as well as in respect of its position, may well be considered a serious obstacle to navigation. Black Jack Rock with its beacon is an excellent guide. There is only one obstruction after the bar is crossed; this is Mussel Bank, a bed of stones and gravel 300 feet long by 30 broad, dry at low water.

Range lights.—Two range lights, situated on the western side of the entrance to the river, in range 191°, lead over the bar. The light towers are painted white.

Ulverstone.—This post and telegraph township is situated on the east bank of Leven River, a short distance within the entrance. It is in a good pastoral and agricultural district, and there is very picturesque scenery up the river. It is on the Western Railway. Small steam vessels from Launceston make this a port of call. The population is about 2,000.

Pilot.—Vessels of 80 to 100 tons frequent Leven River, it being commodious and safe for vessels of light draft to go in or out, but strangers should make a signal for assistance before entering.

Directions.—In coming from the eastward, make for a gap in Dial Range, on the west side of Leven River, and when off the entrance obtain a pilot. To enter, steer for Black Jack, a large isolated round rock having a white beacon on it, which is the second beacon observed in standing in. When Black Jack is covered there is said to be 10½ feet water on the bar. East of this is Half-Tide Rock, uncovered at half tide, which will be avoided by keeping Black Jack Rock bearing 189°. Leave Black Jack Rock on the starboard hand and anchor off Macdonald's public house, which is situated in the township of Ulverstone.

Exports.—The exports consist of split timber of every sort; the timber is of good quality and well adapted for shipbuilding and railway sleepers. Farm produce of various kinds is also exported from Leven River, which, from the fertility of the adjacent land, is likely to increase.

The climate, like that of the other rivers on this coast, is salubrious and admirably adapted to invalids.

Tides.—It is high water, full and change, in Leven River at 11h. 45m.; springs rise 9½ feet.

Coast.—**Dial Range** is a ridge of mountains 1,590 to 2,100 feet high, some 5 miles westward of Leven River mouth, and terminating to the northward, in two headlands lying northwest and southeast distant 2½ miles from each other, the southeastern projection being Dial Point, the western entrance head of Leven River. Both heads are fronted by dry and covered rocks, some of which appear to extend above ¼ mile from the shore, with depths of 6 to 7 fathoms close outside them. From the northwestern of these two headlands the coast trends west-northwestward 6¼ miles to Blythe River.

Penguin Creek (lat. 41° 05′, long. 146° 03′).—The coast between Leven and Blythe Rivers is intersected by Penguin Creek, which is merely a boat harbor.

The mouth of this creek is sheltered from westerly and northwesterly winds by a small headland, and a ledge of rocks which covers at half tide partially protects it. The ledge extends in a northeasterly direction ⅛ mile from the point westward of the creek

A wooden jetty extends 142 yards in an easterly direction from the western side of the creek's mouth; the jetty was built to shelter the corner in which the creek's mouth is situated. Vessels of 8 or 6 feet draft run in alongside the jetty at high water, and lie aground while taking in or discharging cargo. The scend of the sea toward high water is at times considerable. At about 100 feet off the end of the jetty there is a depth of 14 feet at high water. There is no bar at Penguin Creek; from the end of the jetty seaward the water first deepens gradually and then more rapidly; at ¼ mile off there are depths of 4 to 5 fathoms.

Penguin, a seaport with a post and telegraph station, is on Penguin Creek. There are lodes containing silver and copper in the locality; also iron mines. The district possesses a considerable area of first-class agricultural land. There is railway communication with Launceston and Hobart; also with Emu Bay and Macquarie Harbor.

Population is about 1,200.

Blythe River is only accessible to small vessels, the entrance being narrow with a dangerous rock in it, which might be removed, as it is a flaky, rotten stone. No vessels should attempt to enter without the assistance of a person acquainted with it. There is a well-constructed bridge over the river, 1 mile above the heads, for the Circular Head Road.

Small vessels frequent Blythe River for paling, posts, and rails; and there is a large quantity of good splitting timber in the vicinity. Good fishing and shooting can be obtained.

Round Hill Point.—Round Hill Point, west-northwestward 1¾ miles from Blythe River, is backed by a hill 717 feet high, after which it is named.

On the hill are two white beacons, which, when in range, lead eastward of the rocky reef extending from Blackman Point. Between Round Hill Point and Blackman Point, 2¼ miles westward of it, is Emu Bay, into which flows the small river of that name.

Emu Bay (lat. 41° 03′, long. 145° 57′) is 2 miles broad by 1,500 yards in depth, and is open to winds from north round to east;· it has a very even bottom and good holding ground. Its disadvantage is that during easterly winds embarkation or disembarkation is not always possible. Emu Bay is reported to afford safe anchorage in all weathers to seagoing vessels of any size possessing good ground tackle. The water in the bay deepens rather suddenly to 4 fathoms at low water. ·The breakwater at Emu Bay, extending eastward from Blackman Point, is formed of concrete, is 243 yards in length and has 27½ feet at low water at its extremity; small coasting craft and also the interstate steamers lie alongside.

Emu Bay is the outlet for the minerals obtained from the country extending about 40 miles to the southward, 25 miles to the westward, and 12 miles to the eastward of the bay. The minerals are tin, gold, silver, copper, iron, bismuth, zirconia, corundum, nickel and asbestos.

Pier.—A pier 600 feet long extends from the shore 160 yards southward from the breakwater, and has depths of 22 feet at its outer end on the northern side, and 20 feet at its outer end on the southern side, thence decreasing gradually toward the shore. Vessels berth on both sides of the pier.

Lights.—A flashing white light with red sector, visible 10 and 5 miles respectively, 39 feet above water is exhibited from the head of the breakwater.

A light is also shown from the outer end of the pier, to assist vessels going alongside, a green sector showing the direction of the pier.

Shoals.—Northward of Blackman Point are rocky patches with depths of 15 feet and 30 feet, respectively, over them. These patches are situated nearly 1,600 yards 338°, and 1 mile 352°, respectively, from Burnie Breakwater Lighthouse. Rocks which dry extend 800 yards northward from the point. Vessels rounding Blackman Point should not pass southward of the 30-foot patch.

Anchorage.—Anchorage may be had anywhere in the bay, according to the chart, but as a swell is nearly always setting in around the reef off Blackman Point, it is advisable to anchor in a position about 900 yards eastward from the breakwater lighthouse, where a vessel would be in 7 fathoms, sand and clay bottom, and where the swell would be less felt.

Tides.—It is high water, full and change, at Emu Bay at 9 h. 6 m.; springs rise 11 feet, neaps 8 feet, neaps range 6 feet.

Fish.—Very good seining can be obtained (except when the wind is from seaward on account of the surf on the beach). The best place is off the mouth of the Emu during the first of the flood, when fine gray mullet may be caught in abundance in the season.

Burnie, the town in the northwest corner of Emu Bay, has a population of over 3,000; it is rapidly increasing, principally as it is the port at which most of the minerals are shipped from the many valuable gold, silver, and tin mines in the northwest of Tasmania. It is a watering place with a fine sandy beach, which extends nearly 1½ miles, and is much frequented by visitors from Melbourne during summer.

There is biweekly steamer communication with Melbourne, weekly with Sydney, and more frequently with Hobart, Launceston, and Strahan. Coaches ply twice daily to and from Wynyard and once to and from Stanley. A well-constructed bridge carries the main road over the river Emu.

It is the junction of the Western Railway connecting with Launceston and Hobart on the main line, and Zeehan and Macquarie Harbor via Waratah. There is a post and telegraph station here.

Supplies.—About 200 tons of coal is usually kept in stock, but vessels can not depend upon obtaining any supply. Provisions are easily obtained at moderate prices.

Water.—Fresh water is laid onto the breakwater and pier, the charge being 3s. for the first 1,000 gallons and 1s. 6d. after.

The coast, from Blackman Point, curves west-northwestward 7 miles to a sandy projection, between which and Table Cape, 3¼ miles northwest from it, is a bay with reefs extending about ½ mile from its southern shore. The Inglis River flows through the reefs into the bight of the bay. The coast between Blackman Point and the bay is bordered by a reef, and is intersected by several streams of which the largest is Cam River, 3¼ miles westward of Blackman Point.

The light on Table Cape shows red along this coast to 1 mile seaward of Blackman Point, see next page.

Inglis River (lat. 40° 59′, long. 145° 46′).—The mouth of the river, about 3 miles southward of Table Cape, is open to the northeastward and is protected from all winds west of north-northwest by the cape; it is further protected by a ledge of rocks extending to the northeast from the left bank of the river, upon which ledge an embankment of stones has been formed; it is also protected from the eastward by ledges of rocks extending from the shore in that direction. There is said to be no outer bar at the Inglis; the depth over the inner one is about 8 feet.

At the wharves in the river there is a depth of only 5 feet at low water and not much room in the stream. Small craft of 8 feet draft use the Inglis and lie aground while taking in cargo. On the right bank of the river a wooden embankment has been formed, 450 feet long, which is said to have had the effect of driving the bar seaward for a considerable distance.

Lights.—From two piles on the south side of the entrance to Inglis River are exhibited two lights, which in range bearing 249° lead over the bar.

Wynyard.—The post and telegraph township of Wynyard is at the mouth of the river. Two coaches run daily to and from Burnie Station, and there is a daily coach to Stanley. In the neighborhood are well-appointed sawmills, and the land is some of the best in the island and is well watered. Rich indications of gold, silver, and copper have been found. Population is about 1,400.

Table Cape is the cliffy extremity of the woody flat-topped land, 380 feet high.

Light.—A fixed white light with red sector, visible 27 and 10 miles, respectively, 390 feet above water is shown from a white brick lighthouse on Table Cape, 50 feet high.

Caution.—The seaward edge of a red sector of the light passes 1 mile off the reef off Rocky Head, and the same distance off the reef extending from Blackman Point; therefore, keep in the white light.

Lloyd's signal station.—There is a Lloyd's signal station at Table Cape Lighthouse; it is a telegraph station in telephonic communication with the system of the state, and is easily accessible by boat. Communication can be made by the International code.

From Table Cape the coast extends westward 7 miles to a low point surrounded by the Sisters, two remarkable round hills 870 feet high. A reef, with a small islet on it, projects northward nearly 1 mile from the point; and a detached patch lies east-northeastward about ½ mile from the islet.

There is a small boat harbor 5 miles westward of Table Cape.

On the west side of Sisters Point is a sandy bay, with 2 fathoms water near the shore, and a small stream flowing into it; this bay is apparently protected from the eastward by the reef, with the islet on it, which projects from the point. The coast from Sisters Point to Rocky Head, 5 miles northwest from it, is bordered with rocks, but it may be approached to a mile in from 9 to 10 fathoms.

Rocky Head (lat. 40° 51′, long. 145° 32′) has a high pointed summit, with other peaks inland, rising to the height of 1,000 feet. The head is bordered with rocks; and a rock, 2 feet dry at low water, surrounded by a reef, lies northeastward nearly 1¼ miles from the head. A red sector of the light on Table Cape shows over it.

Sawyers Bay extends from Rocky Head west-northwestward for 11 miles to Circular Head, and is 3½ miles deep, with low sandy shores, except between Detention River, 3 miles southwestward from Rocky Head, and Black River, 5 miles southward from Circular Head, where the shore is rocky, with hills rising behind it. From a depth of 19 fathoms 2 miles northwestward of Rocky Head there are from 16 to 8 fathoms across the bay to about 2 miles southeastward of Circular Head, with 3 fathoms close off the beach near Detention River, and 3 to 7 fathoms in the bight close to the southward of Circular Head.

Anchorage in Sawyers Bay can be obtained in the bight southeastward of Circular Head, in depths of 5 to 8 fathoms, sheltered from westerly winds.

Tides and tidal current.—It is high water, full and change, in Sawyers Bay, at 11 h. 40 m.; springs rise 9 feet. The northwest going current begins two hours before high water.

Circular Head is the east point of a peninsula which projects 4½ miles northward from the coast, and is ½ mile to 1½ miles broad; the isthmus which connects this peninsula with the mainland is low and narrow, with an inlet on either side. The head, which appears from the eastward like a small flat-topped island, is a singular mass of trappean rock, rising abruptly from the sea to the height of 460 feet, and is visible in clear weather from a distance of 30 miles. A slight covering of grass, with some bushes, gives it a smooth appearance. The head is connected with the peninsula by a narrow neck of lower land. See view on H. O. Chart No. 3441.

Signal station.—There is a signal station on Circular Head; it is connected by telegraph and is easily accessible by boat.

North Point (lat. 40° 42′, long. 145° 17′), situated 3¾ miles northwestward from Circular Head, is a low shingle point with a dangerous rocky ledge, drying 3 feet at low water, extending east-northeastward 1,500 yards from it, on which several vessels have run. This ledge may be avoided by keeping the bluff extremity of Circular Head open of an intermediate projection of the land.

The light at Hyfield (see next page) shows red over this ledge.

Beacon.—A circular concrete column, surmounted by a red staff and ball, the whole 19 feet high, stands 300 yards from the eastern extremity of the reef extending from North Point.

Depths of 10 fathoms and over will be found 1 mile to the eastward of this beacon. There are heavy tide rips off this reef with easterly winds.

Shoal water also extends 1 mile nortwestward from North Point.

Tidal currents.—Both the flood and ebb set over the reef eastward of North Point at the rate of from 2 to 3 knots an hour.

Coast.—Between Circular Head and North Point is a rocky point, Hyfield, 134 feet high, with dry rocks lying 1,500 yards to the northeastward.

Between Circular Head and Hyfield is a small sandy bight with 4 to 6 fathoms water in it.

Half-moon Bay is between Hyfield and the reefs off North Point.

Light.—A light, elevated 50 feet, is exhibited from a white hut situated on the northern extremity of the rocky point (Hyfield) midway between Circular Head and North Point.

Stanley is a rather important seaport town. It was laid out by the Tasmanian Land Co., and is situated on a flat facing the bay on the south side of Circular Head.

Communication.—It is the nearest port to Melbourne, with which port there is frequent steam communication, and there is regular communication weekly with Launceston and Macquarie Harbor by steam vessel, and also by traders, and a daily mail service by coach to Wynward and Burnie, thence by rail. The district is agricultural and contains some of the finest grazing land in the island; there is a large trade with Sydney, Adelaide, Melbourne, and Tasmanian and Queensland ports, consisting of cereals, potatoes, apples, pears, horses, cattle, sheep, and pigs. There is a post and telegraph station here. The population of the district is about 5,000.

Breakwater—Pier (lat. 40° 46', long. 145° 18').—A breakwater extends from the south side of Circular Head in a southerly direction for a distance of 215 yards, having a depth of 29 feet at the outer end and 15 feet at the inner end at low-water springs. The outer 70 yards, broken up by a gale in 1914, now consist of detached blocks of concrete, which, however, are not dangerous to shipping.

A pier about 200 yards westward from the breakwater extends from the shore in a southerly direction for a distance of 100 yards, having a depth of 22 feet at the eastern end of the head and 19 feet at the western.

Lights.—A light is exhibited 71 yards from the outer end of the breakwater and a similar light at its inner end; both these lights are stated to be obscured from seaward, but visible over the anchorage.

A light is exhibited from the eastern head of the pier.

A light is also shown from a small jetty, not shown on the chart, to the westward of the above pier.

Coast.—From the western side of North Point the coast, with a beach of sand and shingle, trends southward 4 miles to West Inlet; thence westward, with a sandy beach, 6 miles to the opening to Duck Bay. Westward of the opening is Perkins Isle, the sandy coast of which trends northwestward 3½ miles to Robbins Passage. Cape Elie, on the northern side of Robbins Passage, is 2½ miles north-northeastward from Perkins Isle.

Perkins Bay, between North Point and Cape Elie, is 8 miles wide and 4½ miles deep. There are from 8 to 4 fathoms of water across the entrance of the bay, with regularly decreasing depths toward the shore; the bottom being sand over clay affords good holding ground and good anchorage with easterly winds. A heavy swell runs into this bay with strong northerly and northwesterly winds.

West Inlet, in the southeast corner of Perkins Bay, extends for a distance of 2½ miles in a west-southwesterly direction parallel to the south shore of the bay, forming a point half a mile wide, which is covered with dense timber. At low water this inlet dries, leaving only a narrow channel running down the center of it. The entrance to the inlet is 2½ miles southward from North Point and is a narrow channel with depths of 8 to 18 feet, running eastward for ½ mile and then southward for 1¾ miles, close to the shore.

Duck Bay and River—Anchorage.—The southwestern extremity of Perkins·Isle is separated from the mainland by a narrow channel 300 yards wide, which dries at half ebb; this, and the other opening between the southeast point of the isle and the spit of the sandy beach, communicate with Duck Bay, a land-locked inlet, 5¼ miles long, east and west, and 2 miles wide at its broadest part, with Duck River flowing into its south and Deep Creek into its east corner.

Inside the entrance of Duck Bay there is safe anchorage, in any winds, in from 3 to 4 fathoms water, sandy bottom.

Duck Bay nearly all dries at low water, leaving only two narrow channels running to Duck River and Deep Creek.

Smithson, a small township at the mouth of the Duck River, is 13 miles from Stanley, to which it is connected by road. A coach runs daily between Smithson and Stanley. Population is about 700.

Pier.—A wooden pier 1,750 yards long extends from the western point of entrance of the Duck River in a northerly direction toward the entrance of the inlet. The pier dries for most of its length, but has depths of from 9 to 18 feet at its extreme end. A tramway connects the pier to Smithson.

Small vessels may anchor off the pier in 3 to 5 fathoms of water, sandy bottom, protected from all winds.

Beacons.—To mark the channel into Duck Bay two sets of beacons have been erected, one pair on the eastern point of entrance bearing, when in range, 163°, and the other pair near the pier bearing, when in range, 200°. As the entrance to the bay is continually shifting, these beacons do not now lead over the deepest water, and the channel should therefore be only used by those possessing local knowledge.

Depth on bar.—There was a least depth of 7 feet at low water ordinary springs on the bar in 1912. The channel, however, is very narrow. The channel from the end of the pier to Smithson is continually altering, and is marked by piles as necessary.

Coast.—From the southwestern extremity of Perkins Isle the coast extends 12 miles in a westerly direction to the east entrance point of Welcome River. This coast consists of small inlets and points, and is intersected by two streams, the Montagu and Harcus Rivers, at 5¼ and 9¼ miles from Perkins Isle. A small islet from 1 to 3 feet high lies close off the mouth of each stream, and Long Islet lies 1,500 yards off the west point of entrance to Harcus River.

Robbins Passage, which separates Robbins Island from the mainland, is bounded to the southward by Perkins Isle and the coast from thence to Welcome River, and to the northward by Robbins Island. The eastern entrance to Robbins Passage, which appears like the mouth of a river, is 2 miles wide between the north point of Perkins Isle and Cape Elie, the southeast point of Robbins Island. A bank of sand and shingle, 600 yards in extent, drying 3 feet at low water ordinary springs, lies in the middle of the eastern entrance to the passage, and a sandy spit extends to a distance of 1,400 yards off the north point of Perkins Isle. There is a least depth of 9 feet between the middle bank and Cape Elie, the depth inside then increasing to 2 and 3 fathoms.

Two miles within the eastern entrance the channel, which is here only 800 yards wide, divides into two narrow gutters with depths of from 9 to 18 feet, the banks on either side being sand and mud, drying at quarter ebb.

At Stony Point on the mainland, 3 miles within the entrance, there is a small pier with a depth of 4 feet alongside it at low water. A road runs from this pier to the township of Montagu. A rocky spit extends 1¼ miles to the northward from Stony Point, being broken in the middle by a narrow channel with a least depth of 3 feet in it. Goat Islet, 27 feet high, lies on this spit on the northern side of the channel.

To the westward of Stony Point the channel narrows considerably, with depths of from 6 to 9 feet in it, and runs toward Kates Point, the southern point of Robbins Island, close to the westward of which it stops, a bank of sand and mud extending right across between Robbins Island and the mainland of Tasmania. The channels in Robbins Passage are continually shifting, and should only be used by small craft with local knowledge.

The village of Montagu lies on the side of a bare hill, 1¾ miles inland from Stony Point. It is connected by road with Smithson and Stanley, and also the west coast of Tasmania. A coach runs daily between Montagu and Stanley, passing through Smithson and running to the west coast.

Coast.—Between Montagu River and Woolnorth Point the land is low and thickly wooded; a flat of sand and mud, which drives in patches, extends 2 to 3 miles off the shore.

Woolnorth Point (lat. 40° 38', long. 144° 44'), the northwestern extremity of Tasmania, is low and rocky, with low sandhills partially overgrown with coarse grass and scrub. Two miles southwestward from the point the land rises to a height of 269 feet, and becomes, toward the west coast of the island, open undulating land from 300 to 400 feet in height. At 3 miles from the point is an out station of the Tasmanian Agricultural Co.

From Welcome River the general trend of the coast is northwest for 6 miles to Woolnorth Point. There are several islets and rocks close to the east side of Woolnorth Point, the largest two being Murkay and Harbor Islets, the former lying 1 mile to the southeastward, and the latter close to the northeastward of the point.

Hunter Group consists of three principal and many small conspicuous islands, extending 28 miles northwestward from the southeastern extremity of Robbins Island to Albatross Islet and 18 miles north-northeastward from Woolnorth Point to the northeastern extremity of Three Hummock Island, and includes Black Pyramid, which lies 20 miles west-southwestward from Cape Keraudren, the north point of Hunter Island.

Robbins Island, the southeastern and second in size of the Hunter Islands, is a sandy island of a somewhat triangular form, 7 miles long in a north and south direction and 8½ miles wide on its southern side. The island is generally flat and swampy, but a ridge of hills, the summits of which are bare and from 166 to 228 feet high, lies at the southwest end of the island, and a ridge of timbered hills, 250 feet high, lies near the east point of the island. A belt of thick timber extends for 1,500 yards from the southern shore of Robbins Island.

Cape Elie, the southeast point of Robbins Island, is low and sandy. Fresh water may be obtained from wells at this point, and there are also two small fresh-water swamps. There is a farmhouse at Cape Elie, from which fresh meat may be obtained at a reasonable price.

From Cape Elie the east coast of Robbins Island extends for a distance of 2¼ miles in a northwesterly direction to Guyton Point, thence in a westerly and northwesterly direction for 5½ miles to the northeast point of the island.

Guyton Point divides the east side of Robbins Island into two sandy beaches, the northwestern and more extensive of which forms a slight indentation, called Ransonnet Bay, with a depth of over 3 fathoms of water at a distance of ½ mile off it. Between Guyton Point and Cape Elie the bottom is foul for a distance of 600 yards offshore, but depths of 5 fathoms and over will be obtained at a distance of 1¼ miles off it.

175078°—20——18

Between the northwest end of the largest of these sandy beaches and the north point of the island is an inlet about 1,200 yards in extent, locally known as Mosquito Cove; this, however, all dries at low water.

Walker Island, only separated from Robbins Island by a narrow winding channel which dries in places at low-water springs, is 3 miles long, north and south, and 1 mile wide in its broadest part. A fringe of low hills borders the east and west sides of the island, rising to a height of 88 feet at its southeast point.

A chain of small islets and rocks, the highest 58 feet in height, extends eastward for ⅓ mile from the north point of Walker Island, and a small rock, 15 feet in height, lies 1,200 yards southwestward from the same point and 400 yards offshore.

The west sides of Robbins and Walker Islands are bordered by a sand flat which dries at low water and extends from ½ to 1 mile from the shore.

Petrel Islets, a cluster of four principal islets, lie off the north end of Walker Island.

The largest islet lies ½ mile northward from the north point of Walker Island, and rises in a sandhill, the top of which is covered with scrub to a height of 73 feet. The outer, and northeastern, islet is a reddish-colored rock, and lies about ⅓ mile northeastward from the last-mentioned islet, and a rock 1 foot high lies off its northern end. The other two islets lie respectively close northwestward and southwestward from the main islet, to which they are almost joined at low water. They are of the same character as the northeastern islet, and are 69 feet and 65 feet high, respectively.

There is a channel for quite small vessels between Petrel Islets and Walker Island, but it is not recommended.

Petrel Bank, of mud and sand, with a least depth of 1¼ fathoms over it, extends to the eastward of the northeastern Petrel Islet. The eastern extremity of this shoal, as defined by the 3-fathom curve, is situated in a position 3½ miles, 100°, from the northeastern Petrel Islet; the northern edge of the same shoal, which is steep-to, is situated 1½ miles, 66°, from the northeastern Petrel Islet. The shoal is dangerous to vessels rounding the Petrel Islets, as no clearing marks can be given and the tidal currents are strong.

Middle Bank is a sand bank, 4 miles in length in a northeast and southwest direction, lying between Walker and Hunter Islands, and 2 miles in breadth. It is separated from the banks extending from Walker Island, by Walker Channel; its southern part dries 6 to 7 feet. Its east and west edges are steep-to, but it shoals gradually to the northward.

Walker Channel, about ¼ mile wide, has average depths of 7 to 8 fathoms.

Three Hummock Island, the south point of which lies north-northwest, distant 6 miles from the north point of Walker Isle, is the northeastern island of the Hunter group, being 6 miles long, north-northeast and south-southwest, and nearly 5 miles broad. It is of an oval form, with a bay on its northwest side, and a coast ridge of moderately elevated land, partly bare of vegetation, extending from the south to the northeast point of the island. Three hills, from which the island derives its name, rise gradually from this ridge, the southern, a conical peak 784 feet high, 1 mile northeastward of the south point, is the most elevated part of the island.

The northern hill, 1 mile southwest of Cape Rochon, the northeast point of the island, is 551 feet above the sea and densely wooded. At a distance of 1½ miles south of this hill is the third and intermediate hummock, 380 feet high.

Between the south and northeast points of Three Hummock Island the coast consists of sandy bays and rocky points. Cape Adansan is the name given to the rocky point on the east coast of the island to the southward of the sandy beach.

On the northwest side is Coulomb Bay, a broad shallow bay, with a long sandy beach. The shores of the island generally consist of a number of rocky points with several off-lying bowlders; one 10 feet high lies 300 yards off the south point of the island, with 15 fathoms water at a short distance. The largest and most conspicuous of the several bowlders on the western point of the island is 25 feet high. The tidal currents are strong on the south side.

Anchorages.—There is good anchorage off both the northwestern and eastern sandy beaches with offshore winds.

Tides.—It is high water at full and change at Three Hummock Island at 11 h. 30 m.; springs rise 8 feet.

The channel between the Petrel Islets and Three Hummock Island has depths of 12 to 15 fathoms; the tidal current sets east and west, at from 1 to 2 knots.

Taniwha Rock, with a depth of 5 feet over it at low water springs, is situated ½ mile, 74°, from the east point of Three Hummock Island, with deep water around it. A rock with 7 feet water on it lies 400 yards northwestward from it, and some rocks, 1 to 3 feet above water, lie between it and the shore.

Mermaid Rock is a cluster of several rocky heads which dry from 2 to 3 feet, situated ½ mile northward from Cape Rochon; there is deep water close to the rock, and a passage a little more than 200 yards wide between it and the shore.

From Cape Rochon, the northeast point of Three Hummock Island, its rocky coast trends west 3½ miles to the northwest point, and thence southeast 1 mile to the northeast point of Coulomb Bay, which extends 2½ miles in a southwest direction and is 1 mile deep; between the points of the bay there are from 4½ to 7 fathoms water. A short distance behind the beach is a small lagoon of fresh water. A projection 1,500 yards south of the southwest point of this bay forms the west point of the island which, although rocky, may be rounded at the distance of ½ mile, in 11 to 7 fathoms water.

Water is plentiful on Three Hummock Island. The island is covered with an impervious scrub, the trees being small and stunted.

Hope or Peron Channel, between Hunter and Three Hummock Islands, is 2 miles wide, with depths of 8 to 20 fathoms water, and apparently no other hidden dangers than the rocks which closely border the west point of Three Hummock Island.

Hunter (or Fleurieu) Island, the western and largest of the Hunter group, and formerly known as Barren Island, is 12 miles long, north and south; at the middle and broadest part of the island it is 4 miles across, with a small rocky bight on the west side. The southern part of Hunter Island is 2¼ miles broad between Renard and Perigo Points, but the northern part, from 1¼ miles in breadth near the middle, narrows gradually for a distance of 5 miles to Cape Keraudren, the north point of the island. It is moderately elevated, the highest part, Chase Hill, being 300 feet high at 3¼ miles southward from Cape Keraudren. Its northern part has a most barren and sterile appearance, but its southern coasts are formed by wooded hills of moderate height.

Cape Keraudren (lat. 40° 24′; long. 144° 47′) is a low, sloping, rocky point, with 23 to 24 fathoms within ½ mile of it.

The east side of Hunter Island is nearly straight, north and south, and has small sandy bays between its slightly projecting points, off which there is a good anchorage in a moderate depth, with shelter from all but easterly winds. A bank, with 2 to 3½ fathoms on it, borders the east side between 2 and 5½ miles from the south point, with its north end extending 1½ miles from the shore. Depths of less than 5 fathoms extend off this bank to a position 2½ miles eastward of the middle of the eastern coast of Hunter Island. From the northern extremity of this bank to Cape Keraudren, there are 12 to 7 fathoms water within ½ mile of the shore.

Anchorage.—The best anchorage on the east side of Hunter Island is in about 8 to 10 fathoms, from 1 to 1½ miles off-shore, with the west point of Three Hummock Island 19° distant 3½ miles. The approach to this anchorage between Hunter and Three Hummock Islands has a least depth of 8 fathoms.

Dangerous Bank, with a least depth of 3½ fathoms, and 1,400 yards in extent, northeast and southwest, within the 5-fathom curve, and upon which the sea breaks mainly with a moderate swell, lies westward of Cape Keraudren with its shoalest spot 256° distant 1.9

mile from the cape. The tidal current sets strongly over the bank and round the cape. There is deep water between the cape and the bank, but the channel is not recommended.

Cuvier Bay, which extends 6 miles south-southwestward to the west point of Hunter Island from Cape Keraudren, is 1¾ miles deep in its southern part. As this bay is entirely exposed to westerly winds, it can not afford desirable anchorage; but there is temporary anchorage during south and easterly winds in the south part in 8 to 10 fathoms, mud, 1,500 yards off shore. The coast in the bay is steep, except near the northern part, and rocks extend ½ mile northward from the western point of the bay, which should be given a berth of a mile.

Cutter Rock, with a depth of 18 feet over it at low water, is situated 1,400 yards westward from the west point of Hunter Island. A rock which dries 1 foot at low water is situated ½ mile southward from the above. These should be given a wide berth.

The west side of Hunter Island is rocky, and as dangerous rocks and breakers extend considerably to seaward from the west point of the island, it should be carefully avoided.

From the west point of Hunter Island the coast trends 1½ miles southward to a point off which is an islet and several rocks, also a shoal spit extending ½ mile to the southwestward. From this point to Perigo Point, 4 miles to the southward, the coast is foul to a distance of ¼ to ½ mile from the shore.

Weber Point (lat. 40° 36′, long. 144° 45′).—From Perigo Point the rocky coast, with foul ground 200 to 600 yards off it, trends 2¼ miles southward to Weber Point, the southernmost point of Hunter Island. Between Perigo Point and Weber Point are Logan and Keafer Points. Weber Point is rocky, and shoal water extends 300 yards off it.

From Weber Point the coast of Hunter Island trends north-northeastward 1¾ miles to Renard Point, with foul ground extending 300 yards off the rocky shore between these points. Nearly ½ mile northward from Renard Point is the south end of Ainslie Beach, a sandy beach 1¼ miles long, with shoal water extending 1 mile eastward from its northern end.

Albatross Islet (lat. 40° 22′, long. 144° 40′), the northwestern of the Hunter group, lies 6 miles westward from Cape Keraudren; the islet is 1,500 yards long, north and south, ¼ mile broad, and 125 feet high, being visible in clear weather at a distance of 16 miles. The eastern side is an almost perpendicular cliff with deep water close to. Both sides are steep, but rocks extend 400 yards from the southern point of the island. When seen from a southwest by west or northeast by east direction a deep notch in the middle of the island appears to divide it.

There are strong tide rips over both ends of the island during the strength of the currents.

North Black Rock, 33 feet high, is about 50 yards in extent and lies 6 miles southward from Albatross Islet; it is steep-to.

South Black Rock, nearly 5½ miles south-southwestward from North Black Rock, is a round mass 127 feet high, and 300 yards in extent, with a rock which dries 7 feet, situated 900 yards southwestward of it; with this exception it is steep-to.

Tidal currents.—The tidal currents westward of these islets and reefs run at the rate of 2 to 3 knots, the flood to the northeastward.

Caution.—Soundings give no indication of approach to Albatross Islet, there being 25 to 34 fathoms within 1½ miles of its west side, and 31 fathoms at 1,500 yards from its north end, over a coarse ground with sand and shells. These depths correspond so nearly with those toward King Island and for several miles toward the westward of it, that in the night or in thick weather it should be approached with caution.

Steep Islet (lat. 40° 34', long. 144° 41'), 214 feet high, situated between South Black Rock and Hunter Island, is 600 yards in diameter, and its summit is covered with grass; its coasts are mostly steep cliffs. Rocky ledges extend off its north and south sides and a bank extends 600 yards off the east side. A rock, 10 feet high, lies 400 yards southwestward from its west point.

A rock, with a depth of 1¼ fathoms, is situated 84°, distant 1½ miles from the northern extremity of Steep Islet.

Roller Shoal, a bank of sand, rock, and gravel 400 yards in extent, north and south, with a least depth of 3 fathoms at low water, is situated 1¾ miles southwestward from Steep Islet. It breaks, except in quite calm weather.

Nares Rocks, situated nearly 1 mile southward from Steep Islet, consist of four small rocky islets, the largest of which is 29 feet high. A rock which dries 6 feet lies 800 yards west-northwestward, and another drying 8 feet, 200 yards eastward from the highest rock.

Delius Islet, 21 feet high, is the largest of a group of small rocks, situated 1¼ miles eastward of Nares Rocks, on a reef about 400 yards in extent. Rocks lie 100 yards to the southward and to the westward of the reef. A rock which dries 7 feet lies 600 yards to the northward from the islet. Two other rocks lie between it and the islet.

Bird Islet, 1½ miles westward from Weber Point, is 1,200 yards long, north and south, and 50 feet high at its southern end, with a channel ½ mile wide between it and Keafer Point, the southwestern extremity of Hunter Island, in which the depths vary from 7 to 23 fathoms.

Between Bird Islet and Woolnorth Point, 2 miles to the southward, the ground is foul.

Brown Rocks, 28 feet high, are situated nearly 1¼ miles to the westward of Bird Islet. They are a chain of bare rocks extending for 600 yards east and west. A rock with less than 6 feet of water on it lies 200 yards northward from the easternmost of the Brown Rocks.

Trefoil Islet, 1¼ miles westward from Woolnorth Point, is nearly 1 mile in extent, 276 feet high, and receives its name from its resemblance to a clover leaf. The southern and western sides are abrupt cliff; it is covered with grass, but bare of trees. The highest part of the island is over the southern point. Reefs extend from its northwest and south sides. There is landing on its eastern side on a shingle beach.

A bank of sand, with depths of 1½ to 3 fathoms, joins Trefoil Island with the mainland. It breaks heavily with westerly winds.

Little Trefoil Islet, a small rocky islet 25 feet high, is nearly midway between Trefoil Islet and the western side of Woolnorth Point.

Henderson Islets, situated about 1½ miles north-northeastward of Woolnorth Point, consist of a group of small islets and rocks occupying a space 1,500 yards long north and south; the northern islet is 15 feet high. They are situated on the foul ground which extends from Woolnorth Point to Bird Islet.

Crescent Bank is 4½ miles long and about ½ mile wide. It extends eastward, from a position 2 miles eastward from Woolnorth Point, as far as Middle Bank. Its shoalest part is near the western end, 2½ fathoms; the eastern part has depth of 3 to 4½ fathoms on it.

Stack Islet, 127 feet high, about ½ mile eastward of the south point of Hunter Island, is small and rocky, with Eagle shoal projecting about ½ mile from its northeast point. Rocks project 300 yards from its southwest point, and rocks and shoals ¼ mile from its west side, on the latter of which, Blanchard reef, there is a beacon with cage. Two small rocky islets, Dugay and Edwards Islets, lie off the northern end of Stack Islet to which they are almost joined at low water. There are 27 fathoms at ½ mile and 18 fathoms at 1,500 yards from the south side of the islet, with quickly decreasing depths to the southeastward, in the direction of the banks which extend from the mainland.

Between Stack Island and Hunter Island is Hunter Passage.

Penguin Islet, 45 feet high, is small and rocky. It is situated on the western part of Holbrow bank at about 2¼ miles northeastward from Stack Islet.

Holbrow Bank consists of two narrow parallel shoals divided by a gully, ¼ mile wide, in which are depths of 3½ to 7 fathoms. The eastern shoal is 4 miles long in a north-northeasterly direction and ¼ mile wide, with depths of 3 to 15 feet on it. Its southern end is 1

mile eastward from Stack Islet. · There is a space 1 mile wide with
depth of 7 to 8 fathoms between Holbrow Bank and Middle Bank
except at their northern ends which almost meet. The western shoal
of Holbrow Bank, on which is Penguin Islet, is 2½ miles long and
¼ mile wide. It extends 1 mile northeastward from the islet with
depths of 2¼ fathoms. To the southwestward of the islet it extends
·1½ miles, with 4 feet water at that distance and also at other places
between it and the islet. Rocks which dry at low water extent 600
yards westward from Penguin Islet.

Clearing mark.—The western extremity of ·east Doughboy Islet,
in range with Weber Point, 223°, leads westward of the rocks off
Penguin Islet and also of the northwestern edge of Holbrow Bank.

Hunter Passage.—The passage to the southward of Hunter
Island is from ¼ to 1 mile in width, the shoalest part is between
Weber Point and Stack Islet, where a bank with a least depth of 3¾
fathoms projects 700 yards westward from the islet, narrowing the
channel between the 5-fathom curves from each shore to about 100
yards. It is not recommended without local knowledge as the channel is tortuous and the tidal currents strong. The course of Hunter
Passage is between Hunter and Bird Islands, and thence westward
between Delius Islet and Nares Rocks on the north, and Brown Rocks
on the south.

. **Directions.**—Hunter Passage should be used with great caution,
as the course is tortuous and the tidal currents are strong.

For the western part of the channel, when approaching from the
westward, South Black Rock showing midway between the two
highest Nares Rocks, 288°, astern, leads between Bird Islet and
Delius Islet, and Delius Islet twice its own breadth open to the northward of the north point of Steep Island, 316°, astern, will lead in
mid-channel between the southwest side of Hunter Island and Bird
and Henderson Islets.

The only further directions that can be given are to proceed in ·
midchannel, rounding the south point of Hunter Island at a distance of not less than 600 yards, and to pay great attention to the
steering and to the set of the tides and tidal currents.

Tides and tidal currents.—It is high water, full and change, at
Stack Islet, at 0 h. 15 m.; springs rise 8½ feet, neaps 6½ feet; neaps
range from 5 to 6 feet. At and near springs, the tide ebbs for from
half an hour to an hour longer than it flows. At neaps the ebb and
flow are equal.

Eastward of Hunter Island the west-going or flood current begins
at from 4½ to 3½ hours before high water. It sets to the westward
between the Petrel Islets and Three Hummock Island, when it
divides, part running to the southwestward through the channels
between Walker Island and Hunter Island, and thence westward

between the islets southward of the atter; the other part sets northward up Hope Channel, turning sharply to the westward and southwestward round Cape Keraudren and racing with great strength
over Dangerous Bank. . The east-going or ebb current begins about
from 3 to 3½ hours after high water and sets in the opposite direction.

Westward of Hunter Island, the flood or west-going current is
met by the flood current setting to the northward up the west coast
of Tasmania.   In general, within a distance of a few miles westward
of Hunter Island the current runs in accordance with the currents
on the east side of the island, viz, flood to the southwestward and ebb
to the northeastward, and turn about the same time.   These currents,
however, are complicated by the currents on the west coast of Tas-,
mania, which often results in a preponderance of set to the northeastward.   The westerly and southwesterly gales of winter also
greatly affect the currents westward of Hunter Island.

The tidal currents set with great strength through Hunter Passage
and the chain of islets and rocks between Hunter Island and Woolnorth Point, attaining a velocity of 5 knots in places at springs and
forming heavy races off the points of the islands and many swirls
and eddies in the channel.

Directions from Stanley to Hunter Passage.—After leaving
Stanley and clearing North Point Ledge, a vessel bound westward
through the Hunter group should steer for the south hummock of
Three Hummock Island on a 307° bearing until the northeastern Petrel island bears 240°, so as to clear the Petrel Bank.   From this
position, if intending to proceed altogether by Hunter Passage, a
vessel should steer 268° until Stack Islet is open its own breadth
westward of Penguin Islet in order to clear the banks northward of
the latter, then alter course gradually to the southwestward and steer
in mid-channel between Stack Islet and Hunter Island.   If intending to use Walker Channel, from a position 60°, 3¼ miles from the
northeastern Petrel islet, steer to pass not less than 1,500 yards northward of that islet; then alter course toward the western Petrel islet,
and passing 400 yards westward of it, steer 198° until the southern Petrel island bears 43°, when proceed 223° with the islet astern, down
Walker Channel.   When the summit of Trefoil Island bears 260°,
steer for it between Crescent and Middle Banks, and when Stack Islet
comes in line with the highest hill (200 feet) on the southern part of
Hunter Island, 337°, proceed as requisite through the western part of
Hunter Passage, the ranges for which are given in the direction for
Hunter Passage.

Black Pyramid, situated 15 miles west-southwestward from
Albatross Islet, is the most prominent of the islets west of Hunter
Island, and is the first seen from a vessel approaching the Hunter
group from the westward.   Black Pyramid is a small, dark-looking

islet, with a round summit, 240 feet above the sea. It appears bold
to approach, there being depths of 24 to 25 fathoms within a mile
south, east, and north of it. See view on H. O. Chart No. 3441.

Channel between Tasmania and King Island.—The channel
between Hunter and King Islands is 38 miles wide and the position of the islets in it are known, but it is little used by vessels
going through Bass Strait, as the safer entrance between King
Island and Cape Otway is generally preferred.

Reid Rocks (lat. 40° 15′, long. 144° 10′).—The northwestern and
highest of these rocks, which lie in the northwest part of this channel, is a small, dark mass 40 feet in height; it bears 116°, distant
11.5 miles from Stokes Point, and has a rock dry at low water
eastward half a mile from it. The other patches of this cluster lie,
respectively, 1 mile southeastward and 1¾ miles southward from
the northwestern rock, and on the latter patch is the south Reid
Rock, 6 feet high. The space between and immediately around
these rocks is dangerous ground.

There are depths of 27 fathoms 1 mile northeastward of Reid
Rocks, and 35 fathoms between 3 miles to the southeast and southwest of them, showing that the lead is no certain guide for approaching these dangers at night or in thick weather; and as the
tidal currents here are rapid, this vicinity should be avoided at such
times, unless the position has very recently been well ascertained by
bearings of the land.

Bell Reef, bearing 202°, distant 8½ miles from the northwestern
Reid Rock, and 151° 14½ miles from Stokes Point, is about 1¼ miles
long, in a north and south direction, and ¼ mile broad, with 33
fathoms 1 mile east-southeastward from its southern extremity, and
36 fathoms, sand and shells, midway between it and Reid Rocks.

This reef lies much in the way of vessels using the passage south
of King Island, and is the more dangerous as the sea only breaks
at intervals on it, even with a heavy swell.

Clearing marks.—Black Pyramid, bearing 98°, leads 2½ miles
southward of Bell Reef; and the northwestern Reid Rock bearing 8°
leads 2 miles eastward of the reef.

Soundings.—There is no bottom in 220 fathoms at 47 miles southwestward of Black Pyramid, but at 35 miles from it in the same direction there are depths of 70 fathoms, sand and shells, with regular
soundings in 44 to 35 fathoms, between that depth and Black
Pyramid. In the channel between the Hunter Group and King
Island the soundings generally range from about 24 to 36 fathoms,
the deepest water being 44 fathoms, at about 6 miles to the westward
of Albatross Islet.

Directions.—The channel between the Hunter Group and King
Island, as before stated, is not recommended; and as there is a pos-

sibility that some dangers are still undiscovered between King Island and the northwest coast of Tasmania, the safer passage between King Island and Cape Otway should be preferred. But should it be necessary to enter Bass Strait by this channel, keep well to the southward of Bell Reef, observing the clearing marks given before, and pass close to Black Pyramid. Or, with a commanding breeze, a vessel may pass between King Island and Reid Rocks, without danger, by keeping well over on the northwestern side and paying attention to the tidal current, which sets across the channel, occasionally with some strength.

Tidal currents set through mid-channel between King Island and Hunter Group from 1 to 3 knots, the flood to the northeast and the ebb to the southwest.

CHAPTER VI.

WEST COAST OF TASMANIA.

The west coast of Tasmania is mostly rocky, of sterile aspect, with reefs fronting it to the distance of 3 or 4 miles in some places, and a heavy swell usually rolling in upon it from the southwestward. The prevailing winds are from the same quarter, and bring much bad weather, especially in the winter months of June, July, and August.

Cape Grim (lat. 40° 40′, long. 144° 41′), the northwest cape of Tasmania, is a bold grass-topped headland of dark-colored rock, 269 feet high, with an almost perpendicular front. Steeple Rock, a fallen fragment from the cliffs above, lies close southward of it, and is 140 feet high.

The Doughboys are two remarkable islets, both 245 feet high, with almost perpendicular sides, lying east and west of each other and close off Cape Grim; their tops are covered with coarse grass. The western Doughboy is situated 1,500 yards westward from Cape Grim. There is deep water close seaward of them.

The coast between Woolnorth Point and Cape Grim consists of a sandy beach and a rocky point, fronted by dry and covered rocks.

Caution.—At a distance of 4 miles south-southwestward from the high conical rocks which lie close to Cape Grim, and 3 miles from the cliffy land abreast there are depths of 120 fathoms, on a sandy bottom. A coast so steep should be avoided in the night, or in thick weather, especially with the wind blowing from the westward.

Tides and tidal currents.—It is high water, full and change, at Cape Grim, at 10 h. 30 m.; springs rise 8 feet; the southwest going current has a velocity at springs of 5 knots, and at neaps of 3 knots.

Studland Bay.—To the southward of Cape Grim black cliffs extend 5 miles to the northern Bluff Point, on the east side of which is Studland Bay, a small exposed sandy bight with an islet in it.

Boat Harbor.—From Studland Bay the coast trends southward 10 miles to the northern boat harbor, from the bight of which Green Point stretches out nearly 1½ miles to the northwest. There is a rock close off Green Point, and a reef lies nearly 1 mile to the southwest of it. Within the reef is a small bay, from the inner part of which the coast trends southwestward for 2½ miles to West Point.

Hally Bayley Rock, on which the sea breaks heavily in bad weather, has a depth of 3 fathoms on it, and is situated about 15 miles off the west coast of Tasmania. It bears 240°, 16 miles from Cape Grim, and 306°, 14 miles from West Point.

This rock was reported in 1872.

West Point (lat. 40° 57′, long. 144° 38′) lies 17 miles southward from Cape Grim, and is a sandy projection, inclosed by dry and covered rocks.

Light.—A flashing white light, visible 17 miles, 118 feet above water, is shown from a white steel skeleton tower 75 feet in height on West Point.

Porpoise Shoal, which breaks occasionally, extends about 2 miles in a north and south direction. It is situated at about 6¼ miles, 278° from West Point. Depths of from 34 to 39 fathoms were obtained between these breakers and the shore.

Between West Point and the southern Bluff Point, which lies south-southeastward 3½ miles from it, is a bight with an islet near its southeastern shore, and 1½ miles southeast of the latter point is a small opening, close off the entrance of which is Church rock. From Church rock the coast trends east-southeastward 4 miles to Arthur River.

Arthur River is about ½ mile wide at the mouth, and at 17 miles above it, in a southeast direction, it is joined by Hellyer River, a small stream which rises near Valentine Peak.

Southern boat harbor.—From the mouth of Arthur River the general trend of the coast is southward for 12 miles to Ordnance Point, the southern boat harbor being an inlet, with a narrow entrance, 7 miles to the southward of the river. Both entrance points of the harbor are fronted by rocks and Ordnance Point has dry and covered rocks lying about 1½ miles off it.

Sandy Cape, situated south-southeastward 11 miles from Ordnance Point, projects 2 miles from the line of coast. The cape and the exposed bight between it and Ordnance Point are bordered by reefs of dry and covered rocks. Between the southern boat harbor and Sandy Cape there are depths of 44 to 26 and 45 fathoms at 4 to 7 miles from the shore, with irregular depths of 35 to 5½ fathoms between the former soundings and the reefs.

Asbestos is found to the eastward of Sandy Cape.

Between Sandy Cape and another projection 13 miles southeastward from it the coast forms an exposed bight, having an inlet about 7 miles southeast of the cape. From the southeast point of the bight the coast trends east-southeastward 5 miles to the entrance of Pieman River, 2 miles to the northwestward of which is a small inlet or creek.

There is a patch of dry and covered rocks close off the mouth of Pieman River, and 2 miles to the southward of it are two conical rocks standing on a reef of dry and covered rocks extending along a projecting part of the coast.

Pieman River (lat. 41° 40', long. 144° 58')—**Bar.**—There are three bare rocks on the north side of the entrance to this river, the least water on the bar, 10 to 12 feet, being south of the eastern one. The sand forming the bar is continually shifting; in fine weather the entrance to the river is contracted both in width and depth; after heavy rains the scour of the current deeps and widens the channel. With southerly winds the conical rocks south of the river entrance partly break the sea on the bar; with northwesterly or westerly winds the bar is very dangerous with heavy breaking rollers.

The channel over the bar takes a northeast direction and passes close south of the eastern of the three rocks north of the river entrance, thence along the north side of the river until the sandy beach on the south side is passed, after which it is midway between the river banks. Inside the bar there is deep water and a fine river extending many miles into the interior. The township of Corinna is situated about 12 miles from the entrance.

Buoy.—A buoy, painted red, has been placed near a rock with 5 feet of water on it, in the mouth of the Pieman River, about 30 feet from the north bank.

Coast.—From the two conical rocks the coast is rocky, with high cliffs, and takes a south-southeasterly direction 13 miles to Heemskerk Point, 1 mile within which is a stream, whence Long Sandy Beach curves southeast and southward 20 miles to the entrance of Macquarie Harbor. At 8 and 14 miles northward of Macquarie Harbor the beach is intersected by two small streams. There are depths of 12 and 13 fathoms water, at 1¼ miles, and from 20 to 28 fathoms between 4 and 5 miles from the beach.

Trial boat harbor is a small bight, sheltered from winds north of northwest by Heemskerk Point, and from the southward by a low reef of rocks that extends in a southwesterly direction, but the harbor is wholly exposed between northwest and southwest. In fine weather a small vessel or boat, though there is room for only one at a time, may run in and land cargo, but it is a dangerous place: great caution is necessary, and no one should approach Trial Boat Harbor except in very fine weather.

Ringbolts have been let into the rocks for warps. Any craft using this landing place is recommended to let go an anchor and run warps to the ringbolts, then hauling into a position with a depth of 10 or 12 feet, sandy bottom. The water space is very confined, the distance between the rocks north and south being less than 150 feet.

Signals.—The following signals are made from a mast which has been erected on the shore of the harbor:

Two red flags_____The harbor is unsafe, vessels must not enter.

One red flag_____No more vessels must enter at present.

A red and blue flag_____Vessels in the harbor must put to sea immediately.

A red and white flag_____The harbor is unsafe, the sea moderate.

A white flag_____The harbor is safe.

Signal station (lat. 41° 58′, long. 145° 04).—There is a signal station at Remine, a small post town, 5 miles southward of Trial Boat Harbor, and communication can be made by the International Code. Landing can only be effected in fine weather.

Aspect.—From West Point to about 60 miles southward of it the country is low for 2 or 3 miles inland; it then rises gently to a chain of low barren hills, behind which there is a second chain much higher and better wooded than the first.

Mount Norfolk, situated 10 miles eastward of Sandy Cape, is the northern and higher of two hills near each other, which are conspicuous from the offing, and in clear weather are visible before the coast abreast of them.

Mount Heemskerk and Eldon Range.—Mount Heemskerk, situated northeastward 4 miles from the north end of Long Sandy Beach, is the western summit of a ridge extending thence nearly 26 miles eastward to Eldon Range, 4,739 feet high; the former is visible at a distance of more than 30 miles.

Soundings.—From 35 fathoms, rocky bottom, 17 miles westward of Sandy Cape, the depths increase to 106 fathoms, fine white sand and shells, about 30 miles westward of Mount Heemskerk, 5 miles outside which the depth decreases to 66 fathoms, rock. About 27 miles southwestward from Mount Heemskerk there is no bottom at 120 fathoms, the intermediate depths being 95 and 91 fathoms; and there are from 85 to 91 fathoms between 11 and 17 miles from Long Sandy Beach.

Cape Sorell (lat. 42° 11′, long. 145° 10′) is a rocky projection of moderate height forming between it and the north end of Long Sandy Beach an extensive bay, in the southern part of which is the entrance of Macquarie Harbor, Cape Sorell being the western head of the entrance.

Its extremity is low, terminating in straggling bare rocks of brown appearance, and the coast on each side is very rocky and sterile. Many patches of breakers and rocks above water lie detached from the shore; and there is one small rock just above the water's surface, lying 400 yards northwestward from the cape, with apparently no safe channel inshore of it.

Light.—An alternating white and red light, visible 20 and 12 miles, respectively, 186 feet above water, is shown from a white cylindrical tower 100 feet high on Cape Sorell.

Lloyd's signal station.—There is a Lloyd's signal station at Cape Sorell, and communication can be made by the International Code. It is not connected by telegraph, but the pilot station at Macquarie Heads, distant 3 miles, is connected with Strahan by telephone, and urgent messages are conveyed by messenger to and from the signal station.

Watts Hill, eastward 1½ miles from Cape Sorell, is a conspicuous lump of rock on the northeastern part of the cape. A rock above water connected with the coast by a reef lies 300 yards northwestward from the foot of the hill. There is a small rocky islet eastward from the hill and about 100 yards from the coast, the least depth of water between them being 3 fathoms on a sandy bottom, with somewhat less close to the northward of the islet, in a small bight formed in the southern edge of the shoal which extends from the shore.

This small nook, although scarcely an eighth of a mile across in any direction, would nevertheless afford shelter in very smooth water, to a vessel caught suddenly by a northwester in the outer road and unable to cross over the bar of Macquarie Harbor.

Macquarie Harbor is an extensive sheet of water with an area of 112 square miles, trending from its entrance 19 miles southeastward, and is from 2 to 4 miles wide, with regular depths within the entrance, ranging from 5 to 20 fathoms.

Depths.—The entrance is narrow and obstructed by an 8-foot bar between the outer and inner roads.

It must be borne in mind that the channels are liable to alter in position and depth, owing to the occasional great rush of water out through the banks and shoals which, being composed of sand, are of a shifting nature.

Frenchman Cap, 4,756 feet high, distant 30 miles eastward from Cape Sorell, would probably serve in clear weather to point out the entrance of Macquarie Harbor.

Pilot Bay extends from the foot of Watts Hill southeastward about 1 mile to the western entrance point of Macquarie Harbor and has a sandy beach, in the western bight of which, behind some dry and covered rocks, is a small run of fresh water flowing from the swampy land behind it; but this bay is only accessible to boats. on account of its being filled by the western sands of the bar, there being only 6 feet water on their outer spit, about 1,340 yards to the eastward of Watts Hill.

Mount Antill, south-southeastward about 1 mile from Watts Hill, is similar to it, but has a remarkable double summit; Mount

Antill is situated about ¼ mile southward of the beach of Pilot Bay, and is about the same distance from the sea to the westward; there is abundance of fresh water near the mount.

Entrance Islet lies about 85 yards eastward of the steep rocky projection which forms the southeastern point of Pilot Bay and the west entrance point of Macquarie Harbor. The islet is a mass of rock, having some small detached rocks extending about 100 yards from its north point. The proper channel into Macquarie Harbor is between this islet and the western entrance point, where there are depths of 14 fathoms.

Breakwaters.—From a point situated about ¼ mile westward of Entrance Island, a breakwater extends from the shore in a 359° direction for a distance of 920 yards, thence for a further distance of 330 yards in a 336° direction.

At ¼ mile northward of Sandy Point a breakwater is in contemplation, which will be 1,760 yards long, 320°, and 500 yards long, 339°, having an entrance channel 400 yards wide between the breakwaters.

Lights.—A light is shown 25 yards from the end of the west breakwater.

From a white, wooden, hexagonal lighthouse on the western side of Entrance Islet a light is exhibited at 34 feet above high water.

Entrance Islet Lighthouse in range with the lighthouse on Bonnet Islet, or the lights in range at night, bearing 158°, lead over the bar in 8 feet at low water.

A sector of red lights is intended to cover the north spit, which extends from Entrance Islet to the bar, and to guide vessels approaching from the northward, when the inner range light on Bonnet Islet is obscured by Entrance Islet.

The bar, which has only a depth of 8 feet at low water springs, lies 1,340 yards outside Entrance Islet, and separates the outer from the inner road. The depths outside the bar, from 14 fathoms at 2 miles northeastward of Cape Sorell, decrease irregularly toward the bar. At ¼-mile within the shoalest part of the bar the channel is about 400 yards wide, with 3 to 5 fathoms water, whence it narrows toward Entrance Islet, and the depth increases. It is proposed to deepen the bar to 20 feet at low water.

Caution.—As the bar shifts occasionally to a greater or less distance from the north spit, mariners are cautioned that they should not depend on being clear of the spit, and in the channel, directly the light changes from red to white, but should bring the range lights in line. It is advisable not to enter without local knowledge.

Sandy Point.—From about ½ mile eastward of Entrance Islet the sandy beach which forms the eastern side of the entrance to Mac-

quarie Harbor trends southward ¼ mile to Sandy Point, on each side
of which the land is low and sandy for several miles, and covered
with shrubs. The land which forms the western side of the channel
is steep, and rises to irregular ranges of rocky hills of quartzite and
sandstone.

The western side of Sandy Point is fronted by a bank, the outer
edge of which extends westward from about 600 yards off the point
nearly to the opposite shore, then passing close to Entrance Islet,
and northward ½ mile to a spit which always breaks, forming the
eastern part of the bar. From this spit the northeastern edge of this
bank trends southeastward to within ¼ mile of the beach. There is
said to be a narrow channel, with 12 to 18 feet water, close to the
eastern side of Entrance Islet.

Buoy.—A black buoy is placed on the western edge of the shoal
extending from Sandy Point in 3 fathoms at 700 yards westward
from the point.

Pilots—Tidal and pilotage signals.—Tidal signals are shown
from the Bluff flagstaff on the western side of the entrance to Macquarie Harbor.

The figures signaled are those indicated by the tide gauge inside
the heads. Mariners take the bar at their own discretion, as it is
impossible that the exact depth signaled should be guaranteed.

Pilots will, if the weather permits, board vessels outside the bar;
when this is not the case, vessels should be steered with the range
lights in line, and they will be directed by signals from the flagstaff, as follows:

(*a*) A pennant at east yardarm indicates—Alter course to the
eastward.

(*b*) A pennant at west yardarm indicates—Alter course to
the westward.

The pennant will be kept hoisted until it is observed that a safe
course is being steered; no signal will otherwise be made.

When unable to go outside, the pilot will, if practicable, come
out in a boat and direct the vessel's course by a flag waved on that
side of the boat to which the course is to be altered.

Strangers should not attempt to enter at night, and no sailing
vessel should cross the bar on an ebb tide without a commanding
breeze.

Wellington Head.—From the western entrance point the western
coast of the bay trends southward ½ mile, and then southeastward
1,500 yards to Wellington Head, a conspicuous hill, situated 1,400
yards southward from Sandy Point. This head rises rather abruptly from the west side of the harbor, and is easily distinguished
by its table top, which is 260 feet above the level of the sea, and
is separated from the other hills to the westward by a deep notch

that gives it the appearance of being isolated, before the connecting land becomes visible. There is a white mark about halfway up the hills on the south side of the head.

Bonnet Islet lies close to the shore, ⅛ mile northward of Wellington Head; it is small with a round bushy summit. There is a narrow channel with 22 feet water between this islet and the western shore. A wire hawser to facilitate communication has been stretched from Bonnet Islet to the mainland; there is, therefore, no channel for vessels southwest of it. Cap Islet is a small rock about 100 yards southeast of Bonnet Islet.

Light.—A fixed green light, 45 feet above water, visible 10 miles, is shown from a white, hexagonal, wooden lighthouse on the north end of Bonnet Islet.

Bowra Rock, with a depth of 3 feet on it, is situated on the northern side of the channel at a distance of 250 yards northward from Bonnet Island Lighthouse.

Telegraph cable—Beacons.—A telegraph has been laid across Macquarie Harbor 200 yards above Bonnet Islet. The shore ends of the cable are marked by a pair of beacons on each shore.

Prohibited anchorage.—Vessels are prohibited from anchoring in the vicinity of the cable. It is advisable that vessels should not anchor in the space inclosed between lines drawn from Wellington Head and Bonnet Islet to the telegraph beacons on the north shore.

The channel.—From Entrance Islet to Bonnet Islet the channel is from 200 to 300 yards wide, with from 5 to 6 fathoms water in the fairway; but from Bonnet Islet to Wellington Head, between which and the edge of the eastern shoals the channel is only 200 yards wide, the depth of water varies from 13 to 23 feet.

Range beacons.—Three beacons stand on the western shore of the entrance to the harbor. When in range 322°, lead through the fairway above Bonnet Island, as far as the first light beacon, in a least depth of 12 feet at low water.

Channel Bay, training wall.—Channel Bay extends from Wellington Head southeastward nearly 1 mile to Spur Point, and is half a mile deep. Its northern entrance is blocked by the training wall built from Wellington Head across the northern entrance to Channel Bay and on the extensive shoal fronting Channel Bay. The wall extends 1½ miles in a curve, parallel to the eastern sand banks at a distance of about 400 yards. The channel between the wall and the sand banks has been dredged to a depth of 12 feet at low water.

Channel Bay may be entered from the southward; it has depths of about 8 to 12 feet.

Mosquito Cove is a small sandy bight on the south side of Wellington Head with a run of fresh water, and good anchorage in 12 to 20 feet water, on a sandy bottom, within 40 to 50 yards of the beach. It can only be approached from the southward, between the training wall and Spur Point, and is probably not now used as an anchorage.

Round Hill is a high, steep projection, with depths of 2 to 4 fathoms between Spur Point, its extremity, and the training wall.

The north shore from Sandy Point trends east-southeastward 2¼ miles to River Point, and then sweeps round in a northeast direction about 1 mile to Yellow Bluff, and is fronted by extensive sandbanks, nearly dry at low water, which form the eastern and northern sides of the channel leading into Macquarie Harbor.

Backagain Point.—From Round Hill the southwestern shore forms a bight extending 1½ miles southeastward to Backagain Point, a high projection, having 4½ fathoms water close to it. The steep elevated shore of this bight is separated from the southern extensive sandbanks in front of it by a narrow channel, which is said to be finally lost among the shoals to the eastward.

Table Head and Liberty Point.—Between Backagain Point and Liberty Point, the northern extremity of a narrow sharp ridge of moderate elevation, lying 2¾ miles eastward from Backagain Point, the coast forms two bights separated by Table Head, a high, steep, flat-topped point, 1⅝ miles east-southeastward of Backagain Point. Each of these two bights is about 1,500 yards in extent, very shallow, the 2 fathoms curve being some 2 miles northward of the point.

Betsy and Bird Islets lie eastward 1,500 yards and 1,340 yards, respectively, from Backagain Point; the former, though little more than 200 yards in extent, is conspicuous; but the latter is a mere rock. Both islets, together with the rocks about them, are connected with and surrounded by the extensive sandbanks which stretch 2 miles to the north and northeast from Table Head, and which are generally covered, and above referred to.

Kelly Channel, the passage from the entrance channel into the deep water of Macquarie Harbor is about 300 yards wide, with from 3 to 2 fathoms water at its western end, north of Round Hill. Thence the channel takes an easterly direction between the sandbanks for 1¼ miles, with depths of 1 to 3 fathoms. Kelly Channel then gradually widens in an east-northeast direction, to more than a mile in width at its eastern entrance between the sands, where the depths increase to more than 12 fathoms.

Light beacons—Beacons—Buoy.—The northern side of Kelly Channel is marked, from abreast Spur Point for a distance of 1 mile, by four light beacons; the rest of the northern side by three black beacons. The southern side is marked by a white beacon 1¼ miles southeastward from River Point, and by a black conical buoy midway between this beacon and the end of the training wall.

Sophia Point, a low projection of the northeastern shore of Macquarie Harbor, lying 2¼ miles northeastward from Liberty Point, is inclosed by a reef, with straggling rocks extending about 400 yards from it. Sophia Point, and Yellow Bluff, about 3 miles westnorthwestward from it, form the entrance points of the north arm of the harbor, which extends nearly 5 miles in a northerly direction.

Pine Cove is a bight in the eastern shore of the north arm of Macquarie Harbor, lying north between 1¼ and 2¼ miles from Sophia Point. In proceeding from Kelly Channel to Pine Cove the steep south side of the spit which projects from Yellow Bluff, marked by a beacon, must be approached with caution, as the soundings are very irregular, but thence the depths gradually decrease to 3 fathoms within the cove, where there is good anchorage for small vessels, with muddy bottom.

Tide and tidal current.—There is little or no tidal current in Pine Cove, and the rise and fall does not usually exceed 1½ feet.

Kings River.—From the north entrance point of Pine Cove a narrow peninsula extends northwest for 1,340 yards to the south point of the mouth of Kings River, which is 670 yards wide, but it is encumbered by two islets, from the outer and smaller of which a shoal extends at least 670 yards to the southwest, as another does also from the northeast entrance point of the river. Kings River takes its rise among the mountains to the eastward.

Strahan, a post and telegraph town and the principal port on the western side of the island, is situated nearly 2 miles northward of the entrance to Kings River; it is connected with the railway system of the island and a line runs to Kelly Basin. Steamers call here regularly from Hobart and Launceston, and there is direct communication twice weekly with Melbourne. Small local steamers ply to Kelly Basin. Population is about 1,000.

Strahan is situated in a mining district, and the population in the locality is rapidly increasing. Gold exists in the district and rich discoveries at Kings River, Mount Lyell, Mount Zeehan, and Mount Dundas of gold and silver lead have caused very many mining leases to be taken up and worked. A good seam of lignite has been found in the immediate neighborhood of the township.

Signal station.—There is a signal station at Strahan, and communication can be made by the International Code. It is connected by telegraph.

Smith Cove is a small indentation forming a natural dock, situated 2¾ miles northward of Yellow Bluff. The entrance has a depth of 10 feet, and there are upwards of 2 fathoms over the greater part of the cove inside. A ledge of sunken rocks extends east-southeastward from the west entrance point; and rocks extend a short distance southward of the eastern point.

Buoys.—A black buoy is moored at the extremity of the western reef of Smith Cove which buoy should be left close on the port hand in entering, thence steering toward the middle of the cove. A black buoy is moored at the extremity of a rocky ledge, extending nearly 400 yards eastward of the point situated 200 yards eastward of Smith Cove.

Swan Basin, on the west side of the north arm of Macquarie Harbor, extends from 1 to 2¼ miles northward from Yellow Bluff. From the southern extremity of this basin a narrow neck of land sweeps round northeast and north nearly a mile, and terminates in a peninsula, half a mile long, east-northeast and west-southwest between which and a small island to the northward of it is the narrow and only entrance into the basin. This small island, and the rocks northward of it, are connected with the north part of the basin by a dry sandy flat, which lines its shores. A vessel may lie completely land-locked in Swan Basin; but from the narrowness of its entrance and the confined space within, it can scarcely be called a port.

The head of the north arm of Macquarie Harbor, above King River and Swan Basin, is formed by numerous points and bights, affording several sheltered anchorages, secure from all but southeast and southerly winds.

The southwest shore of Macquarie Harbor from Liberty Point trends southward 2¼ miles, and eastward 1,500 yards to a projecting head, forming the northwest entrance point of Double Cove.

Double Cove is 670 yards wide at its entrance between two projecting points, within which it is little more than ½ mile in extent, with only from 3 to 6 feet water, and is much contracted by a projection near the middle of it, which renders the anchoring space very confined, even for the small vessels which are enabled to cross over the bar at the mouth of the cove.

Good shelter for boats may, however, be found here; and there are several runs of fresh water crossing over the beach from the higher land behind.

From the southeast entrance point of Double Cove the southwest shore of Macquarie Harbor extends in an east-southeast direction 4¼ miles, and thence southeastward 2 miles to the northwest entrance point of the south arm of the harbor. It consists of rocky points and small bights mostly fronted by sunken rocks, none of which appear to extend more than ¼ mile from the shore. Inshore, from Double Cove to the south arm of the harbor, the land chiefly consists of yellow loam, and is thickly wooded.

Headquarters Island.—Between the northwest entrance point of the south arm of Macquarie Harbor and the projection at 2 miles to the northwest of it, the shore is fronted by a reef extending about ¼ mile from each point and 1¼ miles from the shore midway between

them. Headquarters Island, the central and largest of the islets and
rocks on this reef and which lies 8½ miles southeastward from Liberty Point, is ¼ mile long in a northeast and southwest direction, but
is only 200 yards broad. It has dry and covered rocks close to each
end, and there is a small islet on the spit of the reef 600 yards east-
northeastward from the northeast point of Headquarters Island.
There is anchorage in 4 to 6 fathoms water in the bight of the reef
about ½ mile to the southeast of the island.

Birch Inlet.—The south arm of Macquarie Harbor is 2 miles
wide at its entrance, whence it gradually narrows for about 2 miles
to the southwestward, where it is only ⅛ mile wide, and after continuing this width nearly a mile to the southward the channel opens into
Birch Inlet, a sheet of water above 1 mile wide and extending 3
miles, and probably more, in a southeast direction.

Gordon River flows into the southeastern end of Macquarie Harbor between the southeast entrance point of the south arm and another point at 1¼ miles to the northeast of it. Both entrance points
of the river have rocks projecting about 200 or 400 yards from them,
between which is a bar with 12 feet water on its deepest part, upon
the southwest side of the entrance of the river. Thence Gordon
River trends southeast 2 miles, and after turning to the northeast
for 1 mile it winds nearly 6 miles in an easterly direction, and then
trends 4 miles southward to some marble cliffs on the west side, above
which the river is formed by several streams flowing from the interior mountains. From 2 fathoms on the bar the depth of water increases to 10 fathoms 2 miles within it, with navigable water to
within ½ mile of the falls.

The northeastern shore of Macquarie Harbor, from a small
bight on the southeast side of Sophia Point, trends southeastward
4 miles and thence southward nearly a mile to Coal Head; there
is a small creek or rivulet 1¾ miles from Sophia Point. The land
inshore, although poor, is thickly wooded.

Between Coal Head and a projecting point southeast 3¾ miles
from it are two bights, the northwestern one being filled by a shoal
flat which extends 1,500 yards from the shore; but the southeastern
bight may be approached within ¼ mile of the shore in from 2 to 4
fathoms. The land behind these bights is poor and heathy, rising
inland to Mount Sorell, situated about 7 miles eastward of Coal
Head.

Phillip Isle, situated 1 mile southeastward of Coal Head, is 1,200
yards long, about 400 yards broad, and situated on the edge of a
rocky shoal, which extends about ½ mile from the broad projection,
which separates the two bights just noticed. A rock above water
lies between the island and the shore.

Pine Point, 1,500 yards eastward of the southern extremity of the southeastern bight before mentioned, is the extremity of an irregular projection of the northeastern shore, stretching out nearly a mile in a southwest direction, and separating a kind of basin on its northwest side, from the northeast arm of Macquarie Harbor. This basin is more than a mile across each way, with an island in the center; but its entrance has a reef stretching nearly halfway toward Pine Point from the western side.

Kelly Basin is a sheet of water 1¼ miles in length by 1,500 yards in width, forming the head of the northeast arm of Macquarie Harbor, which from its entrance between Pine Point and the northeast entrance point of Gordon River extends about 2½ miles in a northeast direction to the entrance of this basin, which is only ½ mile wide.

Nothing is known of the depth or capabilities of this branch of the harbor, nor of the basin on the northwest side of Pine Point.

Mountains and rivers.—To the northeastward of Kelly Basin are some high ridges of white-topped mountains, which are visible from the borders of the River Derwent. The Gordon River, flowing from Lake Richmond, and receiving in its course the Wedge, Denison, Serpentine, and Franklin Rivers, falls into the southeastern part of Macquarie Harbor.

Kings River, with its tributaries, the Queen and Eldon, flow into the northeastern part of Macquarie Harbor as before mentioned.

The depths in Macquarie Harbor, between the spit off River Point and the reef projecting from Headquarters Island, range from 13 to 20 fathoms in mid-channel, and thence generally decrease to 10 and 6 fathoms within ½ a mile of the shore on either side. From 8 fathoms at a mile southeastward of Headquarters Island the depths decrease to 2 fathoms on the bar of Gordon River.

Directions.—The northwest and westerly gales which frequently blow with great violence on the west coast of Tasmania, not only influence the tides in Macquarie Harbor very considerably, but render it unsafe for any vessels to anchor outside the bar when there is a prospect of the wind blowing from those quarters, as there is no shelter between north and west in the outer road, for any but small vessels.

Vessels bound for Macquarie Harbor should make Cape Sorell, and it is recommended that the services of the pilot be obtained.

Waiting for tide.—In fine weather vessels waiting for the tide to cross the bar should anchor about ½ mile from the nearest part of the bar and the same distance northeast of Watts Hill, with the northern extremity of Cape Sorell bearing 271°, and the range lighthouses in line (158°) in depths of 6 to 7 fathoms, sand. Vessels of 6 to 7 feet draft can generally enter Macquarie Harbor, except with strong westerly winds, the least depth on the bar at low water with the lighhouses in line being 8 feet, which depth continues for about 100 yards.

To cross the bar, the range lighthouses should be brought in line, bearing 158°, before the northern extremity of Cape Sorell bears 279°. Keep the lighthouses in range across the bar, and when the water deepens to 2¼ fathoms alter course to the southward until Bonnet Island Lighthouse is midway between Entrance Islet and the bluff rocky point immediately west of that islet. Then steer toward the Bonnet Islet Lighthouse. After passing Entrance Islet keep near the western shore, and leave the black buoy on the port hand and Bonnet and Cap Islets on the starboard hand. Bring the three range beacons (on the western shore near the entrance) in range astern until the first light beacon is passed, then steer to leave the light beacons and black beacons on the port side, and the conical buoy and white beacons on the starboard side.

If bound to Smith Cove, Strahan, or the north part of Macquarie Harbor, the black beacon eastward of Yellow Bluff may be rounded at the distance of 400 yards, leaving it on the port hand.

Sailing vessels, inward bound, should not attempt to cross the bar, against the ebb tidal stream, without a strong commanding breeze.

Anchorage.—There is good anchorage in the inner road between the bar and Entrance Islet, in 14 to 35 feet, clear, sandy bottom, with the summit of Wellington head over the west point of the narrow entrance; but the breadth between the west breakwater and the breaking water on the east side is, in some parts, only ¼ mile.

In working through the inner road the shoals on the east side should not be approached nearer than to bring the east pitch of the summit of Wellington head over the west end of Entrance islet; nor the shoals on the west side at the inner end of the West breakwater nearer than to bring the lighthouse on Bonnet Island in line with the Bluff flagstaff on the southwestern side of the narrow entrance.

Caution.—Great attention must be paid, not only to the ranges and to obtaining quick soundings, but to the tidal currents, which run here with great strength, and during freshets, sometimes at the rate of 5 and 6 knots. In the narrow channel between Entrance Island and the western rocky shore, the ebb tide runs at times like a cataract, with a velocity of 10 miles an hour.

In sailing against the ebb between Entrance Islet and the steep rocky point to the westward of it, favor the western shore while passing the islet, as the tidal current sets strong out of a bight just within it, and is likely to drift a vessel upon the islet.

At night.—No stranger should attempt to enter Macquarie Harbor at night, unless on a lee shore and unable to keep outside.

Productions.—The land in the vicinity of Macquarie Harbor and the rivers which flow into it is said to be wholly unfit for cultivation, but the forests abound with various kinds of timber, fit for spars, boat building, cabinet work, and architecture. The Adventure Bay pine, which is fit for small spars and a variety of other purposes, grows about Kings River, and in the southeast as well as the northwest parts of the harbor; it ordinarily grows to the height of 40 or 50 feet, and is from 12 to 16 inches in diameter, with leaves resembling parsley. These spars are generally rafted over the bar and taken on board in the outer road.

Supplies.—Fish may also be procured in plenty near the rocky parts of the shore of the harbor, and fresh water almost everywhere.

Tides—Tidal currents.—The time of high water, full and change, on the bar at the entrance of Macquarie Harbor, is 7 h. 30 m.; springs rise 3 feet; but the time of high water and rise are both influenced by westerly and northwest gales, and by great freshets that, during the prevalence of rainy or thick, cloudy weather, flow into the harbor from the high mountains in the interior, at which periods the channels between the shoals are deeper than usual. During a fortnight's observation, the tides were irregular, making high water sometimes twice, and at other times only once in 24 hours, and in both cases the ebb ran twice as long as the flood, producing a difference in the level of the water, which on several occasions did not exceed the average fall of 18 inches.

An ebb for nine days together, without the water rising or falling so much as 1 foot, has been experienced, although at other times, during northwest gales, the inundations were great, frequently overflowing the adjoining lowlands.

The tides at Macquarie Harbor are very irregular, partly owing to their being disturbed by the winds, which have an extraordinary influence on the height of the water in the harbor. The extreme range of tide seems to vary from 1½ to 3 feet, but this is so irregular that it is impossible to predict anything about the tides, except that in fine weather and with southwesterly winds the tides are low, but with strong northerly winds the tides are highest. An on-shore gale, or even a fresh breeze, may raise the level of the lagoon to the extent of completely masking the tides.

At the entrance there is usually only one tide in the 24 hours, but a false tide often follows a short time after high water, the effect of which is that after the water has begun to ebb strongly the tide again rises, the ebb slackens for an hour or more, then finally the tide begins to fall, and the strong ebb sets out again.

The extensive shoals which obstruct the entrance prevent the full effect of the sea tide reaching into the lagoon; thus, in fine weather, with a range of tide varying from 1 inch to 2 feet, there is a strong flood and ebb tide into and out of the harbor, the effect of which extends even 20 miles up the Gordon River; but this range of 18 inches or 2 feet causes a range of only 9 to 15 inches at Strahan, and the same at the mouth of the Gordon River.

The tides are higher in winter than in summer to the extent of nearly 1 foot; about the time of high and low water the tide is slack for more than an hour, while the water slowly rises for some time before the flood tide makes; this prenomenon is said to account for the extraordinary difference observed in the length of the ebb as compared with the flood tide, the ebb being frequently 18 hours and the flood only 6. With a northerly gale coming on the tide flows into the harbor very strongly, and often for 24 hours continuously; the harbor then fills up from 3½ to 4 feet, and even 5 feet above low water. As soon as the gale begins to abate, or even during its height, if it shifts to the westward, the water of the harbor ebbs out with great force, the duration of the ebb being often 18 hours, and only checked for 5 or 6 hours as the tide rises at sea. In fine weather there are often days when there are no tidal currents either in or out; on the other hand, the flood tide is longer and the ebb shorter in fine weather. In fine weather, undisturbed by approaching bad weather or by floods in the rivers, the flood and ebb tides are of nearly equal duration; occasionally the flood tide attains a velocity of 4¼ miles an hour, and this generally indicates the approach of bad weather.

During a very heavy gale from west, the range of tide varied from 10 to 13½ feet on the bar; the tide ebbed and flowed all through the gale, with very short flood, about 7 hours, and long ebb, about 17 hours.

The west coast of Tasmania from Cape Sorell (lat. 42° 11′, long. 145° 10′) extends south-southeastward 26 miles, and thence southwest 3½ miles to Point Hibbs, and consists of a series of rocky bights and projections. For the first 12 miles from the cape the coast is fronted by rocky ledges and rocks above water, generally extending about 1¼ miles from it. The land behind the whole of this coast rises by a gentle ascent, for a distance of 2 or 3 miles, and is apparently smooth and uniform, but destitute of wood and almost of other vegetation.

Sloop Rock, lying 10½ miles southward from Cape Sorell, is a small islet about 2¼ miles from the shore with some sunken rocks at 1½ miles to the northward, and others to the southeastward of it.

It is reported that this rock lies about 3 miles to the northward of the position shown on the charts.

Breakers are reported to exist in two places between Sloop Rock and Point Hibbs.

Point Hibbs (lat. 42° 38′, long. 145° 15′) **and Pyramid Rock.**— Point Hibbs projects southwestward about 3 miles from the line of coast, and is higher than the neck by which it is joined to the back land. A remarkable pyramidal rock lies northeastward nearly 2½ miles from Point Hibbs, which rock may be seen, appearing like the crown of a hat, when bearing north-northeastward over the extremity of the point. A ledge of rocks project about 1½ miles from Point Hibbs, and along the south side of the point, some of the rocks on the eastern part of the ledge being above water. There is a fresh-water pond near the shore abreast of Pyramid Rock, and at 1½ miles southeast of the pond a small stream flows into the bight on the south side of Point Hibbs.

Small-craft anchorage.—From the small stream which flows into the bight on the south side of Point Hibbs, the coast trends southeast 8½ miles to a headland projecting 1½ miles from the coast line; between this headland and a cliffy peninsular head southward of it extending 3 miles from the coast is an inlet half a mile wide, said to afford anchorage for small vessels. A reef with a rock above water on it extends from one mile southwestward to 1½ miles northward of the north head, and a larger reef with high rocks on it, 1 mile off shore, projects southwest 2½ miles from the cliffy peninsular head, which forms the south side of the reported anchorage.

Mainwaring Cove is the bight formed on the south side of the cliffy peninsular head just noticed, and Mainwaring Inlet, which has a reef projecting from each side of its entrance, lies southeastward 3½ miles from the head.

Rocky Point.—From Mainwaring Inlet the coast curves slightly in a south-southeasterly direction for about 8 miles to Rocky Point from which reefs extend about 1½ miles to the southwest and nearly a mile to the northwest. The land between Point Hibbs and Rocky Point is somewhat more elevated, and not so destitute of wood as that northward of Point Hibbs; the summit of Junction Range, 1,210 feet high, is situated 8 miles northward of Rocky Point.

Relief station.—A station for the relief of shipwrecked mariners has been established about 2½ miles inland from Rocky Point, west coast of Tasmania.

The relief station is a frame structure, painted red, plainly visible from seaward; a signal staff for signaling to passing vessels has been erected on Rocky Point.

A track has been cut from the signal staff on Rocky Point to the relief station.

From the relief station a track has been cut to the northward for a distance of about 8 miles, and to the southward there is a track as far as the north head of Port Davey. These tracks follow the coast line, and are nowhere more than 1 mile distant from it; in 11 places junction tracks have been cut from the shore to the main track.

At every quarter of a mile on the main track iron index hands have been erected, pointing in the direction of the relief station and indicating the distance from it. A raft has been constructed at the Giblin River, where the depth of water is too great to permit fording.

In the building a sufficient supply of clothing, provisions, and cooking utensils has been placed, with all necessary directions, and mariners wrecked on the coast in this vicinity will have no difficulty in finding their way to the relief station, from whence by means of signal station any passing vessel may be notified of their presence.

Black Rock, situated 7 miles northwestward from Rocky Point and nearly 3 miles from the shore, is 20 feet high and surrounded by rocks and breakers, with another patch of rocks and breakers about 2 miles southeast of it.

Elliot Cove.—Between Rocky Point and a roundish projection of the land 14 miles southeast of it, the coast forms a bay 5 miles deep, the head of which is Elliot Cove.

From the roundish projection which forms the southeastern extremity of the bay just noticed, the general trend of the coast is southeast 8 miles to Point St. Vincent, between 2 and 3 miles to the northwest of which is a small bight having two islets or rocks 1 mile off its entrance; they lie close together and are connected by a reef.

Aspect.—The coast for about 18 miles to the southeastward of Rocky Point is high, and at the back are several bare white peaks, as if covered with snow; De Witt range, 2,445 feet high, the most elevated of these peaks, is situated 17½ miles east-southeastward from Rocky Point.

Point St. Vincent (lat. 43° 17', long. 145° 50'), **North Head, and Dock Islet.**—Point St. Vincent and North Head, at 2 miles to the southeast of it, are each fronted by a reef with dry rocks on it. Dock Islet lies about 1 mile off the bight between the point and the head, and there is a detached reef about ½ mile southward of North Head.

Pollard Head.—From North Head the coast trends east-southeastward 1¼ miles to Pollard Head, the northwest entrance point of Port Davey; there are some sunken rocks close to Pollard Head, but there is a depth of 5 fathoms 200 yards off it.

Port Davey.—When nearing this port the land on either side presents a most rugged and barren aspect, and is steep and mountainous to the eastward. The entrance, which is easily known by the high pyramidal rock on the south side of the entrance is 3¾ miles wide between Pollard Head and Hilliard Head southeast of it; it has a bold approach, and is easy of access.

Port Davey extends 10 miles from its southeastern to its north-western extremity, and has several branches; that which affords the most secure anchorages being apparently on the east side, which includes Bramble and Schooner Coves.

Danger.—The chief danger to be avoided is a sunken rock reported to lie nearly midway between Pollard Head and Pyramidal Rock.

Depths.—The. soundings across the entrance gradually increase from 5 fathoms at 200 yards off Pollard Head to 27 fathoms near the reported sunken rock, and from thence decrease to 9 fathoms close to Pyramidal Rock. From 25 fathoms midway between Pyramidal Rock and Garden Point, the soundings gradually decrease to 12 fathoms within 200 yards of the rock, and to 9 fathoms ¼ mile from Garden Point. From 5 fathoms close to Nares Rock, the soundings increase to 23 fathoms 1¼ miles in a northerly direction, and thence decrease to 10 fathoms at 1 mile eastward of Garden Point. From a line between this point and Kathleen Isle, where the depths increase regularly from 4 to 10 fathoms, the soundings up the harbor to Payne Bay decrease regularly to 4 fathoms. The shores on either side of the harbor, as far up as Earle Point and Bluff Head, may be generally approached within ¼ mile in from 5 to 6 fathoms; but off Earle Point, on the west side, and between Bluff Head and Woody Points, on the eastern side, there are only from 3 to 4 fathoms at that distance from the shore.

Hilliard Head, the south point of the entrance to Port Davey, is a high craggy projecting point, with some sunken rocks close to, and a group of islets and rocks southeastward of it.

Chatfield Islands.—These islands, which are peaked, extend about 1 mile distant from Hilliard Head in a southeasterly direction, and are more or less connected by rocks to the highest Chatfield Island. No dangers are known to exist outside the western peaked rock of this group, which are locally known as the East Pyramids.

Sugarloaf Rock, about 250 feet high, the southern and highest of the Chatfield Islands, is somewhat similar in appearance and height to Pyramidal Rock.

Stephens Island, 1¼ miles southeastward from Hilliard Head, is low, with no definite summit, and from seaward looks like part of the coast. There are sunken rocks between it and the Chatfield Islands.

Pyramidal Rock, about 250 feet high, 1,500 yards northwest of Hilliard Head, is locally known by the name of Caroline Rock.

Southeast shore of Port Davey.—From Hilliard Head to Forbes Point, 1,600 yards northeast of it, the shore forms a bay, between which and Pyramidal Rock is Swainson Islet, about 200 feet high, with some sunken rocks close round it and a dry rock near its northwestern extremity. There are depths of 7 to 10 fathoms between Hilliard Head and Swainson Islet, and from 8 to 15 fathoms between the head and Pyramidal Rock.

On the east side of Forbes Point is Norman Cove, about ¼ mile in extent, having from 4 to 5 fathoms water in it, from the east side of which the shore sweeps round ¼ mile to Knapp Point, close off which is Hay Islet, lying northeastward about ½ mile from Forbes Point.

Hannant Point, which lies in line with Hay Islet and Forbes Point, is a narrow projection separating Spain Bay on the southwest side from Hannant Inlet on the northeast side of the point. Spain Bay has depths of 8 to 11 fathoms across its entrance, close within which there are two small rocks. This bay, which runs in about 1,500 yards from its entrance, has not been sounded inside the small rocks.

Hannant Inlet.—The entrance of this inlet is barely 400 yards wide between Hannant Point and O'Brien Point to the northward of it, and is nearly barred across by a narrow islet close within the entrance. Thence the inlet trands 3 miles to the southward, but is useless for any vessels but the smallest, the water in the inlet eastward of the island being shoal. It has a hard sand bottom; the southern end dries at low water and is of mud and sand.

Nares Rock, lying 52° true, 1¼ miles from Pyramidal Rock, is awash at low water. There are depths of more than 17 fathoms between Swainson Islet and Nares Rock, and from 18 to 7 fathoms between the rock and Norman Cove.

Landing.—There is invariably a heavy swell settling into Norman Cove, which would make it impossible for a boat to land here at any time.

With the wind between south and east a boat may land on the southwest side of Spain Bay, but with the wind with any westing in it this is impossible.

Shanks Islets, eight in number, the highest and largest of which is 995 feet high, lie 1,500 yards northwestward from Hannant Point. These islets, which extend 800 yards north and south, have sunken rocks close about them, but there is a clear channel, with from 6 to 18 fathoms water between the shore, about Knapp Point and a line from Nares Rock to Shanks Islets, and from 12 to 9 fathoms from Spain Bay to within 400 yards eastward of Shanks Islets.

The east shore of Port Davey, from O'Brien Point, trends northward 1¼ miles to Turnbull Head, which forms the southeast side of the entrance to Bramble Cove. There are from 6 to 10 fathoms water 200 yards offshore, except between Shanks Islets and the mainland, where there is a depth of 4¾ fathoms 400 yards from the shore, and at nearly ¼ mile southward of Turnbull Head, where a rocky ledge projects nearly 200 yards from the shore.

Breaksea Islands extend from ¼ mile northward of the Shanks Islets to nearly 1 mile west-northwest of Turnbull Head. They are three in number, the middle and longest island being 250 feet high near the center and over ¼ mile long; the northern island is over ¼ mile long and 255 feet high, while the southern, which is almost joined to the middle island, is about 400 yards long and 175 feet high, but neither of them exceeds 350 yards in width.

There is a rock 13 feet high near the south end of the southern island, and the coasts of all have dry and sunken rocks close along them, but there are from 13 to 9 fathoms water within 200 yards of their west sides, and from 3 to 13 fathoms at the same distance from their east sides, between which and the mainland there are depths of 5 to 11 fathoms.

These islands are joined to the mainland near Milner Head by a 5-fathom bank.

South Passage, the channel between Shanks Islets and Breaksea Islands is ¼ mile wide, with from 16 to 12 fathoms water; and there are from 14 to 5½ fathoms from the middle of the passage to within 200 yards of the ledge of rocks south of Turnbull Head.

North Passage, between Breaksea Islands and Boil Rock, 5 feet high, northward of it, and Milner Head, is ⅛ mile wide, with a least depth of 5 fathoms.

Bramble Cove is a safe and commodious harbor within Port Davey, having an entrance 600 yards wide, with from 4 to 12 fathoms water, between Turnbull Head and Milner Head, 800 yards northward of it. There is a rock 6 feet high close to Turnbull head, and the edge of the 5-fathom bank extends 200 yards northward of the head.

Depths.—Within the entrance Bramble Cove forms a basin extending 1 mile east and west and nearly 1,500 yards north and south, with regular soundings, decreasing from 14 fathoms in the entrance to about 4 fathoms about 200 yards off the shores, except to the eastward of Sarah Island, where 4 fathoms will be found 300 yards from the shore, and to the eastward of Turnbull Head, where there are from 12 to 22 fathoms.

There is a rock 2 feet high, with some sunken rocks, close to the eastern shore; and on the south side is Sarah Island.

Sarah Island, 100 feet high, is known better as Tonguers Island; it is cleared of trees to seaward, and has a cairn on its summit. A rocky ledge with 2 fathoms of water on it, extends to a distance of 155 yards from its north end.

There is a boat passage only between Sarah Island and Hixson Point.

Anchorage.—In Bramble Cove the best anchorage is with the north end of Breaksea Islands in line with the southern extremity of Milner Head, with the highest summit on the land between Pollard Head and Kelley Basin showing over these points; and the west end of Sarah Island bearing 190°. This position gives good shelter from the northwest winds, which are reputed the strongest hereabouts. The bottom is hard sand.

Mount Misery, which is 1,570 feet high and very precipitous on its southern side, directly overlooks Bramble Cove; it is the southern summit of a conspicuous range running about north and south; Mount Berry or Erskine Hill is the central or highest part of this range, being 2,200 feet high and completely hidden from Bramble Cove by Mount Misery.

Tides.—It is high water, full and change, at Bramble Cove at 1 h. 00 m. (about); springs rise about 3¼ feet.

The tides are very irregular.

Directions are given later.

Bathurst Channel is the narrow portion of the channel connecting Port Davey and Bathurst Harbor. Bathurst Harbor is the broad expanse of water some 5 miles from Bramble Cove.

There seems but little doubt from a cursory examination that ships of fairly deep draft will eventually be able to go into Bathurst Harbor itself; boats on several occasions sounded in passing up and down to Bathurst Harbor and found no dangers while keeping a fair midstream course.

The entrance into Bathurst Channel from Bramble Cove is 300 yards wide, with depths of from 9 to 22 fathoms, between a projection ¼ mile eastward of Turnbull Head and the southwestern extremity of Sarah Isle.

The bottom in the channel was found to be hard sand.

The peculiar color of the water which prevails here—a light coffee tint—entirely hides any signs of uneven bottom; this color is evidently due to the fact that the fresh-water streams all flow through peaty soil in their course down to the sea; the color is always noticeable, under all conditions of rain, etc., and is to be seen in the smallest mountain stream.

It is said by some that there is always a strong current setting out of the channel, but this was not found to be the case during the stay of the *Dart*, there being very little current of any sort discernible, tidal or otherwise, even after heavy rain; after some weeks of heavy rain this reported strength of current may possibly be true.

175078°—20——20

From its entrance, Bathurst Channel trends eastward 1,500 yards to Mundy Isle, and is ¼ mile wide, with 11 to 20 fathoms water in mid-channel.

Mundy Isle, which is about ¼ mile long and very thickly wooded, lies midway between Helby and Forrester Points, the projections of the north and south shores leaving a channel nearly 200 yards wide, with from 10 to 25 fathoms, between the southwest end of Mundy Isle and Forrester Point, and a channel with from 4 to 10 fathoms between the north side of the isle and Helby Point.

From Mundy Isle, Bathurst Channel trends one mile eastward, with an average width of ¼ mile, and depths decreasing from 19 to 12 fathoms in the center to 5 and 4 fathoms 200 yards from either shore. Deep point lies 300 yards from the northeast end of Mundy Isle, and Noon Point ¼ mile southward of Deep Point; Night Islet near the shore, lies ¼ mile westward of Noon Point.

Schooner Cove is about 700 yards long between Forrester Point and Night Island, and about the same in depth; it is only fit for small craft, the south side of the cove being shoal.

At a mile above Mundy Isle, Bathurst Channel is only 400 yards wide; it increases afterwards to Bathurst Harbor.

Spring River.—At 1½ miles eastward of Noon Point is a thickly wooded islet which lies off the mouth of Spring River, a stream flowing into Bathurst Channel from the northward; between this islet and Horseshoe Inlet, an opening in the south shore nearly ½ mile to the southward of it, there are depths of 5 to 11 fathoms.

Mount Rugby is a rugged-topped mountain, about 2,300 feet high on the north side of the channel leading out of Bathurst Harbor, and is on the east side of Spring River, being separated from it by an inlet extending about a mile to the northward; Mount Rugby lies about 4 miles eastward of Mount Berry.

Bathurst Harbor.—From the mouth of Spring River Bathurst Channel winds about 3 miles in an easterly direction, when it opens into an extensive sheet of water, forming Bathurst Harbor. It has a cluster of small islets in its southwest corner, and a narrow branch extends above 4 miles to the southward; the harbor has only been partially examined.

All winds with any westing in them apparently tend to follow the line of the channel toward Bathurst Harbor.

Opposite the entrance to Spring River is a hill named Balmoral, about 450 feet in height; it has a rounded top, and is easily recognized. Immediately westward of Balmoral Hill is the entrance to Horseshoe Inlet; this narrows to a small distance across about a mile above the entrance, and then widens out into a circular-shaped piece of water, which is, however, very shoal and of no practical use to any but small boats.

Kathleen Isle.—From Milner Head to Ashley Head, 2¼ miles northwest from it, the northeast shore of Port Davey forms a bay, fronted by Kathleen Isle, 360 feet high, which lies equidistant from the two heads and about ¼ mile from the shore. There is a cluster of islets and sunken rocks, named Needle Rocks, between the northern extremity of the island and the shore to the northward; and between Kathleen Islé and Boil Rock to the southward of it is a channel 600 yards wide, with from 5 to 10 fathoms water.

There is a clear passage between Kathleen Island and the southern Needle Rock, which is useful for a boat.

Ashley Head and Bluff Head, 1,500 yards north-northwestward of it, are each bordered by a rocky ledge but may be approached within 200 yards, in from 6 to 5 fathoms water.

Pym Point.—From Bluff Head the shore curves northward nearly 1¼ miles to Pym Point; it is intersected nearly midway by an inlet, close off which are three small islets, with some sunken rocks.

The west shore of Port Davey from Pollard Head curves northeast 1¼ miles to Garden Point, and thence forms another curve extending north-northwestward 2 miles to Earle Point. Between Garden and Earle Points the shore is lined with rocks, and a shoal with sunken rocks extends ¼ mile northward and eastward from Earle Point.

Whaler Cove is a slight indentation of the coast between Garden Point and a small islet near the shore 1 mile northwestward of it. From this islet a rocky reef extends about half way to Garden Point, and nearly 400 yards from the shore. There are from 4 to 7 fathoms water within 400 yards of the shore in the southeast part of the cove, where there is tolerably sheltered anchorage in northwest or westerly gales; but it is exposed to the wind and sea, if blowing hard from the southwest.

Bond Bay lies between Earle Point and Curtis Point, 1¾ miles northward of it, and is 1¼ miles deep; but nearly the whole bay is occupied by a flat having about 8 feet water on it, except in the entrance, where there are from 2 to 2¾ fathoms, between ½ mile northward of Earle Point and ¼ mile southward of Curtis Point.

Kelly Basin.—In the bight of Bond Bay, at 1¼ miles westward of Earle Point, is an opening about 300 yards wide, having from 9 to 12 feet water, which leads into Kelly Basin, a circular sheet of water 1¼ miles in diameter; it is filled by a shoal flat, except for about 1,500 yards to the southwestward from its entrance, where there are from 12 to 6 feet water.

Payne Bay, the northern part of Port Davey, is a little more than 2 miles wide, east and west, at its entrance between Pym and

Curtis Points, whence the bay extends, 1⅔ miles to the northward. The east shore of Payne Bay from Pym Point trends northward 1 mile to Woody Point. Two small islets, with sunken rocks about them, lie ⅓ mile northwestward from Pym Point, near 400 yards from the shore; and from Woody Point three similar islets, with sunken rocks extend nearly 400 yards.

Between Woody Point and Fitzroy Point, 1¾ miles northwestward from it, the north shore of Payne Bay forms a bight, having two small islets near the shore, ½ mile northwestward of Woody Point, and another islet 1,500 yards eastward of Fitzroy Point, the shore being mostly lined with sunken rocks.

Fitzroy Islets, which are four in number, with sunken rocks about them, extend ⅓ mile southward and ¼ mile southwestward from Fitzroy Point. The northwest islet of Fitzroy group is the only one that is always above high water; it is about 10 feet high. The others are rocks that are awash at high water.

Stephen River.—Above Payne Bay, the north part of Port Davey, from the width of nearly 1½ miles between Curtis and Fitzroy Points, contracts to ¼ mile across, at the mouth of Stephen River, which flows from the northward into the head of the port, 1¾ miles west-northwestward from Fitzroy Point. Sunken rocks lies close along the north shore, and others extend about ¼ mile from the bight on either side of Observatory Point, which lies 1½ miles north-northwestward of Curtis Point.

The north part of Port Davey is filled by a shoal flat, having generally 6 to 8 feet of water on it, the 2 fathoms edge of which from ¼ mile off Curtis Point, trends northward to about ½ mile westward of Fitzroy Point. At nearly ½ mile northward of Observatory Point a ridge, with from 3 to 4 feet water on it, stretches east and west nearly across from shore to shore.

Small vessels can anchor and get good shelter from northwest or west winds, under the land to the north of Curtis Point.

The depth in the channel is about 3 feet at low water, but when there is more water coming down Stephen River, this depth is somewhat increased; the bottom is hard sand.

Directions.—With the assistance of the chart there is no difficulty in entering Port Davey, by passing between Pollard Head and the Pyramidal Rock, taking care to avoid the sunken rock reported to lie nearly in mid-channel. In entering from the southward a good offing must be kept until Pyramidal Rock bears 54° to clear the high-peaked Chatfield Islets.

Working into Port Davey, the west shore between Pollard Head and Garden Point may be safely approached until the water shoals to 8 fathoms; but in standing toward the southeast shore, care must be taken to tack in time to avoid Narves Rock.

In the event of being obliged to run into Port Davey through stress of weather, and unable to get into Bramble Cove, when blowing from northwest or west, having cleared Pollard Head and the reported sunken rock to the southeastward of it, haul round Garden Point and anchor in 5 to 7 fathoms, in Whaler Cove; but it is an exposed place with the wind from any quarter with southing in it, and is not recommended, except for vessels taking shelter only from a northwest wind.

If compelled by southerly gales to leave Whaler Cove, and unable to fetch Bramble Cove, run to the northward for Bond Bay, taking care not to shoal the water to less than 3½ fathoms, and to give Earle Point a good berth, to avoid the sunken rocks which project east and north from it. Having passed Earle Point, and brought Bluff Head to bear 127°, haul into Bond Bay till the peak of Pyramidal Rock is just shut in with, and visible over Garden Point, bearing 178°, and anchor in 3½ to 3 fathoms, ½ mile off the northern extremity of Earle Point. Small vessels might run further up the bay, and anchor in 3 to 2½ fathoms, and be more sheltered from the sea that runs up the port.

To enter Bramble Cove or Bathurst Channel, run in for the North Passage, by steering for the northern extremity of Breaksea Islands; pass between them and Boil Rock to the northward. Or steer for the South Passage between the south end of Breaksea Islands and Shank's Islets, to the southward, and then into Bramble Cove, which is easy of access either from the North or South Passage; both sides of the cove are bold, and may be approached within 200 yards in 4 fathoms, and the entrance is well protected by Breaksea Islands from the heavy sea which rolls into Port Davey. Bramble Cove, and Bathurst Channel within it, are perfectly secure in the most boisterous weather.

For Bramble Cove from the southward, after rounding Pyramidal Rock, steed 65° for the South Passage, and, having passed between Shank's Islets and Breaksea Islands, enter Bramble Cove as previously directed.

In working out from Bramble Cove, if the wind be from the north or northwest, the South Passage between Breaksea Islands and Shank's Islets is the most practicable; but if from the west or southwest, the North Passage is the more safe and convenient one for going out, leaving Boil Rock on the north side of that passage, on the starboard hand, and giving it a good berth. If necessary to tack when in this passage, do not stand within 400 yards of Boil Rock or of the Breaksea Islands, as the heavy swell which sets in may cause the vessel to miss stays; then, if not nearer than that distance, there is sufficient space to bear up and go to leeward of either Boil Rock or the islands, where there are from 8 to 10 fathoms water at 150 yards distance from either, and ample room to get the vessel again under command.

As the chart is a sufficient guide for entering Bathurst Channel from Bramble Cove, it is only necessary to state that the proper channel is between Turnbull Head and Sarah Island.

Anchorage anywhere seaward of Bramble Cove is to be avoided if possible, as a shift of wind from south or southwest will bring a heavy sea right into and up the port.

Sailing vessels can only be absolutely safe from all winds in Bramble Cove or in Bathurst Channel, as any position outside in the port is exposed to wind from some unsafe quarter.

Tides.—From what was observed during a short period in Port Davey, there appears to be no uniform motion in the tides, neither in their ebbing nor flowing, nor in their rise. It seems, however, that they are greatly influenced by the force and direction of the winds, for previously to a strong westerly breeze the water rose from 4 to 5 feet and fell but 2 feet. When the fine weather returned, 2 feet appeared to be the extent of the rise, and this was about the time the moon changed.

The coast from 1¼ miles eastward of Hilliard Head trends south-southeastward for 11 miles to the Southwest Cape of Tasmania; the land is mountainous and presents a barren and desolate appearance.

CHAPTER VII.

TASMANIA.—SOUTH AND EAST COASTS.

Southwest Cape (lat. 43° 33′, long. 146° 02′) is bold and remarkable, with a sharp and rugged outline. Approaching it from the westward, no danger is to be apprehended; but from the eastward it is necessary to keep a good offing as the prevailing winds are from the westward, and the long westerly swell which rolls in with great force, in conjunction with the current which generally sets to the eastward and toward the cape, throws a vessel very fast to leeward.

The south coast of Tasmania extends from Southwest Cape, nearly east, 36 miles to South Cape, and, as might be expected from its exposed situation, is rugged abrupt, and barren. Some small islands lie from 3 to 12 miles off it. The projecting heads of land are supported by basaltic columns, like the Giant's Causeway of Ireland, and it is without any known places of shelter from onshore winds, although it contains two or three sandy bays.

Between two steep rocky heads, distant 3½ and 7 miles east-north-eastward from Southwest Cape, is a sandy bay divided into two bights by a rocky point, with two clumps of rocks in the entrance.

Cox Bight.—From the east point of the sandy bay just described to the northwest point of Louisa Bay, 6½ miles east-northeastward of it, is an indentation, of which the western corner forms Cox Bight, a deep, sandy, but exposed bay. From Cox Bight to Louisa Bay the coast rises to Bathurst Range, which attains an elevation of 2,626 feet.

Louisa Bay and High Bluff.—Louisa Bay extends about 1 mile from northwest to southeast, and has an islet in its entrance. From the southeast point of the bay the coast trends east-southeast 3 miles to High Bluff, the appearance of which may be inferred from its name.

Maatsuyker Isles—Needle Rock.—Maatsuyker Isles consist of two large and several smaller isles lying between 3 and 9 miles off High Bluff; the southwestern of the two principal isles, which lies 13 miles east-southeastward from Southwest Cape, is 920 feet high, and has a reef projecting to the southwest, on which is Needle Rock.

There are several islets and rocks on a reef which extends northward from the island. The northeast Maatsuyker Isle is 1,160 feet high, and lies midway between the southwest isle and High Bluff. There is a sunken rock midway between Louisa Bay and the inner isle, and 2¼ miles southward from the latter is a cluster of rocks, the highest being 540 feet above the sea.

Light.—A group flashing white light, 350 feet above water, visible 25 miles, is shown from a white, circular lighthouse 42 feet high, on the south end of the southwest Maatsuyker Isle. The light is obscured by land between the bearings of 185° and 263°, and by Needle Rock, when within 6 miles, between 72° and 78°.

Mewstone (lat. 43° 44', long. 146° 23'), 6¼ miles southeastward from the southwest Maatsuyker Isle, is a cliffy islet 440 feet high; there are rocks close to the eastward and westward of it. The Mewstone swarms with birds.

Soundings.—There are from 61 to 45 fathoms between Maatsuyker Isles and Mewstone; but vessels are recommended to pass south of Mewstone, 6 miles southwest of which there are depths of 85 fathoms, coral and fine brown sand.

The coast from High Bluff trends northeastward 7 miles to an inlet, and thence extends southeastward 12 miles to the west entrance point of South Cape Bay, 2 miles northwestward of which is Fluted Point. Two rocks above water lie close off a cliffy point 3 miles to the southeastward of the inlet just noticed, and 2 miles south-southwestward from the outer of these two rocks is Isle du Golfe; there is also a small islet or rock near the shore 2 miles to the northwest of Fluted Point.

La Perouse—Aspect (View B on H. O. Chart No. 3570).—From 3 miles southeastward of the inlet just noticed to South Cape Bay the coast consists mostly of high cliffs, from which the land rises to the lofty La Perouse Range. ' La Perouse, 3,925 feet high, is a remarkable table-topped summit, with precipitous cliffs along its south and southeast sides; it bears northward, distant 8 miles from the west entrance point of South Cape Bay. The summit of this range of mountains is a conspicuous thumb-shaped peak, 4,200 feet high, lying west-southwestward, 2½ miles from La Perouse. A sharp remarkable conical apex, 2,630 feet high, rises from a spur trending from La Perouse toward Recherche Bay; this apex usually shows out clearly when the higher mountains to the westward are obscured. Another spur trends east-southeast from the summit of the above range and joins the hills above Three Hillock Point and Whale Head. On this spur the most remarkable part is a dome-shaped wooded summit 1,600 feet high, which rises abruptly from the flat country surrounding it. The Catamaran River flows through the valley between these two spurs into Recherche Bay. The higher portions of the mountain ranges above 3,000 feet elevation are usually rocky and precipitous, but below that altitude the mountain sides and valleys are very thickly wooded.

From La Perouse a ridge trends to the northwest, and the main range runs 25 miles to the north, the most elevated part of it being Adamson Peak, 4,085 feet high and situated 10 miles northward from La Perouse. From Mount Alexander, a not very remarkable peak 3,446 feet high and situated 2 miles northward from La Perouse, a ridge trends east-northeast and termintes in Wooded Hill, 1,835 feet high, a peak which shows as a perfect cone from all directions and is quite unmistakable. This ridge separates the valley of d'Entrecasteaux River to the southward from that of the River Lune to the northward, the former flowing into the Pigsties at the head of Recherche Bay and the latter into the Left Hand Narrows northwestward of South Port.

South Cape Bay extends eastward 4½ miles across from its west entrance point to Three Hillock Point, and is 3 miles deep, but it is too open and exposed to deserve further notice. There are some ponds of fresh water behind the eastern bight of the bay, 1½ miles to the northward of Three Hillock Point.

South Cape (lat. 43° 38', long. 146° 51').—Three Hillock Point, about 500 feet high, forms the southwest extremity of South Cape, which is a broad projection terminating eastward at Whale Head, 155 feet high, 2 miles eastward of Three Hillock Point. Two miles northward of South Cape the land rises to Bare Hill, which is 905 feet high.

Soundings.—From a depth of 80 fathoms, rocky bottom, at about 2 miles southward of the Mewstone, to 2 miles southwest of South Cape, there are from 74 to 48 fathoms. A bank with depths of less than 20 fathoms extends 1,500 yards southeast of Three Hillock Point; and there are depths of 30 fathoms, sand, ½ mile south of that point.

Piedra Blanca and Eddystone, situated nearly 15 miles southeastward from Three Hillock Point, are two cliffy islets connected by a rocky reef, and lying in an east-northeast and opposite direction, 1¼ miles from each other; the former is about 150 feet high; the Eddystone, the eastern one, resembles an ill-shaped tower, and is about 200 feet high.

Sidmouth Rock, distant 5 miles northeastward from Eddystone is about 100 yards in diameter, and awash, with a reef projecting about ⅓ mile to the northeast of it. There is no bottom at 20 fathoms close round this rock and reef, and the passage between it and Eddystone seems to be free from danger.

D'Entrecasteaux Channel (lat. 43° 36', long. 147° 09') is a smooth water passage between the southeast coast of Tasmania and Bruny Island, leading from the southwestward to the Derwent River. The south entrance of this channel extends from South Cape east-northeastward 20 miles to Tasman Head, the south point of Bruny Island, with soundings in 40 to 60 fathoms, for the greater part of the distance across. The channel, about 35 miles long, is slightly winding, the general direction being north-northeast; but its width is irregular, varying from 5 miles within the south entrance to little more than ¼ mile in the north entrance.

The depths range from 40 to 6 fathoms in the fairway.

Directions for D'Entrecasteaux Channel are given later.

The coast from Whale Head trends northeastward 3¼ miles to Second Lookout Point, a rocky projection, on the north side of which is a landing place; this coast has from 22 to 4 fathoms at ¼ mile off. From Second Lookout Point, a rocky indentation, with deep water in it, trends northward one mile to First Lookout Point, and thence a bold rocky coast extends 1,500 yards to the south entrance point of Recherche Bay, which is low and grassy.   .

Recherche Bay.—The north entrance point of this bay is cliffy and 20 feet high. Mutton Rocks extend ¼ mile to the eastward of it, the largest of them being 700 yards eastward from the point. This rock is about 200 yards in length north and south, and 20 feet high; the outer Mutton Rock is small, 5 feet high, and situated 200 yards southeastward of the largest rock. Shoal water, marked by kelp, extends nearly 200 yards southward of the outer rock.

A small and dangerous patch, marked by kelp, is situated 700 yards east-southeastward from the largest Mutton Rock. In ordinary weather this patch has sometimes no breaker on it for hours, when suddenly a tremendous roller sweeps over it. A depth of 4½ fathoms was obtained on it, but there may be less water.

Blind Reef, with a least depth of 7 feet on it at low water, lies nearly 1 mile east-southeastward from the outer Mutton Rock. In moderate weather the sea breaks very little on this reef. Except to the northward the soundings decrease suddenly from deep water to the 3-fathom edge of the reef, and there are patches of kelp northwest of it.

Denmark Reef is a small patch, marked by kelp, with a depth of 2¼ fathoms on it, situated northwestward of the south entrance point of Recherche Bay. Between this rock and the point west of it are Kelly Rocks, some of which are always dry.

Rocky Bay, the south arm of Recherche Bay, has secure anchorage except with northeast and east winds. The eastern part of the bay is shallow. Two rocks above water are situated in the southwestern part of the bay. The hills bordering the shores are densely wooded, and elevated from 600 to 900 feet.

Depth.—Only 3½ fathoms in the entrance, but 4 to 5 fathoms inside.

Anchorage.—Northeast and east winds throw a heavy swell into Rocky Bay, rendering it an unsafe anchorage; several vessels have been lost in this bay, and whaling vessels do not frequent it very much.

The deep-water area, northeast of the eastern rock above water in the southwestern part of the bay, is fully exposed to the heavy squalls which sweep with great violence down Blowhole Valley, and therefore it can not be recommended as an anchorage. With southerly winds the best berth is in the northwestern bight of Rocky Bay, off Ramsgate, about 600 yards offshore, in 4½ fathoms, sand.

Ramsgate is a small postal township, with telephone office situated in the northwest corner of Rocky Bay. Industries are timber and fishing.

Directions.—Give the south entrance point of Recherche Bay a berth of 400 yards, and avoid the kelp off that point. Pass to the southward of Denmark Reef; the south point of the projection westward of Kelly Rocks bearing 270°, or in range with a remarkable dome-shaped wooded summit, 1,060 feet high, leads between Denmark Reef and the shoal water to the southward. Round the south point of the projection westward of Kelly Rocks at a distance of 200 yards, and then steer for the anchorage.

Catamaran River mouth is 150 yards wide, with a depth of 6 feet in it; a rock awash is situated 100 yards northward of the south entrance point of the river.

Recherche Bay anchorage.—The anchorage in Recherche Bay is so much exposed to easterly and southeasterly gales, which, although not of frequent occurrence, send in a heavy sea, that vessels of sufficiently light draft usually anchor well to the eastward, and on the approach of gales from those quarters slip their cables and run into the Pigsties for shelter. The best anchorage is about 600 yards northward of the north point of the projection westward of Kelly Rocks in 9 fathoms, sand.

Depths in the anchorage, 7 to 9 fathoms.

The Pigsties (lat. 43° 33′, long. 146° 54′), the north arm of Recherche Bay, has a rock above water (Shag Rock) in the middle of the entrance, with another rock awash about 30 yards southeastward of it. The edge of the kelp affords a good guide for navigating the channel eastward of these rocks, as it marks approximately the 3-fathom line, and it may be skirted closely.

The depth in mid-channel eastward of the rocks is from 5 to 6 fathoms, shoaling to 4 and 3 fathoms inside the bay.

Foul ground on which the sea breaks in bad weather, and marked by kelp, extends 300 yards southwestward from the south part of the east entrance point to the Pigsties.

Directions.—To enter the Pigsties, when north of Kelly Rocks, bring Shag Rock in range with Wooded Hill (a cone-shaped mountain), bearing 347°, and keep that range until abreast the south part of the east entrance point; pass about 70 yards east of Shag Lock, and then keep the north part of the east entrance point open on the starboard bow, taking care not to stand so far to the westward as to bring Shag Rock outside the southernmost point of land seen. After passing the north part of the east entrance point, keep westward of the north and south parts of the east entrance point in line, until a red cliff on the eastern shore of the Pigsties bears 77°, when alter course toward the cliff for the anchorage.

Anchorage.—The best anchorage in the Pigsties is in a depth of 3½ fathoms, mud, with the summit of the trees on the point westward of Kelly Rocks in line with the north part of the east entrance point. This is an excellent harbor for small vessels, with very smooth water, and the bottom is such soft mud that a vessel is not injured by it if aground.

This harbor was formerly much frequented by whaling vessels during southeasterly gales. The land around is densely wooded.

Supplies.—Water may be obtained on the western side of the Pigsties; milk and butter may be procured from the inhabitants, and fishing boats frequent the bay.

Telegraph and mail communication.—The telegraph office is on the western shore of the Pigsties, whence there is communication by telephone with Hastings, and thence by telegraph to Hobart. The telephone wire has been carried to Bare Hill above South Cape, where it has been proposed to establish a signal station. The post-office is on the west entrance point to the Pigsties; there is mail communication twice a week with Hobart, but the road is a mere track as far as the Narrows at South Port.

Actæon Isles (lat. 43° 32', long. 147° 01') **and shoals,** which lie about 3 miles northeastward of Recherche Bay, are two isles with numerous rocks and reefs extending from them.

Sterile Isle, the south Actæon Isle, is 25 feet high and covered with grass and bushes; it lies nearly 3½ miles eastward from the entrance of Recherche Bay, and has rocks above and below water close to the eastward and westward. Depths of 2 to 4½ fathoms on the inner part of which the sea breaks, extend 800 yards northwestward from the northwest point; and breakers roll over the foul ground which stretches ½ mile to the southeastward, and 1 mile to the southward from the island. Upon some of these patches the sea does not always break.

Beacon.—A white truncated pyramid, surmounted by a staff with horizontal cross pieces, 30 feet in height, stands on the northwestern end of the island, the top being 55 feet above high water.

Southeast Break is a small detached patch with a depth of 9 fathoms on it, lying upwards of 1 mile southeastward from Sterile Isle. The sea breaks on this patch only in bad weather.

South Break, 196° true 1¼ miles from Sterile Isle, has a least depth of 6 feet on it. Vessels should not attempt to pass between South Break and Sterile Isle. The sea breaks in bad weather in a depth of 8 fathoms on the bank to the southward of Sterile Isle.

Actæon Isle, the north Actæon Isle, is 53 feet high and covered with scrub and grass; it is nearly divided into three parts, the northern narrow neck being dry at low water, and the southern neck always dry. A rock dries close to the north point of Actæon Isle, with a reef and foul ground extending 400 yards northward from it. Rocks above water, on the eastern part of which the *Actæon* was wrecked, stretch 400 yards from the south point. The sea breaks on the shoal extending nearly a mile southward of the south point of Actæon Isle. The reef and spit extending 400 yards from the west point of the isle has kelp beyond the danger.

Deep Water Bank or **Ring of Kelp Patch,** the eastern of the Actæon Shoals, is a small rocky patch, lying nearly 1 mile northeastward from Sterile Isle. A depth of 5 fathoms was obtained on this patch, but there may be less water; the sea only breaks on it occasionally.

Black Reef, 1¼ miles northwestward of Sterile Isle, is a cluster of rocks awash at high water. A bank with depths of 3½ to 5 fathoms extends ½ mile in a northerly direction from Black Reef, and then, with a slight break, curves round to the coast northward of Sullivan Point, its whole length being thickly grown with kelp, which also grows in a southeasterly direction from Black Reef for ¼ mile.

The coast, which is low, with Black Swan Lagoon behind, trends from the north entrance point of Recherche Bay northeastward 1¼ miles; it then becomes more elevated, and trends eastwards ½ mile to Sullivan Point, 400 yards southeast of which is Bowden's Mistake, a reef which in moderate weather breaks occasionally.

South Port Lagoon.—From Sullivan Point a rocky coast sweeps round in a north-northwest direction 1¼ miles to a narrow tongue of land, extending north-northeastward, nearly 2 miles to the entrance of South Port Lagoon, which entrance is upwards of 200 yards wide, and has generally a heavy surf across it.

George III Rock is a small patch with a depth of 8 feet on it lying 2 miles northeastward from Sullivan Point. This rock seldom breaks except in heavy weather, and there is a little kelp round it.

South Port Bluff, 1 mile northeast of the entrance to South Port Lagoon, is 65 feet high, and on it stands a tomb.

Blanche Rock, nearly ⅓ mile southeastward of South Port Bluff, and 40 feet high, is bare, with a few dry and covered rocks extending nearly 200 yards to the southeast from it.

South Port Isle, nearly ⅓ mile northeastward of South Port Bluff, is flat topped, with a cliffy coast; it is 83 feet high, and covered with grass. A few rocks, with shoal water 100 yards beyond them, extend 200 yards southeastward from the east point.

South Port (lat. 43° 27′, long. 147° 01′).—From South Port Bluff the coast trends north-northwestward 1,500 yards to a rocky point, ⅓ mile beyond which is a projection, with 3 fathoms water close to it. The west shore of the port is fronted by a shoal, stretching nearly 1,500 yards eastward of the entrance to the Narrows at the head of South Port; on the eastern part of this shoal is Pelican Islet, a low rock covered with grass and with several high trees on it.

Depths.—Eastward of Pelican Islet are depths of 9 to 12 fathoms. In the bight northward of it are depths of 4 to 5 fathoms. There are depths of 4 to 8 fathoms, fine gray sand, in Deep Hole, the bight to the south of Pelican Islet; and a depth of 4 fathoms can be carried to the head of the pier in the southwest part of that bight. The eastern extremity of the shoal eastward of Pelican Islet is marked by a mass of thick kelp.

Shallow water, the edge of which is steep, but not marked by kelp, extends more than 200 yards north of the low rocky point south of Pelican Islet. The water shoals rapidly from 5 to 2 fathoms on the northwest side of this bight, the edge of the bank being on the line joining the pier and the west end of the rocks westward of Pelican Islet.

Jetty approach—Buoys.—The approach to Deep Hole Jetty is marked by 3 black and 2 white cask buoys, moored in 20 feet of water. When approaching the jetty the black buoys are to be left on the port hand and the white buoys on the starboard hand.

The Narrows is a shallow muddy inlet forming three branches, with the River Lune discharging itself into the middle branch or Left-hand Narrows. The entrance is marked by four beacons.

Hastings, a post and telegraph township, is situated on the eastern side of the north branch of Right-hand Narrows. There is a large sawmill here, with a tramway running into the bush from it, and a good road to Hythe. Population is about 400.

Hythe.—The north shore of South Port consists of two bays, separated by a rocky promontory. The post and telegraph township of Hythe is at the head of the western bay, and has a large saw-mill and tramways. Population is about 200.

Stack of Bricks, a rock 30 feet high, is situated off the eastern point on the north shore of South Port, and is steep-to to the southward.

Anchorages.—The eastern bay on the north side of South Port is very much exposed to easterly and southeasterly gales, and with those winds it would be imprudent to anchor there; but it is well sheltered against northwesterly gales, with good holding ground, the best berth being in 7 fathoms, sand, with the extremity of the promontory bearing 257°, distant 600 yards.

The bay in which Hythe is situated has a thick growth of kelp, in which the depth is 4 fathoms, west of the rocky promontory; kelp also grows for 200 yards south of that promontory, with deep water immediately outside it. This bay affords anchorage in $3\frac{1}{2}$ fathoms, sand and mud, but is exposed to easterly and southeasterly gales, although by anchoring westward of the kelp that weed might break the sea. The best berth is in $3\frac{1}{2}$ fathoms, with Hythe Pier bearing 4°, and Stack of Bricks just open south of the rocky promontory on the east point of the bay.

The bight to the southward of Pelican Islet, Deep Hole, is more protected with winds from the eastward, but the deep-water space is too narrow to admit of anchoring near the pier. The best berth is with the eastern extremity of Pelican Islet in line with the school-house at Hythe (on a point eastward of the township), bearing 160°, and about 350 yards from Pelican Islet, in $7\frac{1}{2}$ fathoms, sand.

Directions.—To enter the anchorage off Hythe, South Port Island, just hidden by the south entrance point of South Port, leads to the southward of the kelp in the middle of the bight; and when the eastern extremity of Pelican Islet is in line with the eastern extremity of the sandy beach in the south part of South Port, keep that range astern, and anchor as directed above.

Communication.—Hythe and Hastings are both in telegraphic communication, and there is a mail from Hythe to Hobtra twice a week overland. Steam vessels from Hobart call at Hythe twice a week.

Timber.—The hills at the back of Hythe rise to a height of 1,000 to 1,520 feet, and are thickly timbered with blue gum and stringy bark. The land on the south side of South Port is thickly wooded and from 200 to 300 feet high.

Ballast.—Vessels may discharge ballast on the seaward or southeastern side of an imaginary line drawn from Pelican Island to the Stack of Bricks. but in not less than 10 fathoms of water.

Coast—Burnett Point (lat. 43° 26', long. 147° 02').—From Stack of Bricks the coast trends northeastward nearly 1,500 yards to Burnett Point, the cliffs being about 100 feet high, and thence north-northwestward over ¼ mile to the south point of Sisters Bay, off which a reef extends 400 yards from the shore, having, on the outer part, a small rocky islet 5 feet above high water.

Sisters and Lady Bays are two indentations of the coast, ½ and ¼ mile deep, respectively, with depths of from 3 to 10 fathoms, sandy bottom, and separated by a rocky promontory, 1½ miles northward from Burnett Point. In the north of Lady Bay there is a sawmill and pier.

There is another smaller bight ¼ mile to the northward of Lady Bay, from whence a rocky, cliffy coast line extends north-northeastward for 2 miles to Scott Point. This coast is fringed with kelp its whole length.

Scott Point (lat. 43° 21′, long. 147° 04′) is on the south side of Port Esperance; from the projection north of it a rocky bank, with from 4 to 5 fathoms, and marked by kelp, extends in a northerly direction nearly 800 yards, having a small patch of 3 fathoms on its outer extremity.

Port Esperance.—From Scott Point the coast trends northward ¼ mile to a projection between which and Esperance Point, a little more than 1 mile northward from it, is the entrance of Port Esperance, which extends thence 2¼ miles in a westerly direction, and is 1¾ miles wide.

About a mile westward of Hope Isle, a point of the south shore projects to the northward, on the west side of which is the entrance of an inlet 670 yards wide, with depths of 8 fathoms in mid-channel, and from 8 to 20 fathoms between it and Hope Isle. From its entrance the inlet winds about 1¾ miles in a west-northwest direction to a point which divides it into two branches, one trending ½ mile to the southward, and the other about the same distance westward to Esperance River. One-third of a mile within the entrance of this inlet is Rabbit Islet, between which and the west entrance point there is a narrow passage, with from 4 to 2½ fathoms, and 3½ fathoms water within the islet, above which the channel appears to be obstructed by small islets or rocks. A vessel may lie in this inlet perfectly landlocked.

Depths.—Northward of Hope Isle the depths are 8 to 4 fathoms; southward of Hope Isle are 9 to 20 fathoms.

Hope Isle, which is about 100 feet high, nearly ½ mile in extent, and with a few trees near the summit, lies 1 mile within the entrance, dividing it into two channels, that on the south side of the island being ¼ mile wide, with from 9 to 20 fathoms, mud and sand, where a vessel may be sheltered from all winds.

Light.—An occulting white light, 37 feet above water, is shown from a square white tower 18 feet high, on the eastern extremity of Hope Isle.

A bank of from 3½ to 4 fathoms extends from Hope Isle to the northward across the bay, on which, and ¼ mile north of Hope Isle, is situated a small islet with a few trees on it, named Dead Islet.

Anchorage.—Between Dead Islet and Esperance Point there are from 6 to 8 fathoms water, sand and rock, affording a convenient anchorage, with Esperance Point in line with Ventenat Point bearing 100°, and Scott Point 168°, the western extremity of South Bruny being open to the eastward of it. There is also good anchorage southward of Hope Isle, as stated above.

Snachall Islets (lat. 43° 19′, long. 147° 02′) are two small rocky islets ½ mile northwestward for Dead Islet, having a reef extending 300 yards from them in a southeasterly direction.

Between this reef and the bank on which Dead Islet stands is a narrow channel with depths of 8 to 9 fathoms. Shoal water extends from Snachall Islets in a north-northwesterly direction to the shore.

Ballast.—Vessels may discharge ballast toward the south or southeast of Hope Isle, but is not less than 10 fathoms of water.

Dover, a postal township in the northwestern part of the port, is in telegraphic communication. There is much fine timber in the neighborhood, which gives employment for several sawmills.

There is regular steamboat communication with Hobart and river ports.

Water.—There is a narrow bight in a sort of ravine formed between the heights of Folkestone on the south shore, southwest of Hope Isle, having a depth of 7 fathoms in the entrance and from 4½ to 2¾ fathoms farther in, and affording shelter for heaving down a vessel. At the bottom of the bight is a rivulet of excellent water.

Roaring Bay.—From Esperance Point the coast trends northnortheastward for ¼ mile to a point having a small bight on its west side; the northern portion of this bight is marked by a conspicuous red cliff, 150 feet high. Between the red cliff and another projection, north-northeast about 1⅜ miles from it, there is an indentation named Roaring Bay.

Coast.—From the northeast point of this bay the coast trends northwards ½ mile to Huon Point, the west entrance point of Huon River.

Mount Esperance, 1,515 feet in height, and situated 3 miles westward from Huon Point, is a summit of the mountain range, with spurs both to Huon River and Port Esperance, and is everywhere densely wooded.

Tasman Head (lat. 43° 31', long. 147° 18').—Tasman Head, the south point of South Bruny Island, forming the northeastern point of the south entrance of D'Entrecasteaux Channel, is high, abrupt, and composed of basaltic pillars, with a bank which should be avoided, and with depths of 12 to 19 fathoms extending south-southwestward 1¼ miles and south-southeastward 2¼ miles from it, on which are several small islets and numerous rocks, some of the former producing vegetation. Arched Rock, 170 feet high, is situated 200 yards offshore, southeast of Tasman Head.

Friar Rocks.—The largest and most conspicuous islet of Friar Rocks is 325 feet high, and is ½ a mile southward of Tasman Head, the passage between being apparently free from danger. It is possible to land on the north side of this islet when the sea is smooth. A chain of islets and rocks extends 1,200 yards southeastward of the largest islet, terminating in a rock nearly awash, on which the sea usually breaks. The two southeastern islets of the group are 90 and 95 feet in height, pyramidal in shape, and, except where whitened by sea birds, have a bleak, weather-beaten appearance. The southwestern islet is 250 feet high, and appears to be split in two when seen from the southeast or northwest. Vessels should not attempt to pass between the rocks and islets of this group.

Bank (lat. 43° 37', long. 147° 16').—A small rocky bank, with from 17 to 20 fathoms water on it and 30 to 45 fathoms all round, is situated southward, distant 4¾ miles from the southwestern Friar Islet.

The coast, for 2½ miles to the northwestward, between Tasman Head and East Head is bold, rocky, and precipitous. The land at the back rises in smooth grass and scrub-covered summits, 800 to 1,000 feet high, culminating in Mount Bruny, a saddle-shaped summit, 1,700 feet high, and covered with thick scrub.

Cloudy Bay, a bight in the southern end of South Bruny Island, exposed to all the fury of southwest gales, is 3 miles wide, east and west, at its entrance between East and West Heads, whence it extends 3¼ miles northward to a long narrow tongue of land stretching westward from the east side, and separating this bay from Cloudy Lagoon. The east shore of Cloudy Bay for the first 1¾ miles is rocky and irregular, the most projecting danger being a reef with dry rocks upon it, extending from a point about midway between East Head and the head of the bay; there is a rock 5 feet high at its extremity, 300 yards westward of the point. Another reef, with thick kelp on it, extends ⅓ mile northward from the northern extremity of this point, on the east side of which is a small bight, with depths of 2 to 1¼ fathoms water, affording complete protection in all weathers to fishing and other small vessels. The east side of Cloudy Bay between this bight and a projection of the north shore appears to consist of a sandy beach. The head of the bay is exposed to a great surf.

Cloudy Lagoon is a shallow sheet of water 1¼ miles long in an east and west direction, 1¼ miles wide, and communicates with the northwest corner of Cloudy Bay by a narrow channel trending north and south 1,840 yards. The land for about 1½ miles northward of the lagoon is low and swampy.

Half Moon Bay.—The western shores of Cloudy Bay fall in steep cliffs from grassy downs, 300 to 680 feet high, with several small open bights, the southern of which, Half Moon Bay, is the most important. It has depths of 7 to 12 fathoms, white sand, with thick kelp ¼ mile from the shore along the head of the bay.

Anchorage.—Except the small vessels before mentioned, no vessel should anchor in Cloudy Bay, unless it is absolutely necessary to do so. In such a case the best position is in Half Moon Bay, as there a vessel is well to windward should southerly winds come on.

The coast between West Head and another point lying west-southwest from it, and ¼ mile from Cape Bruny, forms an exposed bay 1¼ miles wide and 1,500 yards deep, its bight being a sandy beach, with a rocky point, ¼ mile westward of West Head. Chains of high rocks extend to the southward for upward of 200 yards from each point. A bank with from 16 to 20 fathoms extends 1¼ miles southward from West Head.

Cape Bruny (lat. 43° 30′, long. 147° 09′), the southwest point of South Bruny Island, is 291 feet high.

Light.—A flashing white light, 346 feet high, visible 22 miles, is shown from a white, circular stone lighthouse on Cape Bruny, 44 feet high.

Lloyd's signal station.—There is a Lloyd's signal station at the lighthouse, connected by telephone with Hobart.

Courts Isle extends from a few yards to ¼ mile southward of Cape Bruny, and is nearly ¼ mile wide. It is 200 feet high, flat-topped, grassy, and precipitous, with a small islet, 60 feet high, 200 yards southwest of it; there are depths of 10 to 14 fathoms close to the southward of the isle and islet.

Bank.—A rocky bank extends 1¾ miles south-southwestward from Cape Bruny, with depths of 12 to 17 fathoms on it. The light-keeper has reported a very heavy break on this bank in bad weather, but nothing less than 12 fathoms was obtained during a careful search by a government surveying vessel; the bottom is very irregular.

Mount Barren, 1,500 yards northward of Cape Bruny, is 500 feet high, with a cairn on it. The land near Cape Bruny is covered with grass and scanty scrub.

Standaway Bay extends from Cape Bruny northwestward 4 miles to Point la Billardière, 1 mile southward of Hopwood Point. The detached rocks in the bay are from 10 to 30 feet high, and lie from 200 to 500 yards offshore. Mount Bleak is 510 feet high, and its southwestern slopes are covered with scanty scrub. The depth of 20 fathoms is about ¼ mile offshore throughout Standaway Bay.

Hopwood Point is the northwest point of the promontory, from 300 to 500 feet high, extending 4½ miles in a northwest direction from Mount Barren. It is the north end of the cliffy coast line, extending from Cape Bruny, and when seen from the vicinity of the Zuidpool Rock appears as the northern extremity of the promontory.

About ¼ mile northward from Hopwood point is a projection having some rocks above water close to its west side, eastward 1,500 yards from which is a third point forming the west entrance point of Great Taylor Bay. These points are low and rocky.

Partridge Isle, which extends from about 200 yards to 1½ miles northward from the northwest point of this promontory, is ¼ mile broad, with from 18 to 7 fathoms water close to its west shore. It is wooded and 230 feet high.

Pier—Anchorage.—There is a pier on its east side, off which there is anchorage in 10 fathoms, mud and sand, with Hopwood Point in line with the southern extremity of Partridge Isle, bearing 216°, and the northern extremity of the island 317°.

Great Taylor Bay.—The entrance of this bay, eastward of Partridge Isle, is 2½ miles wide, from whence it extends about 3¼ miles in a southeast direction. The western shore for the first 2½ miles is nearly straight, and thence irregular to the bottom of the bay; a patch, which is awash at high water, lies ¼ mile from the shore 2¼ miles southeastward from the west entrance point. The eastern shore of Great Taylor Bay consists of projecting points and bights, the most extensive of the latter being the Bay of Islands, which lies midway between the northeastern entrance point and the southern extremity of Great Taylor Bay. The Bay of Islands is over ½ mile wide at its entrance, whence it extends nearly a mile to the northeastward.

The depths in Great Taylor Bay range from 17 fathoms at the entrance to 8 and 9 fathoms 1 mile from the head of the bay, thence gradually shoaling.

Curlew Islet is 23 feet high, covered with grass, and situated 400 yards from the north entrance point of the Bay of Islands, and there is a smaller islet, 30 feet high, nearly the same distance from the southern shore of this bay, over half a mile within the entrance.

Oak Point.—At 1½ miles southward of Curlew Islet is Oak Point, the southern projection of the eastern shore; the bight to the southwest of the point is surrounded by a sandy beach.

The store houses for the lighthouse on Cape Bruny, with a pier for landing, are built on the western shore, southwestward of Oak Point.

Anchorages.—Great Taylor Bay is too large to afford shelter from gales at all times; although the bottom generally is hard black mud, vessels have dragged their anchors, even with a long scope of cable.

Ventenat Point (lat. 43° 21′, long. 147° 12′).—From the east entrance point of Great Taylor Bay the general trend of the western coast of South Bruny Island, which is slightly embayed, is nearly north 3½ miles to Ventenat Point. This point, which forms the west side of the entrance of Little Taylor Bay, is the northern extremity of a tongue of land projecting in a north-northwest direction 2½ miles, and separating Little Taylor Bay from D'Entrecasteaux Channel. There are depths of 24 to 12 fathoms between Partridge Isle and Ventenat Point; a reef projects a short distance north from the point.

Shoal water.—A small rocky patch of 12 fathoms lies about 2¼ miles northeastward from the north end of Partridge Isle and another of 10 fathoms, nearly 2 miles northward from the same point.

Little Taylor Bay is 1½ miles wide, northeast and southwest, at the entrance, whence it extends about 2½ miles southward. There is a small bight in the western shore of the bay, ½ mile within Ventenat Point, where anchorage may be obtained in from 4 to 5 fathoms, sand and mud, sheltered from south and west winds; and there is a larger, but more shallow one, in the eastern shore, between 1,500 yards and 1¼ miles from the northeast entrance point of the bay, but it is not recommended, being exposed to the northwest, from which direction the strongest gales are experienced, and the sea soon gets up. Little, like Great Taylor Bay, is capable of receiving large vessels, although the anchorage in neither of them can be considered good.

Simpson Point.—The west coast of South Bruny Island from Little Taylor Bay takes a general north-northeast direction 6¾ miles to Simpson Point, the northern extremity of a part of the island projecting 3¼ miles to the northward, and separating Isthmus Bay on its east from D'Entrecasteaux Channel on its west side.

Danger.—From Simpson Point a narrow rocky shoal, with 4 to 5 fathoms water on it, extends to the southward for a distance of 1¼ miles parallel to the coast, with its western edge about 800 yards from the shore.

Satellite Island, 2¼ miles north-northeastward of Ventenat Point, and ¼ mile from the shore, is 170 feet high, cultivated, and has a thick grove of trees on its west side.

The coast is cliffy almost throughout, from 70 to 40 feet in height and accessible. There is a pier on the northeast side of the island.

Between Satellite Island and Bruny Island there is foul ground; a reef of rocks extends 400 yards from Bruny Island, and the depth between is from 2 to 3 fathoms.

There is a pier on the point of Bruny Island about 1 mile northeastward of Satellite Island.

Zuidpool Rock (lat. 43° 20′, long. 147° 10′), about 50 yards in extent with 2 fathoms water on it, lies nearly midway between Ventenat Point and Huon Island, and 1¾ miles from the former.

Within the 5-fathom line the shoal is about 300 yards long, north and south, by 150 yards wide, with from 7 to 10 fathoms around.

Buoy.—A black and white checkered conical buoy, which is, however, liable to drift, is moored close to this danger.

Clearing marks.—Hopwood Point, seen just open east of the south extremity of Partridge Island, bearing 215°, leads over ¼ mile eastward of Zuidpool Rock.

The northeast extremity of Hope Island, in line with Esperance Point bearing 263°, leads 200 yards to the northward of the rock.

Cygnet point, in line with the southern extremity of Huon Island, leads northeastward of the Zuidpool rock.

West shore (continued).—**Huon Island,** which lies close off the entrance of Huon River, about 1.7 miles east of Huon Point, is wooded in the center, the tops of the trees being 250 feet above high water. There are some houses on its north end, and a small pier. The island is conspicuously green.

Huon River.—Huon River is about 2.8 miles wide at its entrance from Huon Point to Ninepin Point eastward from it; a cluster of rocks lies off Ninepin Point, between which and Huon Island there is a channel 1,340 yards wide. The southwest shore of Huon River from Huon Point extends northwestward 6¾ miles to a projecting part of Adelaide, which is a small settlement between Surge Bay on its southeast side and Flight Bay to the northwest of it. The objects along this shore which appear most worthy of notice seem to be Surveyor Bay, ¼ mile within Huon Point; Police Point, northwest 2 miles from Surveyor Bay; Desolation Bay, west 1¼ miles from Police Point; and White Bluff, northwest 1¼ miles from Desolation Bay; close to the westward of the bluff is Flower-pot Rock, above water.

Depths.—The river entrance southwestward of the Butts is 20 fathoms deep. Off the entrance to Port Cygnet, 3 miles above the entrance, the depths are 15 to 16 fathoms. Off Huon Lighthouse, at One Tree Point, 8 miles from the entrance, there are 10 to 11 fathoms, decreasing to 5 fathoms 1¼ miles above the lighthouse.

The Butts.—The northeast shore of Huon River from Ninepin Point trends west-northwest 1¼ miles to a small peninsular point, westward nearly 1,600 yards from which is a rocky patch, covered at high water, named the Butts.

Light.—A fixed white light, visible 8 miles, 26 feet above water, is exhibited from a skeleton iron tower erected on the Butts.

Buoy.—A buoy is moored close westward of the Butts Light Beacon.

Clearing mark.—Mount Windsor, a conspicuous summit, in line with Cygnet Point, bearing 324°, leads 400 yards westward of the Butts.

Garden Island.—Between the small peninsula just noticed and a point west-northwest 2½ miles from it, and ¼ mile southeastward of Cygnet Point, is a bight 2 miles wide and 1¼ miles deep, having in its center Garden Island, which is 1,500 yards long north and south, and about 670 yards broad. This island gives the name to a creek flowing into the bight ¼ mile to the northeast of the north point of the island; there is a small cove, Randall Bay, 1 mile to the westward of the rivulet.

Cygnet Point is a broad projection between South Deep Bay, on its southeast side and Eggs and Bacon (Abel) Bay on its northwest side, and forms the southeast point of the entrance of Port Cygnet. Eggs and Bacon Bay is little more than ¼ mile in extent.

Buoy.—A white buoy has been placed to mark Eggs and Bacon Reef on the south side of South Deep Bay. The buoy is moored in 18 feet water.

Port Cygnet.—Port Cygnet is 1¼ miles wide at its entrance from Cygnet Point to Beaupre Point, in a west-northwest direction and extends 4 miles to the northward. The east shore of the port is broken and irregular, consisting of points and bights. Deep Bay, the southern and largest of these bights, lies about 2 miles northward of Cygnet Point, and extends nearly 1,500 yards in an east-northeast direction. On the north side of Green Point, which is about 1 mile northward of the north point of Deep Bay, is an inlet over ¼ mile wide, extending about ¼ mile to the eastward. This inlet is separated from a similar one at the head of the port by two projecting points.

Depths.—The middle of the harbor has from 3½ to 7 fathoms water, upon a mud and sandy bottom; and, with the exception of the interior of some of the bays, a depth of less than 3 to 4 fathoms is seldom found at a distance of over 100 yards from the shore.

Lovett is a small township with post and telegraph stations on the shores of Port Cygnet. Population of town is about 1,500, of district about 4,000. The country around is agricultural and timber-producing. Fruit is exported in immense quantities. Coal seams in the district are systematically worked. There is daily communication with Hobart by steamer and coach.

The west shore of Port Cygnet from Beaupre Point trends north-northeastward 1¼ miles to Slag Point, projecting $_{r}o_m$ Lymington, between which and another point 1,550 yards northward from it, is a bay, Copper Alley, about 1,340 yards deep, with 3½ fathoms water in its center. The western bight of this bay has a sandy beach, to the northward of which is a small inlet. From the north point of Lymington Bay the west shore of Port Cygnet extends northward 1¾ miles to Lovett, at the northern extremity of the port where it forms a narrow shallow inlet.

Each of the five bights just described receives a small stream flowing from the neighboring hills, of which hills Mount Cygnet, 4¾ miles northeastward, and Mount Morrison, 6 miles northward of Cygnet Point, appear most worthy of notice; but Mount Grey, situated 3½ miles northward from Mount Morrison, seems the most elevated, being 2,713 feet high.

The shores of Port Cygnet are a little elevated, and generally steep; and the remarkable fertility of the soil offers everywhere the most enchanting and varied appearance. In several places natural quays are formed, easy of access for large vessels, or even for the purpose of careening.

The northeast shore of Huon River.—Between Beaupre Point and Poverty Point, ¼ mile to the northwest of it, is a cove, whence the shore trends west-northwestward 2 miles to a small stream, with a rock close off it, and thence westward 1¾ miles to One Tree Point, at Brabazon; midway between the stream just noticed and One Tree Point is Petchey Bay, which is barely ¼ mile in extent.

From One Tree Point, Huon River takes a northerly direction for nearly 10 miles, with an average width of ½ mile. Its east shore from One Tree Point to California Bay, 5 miles to the northward of the point, is irregular, and intersected by several small streams; but for the next 5 miles it is nearly straight.

Light.—From a square white tower 27 feet in height on the extremity of One Tree Point a light is exhibited at an elevation of 25 feet above high water.

The west shore of Huon River from Flight Bay to abreast of California Bay consists of points and bights, the largest two of the latter being Hospital and Castle Forbes Bays. Hospital Bay, which lies north-northwest 1¼ miles from One Tree Point, is ½ mile wide, north and south, at the entrance, whence it trends nearly 1 mile to the westward, the mouth of the Kermandic River being in its northwest corner. The bight of the bay is mostly occupied by an islet and shoal water. From the north point of Hospital Bay the shore trends north-northwest 1¼ miles, and northeast 1 mile, to Bullock Point, the intermediate bight being Castle Forbes Bay, between which and Bullock Point is Fleurtys Bay.

Hospital Bay.—There is a pier in Hospital Bay which has 800 feet of berthing accommodation and depths of 10 to 32 feet alongside. At Shipwright Point there is another pier, alongside of which a vessel drawing 16 feet has been berthed.

There is good anchorage in Hospital Bay, with plenty of room in a depth of 6 fathoms at a distance of 600 yards, 235°, from Shipwright Point. The bottom was very soft thick mud.

Four small pile beacons stand in the northeastern part of the bay.

Franklin.—For about 6 miles above Bullock Point the west shore, which is nearly straight, forms the water frontage of Franklin. There are from 10 to 4¼ fathoms water between One Tree and Bullock Points, but from nearly abreast of, to 8 miles above the latter point, the river is mostly filled by the Egg Islands, and the bank extending southward from them.

The postal township of Franklin is 28 miles southwest of Hobart, with a population of about 1,200. It has a telegraph station and daily communication with Hobart by mail coaches, also by steamer. The country around is thickly timbered, giving employment to several saw mills; much fruit is also grown.

The river is unsurpassed for salmon and trout fishing, and the scenery, both mountain and river, is very beautiful.

Huonville.—At 9 miles above Bullock Point Huon River turns northwestward 2 miles to Huonville, whence, after being joined by a small stream from the northeastward, it becomes a mere rivulet, flowing from the westward.

Huonville is situated on the banks of Huon River, at its junction with Mountain River, and is connected with Hobart by a good road. The principal industries are fruit growing and the timber trade. Coal has been found. It is a post and telegraph station, and mail coaches run daily to Hobart. There is good salmon and trout fishing here. Population is about 500; of district about 4,000.

Ninepin Point is the south point of the peninsula formed by the Huon River on one side and the D'Entrecasteaux Channel on the other, and is situated 1,700 yards east-northeastward from the north point of Huon Island.

Arch Islet is a perforated rock, 50 feet high, lying 1¼ miles eastward from Huon Island and ½ mile from the shore, with depths of 2 to 6 fathoms between.

Three Hut Point.—The west shore of D'Entrecasteaux Channel from Ninepin Point after turning 1,500 yards to the northeast, trends nearly east for 1¼ miles, and then again turns northeastward, 1¼ miles to Three Hut Point, behind which is the village of Gordon.

There are from 3 to 4 fathoms water close along the shore for the greater part of the distance, but at 1 mile southwest of Three Hut Point, the 3-fathom curve is nearly ¼ mile from the beach, therefore this point must not be rounded too closely.

Gordon is a port of clearance and has a post and telegraph office. There is daily steam communication with Hobart. Population is about 200.

Anchorage.—During west and northwesterly winds the anchorage off Three Hut point is sheltered. A good berth is in 4½ fathoms, mud, with the pier bearing 280°, and the southern beacon on Long Bay Bank 359°, with a conspicuous white house on the shore, 1¾ miles to the northward, open to the eastward.

Mount Royal, 1,500 yards westward from Three Hut Point, rises to the height of 1,190 feet, and forms the south end of the mountain range extending 17 miles in a northerly direction.

This range attains a height of about 2,000 feet; 2 miles to the westward of Peppermint Bay, and farther northward, it rises to between 2,400 and 2,500 feet, with long spurs extending to the coast, the mountains and valleys being everywhere densely wooded.

Communication.—There is a good road from Hobart.

Steamers run daily from Hobart, calling at all the principal townships and landing places on the way to Huon River, and twice a week on to Port Esperance and South Port.

Long Bay Bank.—From Three Hut Point the coast trends northward 2¾ miles to Whaleboat Rock, above water, and thence northnorthwest 1½ miles to Flower Pot Rock, 10 feet high, close to the shore, 1,500 yards to the north of which is Fleurtys Point. On the north side of Three Hut Point is a shoal bight about ¼ mile wide. A bank borders the coast between Three Hut and Fleurtys Points, and extends 1,340 yards to 500 yards from the shore, projecting farthest from the land 1 mile northward of Three Hut Point.

Beacons.—Four large beacons erected in from 9 to 18 feet water, and a black can buoy in 16 feet at low water, 2,200 yards northeastward, from Three Hut Point, indicate the general bend of the bank which sweeps uniformly round from Three Hut Point to Fleurtys Point, the 3-fathom curve passing about 200 yards outside the two northern beacons.

Light.—From the beacon, situated at a distance 2 miles northnortheastward of Three Hut Point, a light is exhibited.

The channel between Long Bay Bank and the 4-fathom shoal southward of Simpson Point is 1,340 yards wide, with from 6 to 7 fathoms in the fairway.

Dangers.—There is a small 5-fathom patch detached from Long Bay Bank, 1,800 yards westward from Simpson Point, and another the same distance northward from that point.

Middleton is the name of the post and telegraph township at Long Bay; it has daily steam communication with Hobart and has a population of about 200. There is a good pier, off which anchorage may be obtained in 4½ fathoms, with the pier bearing 270°, and the light beacon on Long Bay Bank 201° ⅓ mile from the shore.

Buoy.—There is a buoy on the edge of the bank off Middleton; it is very small and not easily distinguished.

Tides and tidal currents.—It is high water, full and change, at Three Hut Point, at 7 h. 30 m., springs rise 4½ feet, neaps rise 3½ feet. During spring tides, after a continuation of light winds, the flood current sets to the northward, and the ebb to the southward; these currents are felt most strongly near Three Hut Point and Long Bay, where the velocity is from ¾ to 1 knot an hour.

Current.—The prevailing current in D'Entrecasteaux Channel sets in a northerly direction 1 to 2 knots according to the wind.

Birch, Peppermint, Trial, and Flight Bays.—From Fleurtys Point to Simmonds Point the northeast point of Oyster Cove, 4¾ miles northward from it, the western coast of D'Entrecasteaux Channel is embayed to the depth of a mile, and consists of alternate bays and points, nearly all the bays having a small stream flowing into them.

Birch Bay, the southern and widest of these bays, extends from Fleurtys Point, north-northwestward 1¼ miles, and is 670 yards deep. A long low white jetty projects northeasterly from Fleurtys Point. Peppermint Bay forms a double bight, extending 1 mile northyard from the northern point of Birch Bay. The name of the township is Woodbridge, which has a population of about 300. There is a postal and telegraph office, and biweekly steam communication with Hobart.

The pier at Woodbridge is built on the point dividing the bay into the two bights above mentioned, off which the anchorage is in 6 fathoms, mud, with the pier bearing 263°, distant 700 yards, and the north point of the bay 353°.

The northern portion of the bay is shoal, the 3-fathom curve extending 300 yards to the southward of the north point.

Between the north point of Peppermint Bay and another projection north-northeast nearly 1½ miles from it there are three bights, named, Peach, Trial, and Flight Bays.

Little and Oyster Coves lie between the northeast point of Flight Bay and Simmonds Point 1½ miles northeastward of it, and are separated from each other by a broad projection of the coast, which rises to a height of 500 feet.

Little Cove is ⅓ mile across at the entrance, and 1,500 yards deep, with 5 to 8 fathoms in the outer portion. There is a beacon in 10 feet water, near the middle of the cove. A single white pile beacon is on the eastern end of the Mudflat in Little Cove. It is to be left on the starboard hand by vessels approaching Kettering jetty.

Oyster Cove is 1,340 yards in width and the same in depth, with 7 to 8 fathoms in the middle. The northeast point of the cove is named Simmonds Point. There is a pier projecting northeasterly from the southern entrance to Oyster Cove.

Channel Rock is a small rocky patch 150 yards in extent and having 11 feet least water over it, with from 5 to 7 fathoms around. It is situated 500 yards southward from Simmonds Point.

Northwest Bay.—From Simmonds Point the coast trends northward 1¾ miles to Snug Point, the south side of the entrance of Northwest Bay, which is 1¼ miles wide in a south-southwest and opposite direction; there are depths of 3 fathoms close to Snug Point, and 10 to 15 fathoms thence across the entrance to the opposite point. The two entrance points are high and rocky; but the shores of the bay are much lower and easy of access.

Within its entrance Northwest Bay extends 4 miles in a northerly direction, and 2 miles from its entrance to its west shore. The south shore from Snug Point trends west-northwestward 1¾ miles to Snug Bay, where there is a small jetty 1,500 yards to the northward of which is Snug River, whence the west shore extends northward 2¼ miles to the east point of Margate. Between this point and the north corner of the bay 1 mile to the northward is a shoal bight, with Northwest Bay River flowing into it. From the north corner of Northwest Bay, its northeast shore trends southeastward 3¼ miles to the north entrance point.

The east shore of the bay is clear of dangers, but on the west shore a reef extends for nearly 400 yards to the eastward of the north point of Snug Bay, and there is another reef extending 800 yards from the shore midway between Snug Bay and the east point of Margate.

Depths.—The soundings in Northwest Bay gradually decrease from 10 to 15 fathoms across the entrance to 6 and 7 fathoms eastward of the Chimneys (the landing place for Margate), thence shoaling rapidly to 3 and 2 fathoms off the mouth of Northwest Bay River.

Margate is a post and telegraph township with a population of about 550. The district is agricultural, and great quantities of fruit are grown here.

Piers.—There is a pier at Margate at the Chimneys, and a disused coaling pier about 100 yards long, running out at right angles to the shore, in the small bay about 1,200 yards southward of Margate Pier.

Pierson Point (lat. 43° 03′, long. 147° 21′).—From the north entrance point of Northwest Bay the coast trends northeastward 1¼ miles to Pierson Point, which is high and cliffy, and forms the northwest side of the north entrance of D'Entrecasteaux Channel.

Pilot station.—There is a pilot station at Pierson Point.

Anchorage.—Midway between the above points there is anchorage in 8 fathoms, mud, in Tinderbox Bay, a small indentation of the coast at 400 yards from the shore, sheltered from northwest winds.

At ½ mile west-southwestward of Pierson Point, Mount Louis rises to a height of 694 feet.

Isthmus Bay, on the north side of South Bruny Island, is separated from D'Entrecasteaux channel by a promontory, the summit of which is 1,130 feet high, gradually sloping down to Simpson Point, which is low and rocky.

The bay is 3 miles deep; its west shore from Simpson Point trends south-southeastward for 3½ miles to a small islet in it, whence the southeast shore, which is bordered by shoal flats, curves nearly 4 miles in a north-northeast direction, to a small projecting point with reddish-colored cliffs, named the Bluff, with a low rocky point ¼ mile north of it.

The southeast shore of this bay is only separated from Adventure Bay, on the east side of Bruny Island, by an isthmus, which for a distance of 2 miles is from 200 to 400 yards broad, nearly dividing the island midway between its north and south ends. On the north side of the northern of the two small projecting points just noticed is a cove ½ mile wide, whence the east shore of Isthmus Bay trends northwest, 1,500 yards to a cliffy point which separates Isthmus from Great Bay.

Depths.—Between Simpson Point and the Bluff there are from 5½ to 6 fathoms near the former, decreasing gradually toward the head of Isthmus Bay and the Bluff, off which the 3-fathom curve projects over ½ mile.

Great Bay is 2 miles wide at its entrance and nearly 2¼ deep; 1 mile within its entrance the bay is contracted to 1¼ miles in width by projections of the north and south shores, within which it again expands to nearly 2½ miles, north and south.

There is a small cove on either side of a broad projecting part of the south shore; that to the westward being named Fancy Bay, and that to the eastward Ford Bay; and there is an inlet in the northeastern extremity of the bay with a stream running into it, and apparently a narrow channel to it through the flats which border the east shore. The northeast portion of Great Bay is named Adams Bay.

The depths in Great Bay are 5 fathoms at the entrance, shoaling gradually to 3 fathoms at ½ mile from the head of the bay.

Missionary Bay.—From Stockyard Point, a double projection which separates Great Bay from Missionary Bay, to the northwest

of it, the entrance of Missionary Bay extends westward 1¼ miles to
Soldiers Point, whence the bay runs in about 1 mile to the northeastward.   The general depths are less than 3 fathoms in this bay.

Green Island, situated 1 mile southward from Soldiers Point, is
a small grass-covered islet, 20 feet high, with a few bushes on it.   The
5-fathom curve extends 350 yards from it, northeast and southwestward narrowing the passage to ¼ mile between it and the 5-fathom
curve on the west side of the channel, which here projects nearly 1
mile from the shore.

Snake Islet.—To the westward of Soldiers Point is a bay ¼ mile
deep, in the inner part of which is Snake Islet, 30 feet high, and
covered with grass; 200 yards southwestward of the islet is a rock
that dries 3 feet at low water.

Shoal.—A rocky shoal, 250 yards in length west-northwest and
east-southeast and 100 yards in breadth, with depths of 6 to 10 feet
on it, lies nearly 500 yards, southwestward, from the south end of
Snake Islet.

There are depths of 5 fathoms 100 yards southwestward of the
shoal, and a channel, which is used by small vessels, with about 20
feet water between Snake Islet and the rock 200 yards southwestward
of it.

Also several shoal heads with depths of 6 to 10 feet lie about 400
yards southeastward of Snake Islet and on the 5-fathom line.

Buoy.—A red and white vertically striped conical buoy lies in
6 feet water, near the center of the first-mentioned shoal.

Coast.—From the west point of Snake Islet Bay the coast turns
northwestward ¼ mile to Kinghorne Point, thence north-northeastward 670 yards to the south point of Apollo Bay, which is ¼ mile
deep; from the southeast corner of Apollo Bay the coast trends north-
northwestward 1,500 yards and west-southwestward ½ mile to Roberts
Point.   The channel between Apollo and Peppermint bays is 1¼
miles wide, with from 5½ to 10 fathoms water close to the eastern
shore, and 10 to 11 fathoms in the fairway.

Barnes Bay.—From Roberts Point the coast trends northeastward over 1¼ miles to the southwest entrance point of Barnes Bay,
which is ½ mile wide at the entrance, whence it runs in nearly 2 miles
to the eastward.   Immediately inside the southwest entrance point of
this bay is a small cove ¼ mile in width by the same in depth, in which
small vessels may anchor in 7 fathoms, sand and mud.

At ½ mile southeast of the southwest entrance point of the bay is
Sykes Cove, which is ¼ mile wide at the entrance, whence it trends
southeast 1,500 yards.   From the east point of Sykes Cove the southeast shore of Barnes Bay extends northeastward nearly 1¼ miles, to a
point, between which and a projection of the northern shore the bay

is contracted to a channel 400 yards wide. This channel leads into Simmond Bay, which extends 1 mile north and south, forming two narrow bights, one trending to the north and the other to the south-ward, with depths of 3 to 4 fathoms in each.

Depths.—The depths in Barnes Bay, anywhere beyond 400 yards from the shore, are from 9 to 6 fathoms, mud bottom.

Shelter Cove, which lies between the northeast entrance point of Barnes Bay and another point northwestward from it, is ½ mile wide at the entrance, whence it extends ½ mile north-northeastward, and near the head of which and on the eastern side is the Quarantine Station.

Anchorage.—There is anchorage in 10 fathoms, mud, at the entrance of this cove, partially sheltered from northwest winds, with Woodcutters Point bearing 325° and the Quarantine Station 61°.

Mount Roberts is a double-topped hill, thickly wooded, the southeast and highest peak being 774 feet in height, and it is the summit of the peninsula formed by Barnes Bay on the north and Great Bay on the south.

Coast.—From Woodcutters Point, which lies northwestward ¼ mile from the northwest point of Shelter Cove, the coast, after turning ½ mile to the eastward, trends nearly north 1¼ miles to Bligh Point, whence it curves 1¾ miles in a northeasterly direction to Kelly Point, which forms the southeast side of the north entrance of D'Entrecasteaux Channel. This entrance, which is 1,340 yards wide, has 3 fathoms water close to Pierson Point, and from 7 to 10 fathoms in the fairway. In the bights on both sides of Bligh Point there is shoal water, and the 3-fathom curve extends about 300 yards to the northwestward of that point. There is a considerable quantity of kelp off Kelly Point, and the shoal water extends for 300 yards to the northward, and 200 yards to the westward of that point. There is a small jetty just inside Kelly Point.

D'Entrecasteaux Channel—Directions.—The navigation of D'Entrecasteaux Channel is not difficult either by day or night, if provided with the proper chart, the principal dangers being the Actæon Shoals, Zuidpool Rock, the bank which borders the west shore between Three Hut and Fleurtys Points, and the Channel Rock. In passing the valleys and mountains, strong gusts and contrary winds are met with, and a moment afterwards it falls quite calm, an inconvenience common to land of this description. At the various anchorages much trouble is found in weighing the anchor, in consequence of the tenacity of the muddy bottom which everywhere exists.

This channel, which affords safe shelter for shipping, is not recommended as a passage for sailing vessels bound to Hobart, except in the summer season, when dependence may be placed on the sea breeze.

Vessels from the westward have frequently taken this passage, as affording immediate anchorage, secure from all winds; but they were often several days before they reached Hobart. The detention is caused by the direction given to the wind, by the high hills and deep openings that form the west side of the channel, such as South Port, Port Esperance, and Huon River,. each of which gives a respective or distinct course to the wind, though, at sea, it may be blowing strong from the southwest. The passage to Hobart by Storm Bay is preferable for sailing vessels.

In proceeding through D'Entrecasteaux Channel for Hobart, from the westward, and not having a pilot, on no account pass between Actæon Shoals and the west shore; but having arrived abreast of Whale Head, bring it to bear 246°; and not to the southward of that bearing until Burnett Point bears 353° from that position steer 4°, 11 miles, when the north end of Partridge Isle should bear 75°, distant 1¼ miles nearly; this course leads clear of all dangers.

In baffling or contrary winds, keep on the east shore, which may be approached boldly. It is necessary to approach the west shore with great caution, until abreast Blanche Rock.

When working in the channel be careful to keep the lead going and not approach Actæon Shoals to less than a depth of 20 fathoms. Having passed Blanche Rock, the shore on either side may be approached to ½ mile.

From 1¼ miles, 255°, of the north end of Partridge Island steer 55° for about 5 miles till Partridge Island is just westward of Hopwood Point bearing 215°, then steer 35° until about 1,600 yards eastward of Zuidpool Rock. Then steer 44° to abreast of Three Hut Point.

After passing Three Hut Point, favor Simpson Point, to avoid Long Bay bank, which borders the shore between Three Hut Point and Fleurtys Point, and take care not to close the east shore within ¼ mile, to keep clear of the 4-fathom bank on that side.

The west summit of Mount Roberts, 660 feet in height, just open east of Green Island, 16°, leads in mid-channel between these dangers.

Pass Green Island at a distance of about ⅓ mile, and proceed to the northward through the fairway, slightly favoring Woodcutters Point to avoid Channel Rock.

Bruny Island, the west coast of which has already been described as forming the east shore of D'Entrecasteaux Channel, is 27 miles long from Tasman Head to Kelly point in a northerly direction, and 9 miles across at its southern and broadest part. A ridge of mountains 1,700 feet high near its south end extends to the north- ward along the east side of the island from Tasman Head to Simpson Point; it rises to a height of 2,010 feet near its center, and thence gradually falls. These mountains are for the most part densely wooded.

Between Little Taylor Bay and Cloudy Lagoon the land is quite low, and to the westward of this neck a coast ridge runs north and south; the summit, 530 feet high, is situated 8 miles south-southeast-ward from Ventenat Point, and is easily distinguished by its position, height, and small saddle-shaped top.

The hills to the southeast and southwest of Cloudy Bay are grass-covered downs, falling steeply in cliffs to the sea, and sloping gradually inshore.

The south portion of North Bruny Island, between Adventure and Trumpeter Bays, is composed of flat-topped wooded hills, 500 to 660 feet high on the east side.

Between Trumpeter and Adams Bays the land falls considerably, rising again 600 to 700 feet in undulating wooded heights to the northward to Cape de la Sortie, at the north end of North Bruny Island.

The east coast of Bruny Island from Tasman Head trends north-east 1¼ miles to a projecting point which forms a bold headland; the coast thence trends northward 3¾ miles, and thence east-northeast ½ mile to a double point, ½ mile broad; close southward of which is Arched Islet, flat-topped, and 80 feet high. A narrow reef borders the coast between 1 and 1¾ miles southwest of Arched Islet, and kelp extends from it there ¼ mile. From the double point to Cape Connella, north-northeastward 2¾ miles, the coast is embayed to the extent of 1,500 yards. Two small islets lie near the shore, at 1,500 yards, and 1¼ miles to the southwestward of Cape Connella.

Fluted Cape (lat. 43° 22′, long. 147° 23′).—From Cape Connella the coast, precipitous and bold, trends north 1¼ miles to Fluted Cape, and thence northwestward 1¼ miles to the south point of Adventure Bay, close to the northeastward of which is Penguin Island. The cliffs of Fluted Cape are composed of basaltic columns, and are from 700 to 800 feet in height. The summit of Fluted Cape, which is well marked, is 1,000 feet high, and thickly wooded, as is all the neighboring country.

Penguin Island, 1 mile northward of Fluted Cape, is 200 feet high and wooded, with a cliffy coast. The island is steep-to on its north and east sides.

Soundings.—Between Tasman Head and Cape Connella the depth is 20 fathoms about 1,500 yards, and upwards of 40 fathoms, 4 to 5 miles from the land.

Adventure Bay.—From Penguin Island the coast, with some slight windings, first trends southwestward 1¼ miles to a long sandy beach forming the head of the bay, and thence curves for 1¾ miles in a northwest direction to Cooktown, having a rocky projection about midway, off which there is kelp and foul ground stretching a distance of 400 yards.

Depths.—The southern bight of Adventure Bay has depths of 13 fathoms, sand and shells, in the center, shoaling gradually to the southward, where the 10-fathom curve is 1,340 yards from the beach. Several fresh-water streams flow into this bight.

From Cooktown a steep and cliffy coast, bordered by kelp, trends 2 miles northward, whence a sandy beach extends north-northeastward 5½ miles, forming the southeast coast of the isthmus between Adventure and Isthmus Bays. The southern part of this isthmus is flat, with scattered trees on it; the northern part consists of sand hills, the southern of which is the highest, 140 feet, and is covered with scrub; there are two lagoons at the back of the north end of the long beach. West of Cape Frederick Henry is an exposed bight, with 10 fathoms 1,500 yards from its head. On the east side of this bight there is foul ground with some kelp growing on it. The soundings in Adventure Bay are regular, over a sandy bottom, with 10 fathoms about half a mile off shore.

Cooktown (lat. 43° 21', long. 147° 20'), on the west side of Adventure Bay, consists of a few houses near the beach; landing there is very bad, being usually through surf, and the bottom of the bay off the township is rocky and uneven with a growth of kelp, rendering the position a bad one for anchoring. There is a disused coal mine, with a tramway leading to it from the coast, about ½ mile northward of Cooktown. There is a pier about 1,400 yards to the southward from Cooktown.

Anchorage.—The best berth is in the south part of the bay, off a small sandy bight, 1,500 yards to the southwest of Penguin Island, in 10 fathoms, sand and mud. This anchorage has good holding ground, and is protected from all but northerly and northeasterly winds. Although northeasterly gales are not frequent, they occasionally blow with great strength and send a very heavy sea into Adventure Bay.

Wood and water are plentiful in Adventure Bay, but are difficult to obtain, owing to the surf on the beach, except in the sandy bight off which anchorage is recommended above.

Cape Frederick Henry (lat. 43° 15', long. 147° 26') is a precipitous grassy bluff, 350 feet high, with a bare rock, of conical shape and 250 feet high, close to it. From the cape a high, cliffy coast, which is steep-to, extends northward nearly 3 miles to the south point of Variety Bay. The coast ridge, 660 to 450 feet high, runs immediately above the top of these cliffs, in some places falling down sheer from the summit. A thickly wooded ridge, of nearly equal height with the coast ridge, extends parallel to that ridge at a distance of 1,500 yards to the westward, the stream in the valley between discharging into Adventure Bay; the above ridges unite in a lower saddle at Variety Bay, whence the coast hills closely follow the coast cliffs as far as Trumpeter Bay, where the land near the coast and across to Great Bay is low.

Variety Bay is an indentation 3 miles northward of Cape Frederick Henry; its shores are thickly bordered with kelp to a distance of 400 yards off, and there are some rocks near the shore in its northern part, but in the southern part of the bay there is room to obtain a certain amount of protection from southeasterly winds, at a distance of 200 yards outside the kelp in a depth of 10 fathoms, sand.

Trumpeter Bay.—From Variety Bay the coast trends north-northwest 1 mile to the southeast point of Trumpeter Bay. This bay does not afford much protection from southeasterly winds. There are a few houses near the shore and the best landing place is on the sandy beach nearly in the center of the bay, off which also is the best anchoring place, in 8 to 10 fathoms.

Yellow Bluff—One Tree Point.—From the north point of Trumpeter Bay an irregular, rocky, cliffy coast trends north 1 mile to Yellow Bluff, distinguishable by its cliffs of that color; thence it trends north-northwestward 1½ miles to One Tree Point, which projects ⅓ mile from the general line of coast, and is low and rocky. North of Trumpeter Bay the land rises to a well-defined summit, elevated 730 feet, and situated ½ mile to the westward of Yellow Bluff; thence there is a gradual slope to the northward, forming a thick-wooded coast range, which rises again to a height of 790 feet, 1,500 yards southwest of Bull Bay.

Kelly and Bull Bays.—Between One Tree Point and Cape de la Sortie, northwest 2 miles from it, are Kelly and Bull Bays, separated from each other by a broad rocky point; there are depths of from 7 to 2 fathoms within ¼ mile of the shores of these bays, but they are mostly bordered with rocks. From Cape de la Sortie the coast trends west-northwest nearly 1 mile to Kelly Point, and is bordered with rocks, outside which, at about ⅓ mile northeast of Kelly Point, there are 7 fathoms water.

Derwent River.—This river, which is 130 miles long, has conspicuous marks at its entrance: Mount Louis, a conical hill 694 feet high, with the signal station on the west side; and on the east side, Iron Pot Islet with the lighthouse, situated east-northeastward distant 3 miles from Kelly Point, and Betsy Isle 3 miles to the eastward of Iron Pot Islet.

Directions given later.

Signal station.—There is a telegraph signal station on Mount Louis communicating with Mount Nelson, which is 7¼ miles farther to the northward, and 2½ miles southward of Hobart. It transmits all necessary information from the entrance of Derwent River. Another line of telegraph extends easterly from Mount Nelson to Port Arthur and Fortescue Bay, by which the approach of vessels between Maria Island and Cape Pillar is made known.

Pilot station.—There is a pilot station on Pierson Point, and the pilots are provided with a motor boat, in which they board inward-bound vessels in Storm Bay, at a distance depending upon the weather. From the station there is telephonic communication with Hobart. A lookout is kept for vessels by day and night.

Depths.—The entrance of Derwent River between Cape de la Sortie and Iron Pot Islet, is 2¼ miles wide, with depths of 10 to 8 fathoms, sand and broken shells. Thence the river retains an average width of about 2½ miles for the distance of 11 miles to Hobart, the soundings in mid-channel increasing to 20 fathoms at 6 miles above the entrance, and from this depth decreasing to 12 fathoms close to the town. There are general depths of 10 to 12 fathoms within ¼ mile of, and at least 3 fathoms ¼ mile from either shore.

Light.—A fixed white light, 65 feet high, visible 13 miles, is shown from a white, square tower, 40 feet high, on Iron Pot Islet ¼ mile southward of Cape Direction.

The channel between Iron Pot Islet and Cape Direction is rocky, only leaving a narrow passage for small vessels. Coast northward and eastward, described later.

Local magnetic disturbance.—To the southeastward of Derwent Lighthouse an area of local magnetic attraction exists; the center lies with Iron Pot Lighthouse bearing 300°, distant 2 miles, and the area extends 2 miles in a north-northwest and opposite direction, and 1 mile east-northeast and west-southwest. The maximum deflection of the compass whilst passing over this area is 13° to the eastward and to the westward of the normal. The locality should be avoided.

A vessel found her compasses sluggish and unsteady when northward of the northwest Point of Betsy Island, which tends to show that the area of local magnetic disturbance extends over a larger area than that given above.

Western shore—Blackman Bay.—The west shore of Derwent River from Pierson Point trends north-northwest 2¼ miles to the south point of Blackman Bay, close off which are some dry and covered rocks. This bay extends ½ mile north and south, and is ¼ mile deep, with 6 fathoms close off its entrance, and 6 feet near the shore.

Kingston—Browns River.—From Blackman Bay . the shore trends north 1,340 yards to the south point of Kingston Bay, which thence extends north 1 mile, and is ¼ mile deep, with from 12 to 9 fathoms water in the entrance, and 9 to 6 feet near the shore. Browns River flows into this bay, at ¼ mile within its north point.

Kingston postal township is on Browns River, in a grazing and fruit district, and has a population of about 600 inhabitants. It is 6 miles southward of Hobart, connected with it by telegraph, and a mail coach runs daily. There is a jetty for landing goods and passengers, and steamers for Hobart ply frequently.

Alum Cliffs—Crayfish Point.—From the north point of Kingston Bay the rocky shore trends irregularly north-northeastward ¼ mile to the southwest point of Alum Cliffs Bay, which continues thence in the same direction about 1 mile to Crayfish Point. It forms a double bight, ¼ mile to ½ mile deep, with depths of from 7 to 4 fathoms across its entrance, and from 3 to 3¼ fathoms 200 yards from the shore, which is partly bordered by rocks. Alum Cliffs are precipitous and conspicuous, when seen from the distance of 1 mile.

One Tree Point (lat. 42° 55′, long. 147° 22′).—From Crayfish Point the rocky shore extends north about 1 mile to Cartwright Point, when a succession of rocky points and small beaches trends slightly more westerly about 1 mile to One Tree Point, a rocky projection between which and Sandy Bay Point, northwest ½ mile from it, is a smooth beach.

John Garrow Shoal, with a least depth of 5 fathoms, sand, lies 750 yards eastward from One Tree Point. The shoal, with depths of 5¼ to 6 fathoms, extends 120 yards to the southeastward from the above position.

Clearing marks.—Mount Direction over Montagu Point 337°, or Betsy Island summit open of Gellibrand Point 160°, clears the shoal water off One Tree Point.

Light.—An occulting white light, elevated 41 feet above high water, visible 8 miles, is exhibited from a wooden structure on One Tree Point. The light is obscured by land when bearing northward of 358°.

Sewer outfall and buoy.—A sewer outfall pipe extends eastward from One Tree Point, and its outer end is marked by a black tub buoy, moored in 4½ fathoms, about 240 yards from the point.

All traffic between the buoy and the shore, with the exception of that of rowing boats, which must keep clear of the pipe, is prohibited.

Coast.—The shore from Pierson Point to Sandy Bay Point, although rocky, is bold, there being general depths of 5 fathoms water ¼ mile from it, except off One Tree Point, where there are 5 fathoms ½ mile from the point.

Mount Nelson (lat. 42° 55′, long. 147° 21′)—**Signal station.**—
From Sharp Hill, on which is a conspicuous white tower, 1,500 yards
northward of Brown River, a range of forest hills extends in a
northerly direction 2½ miles to Mount Nelson, which is 1,116 feet
high, having a telegraph station on it communicating with Mount
Louis and Hobart, the International Code as well as local signals
being used. Gentle slopes and spurs descend from this range to
the coast from Alum Cliffs to Sandy Bay Point.

Sandy Bay Point is the northeastern extremity of low flat land
projecting about ¼ mile from the more elevated, well-wooded, and
partly cultivated land which descends from Mount Nelson.

Sandy Bay extends from Sandy Bay Point northwestward 1.6
miles to Battery Point; a smooth beach trends 800 yards westward
from Sandy Bay Point to a rocky head, whence the shore, consisting of rocky points and sandy beaches, extends west-northwestward
¼ mile to Dunkley Point, which projects 300 yards from the line
of coast, its outer part being closely fringed with dry and covered
rocks with from 1 to 2½ fathoms close to them. From the inner
part of Dunkley Point the shore trends northwest 600 yards to a
small stream, and thence turns northward ¼ mile to Wellington
Rivulet, 200 yards to the southward of which are some bathhouses.
From Wellington Rivulet the southeast frontage of Hobart trends
northeastward 600 yards to a point, and thence nearly north-northwest ¼ mile to Battery Point, the south point of the entrance of Sullivans Cove.

Depths.—From 100 to 800 yards northwest of Sandy Bay Point
there are depths of 5 to 10 fathoms, with uniform soundings in 12
fathoms thence to 200 yards off Battery Point. The shore of Sandy
Bay may be approached to 300 yards in 5 fathoms, and to 200 yards
in 3 fathoms, except at ½ mile southeast of Dunkley Point and at 800
yards nearly northward of Dunkley Point, where the 3-fathom edge
of the bank which borders the bay projects 300 yards and the 5-
fathom edge nearly ¼ mile from the shore. The 5-fathom curve extends from Dunkley Point 150 yards to the eastward, and 100 yards
to the northward. A detached bank, about 200 yards across, with 3½
to 5 fathoms water on it, lies northwestward 800 yards from Sandy
Bay Point; with this exception the depths of water gradually decreases toward the shore.

Between the mouth of Wellington Rivulet and the point, ¼ mile
southward of Battery Point, there are patent slips, wharves, and
jetties. The 3-fathom edge of a bank extends 100 to 300 yards from
the shore. At about 400 yards to the southward of Battery Point
the bank projects 120 yards from the shore to the depth of 5 fathoms.

Ridges of well-wooded and partly cultivated land descend from
Mount Nelson to the coast between One Tree Point and Hobart, with

several small streams flowing into the bay. A road from the southward to Hobart passes by the villages and houses which are situated
near the shore of Sandy Bay. The suburb of Queenborough, westward of Sandy Bay Point, has a population of about 2,000.

˙**Port Hobart.—Sullivans Cove** extends from Battery Point 500
yards north-northwestward to Ocean Pier, its north point. From a
depth of 4 fathoms at 100 yards northeast of Battery Point, the
depths increase to 10 and 9 fathoms in the middle of the cove and
thence decrease to 6 and 4 fathoms at 250 yards to the eastward of
the north point of the cove. The cove is about 400 yards in extent.
From 9 to 10 fathoms in the middle of the entrance, the depths decrease to 5 and 6 fathoms within 50 yards of the shore and wharves,
over a bottom of mud.

Princes Wharf, 431 yards long, forms the south side of Sullivan's
Cove; there is a depth of 7 fathoms at its extremity, and 30 feet
alongside its entire length.

From the southwest corner of Sullivan's Cove to its north point
there is a continuation of wharves from which project eight piers.

Ferry Pier, the southwestern one, is 43 yards long, with a depth
of $4\frac{3}{4}$ fathoms at its extremity.

Brooke Street Pier, 35 yards to the northeastward of Ferry Pier,
is 75 yards long, with 6 fathoms of water at its extremity. This
pier is used by the local steam-vessels running to D'Entrecasteaux
Channel and New Norfolk.

Franklin Pier, 37 yards to the northeastward of Brooke Street
Pier, is 70 yards long, and projects into a depth of $6\frac{1}{4}$ fathoms.

Elizabeth Street Pier, 37 yards northeastward of Brooke Street
Pier, is 113 yards long, and projects into a depth of 8 fathoms. This
pier is used by the Tasmanian Steam Navigation Co.

Argyle Street Pier, 37 yards farther northeastward, is 111 yards
long with a depth of $7\frac{1}{2}$ fathoms at its end. This pier is used by the
New Zealand Shipping Company's steam vessels.

King's Pier, 85 yards northeastward of Argyle Street Pier, is
233 yards long, and 33 yards wide, with depths of 32 to 42 feet
alongside. There is a 25-ton steam crane at its inner end.

Queen's Pier, 70 yards northeastward of King's Pier, is 184 yards
long, with $6\frac{3}{4}$ fathoms at its outer end.

Kangaroo Pier, 48 yards farther northeastward, is 40 yards long,
extends into 18 feet water, and is used by the steam-ferry boat to
Kangaroo Point.

Ocean Pier.—Ocean Pier forms the northern side of Sullivan's
Cove, and is 1,210 feet in length on its southwest side, and 650 feet
in length on its northeast side. The width is 120 feet, and there is
a depth of 30 feet of water at the inner end to 62 feet at the outer
end on both sides. The railway is carried on to the pier.

Close to the northward of the north point of the cove is the mouth of Hobart Rivulet, between which and Macquarie Point, a little more than 400 yards northward from it, there is a depth of 11 feet at 80 yards from the retaining wall, with irregular depths of 2½ to 5 fathoms, extending 300 yards to the eastward of the mouth of the rivulet.

A part of this shoal to the southward of Macquarie Point is being reclaimed. The retaining wall extends 400 yards southward of the point.

Starting at the inshore end of the northeast face of the Ocean Pier, the shoal ground being reclaimed runs to the northeastward for 100 yards, and then to the northward.

Macquarie Point, over which is Queens Battery, may be approached to 100 yards distant in depths of 5 to 6 fathoms; the ebb tidal current sets strongly around and toward the point.

Sewer outfall buoy.—The sewer outfall pipe extends 143 yards from Macquarie Point, and a black tub buoy has been placed at the end of the pipe. Vessels must pass to eastward of the buoy, and must not anchor in the vicinity of it.

The Domain.—From Macquarie Point the river frontage of the Domain curves northwest and northward 1,800 yards to Pavilion Point, and has landing places 200 yards, and a patent slip 600 yards from the former point. There are depths of 3 fathoms close to the landing places, but a flat, with depths of 3½ to 4½ fathoms extends above 200 yards from them.

The Powder Jetty is on the shore 850 yards northward from Macquarie Point.

Buoys.—Three conical buoys, painted in red and white vertical stripes, are moored to the eastward and northward of the Powder Jetty. Inshore of them is a dolphin.

Bank.—For about ⅓ mile southward of Pavilion Point a bank having 4 to 5 fathoms water on it extends ⅓ mile from the shore. Between this bank and that which projects from the landing places to the northward of Macquarie Point there are regular depths of 6 to 9 fathoms. On the northern part of the Domain are Government house and the botanical gardens.

Eastern shore.—**South Arm** is a peninsula from 1 mile to ¼ mile broad, extending from Cape Direction 5½ miles in a north-northwest direction to Gellibrand Point, which forms the south side of the entrance of Ralph Bay. This peninsula is mostly covered with open forest; the land for about 1½ miles northward of Cape Deliverance, and between 1½ and 2½ miles southward of Gellibrand Point, being elevated; there are two hills near Cape Deliverance, and one 400 feet high, 2 miles south-southeastward from Gellibrand

Point; the remaining portion of the arm is undulating with low narrow flats between the higher land.

The east shore of Derwent River is partly formed by South Arm, which extends from Cape Direction 1 mile northwest to the south point of Half Moon Bay, and forms three bights, the northwestern and largest of which has 3 fathoms water; but the south point of Half Moon Bay and Cape Deliverance have only 2 fathoms water at 400 yards from them.

Half Moon Bay extends 1¾ miles north-northwest and opposite direction, and is 1,500 yards deep, with from 5 to 6 fathoms water in the middle, and from 1 to 4 fathoms close along shore.

From the north point of Half Moon Bay the shore trends northward nearly 1 mile to the south point of Opossum Bay, close to the southward of which is a small cove with depths of 2 fathoms.

Shoal.—At ¼ mile northwest of the north point of Half Moon Bay is a 4-fathom bank ¼ mile off shore, with 5 fathoms inside it.

Opossum Bay is 1,500 yards wide and ½ mile deep, with depths from 6 to 2 fathoms. This bay is separated from a bight to the northward of it by a broad hilly point partly fringed with dry and covered rocks, whence the rocky shore of the bight extends north-northeastward 1,500 yards to Gellibrand Point.

Ralph Bay, which is separated from the east side of Derwent River by North and South Arms, extends 7½ miles in nearly a parallel direction with the river, and 1¼ to 3½ miles east and west. The entrance of the bay between Gellibrand and Trywork Point is 1¼ miles wide with 7 to 14 fathoms water, but within the entrance the water is mostly shallow. Mortimer Bay is an indentation of the east shore of Ralph Bay, between 2 and 3 miles east-southeast of Gellibrand Point. From Maria Point, the northwestern extremity of Mortimer Bay, the east shore of Ralph Bay trends northward 3½ miles to the foot of Mount Mather, which is 575 feet high. Between Mount Mather and the north shore of Ralph Bay, a creek ½ mile wide at its entrance, trends about 1½ miles to the eastward and southward; this creek and Mortimer Bay are both very shallow.

Trywork Point is 1½ miles north of Gellibrand Point. Trywork Point and the rocky shore, extending ¼ mile to the eastward from it, form the north side of the entrance of Ralph Bay and the south end of the north arm. This arm, which separates the northern part of Ralph Bay from Derwent River, is 2 miles long, north and south, 1 mile to ½ mile broad, and consists of a series of undulating grassy hills, with patches of cultivation.

From Trywork Point the east shore of Derwent River curves northward 1¼ miles and thence north-northwestward 1¼ miles to a projection of the shore between which and Kangaroo Bluff, 2¾ miles northwest of it, are two small bays, of about equal extent, separated from each other by a broad rocky point, at the foot of a hill close to the northward of it. The shore from Trywork Point to Kangaroo Bluff, although rocky, has depths from 2 to 6 fathoms about 200 yards from it, and may be approached to ¼ mile in 10 fathoms.

Kangaroo Bluff is the cliffy south point of an elevated peninsula extending north-northwest ½ mile to Bellerive. From the bluff the shore sweeps around in a northerly direction, and, although rocky, it may be approached to the distance of 100 yards in 3 fathoms water.

Kangaroo Bay lies between Kangaroo Bluff and Montagu Point, 1,500 yards northwestward from it; there are depths of 7 to 10 fathoms in the entrance, whence the bay trends 1,500 yards in a northeast direction, gradually decreasing to 200 yards in width between Bellerive and a low point projecting 200 yards from the opposite shore.

From Montagu Point the northwest shore of Kangaroo Bay for about ¼ mile may be approached to 100 yards in 3 fathoms, thence the edge of the northern bank trends eastward to 100 yards north of Bellerive. Above Bellerive the bay expands to 400 yards in width, but it is filled by a flat, on which the greatest depth of water is only 2½ fathoms. From 7 fathoms in the entrance the depth of water decreases to 4 fathoms about 100 yards northwest of the pier.

Bellerive is a rapidly extending suburb of Hobart, on the opposite side of the Derwent, with a population of about 1,000. Steam ferryboats ply across to Hobart. It is the terminus of the Bellerive and Sorell Railway.

Railway wharf.—The railway wharf, 220 yards long, runs out in a southwesterly direction from the shore at the head of Kangaroo Bay. There is a flagstaff on its outer end.

Montagu Point, which has a depth of 8 fathoms within 150 yards of it, is the southwestern extremity of a hilly wooded promontory projecting 1,500 yards from the northeastward, its most elevated part being a hill 316 feet high, situated north-northeast ¼ mile from the point.

Montagu Bay.—From Montagu Point the shore trends north-northwestward ¼ mile to the south point of Montagu Bay, which thence continues in the same direction ⅛ mile. Between the south point of the bay and a rocky spit, 400 yards to the northeastward of it, is a shoal bight with only 2 fathoms water 200 yards from the shore; but the north shore may be approached to about 50 yards in 3 fathoms water. The bay runs in nearly ⅛ mile in a northeast direction, terminating in a small shallow cove, on the northwest shore of which are some smelting works. From 7 fathoms in the entrance the depths decrease to $3\frac{1}{2}$ fathoms at 100 yards south of the smelting works.

Depths.—About $\frac{1}{4}$ mile to the northward of Montagu Point there are depths of 11 to 12 fathoms 150 yards from the shore; but there are only $4\frac{1}{2}$ fathoms at that distance off the south point of Montagu Bay. Between $\frac{1}{4}$ mile northwest of Montague Point and 300 yards southwest of the north point of Montagu Bay there is a gully about 800 yards long and 200 yards wide, having depths of 22 to 26 fathoms.

Directions.—There are two approaches to Derwent River, through D'Entrecasteaux Channel, and through Storm Bay between the north part of Bruny Island and Tasman Peninsula, at about 12 miles to the eastward of it, but the latter approach is much to be preferred by sailing vessels.

From the westward bound into Derwent River through Storm Bay give Tasman Head, the south point of Bruny Island, a good berth, to avoid Friar Rocks. To the northward of Fluted Cape the most remarkable object is Mount Wellington, a table mountain, 4,135 feet high, about 4 miles west-southwest from Hobart, and when Betsy Isle, which is high and wooded, is seen, steer to pass on the west side of Derwent Lighthouse on Iron Pot Islet. In approaching Derwent River the generally strong prevailing westerly winds make it desirable to keep within a mile of Bruny Island. Remember the local magnetic disturbance in that region.

Enter Derwent River between Cape de la Sortie and Derwent Lighthouse, and favor the west shore, passing $\frac{1}{2}$ mile off One Tree Point. From $\frac{1}{2}$ mile off One Tree Point steer to the northwest for the usual anchorage off Hobart, but one anchor should be laid out well to the southeast for convenience in getting under way. There is no danger all the way up, so that the vessels may work in or out without a pilot, tacking at about $\frac{1}{4}$ mile offshore, and may anchor anywhere, on muddy bottom. Avoid the prohibited anchorage shown on H. O. chart No. 3583.

Anchorage will be found in any part of Derwent River, but the safest on all occasions is on the west side, the east being unsafe, especially for small vessels, several having been lost by anchoring near it.

Hobart approach—Examination anchorage.—The examination anchorage is an area 1,500 yards square, situated between Crayfish Point and Cartwright Point, on the western side of the river, in 11 to 15 fathoms water.

Quarantine.—Vessels liable to medical inspection must not proceed higher up the River Derwent, until pratique is granted, than an imaginary line drawn 112° from the seaward end of Prince's Wharf, and intersecting a line drawn 224° from Kangaroo Bluff.

Explosives anchorage.—No vessel arriving in the River Derwent, having explosives on board (other than ship's stores), shall, except with the written permission of the chief inspector and harbormaster, anchor above an imaginary line from the signal staff at Battery Point to Montagu Point, nor within ¼ mile from the shore, but they may, nevertheless, anchor opposite the Powder Jetty in the Queen's Domain at not less than ¼ mile from the shore.

Signals to be made by vessels approaching ports when inconvenienced by searchlights have been described.

Prohibited anchorage.—No vessel shall be allowed to anchor in the port above Sandy Bay Point within—

(1) A fairway between the following two lines: One being a 335° bearing of the southwestern end of Government house, and the other a 338° bearing of the point by the semaphore, and which lines may be stated as—

    (*a*) The southwestern extremity of Government House in line with eastern extremity of Cattle Jetty at Macquarie Point bearing 335°.

    (*b*) The western extremity of the Hobart Gas Works in line with Battery Point bearing 338°.

(2) Sullivan's cove.

The port authorities are very particular regarding the observance of this rule, on account of the hindrance of vessels in the prohibited area to long steamers going alongside wharves.

This prohibited area is shown by dotted lines on the chart.

No vessel may anchor, or weigh anchor (except in case of necessity), above a line drawn from Sandy Bay Point to Kangaroo Bluff, except by permission or direction of the harbor master. Penalty for disregarding this by-law £10.

Pilots.—The only pilot station for Derwent River is at Pierson Point.

Pilotage.—Pilotage payable by any vessel liable to pay pilotage, whether upon entering or leaving the port of Hobart, shall be paid by every such vessel which shall enter or depart from the waters comprised within an imaginary line drawn from Cape Raoul to Southport Isle.

All navigable waters within the said imaginary line and all bays and estuaries to the northward therefrom shall for pilotage purposes be deemed to be the port of Hobart.

Vessels entering must, if not exempted, pay the pilotage fees whether a pilot is taken or not, but there is no penalty for not taking a pilot on board.

Vessels leaving or shifting, if not exempted, must take a pilot on board or they become liable to a fine of £50 in addition to the pilotage fees.

Vessels claiming exemption must fly at the foremast head a white flag 2 yards square at the least.

Pilotage rates.—The pilotage rates payable by vessels entering, leaving, or proceeding from one place to another, within the port, is as follows:

Inward.—Sailing vessels 6 pence per ton, steamers 4 pence per ton; the maximum rate payable in either case is £15, and the minimum £4.

Outward.—One-half of the above rates, with a maximum payment of £5.

Within the port generally.—Sailing vessels 3 pence per ton, steamships 2 pence a ton; maximum rate in either case £5.

Within the port, above a line drawn from **Trywork Point** to **Sandy Bay Point**, or within any outport:

Vessels of 200 tons and under-- £0 10s.

Over 200 tons, and up to 1,000 tons----------------------------------- £1 0s.

Over 1,000 tons-- £2 0s.

Vessels under 50 tons are liable to the above rates only if a pilot is actually employed.

Vessels in ballast, or which do not break bulk, pay one-half only of the port pilotage dues.

Port regulations.—A copy of the " Rules and Regulations of the Port of Hobart " will be delivered to the master of every vessel arriving in the port, and a receipt in writing is to be given for it.

No master of a steamship shall navigate his vessel in Little Sandy Bay to the shoreward of lines drawn from Sandy Bay Point to the tide gauge beacon, and thence to a point 15 yards farther out than the beacon carrying a springboard, and thence to a point at the southeastern corner of the bay, where a diamond-headed mark is fixed, the bearings of such lines being 166°, 177°, and 149°, respectively.

The bridge at Victoria Dock will be opened for fishing and pleasure vessels passing out at the following times and during the following hours: At half past nine and half past eleven in the morning, and half past three and half past six in the afternoon, and between midnight and half past seven in the morning.

Caution.—Vessels are forbidden to sound steam whistles or horns within the port when above a line drawn from Secheron Point and Kangaroo Bluff, except in case of necessity from fog or other cause, or to prevent collision.

Lights.—A light is exhibited from the outer end of Princes Wharf, on the south side of Sullivan's Cove; it is 18 feet above high water.

A light is shown from the end of Brooke Street Pier.

Two lights, placed vertically, are shown from Franklin Street Pier.

Two lights, placed horizontally, are shown from the end of Elizabeth Street Pier.

A light is shown from the end of Argyle Street Pier.

Two lights, placed horizontally, are shown from the end of King's Pier.

Three lights, triangular, are shown from Queen's Pier.

Two lights placed vertically are shown from the center of the end of Ocean Pier.

Tides.—It is high water, full and change, at Hobart at 8 h. 15 m.; spring rise 4½ feet, and neaps 3½ feet. The tides here are irregular, and frequently almost stationary for days.

Tidal currents.—The flood current is barely perceptible between Iron Pot Islet and Kelly Point, but it runs stronger under Mount Louis, and thence parallel to the shore, following the course of the river at the rate of half a knot. Between Macquarie and Montagu Points the ebb current runs south 1½ knots at half tide; off Battery Point it runs about 170°, sweeping around Sandy Bay to the southeastward, at the rate of ¾ knot, and after passing Sandy Point, its strength gradually decreases to ½ knot at the entrance of the river.

Hobart, the capital of Tasmania, has numerous handsome public buildings which cover a large area, the streets are wide, well laid out, and intersect each other at right angles; it is picturesquely situated on a gently sloping plain at the foot of the hills that descend from Mount Wellington, which lies 5 miles westward of Mount Nelson, and is 4,135 feet high, with Collins Bonnet, another mountain, 4,131 feet high, 4 miles to the westward of Mount Wellington.

The population on December 31, 1917, was estimated to be 40,350.

The city is lighted with gas and electricity, and plentifully supplied with water.

Hobart, on account of its invigorating climate, is largely visited during the summer season by visitors from Sydney, Melbourne, and Adelaide and other places in Australia.

Communication.—There is regular steam communication to Melbourne, Sydney, and New Zealand; the Shaw, Savill & New Zealand Shipping Co.'s steamers from London to New Zealand make Hobart a regular port of call; local steamers also trade regularly to east and west coast and river ports, and during the fruit-shipping season large ocean steamers call here, taking shipments for London. Hobart is connected with the railway and telegraph systems of the State.

Radio.—A radio station is established in the Domain. It is open to the public from 9 a. m. to 11 p. m. Call letters V I H. Weather signals are received and transmitted at 10 p. m.

Trade.—The principal imports are manufactured goods, tools, tea, sugar, stores, etc.; and the exports—wool, grain, hops, sperm oil, timber, vegetables, and fruits.

Shipping.—Port Hobart is the principal port of Tasmania and has considerable shipping.

Fort Mulgrave—Signal station.—Fort Mulgrave, on which is a signal station, is situated about 200 yards to the southwest of Battery Point, at an elevation of 85 feet above the sea.

Time signal.—A time ball at the flagstaff on the site of Fort Mulgrave, 85 feet above high water, is hoisted to half-mast about 10 minutes before signal, as preparatory, to about 6 feet from the masthead, at 5 minutes before signal, and is dropped by electricity, at noon and 1 h. 00 m. 00s. p. m. standard time of Tasmania, equivalent to 14 h. 00 m. 00 s. and 15 h. 00 m. 00 s. Greenwich mean time between the last Sunday in September and the last Sunday in March and at 15 h. 00 m. 00 s. Greenwich mean time only for the remainder of the year.

The time signal at 15 h. 00 m. 00 s. G. M. T. is made by electric communication with Melbourne observation.

Patent slips.—See Appendix I.

Repairs to small engines and boilers can be made by the Derwent Engineering and Shipwright Works, where cylinders of 25 inches diameter can be cast and 36 inches bored; boilers can be made to 10 tons in weight, and repaired to any size, shafts of 5 inches in diameter and 18 feet in length can be forged and turned, and a piston of 56 inches diameter has been made. There are sheers at the works capable of lifting 9 tons. There are several other firms at Hobart capable of executing minor repairs to engines and boilers of small craft.

Coal.—About 29,000 tons of coal are imported annually, and from 5,000 to 6,000 tons are usually kept in stock; coaling is done at the piers.

Supplies.—Stores of all kinds, provisions, fruit, water, and firewood, are easily procured. The country in the immediate neighborhood is rich in natural productions, such as coal, iron, black lead, alum, mica, precious stones, and gums.

Water is laid onto all the wharves, and is supplied by the city council at a reasonable price. There is also a small tank boat carrying about 12 tons, available for shipping.

Cornelian Bay.—Cornelian Bay Point is the southeastern point of a peninsula, occupied by the public cemetery, separating Cornelian Bay on the south from Newtown Bay on the north side of it. It is situated on the western shore of the River Derwent, northwestward distant 1,800 yards from Pavilion Point.

The eastern face of the peninsula is cliffy, and extends in a northerly direction for 750 yards to Cemetery Point, whence the coast trends to the southwestward, forming Newtown Bay.

Cornelian Bay, situated on the south side of the above-mentioned peninsula, is ⅓ mile in length and about the same in breadth. It is not available for anchorage for vessels of large size. The 3-fathom curve projects to a distance of 350 yards from Cornelian Bay Point, and on the opposite side of the bay it extends nearly 500 yards from the shore, skirting the head of the bay at the same distance.

Anchorage for vessels of moderate size in 5½ fathoms, mud, may be obtained 350 yards 179°, from Cornelian Bay Point, with the end of the cliff near Cemetery Point, just open to the eastward of the former point.

This berth must be approached from the eastward in order to avoid both the 3-fathom spit projecting from the southern shore of the bay and the shoal ground extending from Cornelian Bay Point.

Newtown Bay.—Between Cemetery Point, the southern entrance point, and Woodman Point, situated 900 yards north-northwestward from the former, the coast forms Newtown Bay 1,500 yards deep, narrowing toward its head. The 3-fathom curve extends nearly 1,200 yards from the head of the inlet.

Rocky and foul ground, with from 1½ to 2 fathoms of water on it, extends to the northeastward from Cemetery Point to a distance of over 500 yards, and shoal ground also extends to a distance of upwards of 200 yards from Rock Cod and Woodman Points on the north side of the bay, thus narrowing the channel into the bay to barely 200 yards in width.

Buoys.—A black can buoy, moored in 4 fathoms water, marks the edge of the bank extending northeastward from Cemetery Point. A black cask buoy marks the shoal ground off Rock Cod Point.

Anchorage for a small vessel may be obtained in 3½ fathoms with Cemetery Point bearing 117° and Woodman Point bearing 21°.

Coast.—Froom Woodman Point a rocky and cliffy coast curves round to the northwestward for 1 mile to Prince of Wales Bay.

Eastward of Woodman Point, and thence to Stanhope Point 400 yards to the northward, the 3 and 5 fathom curves of soundings approach one another closely and skirt the coast at a distance of 250 yards, thus projecting more than half way across the river, which here narrows to a width of 450 yards.

Buoy.—A black can buoy, moored in a depth of 4 fathoms, marks the edge of the shoal water extending eastward from Stanhope Point.

Risdon Ferry (lat. 42° 50′, long. 147° 19′).—A ferry crosses the river from a point, on the western shore about half-way between Prince of Wales Bay and Stanhope Point, landing on the opposite or Risdon side, close to Restdown Point.

Prince of Wales Bay.—The entrance to this bay, ¼ mile westward of Risdon ferry, is about 300 yards wide. The bay turns westsouthwestward about 600 yards and then continues 600 yards to the southward with a width of between 200 and 400 yards.

The depths are between 2 and 3 fathoms, mud bottom, generally over the central portion of the bay, shoaling gradually in the southern end and increasing to 4 fathoms in the entrance about 100 yards from, and parallel with, the northern shore.

Lindisferne Bay.—From Smelting Works Point, the northern entrance of Montagu Bay, on the east side of the river Derwent, the coast takes a north-northeastward direction for upward of 1 mile to the head of Lindisferne Bay, an inlet ¼ mile wide at its entrance abreast Lindisferne Point, which forms the northern entrance point.

There are depths of 4 to 5 fathoms at the entrance of this inlet, and a vessel of moderate size can anchor in 5½ fathoms, mud, 250 yards 173°, from Lindisferne Point, but anchorage on the eastern side of the river Derwent is not recommended, the western side being the safer. The water shoals gradually up to the head of the bay, which the 3-fathom curve approaches to within 400 yards.

The suburb of Beltana is situated on the northern shore of Lindisferne Bay, frequent communication with Hobart being maintained by ferry steamers.

From Lindisferne Point the coast, with two slight indentations, trends west-northwestward for 1,500 yards to the entrance of Geilston Bay.

Off this portion of the coast the 3 and 5 fathom curves project to distances of nearly 200 and 250 yards, respectively.

Geilston Bay is an inlet 400 yards wide at the entrance, extending in a northeasterly direction for ¼ mile, narrowing toward its head. There are depths of 3½ to 5 fathoms across the entrance.

From Geilston Bay a high cliffy coast called Bedlam Wall trends northwestward for about 1,340 yards, and thence curves around to the northwestward for ¼ mile to Restdown Point. Bedlam Wall is steep-to within 30 or 40 yards for the greater part of its length, except just off the northern entrance point of Geilston Bay, where the 5-fathom curve projects to a distance of 70 yards from the coast. In passing up and down the river Bedlam Wall should be passed within 200 yards, in order to avoid the foul ground projecting from the opposite side of the river.

Shag Bay is a small inlet in Bedlam Wall, 450 yards deep, ¼ mile northwestward of Geilston Bay. It is barely more than 100 yards wide at its entrance, across which are depths of 3½ to 4 fathoms, gradually shoaling toward its head.

175078°—20——23

Coast.—Broken cliffy coast extends northwestward for 1,600 yards from Shag Bay to Restdown Point.

From Restdown Point the coast trends northward ½ mile to Risdon Cove, from which Derwent River extends northwestward 2½ miles to a narrow point projecting 1 mile from the west shore, the intermediate portion of the river being nearly 1½ miles wide, and forming an extensive bay on the southwest side. There are from 5½ to 3½ fathoms water in mid-channel.

Bridgewater (lat. 42° 44′, long. 147° 13′).—Between the east shore and the point which projects from the opposite side the river is contracted to ¼ mile in width, and turning thence about 1 mile to the westward it trends northwestward 1¾ miles to Jordan River, which flows into Derwent River from the northward. Between the mouth of Jordan River and Bridgewater, 1¾ miles to the northwest of it, and for about 2 miles to the westward of the bridge, Derwent River is nearly ½ mile wide, above which it is much smaller, with branches flowing into it, mostly from the northward and northwestward.

Bridgewater is 12 miles from Hobart on the north side of the Derwent, which is here crossed by a causeway and bridge. The main line railway also crosses here, a drawbridge being maintained for the convenience of navigation. There is a post and telegraph station.

Mount Dromedary.—The land on either side of the Derwent River consists of hills and fertile valleys, with numerous small streams flowing into the river. The principal summits of these ridges above Hobart are Gunners Quoin and Mounts Direction, Faulkner, and Dromedary, which lie northward and northwestward from 7 to 14 miles from Mount Wellington. Mount Direction is 1,460 feet, Gunners Quoin 1,390 feet, and Mount Dromedary 3,245 feet high. Mount Direction has a nearly flat top with gently sloping shoulders, and Gunners Quoin has a shoulder and sharp fall on the west side.

Winds.—During summer, or from December to March, the winds are generally land and sea breezes, which blow from north-northwest and south-southeast, but with no degree of certainty, for frequently a sudden change takes place in the middle of a fine sea breeze, a violent gale coming on from the westward, which usually lasts 3 or 4 days.

In January and February the weather is generally fine with light northwest to southwest winds, a sea breeze blowing up the harbor in the afternoons; the barometer stand at about 30 inches, and the mean temperature is between 72° and 53° F. Now and then the sea breeze is interrupted by fresh north to west winds, with squalls off the high land, when the barometer is usually below 30 inches. In March, the sea breeze is not so regular, and the mean temperature is between 68° and 51° F.

Winter may be said to commence in April, which month and May are regarded as the finest in the year, the weather being cold, bright, with light winds and occasional spells of heavy gales, generally from the westward, which last 3 to 4 days at a time. During the winter gales are frequent. In June the mean temperature is between 53° and 41° F. There is much rain and sometimes fog. A northwest to west wind at Hobart is often a southwest wind outside.

Hope Beach forms a slight curve extending from Cape Direction east-northeastward 3 miles to Goat Bluff, a small cliffy point, on which there is a hillock 85 feet high. This beach, which is low and narrow, is the only barrier between Ralph Bay and the sea.

There are regular soundings in 11 to 14 fathoms water between the entrance of Derwent River and Betsy Island, with 6 and 7 fathoms close to the sides of the island, and within ½ mile of Hope Beach.

The coast, a sandy beach, from Goat Bluff curves northeastward 1¾ miles to Cape Contrariety.

Betsy Island or Willaumez Island is 1¼ miles long north and south, ½ mile wide, and 470 feet high. A dark red sandy soil shows itself in numerous landslips on its eastern side. The island is bold and precipitous, and landing can only be effected on the north side on a cobblestone beach near a ruined hut.

Little Betsy is a small rocky islet, 60 feet high, about 400 yards southward of the island; 500 yards southeastward of Little Betsy are two small rocks, 3 feet above high water, which nearly always break heavily, and about 200 yards farther to the southward is a rock under water, which only breaks in heavy weather.

Passage.—Between the north end of Betsy Island and Goat Bluff is a passage 1,500 yards wide, with a general depth in the center of 5½ fathoms, but halfway over the Goat Bluff from the northeast end of Betsy Island are the Black Jack Rocks, two flat rocks about 3 feet high, usually covered with sea birds. Thick kelp extends to the southward of these rocks, but there is a good passage on the north side, carrying not less than 4½ fathoms and about 200 yards wide. The range through is the northern or lower summit of Mount Forestier in range with the extremity of Cape Contrariety bearing 65°. This forms a convenient short cut for small vessels proceeding between Hobart and Frederick Henry Bay.

Beacon.—A white hexagonal beacon 15 feet in height stands on the northern head of the Black Jack Rocks.

Anchorage.—There is good anchorage for a small vessel in ordinary weather, out of the swell, off the beach on the north side of

Betsy Island, in from 5 to 5½ fathoms. A heavy surf runs on all the beaches facing to the southward on this part of the coast.

Magnetic disturbance in this locality has been described.

Storm Bay, the west shore of which is formed by the coast of Bruny Island, from Cape Frederick Henry to Cape de la Sortie and the north shore by the coast from Cape Direction to Cape Deslacs, is 16 miles wide from Cape Frederick Henry to Cape Raoul, and extends from its entrance for a distance of 15 miles northward to Cape Deslacs.

Soundings.—From 50 fathoms 1½ miles southwest of Cape Raoul the depth of water across Storm Bay gradually decreases toward Bruny Island, over a bottom of fine red sand, with black specks and small broken shells. The 20-fathom curve crosses Storm Bay at a distance of about 3 miles southeast of Betsy Isle and thence runs to within 1 mile of the coast at the southeast point of Trumpeter Bay, and within ¼ mile of Cape Frederick Henry. The bottom in the northwest part of Storm Bay is brown sand.

Toward the entrance of Derwent River the bottom becomes muddy, which is generally the case where there is any considerable run of fresh water.

Directions.—Entering Storm Bay from the eastward, after rounding Cape Pillar and Cape Raoul, stand over toward Cape Frederick Henry, and steer thence along the northeast coast of Bruny Island for the entrance of Derwent River. In working against a northwest wind, work up along the same coast, to avoid the strong outset from Frederick Henry Bay.

If, when off Betsy Island, the wind should blow from northwest, so as to prevent a vessel from working into Derwent River, good anchorage may be obtained either in Adventure Bay or Frederick Henry Bay. In calms or light winds vessels may, if necessary, anchor with a stream or kedge in Storm Bay until they get a breeze.

Vessels bound to sea from Derwent River, and meeting a southeast gale in Storm Bay, may find safe anchorage in Northwest Bay, just within the north entrance of D'Entrecasteaux channel.

Winds.—During a great part of the summer season, from November to April, when the weather is fine and settled, sea and land breezes generally prevail, the land breeze coming off between 8 and 10 o'clock; both these breezes are preceded by an interval of calms or light airs for two or three hours. From January to March the northwest winds come in very hard squalls.

Eastern shore of Storm Bay.—Cape Raoul (lat. 43° 14′, long. 147° 48′), the south point of Tasman Peninsula, is formed of high basaltic columns, presenting a very remarkable appearance, and falling in perpendicular cliffs from a plateau about ½ mile long, 1,500 yards wide and 700 to 800 feet high, covered with dense scrub.

Four hundred yards northwest of the cape the basaltic columns are separated from the cliff by a deep crevasse; the first isolated column to the southward of the crevasse is 532 feet high, and between it and Cape Raoul is a narrow neck of land consisting of basaltic columns which become more separated toward the cape.

Hill Rock (lat. 43° 14', long. 147° 47'), 14 feet high, is situated westward, nearly 1 mile from Cape Raoul; there is a channel between it and the shore 250 yards wide, but it has been only partially examined. The sea breaks heavily on this rock, which is steep-to. There are depths of more than 20 fathoms within 200 yards of the rock, and 20 to 30 fathoms everywhere within ¼ mile of the cliffs of Cape Raoul.

Lempriere Bay, between Hill Rock and Southwest Point, is 2 miles wide and 1,200 yards deep, with soundings of from 20 to 13 fathoms sand; it shoals gradually from Hill Rock, toward Southwest Point, and the 10-fathom curve in the northern part of the bay is about 400 yards from the shore.

Bartlet Rock and Burton Islet are situated in the southern part of Lempriere Bay, about 300 yards from the shore; the former is limpet-shaped and 12 feet high, and the latter has two sharp rocky peaks, the northern and highest one being 131 feet high. To the northward of Burton Islet the basaltic formation of the cliffs ceases.

Southwest Point, about 3 miles northwestward from Cape Raoul, 290 feet high, is composed of yellow sandstone. After a continuance of northeasterly winds landing may be effected on a small pebble beach about 300 yards northeastward of the point. From Southwest Point the yellow cliffs continue past Tunnel Point to the beach at the head of Tunnel Bay.

Tunnel Point, situated ½ mile northwestward from Southwest Point, is 160 feet high.

Tunnel Bay is 900 yards wide at the entrance points and ½ mile deep; there is a depth of 14 fathoms at the entrance and 10 fathoms at ¼ mile from the shore, shoaling gradually toward the beach which is formed of large bowlders. Midway between the beach and the north entrance point, which is a round bold yellow cliff about 500 feet high, is a cliffy point, 100 yards southwestward of which is a rock which always breaks. At the northern end of the bay, just inside the beach, is a stream of good fresh water.

From the north entrance point of Tunnel Bay the coast trends north-northwestward 1,200 yards to the head of a small cove about ¼ mile in extent, with a depth of 14 fathoms at its entrance, shoaling gradually toward the shore.

Baker Point, the western point of this cove, is low and rocky, and extends about 200 yards to the southeastward from the line of coast.

From Baker Point the coast trends northwestward 1,200 yards to a rock 35 feet high, at the entrance to Three Beach Bay.

Winsor Islet, a square-topped rock, 148 feet high, is conspicuous from northwest and southeast.

Dart Bank (lat. 43° 13′, long. 147° 42′), within the 20-fathom curve, is about 1,200 yards in diameter, and has a least depth of 13 fathoms, coral, near the center; depths of 25 to 30 fathoms, sand and shell, are found close around. With southerly winds there is always a heavy swell. The shoalest part of the bank is situated 4 miles, 291° from the summit of Hill Rock (a mile westward of Cape Raoul) in line with Cape Raoul.

Three Beach Bay is 1,300 yards wide and ½ mile deep, with depths at the entrance of 16 to 18 fathoms and 10 fathoms about 200 yards from the shore at the sides of the bay, and ¼ mile from the pebble beach at the head of the bay. The southern side of the bay is composed of low cliff about 130 feet high; on the southeast side of the bay are two pebble beaches, one close to the eastward of the south entrance point, and the other midway between it and the head of the bay. Kelp extends out to the 5-fathom curve of the beaches in Three Beach Bay.

About 150 yards from the north point of Three Beach Bay there is a rock close off the coast 10 feet high, and from this rock the low cliffy coast trends northwestward 1,500 yards, and then curves to the eastward, forming three small bays, the heads of which are filled with kelp which extends nearly to the 7-fathom curve; from these bays the coast continues low and rocky for 1¾ miles to Low Point, which is very low and shelving, and surrounded by kelp as far as the 5-fathom curve.

Wedge Bay.—The entrance of Wedge Bay extends from Low Point 2¼ miles northward to Lory Point, with depths of 12 to 15 fathoms midway; from its entrance the bay trends 2 miles eastward, where it terminates in a bight about 1 mile wide, with the Sister Islets lying between 200 to 300 yards from its north shore; close behind the low sandy east shore of this bight are two lagoons. With the exception of the inner part of Burnett Harbor or Parsons Bay, which is landlocked, Wedge Bay affords no protection with westerly winds except for small craft. These can find secure anchorage in 3 or 4 fathoms water in the southwest corner of the bay; and in the inner part of Burnett Harbor where the holding ground is good, being of stiff mud.

Mount Clark, eastward of Wedge Bay, 1,796 feet high, and the highest hill on the peninsula, has a broad summit with a long sloping shoulder extending ½ mile to the northward.

Wedge Island (lat. 43° 08′, long. 147° 40′), which lies ½ mile off Low Point, the southern extremity of Wedge Bay, is 1,500 yards long, north and south, and ⅓ mile broad, with Castle Rock, 81 feet high, close off its south point and rocks and shoal water off its east side; the Witch Rock, awash at low water, lying 300 yards eastward from the north point. The west side of the island is formed of perpendicular basaltic cliffs rising to a height of 311 feet a quarter of a mile northward of Castle Rock; the summit is a plateau 400 yards long and 200 yards wide, and from it the land slopes gradually to the northward and more abruptly to the eastward.

The channel between Wedge Island and Tasman Peninsula is only ¼ mile broad between the 5-fathom curve, but it is clear of danger, and has 5½ fathoms least water. During heavy southerly gales the sea often breaks across the channel.

Burnett Harbor or **Parsons Bay** is a considerable inlet on the north side of Wedge Bay, having an entrance ¼ mile wide. The bay extends northeastward 1¼ miles and thence southeastward about the same distance. The outer arm of the harbor is about ¼ mile broad, with depths midway of 10 to 7 fathoms; long kelp grows near the shore on either side.

The southeastern or inner arm of Parsons Bay is 600 yards wide, the depths decreasing from 9 to 1½ fathoms at ⅓ mile from the southeast end of the harbor, over a bottom of stiff mud.

On the eastern side of the bay is a pier with 12 feet at low water alongside its outer face.

Although Wedge Bay and Burnett Harbor are small, yet from their position opposite Derwent River, they may be often found convenient for small vessels when adverse winds prevent their entering that river.

Tides.—It is high water, full and change, at 7 h. 48 m.; springs rise 4½ feet, neaps 3½ feet.

Nubeena is the settlement on the eastern side of the harbor. Population is about 200.

There is telephonic communication between Nubeena, Carnarvon, and Impression Bay, from thence by telegraph to Hobart.

Mails are received and dispatched four times a week to and from Hobart. There is weekly communication between Nubeena and Hobart by schooner.

Fresh water can be obtained from the stream at the head of the bay with difficulty.

The land around Nubeena is good, and much of it is cultivated with potatoes, grain, and apple orchards, which do well, and the pasture in clear places is fairly good for cattle and sheep.

From the north point of the entrance to Burnett Harbor, the coast trends westward one mile to Lory Point, thence about westnorthwest for 3 miles to Outer North Head, with Awk Point about midway between.

Roaring Beach Bay (lat. 43° 06′, long. 147° 40′), situated between Lory Point and Awk Point, has a sand beach ½ mile long, on which there is usually a heavy surf which renders landing impracticable, unless after a continuance of northeast winds. About the center of the beach is a sharp-pointed sand hill 100 feet high.

Hayward Rock, with 3½ fathoms water over it at low water and 9 fathoms close around, which generally breaks with long southerly swells, lies 600 yards southward from Awk Point.

Frederick Henry Bay was discovered in 1792 by D'Entrecasteaux, who named it North Bay (Baie du Nord). Since then, by a series of extraordinary typographical errors, the name North Bay has changed places with the original Frederik Hendrik Bay of Tasman, on the northeast side of Forestiers Peninsula.

Norfolk Bay was much used by government vessels and boats when Port Arthur was a convict establishment, and extensive remains of convict settlements are scattered about round the bay.

Frederick Henry Bay or **Baie du Nord** is 3 miles wide at its entrance between Cape Deslacs and Northwest Point, whence it extends to the northward for 9 miles to Seven-mile Beach and the entrance of Pitt Water.

Depths.—The bay is from 5 to 6 miles wide, and everywhere clear of dangers, with a bottom of sand and shells and depths of 10 to 11 fathoms abreast of Cape Deslacs, shallowing gradually to 4 fathoms at a half a mile off Seven-mile Beach.

Cape Contrariety, 2¼ miles northeastward from Betsy Island, is a bold cliffy point, perfectly steep-to. From this cape the coast is cliffy, the cliffs being from 200 to 300 feet high, rising to 615 feet ¼ mile inland, for 1¾ miles to the northward to the west end of Clifton Beach which extends in an east-northeast direction about one mile to Cape Deslacs.

Cape Deslacs is a long, grass-covered, flat, cliffy point 190 feet high. The cliffs are composed of a whitish yellow mudstone and show conspicuously in all directions.

There is a very good blowhole under Cape Deslacs close to the water's edge. The coast northward of Cape Deslacs is cliffy to the entrance of Pipe Clay Lagoon.

Pipe Clay Lagoon.—One and three-quarter miles to the northward of Cape Deslacs is the entrance to this lagoon, shown by a white sandy beach, with a depth of from 6 to 8 feet of water at the mouth; a narrow shifting channel with 3 to 4 feet of water leads in for about ½ mile to where a pier has been run out for the convenience of ketches which call here for fruit, etc. It then opens out into a shallow lagoon some 2 miles long by 1,500 yards wide, mostly dry at low water; the bottom is composed of a stiff white clay, hence the name.

Calverts Hill (lat. 42° 57′, long. 147° 32′) is a prominent object, 455 feet high, on the north side of Pipe Clay Lagoon, with several large trees on its summit. Behind it are the Muddy Plains, an extensive marsh, where occasionally wild duck may be found.

There are numerous farms about, and an extensive trade in fruit it carried on from this part of the country.

Richardsons Beach is about 2¼ miles northward of Pipe Clay Lagoon; there is a large fruit-growing establishment here and a small pier for shipping the produce.

May Point is just to the north of Richardsons Beach; it has a very conspicuous white patch of mudstone cliff similar to Cape Deslacs.

Ralph Bay Neck.—A stretch of sand northward of May Point called the Two Mile Beach forms the eastern side of Ralph Bay Neck, which in one place is little over ¼ mile wide.

Single Hill is 2½ miles northward of May Point; it is a flat-topped summit 670 feet high. Abreast it to the eastward and about ½ mile offshore is a small sandy patch of 2½ fathoms, but all the northwest corner of the bay is shallow and the 5-fathom curve extends nearly 2 miles from the shore.

Clarence Plains are at the back of Single Hill and extend down to Ralph Bay Neck.

Mount Rumney (lat. 42° 52′, long. 147° 27′).—On the west side of Clarence Plains is a range of hills, averaging 800 to 900 feet in height, the northern summit being Mount Rumney, 1,220 feet high, on which is a large cairn.

Seven Mile Beach forms the head of Frederick Henry Bay and is a curved sandy beach, at the eastern end of which is a narrow opening into the Pitt Water; a sand bank extends out for ¼ mile on the western side of the entrance, leaving a passage about 200 yards wide with from 3 to 4 fathoms in it.

Pitt Water.—This lagoon extends nearly 9 miles in a westerly direction from the entrance.

Within the entrance a narrow channel trends westward between two rocks, above which it passes close to a projecting head on the north side and a point extending from the south shore. There are depths of 3 fathoms between the two rocks, and from 6 to 5 fathoms in the channel to the east and west of them.

Spectacle Island.—One and one-quarter miles to the southward of the entrance to the Pitt Water is a projecting cliffy point, close off which are 3 small islets. Spectacle Island is the southern and largest of these islets, and is 40 feet high.

On the northern islet are two holes right through the rock, which probably gave the name.

Whale Rock, situated 1,400 yards southward of Spectacle Island, is 2 feet high, on which the swell generally breaks heavily. About 200 yards southward of it is a rock under water.

Rocky patch.—At 800 yards westward of Whale Rock is a rocky patch of 4 fathoms, with from 5 to 6 fathoms around.

Carlton Bluff, 2 miles east-southeastward from Spectacle Island, is 370 feet high and looks like an island from the bay. On its north side is the entrance to Carlton River, into which ketches go at high water; it almost dries at low water.

Renard Point is 1 mile farther to the southeast, with Roaring Beach between. It forms the north point of the entrance to Norfolk Bay.

Isle of Caves, 1,700 yards west-southwestward from Renard Point, is a small islet, 35 feet high, bare of trees. At the southeast corner of the islet are two caves in the cliff, which gave the name.

Bass Spit is a spit or bank of sand extending, within its 5-fathom limit, $1\frac{1}{2}$ miles to the southward of the Isle of Caves; it has a breadth of about 1,200 yards, and its north point is northwestward nearly 1 mile from the isle. The least water on it is $3\frac{1}{2}$ fathoms near the southeast corner, which is fairly steep-to.

The eastern side of the entrance to Frederick Henry Bay, extending from Outer North Head $4\frac{1}{2}$ miles northward to Northwest Point, is a bold, rugged, broken up shore, on which the sea always breaks heavily, rendering landing practically impossible. It is steep-to all along and in places lined with kelp.

Outer North Head (lat. 43° 04′, long. 147° 38′), a yellow cliffy point about 250 feet high, which rises to an elevation of 898 feet, is the turning point into Wedge Bay.

Rocks.—At 400 yards southwestward of Outer North Head is Flat Rock, 1 foot high, on which the sea always breaks. Foster Rock with a depth of 9 feet at low water spring tides, and 10 to 12 fathoms close around, lies $\frac{1}{4}$ mile westward from Outer North Head. It breaks only in very heavy weather. A rocky patch with a least depth of $5\frac{1}{2}$ fathoms exists at a distance of 200 yards, 36°, from Foster Rock. The patch is plainly indicated by the kelp growing on it.

Mount Communication, $1\frac{1}{2}$ miles northeast of Outer North Head, is a flat-topped hill 1,120 feet high. The summit is partially cleared, and was formerly one of the convict signal stations. A few hundred yards below the summit on the north side is a well of good water, 40 feet deep; this is somewhat remarkable, as all the water in the district is brackish. On one of the western spurs about 1,500 yards from the summit are the remains of a quarry from which the aborigines obtained green stone for axes, &c.

Storm Cove is a small cove 2¾ miles northward of Outer North Head; there is a patch of sunken rocks here surrounded with kelp, behind which fishing boats obtain shelter in westerly weather.

Northwest Point (lat. 43° 00′, long. 147° 38′), which is surmounted by a summit 690 feet high, named Black Jack, forms the eastern point of the entrance to Frederick Henry Bay. The coast here turns to the eastward for over 1 mile to form a bay 1¾ miles across, with a white sand beach named Sloping Main at its head, behind which is a stretch of marsh land extending back nearly to Coal Mines Summit.

Depths.—Between Northwest Point and Sloping Island, 2½ miles to the northward, the depths in the bay are mostly under 5 fathoms; there is a sandy patch in the center of it, with 4 fathoms water and 5 to 6 fathoms around, lying 1 mile southward from Sloping Island.

Sloping Island, 1 mile in length north and south and ½ mile wide, is a long, low island with a few scattered trees on it. The highest part (220 feet) is near the south end of the island, and it slopes down to its north end, a low point, on the northeast side of which is a small white hut.

At 400 yards off the southwest point of the island is a narrow rocky ledge, 20 feet high, a favorite resort for cormorants. A bank of 1 to 3 fathoms connects Sloping Island with the main, its north edge being steep-to.

On the land eastward of Sloping Island is a sandy beach 1,500 yards long, inshore of which is a shallow lagoon extending nearly to Lime Bay and generally covered with water fowl.

Hog Islet, two-thirds of the way from Sloping Island toward Green Head, is 15 feet high and just at the north edge of the bank.

Green Head (lat. 42° 56′, long. 147° 40′), a yellowish cliffy point, forms the northern extremity of Tasman Peninsula. The land here is 300 to 400 feet high and is rather thickly wooded. A sandy spit runs out under water from Green Head for 450 yards to the 3-fathom line; it is usually plainly visible when the light is good.

Norfolk Bay—Flinders Channel.—Between Green Head and Renard Point, a distance of 2 miles, is the entrance to Norfolk Bay, but the navigable channel named Flinders Channel is only about 1,200 yards wide southward of Bass Spit. Off the north point of Sloping Island, which can be safely passed at a distance of 300 yards, the soundings are over 20 fathoms, the deepest spot in Frederick Henry Bay.

Range.—A red beacon surmounted by a diamond, on Fulham Point, a long, low grassy point about 1,500 yards west of Low Island, in line with Tasman Hill, a well-defined summit on the far side of Blackman Bay, bearing 72°, leads through Flinders Channel in a depth of not less than 7 fathoms and well clear of the 3-fathom spit extending off Green Head.

When the beacon can not be distinguished steer 85° for the northern extremity of Green Head until Cape Deslacs is in line with the north extremity of Sloping Island astern, bearing 235°, then steer 55° until the southern extremity of Spectacle Island is in line with the northern extremity of the Isle of Caves, when course may be altered as necessary.

Lime Bay, 1½ miles to the eastward of Green Head, is very shallow, and at low water dries out for some distance, but it is a convenient anchorage in southerly winds.

Whitehouse Point is a cliffy point on the eastern side of Lime Bay and forms the turning point into Norfolk Bay.

North shore.—From Renard Point, on the north shore, the coast extends to the eastward in a succession of small bays and low grassy points for 5 miles to Dunally Bay. A range of well-wooded hills extends along the coast with occasional cleared patches and farm houses. There is a pier with 5 feet of water alongside it, in Connolly Bay, 2¼ miles eastward of Renard Point. The road from Dunally passes by here and over the hills to a bridge across the Carlton River and thence to Sorell.

Albert Point is the name of the narrow projection ½ mile eastward of Renard Point. A white beacon stands on its southern extremity.

Susan Bay, half a mile wide, is 1,500 yards northeastward of Albert Point.

Dorman Point, the eastern end of Connolly Bay, is nearly 2¼ miles eastward from Albert Point.

Wykeholm Point is 1 mile eastward from Dorman Point.

Boxall Rock, with 5 feet water on it, lies 300 yards southwestward from Wykeholm Point.

Fulham Point is 1,500 yards southeastward from Wykeholm Point.

Low Island, 1,500 yards southeastward from Fulham Point, is about ¼ mile in extent, 60 feet high, and there are a few clumps of trees on it. A reef of rocks extends for a short distance from its north point, and midway between the island and the shore the passage is blocked by a rocky patch, which dries 1 foot.

Beacon.—A white beacon, with diamond-shaped topmark, is situated on the northern part of the reef which dries 1 foot.

Dunally Bay (lat. 42° 55′, long. 147° 47′), which is 1½ miles wide, is very shallow, there being less than 3 fathoms right across the entrance.

A pier, nearly ¼ mile long, extends out beyond the edge of the dry line and has 5 feet of water at its end. On a small rise over the bay is a hotel, a favorite resort in summer months for visitors from Hobart.

The township of Dunally is about ½ mile farther inland, where there is a post and telegraph office. Population of township is about 100; of district about 700.

East Bay Neck is a low neck of land nearly ½ mile wide between Dunally Bay and East Bay, a shallow inlet opening out of Marion Bay on the east coast of Tasmania. It connects Forestiers Peninsula with the mainland. North of Dunally Bay is a high range of hills, with a conspicuous tree on the highest summit which is 1,435 feet high.

Denison Canal.—This canal, which connects Norfolk and Marion Bays, is available for vessels of 10 feet draft. There is a bridge across it, which will be opened for the passage of vessels between sunrise and sunset, from April 1 to September 30, and from 6 a. m. to 6 p. m. from October 1 to March 31; but in any urgent case the superintendent may, in his discretion, cause the bridge to be opened at any other time.

Tide signals.—During the above-hours tidal signals are made from a flagstaff on the western side of the Dunally Bay entrance to the canal.

A red ball denotes flood tide, which runs to the northeast.

A black ball denotes ebb tide, which runs to the southwest.

The canal is in charge of a superintendent, whose directions must be followed by mariners desirous of passing through it.

Beacons and buoys.—The canal is marked by a white buoy on the western side and a black buoy on the eastern side of the approach from Dunally Bay.

East Bay and Blackman Bay channels are marked with 10 single-pile beacons, 4 diamond-headed beacons, and 7 buoys. The entrance at the Narrows (Marion Bay) is marked by two diamond-headed beacons. All beacons and buoys painted white are on the northwestern side of the channel, those painted black on the southeastern side.

There are also two diamond-headed beacons painted white on the northwestern bank of Denison Canal to mark the center of the cutting when approaching from East Bay.

A semaphore at either end of the canal indicates, by a raised arm, that vessels are not to enter.

Dunbabin Channel.—This channel has been dredged in East Bay, between Dunbabin and Green Points. It is about 2,900 feet long and 40 feet wide, with a depth of 8½ feet, and is marked by white stakes, which must be left on the starboard hand in entering the channel from East Bay.

Mount Forestier lies southeast of Dunally, about 1½ miles in land; it has a conspicuous double summit, the northern being 875 feet and the southern 1,050 feet high, on the latter are the remains of a stone hut, one of the old signal stations.

Garden Island, about 1½ miles southward of Low Island, is nearly ¾ mile long north-northwest and south-southeast and ¼ mile wide. It is 140 feet high and clear of trees except a clump on the south end, it contains a few acres of very good pasture land; near the north point are some farm buildings.

On the east side of Garden Island and northward of it to Dunally Bay, the water is shallow; but good anchorage in from 4 to 5 fathoms may be had in northwest winds under the southeast corner of the island.

Norfolk Bay, from a line adjoining Whitehouse Point and Garden Island, extends from 6 to 7 miles to the southward, with a width of about 5 miles. It is perfectly land-locked, clear of dangers and with anchorage everywhere with good holding ground.

Depths.—Off Whitehouse Point, to a distance of 1,500 yards, the depths are 7 to 12 fathoms, shoaling gradually to 4 fathoms near Garden Island. Inside the bay the depths are 10 to 7 fathoms, mud bottom.

Calibration Range.—The Commonwealth Government calibration range is in Norfolk Bay.

The western side of Norfolk Bay has no distinctive feature except the coal mines, where there are a cluster of ruined prison buildings with a summit behind 275 feet high, partially cleared, and on which was a signal station, and the curious underground cells for convicts. The mines were worked by convict labor, an inferior kind of coal only fit for household purposes being obtained.

Close to the water with the remains of a pier in front of it is a large storehouse forming a conspicuous object in the bay.

The soundings are comparatively shallow abreast the coal mines, the 5-fathom curve being about 1,500 yards from the shore.

Ironstone Point (lat. 42° 58′, long. 147° 45′) is a low point covered with trees midway between Whitehouse Point and the coal mines.

Saltwater River, 2 miles southward of the coal mines at the southwest corner of Norfolk Bay, has a bare grassy hill, 175 feet high, with a few houses, some in ruins, on its eastern side; this was a government farm worked by convicts. There is a pier in ruins here with a depth of 7 feet alongside it.

The south side of Norfolk Bay is a succession of jutting yellow sandstone points with small bays between.

Deer Point is the point on the east side of Saltwater River; then come Prices Bay and Primadina Point, 1¼ miles to the southeastward.

Impression Bay is the bay on the east side of Primadina Point. Here is a small township with post and telegraph office. Overlooking the southwest side of the bay are some remains of a small model prison; this was a sanatorium in the time of the convicts.

There is a small pier with 10 feet water at its end extending from a small point in this bay.

Tides.—It is high water, full and change, at Impression Bay at 8h. 15m.; springs rise 6¼ feet, neaps 4 feet.

Kunya.—In the southwest corner of the next deep bay, 1½ miles to the eastward of Impression Bay, is Kunya, formerly a convict station known as the Cascades; two small waterfalls originate the name.

It is a favorite summer resort for tourists; there is a post and telegraph office and a pier with 5 feet of water alongside it.

Single Tree Point, at the southeast corner of Norfolk Bay, is 3½ miles eastward of Primadina Point; it is a jutting rocky ledge.

Woody Island, about 400 yards northeastward of Single Tree Point, is 75 feet high, 400 to 600 yards in extent and covered with clumps of shea oak.' At its eastern point is a ruined signal hut, which formed the connecting link between Signal hill and Eagle-hawk Neck. A few rabbits may be got on the island.

There is very good shelter under Woody Island in from 4 to 5 fathoms for one vessel during heavy northwesterly weather.

Mason Rock, situated in mid-channel between Single Tree Point and Woody Island, has a depth of 5 feet water on it.

Little Norfolk Bay, an inlet about ⅓ mile broad, extends 1½ miles south-southeastward from Woody Island to Taranna, where there is a pier with 9 feet water alongside and to which there is a narrow channel with depths of from 3 to 4 fathoms.

Taranna consists of a few houses and a hotel with post and tele-graph offices. Population is about 200. The road from Carnarvon (distant 7 miles) passes here and then goes by Eagle-hawk and East Bay Necks to Sorell. A steamer runs to and from Hobart three times a week, calling at Dunally and other places en route.

Supplies.—Fresh meat, bread, and potatoes can be obtained at moderate rates, but it is advisable to give a few days' notice. Fish is abundant.

Signal Hill (lat. 43° 05', long. 147° 53') is about 2 miles south-eastward from Taranna; on its shoulder under the summit is a cleared space, where the signal station was that connected with Port Arthur.

Dart Bay is between Woody Island and Heather Point, 1,500 yards to the northward. There are from 7 to 5 fathoms water in this bay to nearly 1 mile eastward of Heather Point. From it Eagle-hawk Bay, a long narrow inlet, extends 2½ miles eastward to Eagle-hawk Neck; this bay is very narrow for the last 1¼ miles, averaging 300 yards in width, with depths of from 1 to 2 fathoms.

Eagle-hawk Neck is a narrow neck connecting Forestiers and Tasman Peninsulas; on it are some sand hills about 60 feet high. On the east side of the neck is Monge Bay, on the east coast of Tasmania. The neck averages about 100 yards broad, but in one place at times there is scarcely a distance of 20 yards between the two high-water levels.

Mount Macgregor, 2¼ miles northward of Eagle-hawk Neck, is 1,925 feet high, the highest summit in this part of the country. It is a well-defined double summit, densely wooded.

Coast.—Three-quarters of a mile to the northwest of Heather Point is a point, close off which is a small islet 10 feet high; 200 yards outside this isl't is a rocky patch about 200 yards in extent with a few bowlders just above water on it.

Flinders Bay, 1 mile farther to the northward, is 1,500 yards deep; there are depths of 7 fathoms in the entrance, and the depth gradually decreases to the head of the bay.

Wattle Point (lat. 42° 58′, long. 147° 49′), 1¼ miles to the south-east of Garden Island and about 3½ miles northwestward from Heather Point, forms the south point of King Georges Sound.

King Georges Sound, a long, shallow inlet, is ½ mile wide at its entrance between Wattle Point and Gull Island to the northeast, a small wooded island 160 feet high and about 800 yards in extent. The sound then extends 1½ miles to the northeast, and has a depth of from 2 to 3 fathoms to a small pier at its head; there is a depth of 5 feet alongside the pier.

Anchorage in Frederick Henry and Norfolk Bays may be had almost anywhere according to the chart. But a swell is nearly always setting in from Storm Bay on the east side of Frederick Henry Bay, and it must be remembered that the heaviest and most sudden blows begin from northwest to southwest. The holding ground is everywhere good, generally sand and shells in Frederick Henry Bay, mud and clay north of Norfolk Bay, and soft mud in Norfolk Bay. In Norfolk Bay, although completely land-locked, a very nasty short sea for boats very quickly gets up in strong winds.

Communication.—A small steamer runs twice a week from Hobart to Norfolk Bay, calling at Saltwater River, Impression Bay, the Cascades, Taranna, and Dunally. Small ketches constantly run from Hobart all round the bays, trading principally in firewood and fruit during the season.

A good road runs from Carnarvon by Taranna, Eagle-hawk and East Bay Necks to Sorell, where there is a railway to Bellerive, opposite Hobart. Another road branches off from Taranna, along the south side of the bay to Saltwater River and from Impression Bay to Wedge Bay.

Supplies.—Fish may be caught in abundance with hook and line, and on many of the sand beaches a good haul of flounders may be had with the seine at night.

The fishing industry is carried on in a somewhat desultory manner by whaler-built fishing boats in Storm and Frederick Henry Bays; the boats usually work in couples, either line fishing, or along the edge of the kelp with " grab all " nets.

Fresh meat and vegetables, dairy produce, etc., may be obtained at Saltwater River and Impression Bay in any quantity on due notice being given. The farmers here are prepared to enter into a contract, if necessary, for a large daily supply.

Tides.—The tides appear to closely follow the tides at Hobart, both as to the times of high and low water and as to irregularity.

There is a large diurnal inequality; with the moon's declination north, the higher high water follows the superior transit of the moon; with the moon's declination south, the higher high water follows the inferior transit.

The greatest range of the tide appears to occur when the moon is at its greatest north or south declination, the least range when the moon is on the Equator. There is therefore no proper establishment.

The height of the tide is usually much affected by the barometer, a low barometer causing an abnormal high tide.

Tidal currents.—The tidal currents are weak and practically imperceptible, but after a heavy gale from the southwest a distinct set was felt into Frederick Henry Bay, and in Flinders Channel toward Norfolk Bay.

Tasman Peninsula, of which the west and north coasts have been described with the shores of Storm, Frederick Henry, and Norfolk Bays, extends northwest and southeast for a distance of 23 miles, and is 13 miles across; it consists of wooded hills and fertile valleys, with numerous streams of pure fresh water.

Cape Pillar (lat. 43° 14', long. 148° 02'), the southeastern extremity of Tasman Peninsula, lies 10 miles east-northeastward from Cape Raoul, and is the most remarkable headland on the coast, being formed of perpendicular columns of basalt rising to a height of 913 feet, and there forming a flat surface, the high land near the cape being mostly without wood.

Tasman Island, 811 feet in height, is rocky, sterile, rugged, and flat-topped; it is close southward of Cape Pillar. Off the southwest end of Tasman Island is a remarkable semidetached rock with two peaks; the gap between the rock and island is perfectly straight and square. On one of the peaks of the rock is a large stone, which.

has exactly the appearance of a lighthouse when on an easterly or
westerly bearing. There is a narrow passage between Cape Pillar
and the island, sometimes available for small vessels.

Light.—A flashing white light, visible 36 miles, 907 feet above
water, is shown from a white iron tower on the southeast part of
Tasman Island.

Port Arthur.—The coast between Capes Raoul and Pillar forms
a bay, in which, midway between the two capes, is the entrance of
Port Arthur, one of the most secure harbors in Tasmania; it extends
in a north and south direction about 5 miles and is clear of all
dangers.

Westward of Cape Pillar the coast continues high and precipitous
for about 5 miles to Arthurs Peak, a conspicuous summit, 1,050 feet
high, which forms the east head of the entrance to the port.

On the point under Arthurs Peak is a remarkable pillar rock
which stands out alone, when viewed from a north or south direction, and close off it is the Budget, a small rock 5 feet high.

Depths.—Over the greater part of Port Arthur the depths are
from 17 to 25 fathoms, but in the bays on its western shore are 8 to
6 fathoms.

Kelp extends off nearly all the points in this harbor to a depth of
from 5 to 10 fathoms. It always grows from a rocky bottom and
should be avoided. `

The entrance (lat. 43° 12', long. 147° 54') is easily made out from
seaward by the dip in the highland, and by Mount Brown, a long
flat precipitous summit 570 feet high, which forms the west head of
the entrance, and appears as an island against the land behind.

The entrance to the harbor is 1,600 yards wide, with deep water
right across between this point and the Budget.

The east shore of Port Arthur is steep-to and extends in a northerly direction for 4 miles to the head of Stinking Bay, 1 mile southward of which is Denmans Cove, an excellent place for hauling the
seine.

Black Rock.—Southwest of Mount Brown, at 600 yards from the
shore, is Black Rock, 30 feet high, with deep water close round it;
between Black Rock and a small point to the northward of it is an
isolated rock, which always breaks.

Half Moon Bay.—Just to the northward of Mount Brown is a
small cove, Half Moon Bay, 800 yards wide by 700 yards deep, in
which there is always too much swell to afford any anchorage.

Black Point, on the north side of Half Moon Bay, is 80 feet high,
flat and cliffy.

Safety Cove is about 1 mile to the northward of Black Point and
is 1,400 yards long, north and south, and 800 yards deep, east and

west; it has anchorage in from 5 to 12 fathoms, sand, but there is always a certain amount of swell here.

Puer Point.—From Safety Cove the coast is about 100 feet high and cliffy to Puer Point, about 1 mile to the northward, 200 yards off which is Dead Island, a small islet 40 feet high, with the tops of the trees 120 feet high. This islet was the burying place of the old convict establishment, and about 1,700 bodies are supposed to have been buried here.

On the northwest side of Dead Island a rocky spit extends for 300 yards toward Frying-pan Point, with from 4 to 5 fathoms water on it; it is distinctly shown by the kelp.

Opossum Bay (lat. 43° 09′, long. 147° 52′), the best anchorage in Port Arthur, is ½ mile wide at its entrance between Dead Island and Frying-pan Point. From its entrance the bay extends about 1 mile to the southward; its south shore is fronted by a sand and mud flat.

Depths.—The depths in Opossum Bay are from 9 to 6 fathoms, sand and mud bottom.

Mason Cove, on the northwest side of Opossum Bay, is 200 yards deep, and forms the water frontage of the settlement of Carnarvon. There are remains of jetties on the south side, and there is a wooden pier on the north side with 18 feet water at its outer end and 12 feet at its inner end. The depths are only 3½ fathoms at the entrance of the cove and it soon shallows.

Anchorage.—There is excellent anchorage off Mason Cove in from 7 to 12 fathoms, sand.

Carnarvon, as the once famous convict settlement of Port Arthur is now called, has a population of about 150; the old prison buildings have fallen into decay.

From its sheltered position a mild climate is experienced all the year round, fruit and flowers grow in abundance, and it is a very pretty little spot.

Communication.—There is communication by coach with Taranna in Norfolk Bay and biweekly steamer to Hobart. A road leads across the peninsula to Wedge Bay (about 7 miles), which brings Carnarvon within 3 hours of Hobart. There is a post and telegraph station here.

Frying-pan Point is a narrow, low point, with tall gum trees on it; from its north end a rocky ledge runs about 70 yards into Stewarts Bay.

Stewarts Bay, 800 yards deep by 600 yards wide, lies between Frying-pan and Garden Points; a small vessel could find anchorage in it in from 6 to 8 fathoms, but shoal water extends around it for about 300 yards from the shore.

Garden Point is covered with trees and forms the extremity of a peninsula with a long flat cleared top 115 feet high and some buildings upon it.

Long Bay extends from Garden Point to the northward for nearly 2 miles; it is 400 to 600 yards wide, with not less than 5 fathoms water for the first 1,500 yards, after which it shallows and the head of the bay dries. Small craft by keeping on its eastern side can get within ½ mile of the head of the bay.

Stingaree Bay, 1,500 yards above Garden Point and on the western side of Long Bay, from which it is separated by a small peninsula extending to the southeast, is shallow and there are many rocks in it.

Oakwood (lat. 43° 07′, long. 147° 52′) is a small settlement at the head of Long Bay.

Supplies.—Port Arthur is situated within an amphitheater of lofty wooded hills (the trees being mostly stringy bark gum), well watered and of a most pleasing aspect. Fish abound and can be caught in any quantity with line or seine. Fresh meat can be obtained at Carnarvon. Nearly all the bays have small streams of fresh water flowing into them. There is excellent timber about the port.

Tides.—From day observations made for a limited period it would appear that spring tides occur at the time of the moon's greatest north or south declination instead of at full and change, so that there is no "age of the tide" and probably no fixed establishment.

The time of high water, full and change, has been given as about 8 h. 20 m., springs rise about 4½ feet.

Rolling Bay.—This bay is 3¾ miles wide between the cliffy point southeast of Mount Brown and Cape Raoul, and is from 1¼ to 1½ miles deep, with depths of from 15 to 37 fathoms.

The coast westward of Mount Brown trends back to the neck of the peninsula on which the mount stands, forming a bay, at the head of which is a remarkable cave, and ½ mile nearer Mount Brown is the "blow hole," which is described as being a magnificent sight in heavy southerly gales.

About 1½ miles westward from Mount Brown is a bold cliffy point, 200 yards, within which is a square cave, 60 feet above high water and conspicuous from the southward and westward; from this point an irregular cliffy coast trends southwestward about 2 miles to a point to the southward of which is a deep cliffy gully with caves at its head; about midway between these two points there is a remarkable waterfall.

From the last-mentioned point the coast trends southwards 1½ miles to Cape Raoul.

EAST COAST OF TASMANIA.

The coast of Tasmania from Cape Pillar (lat. 43° 14′, long. 148° 02′) curves 5 miles in a northerly direction to Cape Hauy, which has a cluster of detached conspicuous rocks close off it, and forms the south point of Dolomieu, or Fortescue Bay.

Hippolite Rocks are situated immediately in front of Dolomieu Bay, from 1 to 2½ miles northeastward from Cape Hauy; they consist of two rocks above water and covered patches, the east and most elevated rock being 216 feet high, of a reddish brown color, and quoin shaped, high at the west end. The west rock is 28 feet high. The passage between the rocks and between them and the shore to the westward is foul and should on no account be attempted.

Dolomieu Bay is 1¼ miles wide, northwest and southwest, and 1½ miles deep, with a white sandy beach; the bay is sheltered only with land winds, the Hippolite Rocks not being sufficient to protect it from seaward.

Monge Bay.—From Dolomieu Bay the coast trends nearly north-northwestward 6 miles to the southeast point of Monge Bay, between 100 and 600 yards northwest of which is the Isle of Fossils, connected with the point by a reef of dry and sunken rocks. From its southeast point Monge Bay extends north-northwestward 1¼ miles to its northwest point, close off which lie the two Clyde Islets. This bay is 1,500 yards deep, with a small sandy beach in its southern bight, and a more extensive one along its western shore; the latter forms Eagle-hawk Neck, the narrow isthmus which connects Tasman Peninsula with Forestiers Peninsula, to the northward of it. From the neck the water in the bay appears very shallow for over ½ mile out and a heavy surf always rolls in.

On the northwest shore of Monge Bay is a curious geological formation named the Pavement, an extensive level of basaltic rock, much resembling a pavement of large flat stones, laid with remarkable regularity between straight parallel lines, which may be seen at low water, and 2 or 3 miles to the southeastward are the Blow Hole and Tasmans Arch.

Signal station.—There is a signal station at Eagle-hawk Neck and communication can be made by the international code; it is connected by telegraph.

The coast from Monge Bay extends 2 miles in a north-northeast direction to Cape Surville. A small islet lies close to a point 1,500 yards northeast of the northeast point of Monge Bay; and ½ mile to the southeast of Cape Surville the Sisters Islets lie within ¼ mile of a projection of the coast. Immediately behind a bay, midway between Eagle-hawk Neck and Cape Surville, Mount Macgregor rises to the height of 1,925 feet.

Between Cape Surville and a projection 1¼ miles northward from it, is a bay ½ mile deep, having two small islets or rocks, one close within the cape, and the other near the shore midway between the two points of the bay. From the north point of this bay the coast extends northward 2 miles to Yellow Bluff, and thence northwestward the same distance to Humper Bluff. The coast is high and bordered with rocks above and under water, and affords neither anchorage nor shelter, as the sea breaks upon every part of it with violence.

Wilmot Cove, locally known as Lagoon Bay, is the west end of an inlet trending westward 1¼ miles from its entrance, between Humper Bluff and Cape Frederik Hendrik, 134 yards northward of it. The south shore of this inlet is fronted by several rocks and small islets, the largest and most distant from the shore being the Kelly Islets, which lie midway between Humper Bluff and Wilmot Cove; some rocks also extend from the extremity of Cape Frederik Hendrik to the southeastward. In this cove is the one solitary dwelling on this part of the peninsula at the time of the survey.

Cape Frederik Hendrik (lat. 42° 52′, long. 148° 00′), formed of basaltic columns, is a narrow point stretching 1¼ miles in an east-northeasterly direction from the line of coast, and forms the southeastern point of Marion Bay.

Marion Bay is an exposed indentation of the coast extending from Cape Frederik Hendrik 8½ miles north-northwestward to Cape Bernier; it is 4 miles deep, but the only part at all available for vessels is the southern portion of the bay from whence is the channel through Blackman Bay, leading to Denison Canal.

North Bay, the original Frederik Hendrik Bay of Tasman, southward of Marion Bay, lies between Cape Frederik and Cape Paul Lamanon; it is 3 miles across, 1¼ miles deep, with a small inlet trending westward for 2 miles from the extremity of the former cape. Green Islet, which lies in front of the bay, 1¼ miles northward of Cape Frederik Hendrik, is too small to afford any protection from seaward.

Cape Paul Lamanon (lat. 42° 50′, long. 147° 57′) is a small projecting point, low, stony, arid, covered with small timber and rough scrub, with the High Rocks, and others above and under water, close to it. From this cape the coast trends westward 1 mile to the northeast entrance point of Blackman and East Bays; it is bordered by a reef of rocks.

Port Frederik Hendrik is an extensive but shoal inlet on the southwest side of Marion Bay, the west side of its confined entrance being formed by a narrow point which projects southeastward 2 miles from the west to within ⅓ mile of the opposite side. The shores are high, and form a projecting double point on the east side, and two long narrow projections on the west side.

Blackman and East Bays are together about 4 miles in length in a northeast and southwest direction, and from ¾ to 1½ miles in breadth. The entrance which leads from the south part of Marion Bay is about 200 yards in breadth, with a depth of 3 to 5 fathoms, but immediately within it is a flat ½ mile across which blocks the passage for any but light drafts, the low-water depth over it being only 1½ fathoms.

Channel to Denison Canal.—Details of the Denison Canal have been given.

Supplies.—Wood can be had in Blackman and East Bays, and plenty of fish may be taken on the large bank at the entrance.

Tides and tidal currents.—It is high water, full and change, in East and Blackman Bays, at 8 h. approx., springs rise from 4 to 5 feet; the tidal current at the entrance runs at the rate of about 2 knots.

The coast.—From the entrance of Port Frederik Hendrik a flat sandy beach curves northward for about 5 miles to Du Ressac Point; landing is at all times dangerous on this beach, and is impossible with winds from the sea, as an enormous surf breaks more than 400 yards from it. A mountain torrent pours through this beach in the rainy season.

Between Du Ressac Point and Cape Bernier the coast, which is of moderate height, forms two sandy bights that may be approached with offshore winds, when that nearest the cape will be found the most convenient.

Cape Bernier (lat. 42° 44′, long. 147° 57′) is high and remarkable on account of its conical shape; there are 6 and 7 fathoms water close to the southward and eastward of it.

Maria Island, which is high and rugged, with sharp irregular peaks toward the north end, is separated from the east coast of Tasmania by a navigable channel from 2¼ to 4½ miles wide; Cape Peron, its south point, lying east 3½ miles from Cape Bernier.

Maria Island is 11 miles long, north and south, and at 5 miles north-northeast from Cape Peron is nearly divided by Riedle Bay on the east, and Oyster Bay on the west side, there being only a low sandy isthmus between them. The southern part of the island is 3½ miles, and the northern 6½ miles broad.

Rocks.—Cape Peron is a bold headland with three rocks situated off it. The one nearest the cape, from which it is separated by only 20 yards, is named Pyramid Rock, and is 118 feet high; the outer rock, which is known to the fishermen as the Boy in the Boat, is a small dangerous pinnacle, steep-to, only 2 feet above high water, and lies 220°, distant 800 yards from Cape Peron, while the middle of the three is 6 feet high, and situated about midway between the other two.

The south coast of Maria Island from Cape Peron trends northeastward 3 miles to Cape Maurouard, on the southwest side of which is a small inlet. From this cape the coast trends irregularly, 1 mile northeastward to Cape Bald, and thence north-northwest 2¼ miles to the southwest point of Riedle Bay. There are depths of 10 fathoms close to Cape Bald; but some rocks lie near the shore 1½ miles to the northward of that cape.

The south and east coasts of Maria Island are all of granite, and rise abruptly like a wall to the height of 200 feet, but gradually descend from Cape Maurouard toward Riedle Bay. There are some caves in which the water breaks with great noise.

Riedle Bay extends nearly 2 miles across from southwest to northeast, and is 1 mile deep, with some rocks projecting from its west and north shores. There are depths of 15 to 9 fathoms, fine sand, in the southwestern part of the bay, but in the northeastern part there are only from 5 to 6 fathoms. Riedle Bay affords but indifferent anchorage for vessels remaining any length of time, being entirely exposed to the wind and sea from south to northeast. Landing may be effected at the southwestern part of the bay, with the wind off the land, but farther to the northward, the approach to the shore is prevented by a dangerous bar.

From Cape Mistaken (lat. 42° 40′, long. 148° 10′), which lies 1¼ miles eastward from the northeast point of Riedle Bay, the east coast of Maria Island trends north-northeastward 1½ miles to Ragged Head, with a small islet or rock close to the east side of cape. The coast from Ragged Head extends northwest 5 miles to Cape Boullanger, with a small inlet midway, and some rocks near the shore 2 miles to the northwest of the head. The land rises from this high steep coast to a lofty ridge; Mount Maria, 3 miles northwestward from Ragged Head is 2,329 feet, and Bishop and Clerk Mount, 1¼ miles southwestward of Cape Boullanger, is 3,000 feet high.

There are depths of 19 fathoms about 1 mile from the shore between Capes Mistaken and Boullanger.   •

From Cape Boullanger the north end of Maria Island forms a bay extending westward 2 miles to the north point of the island, ¼ mile off which is North Islet, with Black and other rocks between it and the shore to the southeast of the islet.

The west coast of Maria Island from Cape Peron trends northnorthwest 1½ miles, and thence north-northeast 3 miles to Point Maugé, the southeast point of Oyster Bay. There are depths of 3½ to 4 fathoms along this coast, but the point is enclosed by a shoal.

Oyster Bay is 1¼ miles wide from southeast to northwest at its entrance, within which the bay expands to 2¼ miles, and is 1¼ miles deep, but its shores are bordered by a shallow flat; the greatest depth of water in the bay is not more than 3½ fathoms, and it generally does not exceed 2 or 3 fathoms, with a white sandy bottom.

The north side of the entrance of Oyster Bay is formed by a low narrow point, projecting nearly 1½ miles to the southwest from the coast-line, from the extremity of which the west coast of Maria Island extends north-northeast 6 miles to its north point, and forms a succession of small bights and points, bordered by a shoal, on which are some sunken rocks. Shoal water extends to about 2¼ miles northward from the north entrance point of Oyster Bay.

The northwest part of Maria Island from Oyster Bay to its north point is low and wooded.

Settlement.—Immediately on the southwest side of the north point of Maria Island there is a small cove, close to which is the settlement. The population is about 100. There is communication by steamer to Triabunna.

Productions.—The soil of Maria Island is excellent and deep in the valleys. Oysters, lobsters, and mussels are abundant, the former of extraordinary size; but other kinds of fish are scarce, particularly in the beginning of winter.

The coast from Cape Bernier trends north 3½ miles to Point des Galets, and is steep with 6 or 7 fathoms water close to. Cockle Bay is merely a slight indentation of the coast, terminating to the southward in a small inlet, on the west side of Pebbly Point, which lies 2 miles north of Cape Bernier.

From Point des Galets the coast trends northwestward nearly 2½ miles to the head of a small inlet formed on its northeast side by a low narrow point stretching out about 1 mile southeastward from the coast line. This point is fronted by a sandy beach, and a small stream flows into the inlet.

Lachlan Islet (lat. 42° 38′, long. 148° 00′), which lies northeastward 1 mile from this point, is of triangular form, 300 yards broad, 43 feet high, and covered with grass. The only conspicuous object on it is a solitary bush, which is nearly in the center. There is a small reef of rocks 300 yards to the northwest of Lachlan Islet, the highest point of which is 1 foot above high water.

The channel west of Lachlan Islet has not more than 12 feet water in it, and only small vessels should attempt it.

Coast.—Between the low, narrow point abreast of Lachlan Islet and a steep cape 2½ miles northward from the point the cost is slightly embayed, with 5½ fathoms water about 1 mile from it. From this cape a high rocky coast extends 3 miles northwest to Luther Point, the south point of Prosser Bay. A range of mountains extends about north and south behind the coast from Port Fredrik Hendrik to Prosser Bay, its principal summits being Gordon Sugar-loaf, 1,350 feet high, and Prosser Sugar-loaf, 2,195 feet high, distant 8¼ miles and 6 miles, respectively, westward of Cape Bernier.

Thumbs, 1,805 feet high, the north summit of this range, is situated about 3¼ miles to the southwest of Prosser Bay.

Between the steep cape just noticed, and a projecting point 2½ miles northward of it, is a deep bay extending 2 miles northwest to Meredith Point, a broad point, which separates Prosser Bay from Spring Bay.

Orford Roads, eastward of Luther Point, is a fair anchorage during the westerly gales which are so prevalent, but can not be recommended in easterly gales, as the bottom is of hard sand, and a swell rolls in round the north end of Maria Island. Depth, 9 to 12 fathoms.

Prosser Bay is about 1 mile across, but it is shallow, and useless for any but small vessels. A vessel may anchor off its entrance, in 8 fathoms water, but would not be sheltered from southerly winds, which are violent in this locality. In easterly gales there is a dangerous sea in this bay.

Jetty.—At ¼ mile westward of Luther Point a jetty 75 yards long projects to the northward.

Rock.—A patch of rock, 30 feet in extent, with a least depth of 2 feet over it at low water, is situated 72 yards to the eastward of the jetty head.

Range beacons.—Two diamond-headed beacons, southward of the jetty, in range, bearing 236°, mark the direction of the rock, and two diamond-headed beacons westward of the jetty, in range bearing 275°, lead northward of the rock.

Prosser River discharges into Prosser Bay. The river is reported to be deep, but the bar has only 2 feet on it at low water.

Orford, situated on the right bank of the Prosser River, is a post and telegraph township, and has a population of about 100. The district is noted for freestone quarries.

Spring Bay (lat. 42° 32′, long. 147° 56′), so called from a chalybeate spring on its east side, is 2¼ miles long in a north-northwest and opposite direction, from 800 to 1,600 yards broad, and has a navigable area of 1 mile, the northern portion of the bay being much choked with sand and mud. It is completely sheltered from wind and sea, and the holding ground of mud is good. The township of Triabunna is on the shore at the head of the bay, 1,500 yards from the nearest anchorage. Two spits jutting out from the east shore narrow the bay, viz., Sappho Spit and Horseshoe Bank. The latter is covered with weed. A bank of 5 fathoms, sand, lies in the entrance at 600 to 800 yards northwestward from Freestone Point.

Depths.—Spring Bay has depths of 6 fathoms and over from the entrance to nearly 1½ miles within the bay.

Anchorage is very good at the mouth of Spring Bay, between Freestone Point and Meredith Point, but this is a long way from the town. To proceed up the harbor as near to Triabunna as pos sible keep well over on the west shore until past Observatory Point, to avoid Sappho Spit, taking care to give a wide berth to the Horseshoe Bank, which shoals very suddenly to 4 feet with Patten Point in line with the courthouse; anchor between Horseshoe Bank and Patten Point, in 20 feet of water, the point being distant 600 yards and in line with the courthouse.

Tides.—It is high water, full and change, at the entrance of Spring Bay at 8 h. 30 m. Springs rise from 4 to 5 feet, according to the wind; neaps 3 to 3½ feet. Neaps range 2 feet.

Triabunna has about 300 inhabitants. The exports are wool, mimosa bark, grain, and fruit. Indications of anthracite coal have been found in the neighborhood.

The bay teems with fish and oysters, and in close proximity native game and birds are abundant.

Communication.—There is postal communication between Spring Bay and Hobart four times a week, the journey by coach and rail occupying from 18 to 20 hours, and there is also a telegraph office. Steamers ply regularly to and from Hobart and Launceston.

The climate of Spring Bay is salubrious.

A tidal wave occurred in Spring Bay on the morning of the 28th August, 1883, a few hours after the volcanic eruption in the Strait of Sunda. The water rose 3 feet higher than high-water spring tides and washed backward and forward many times.

Oakhampton Bay is the name given by the settlers to the deep bight between Home Look-out Point and Lower Look-out Point (Cape Bougainville).

Cape Bougainville (lat. 42° 31′, long. 148° 02′) is a double point about 1 mile broad, projecting 1 mile to the southeast from the line of coast; some sunken rocks extend a short distance from the cape, close outside which there are 17 fathoms.

Directions for Spring Bay.—The direct approach to Spring Bay is northward of Maria Island, and no directions are necessary beyond those previously given with the anchorage.

Vessels from Marion Bay may take the channel between Maria Island and the land if locally acquainted.

The channel is 3½ miles wide at its south entrance, between Capes Bernier and Peron, and there is the same width at its north entrance (reported by the fishermen to be clear of dangers) between Cape Bougainville and the north point of Maria Island, but midway it is contracted to two narrow channels by Lachlan Islet and the rocks

above it. There are depths of 8 to 9 fathoms in mid-channel off
Oyster Bay, 7 fathoms 2 miles northward of the islet, and 19 fathoms
in the north entrance of the channel.

Proceeding to Spring Bay by the south entrance, pass the Boy in
the Boat at a moderate distance, and steer over to 1,500 yards from
Point des Galets, until Lachlan Islet bears 0° distant 2¼ miles, when
haul up to 13° and pass the islet on its eastern side at a distance of ½
mile. A straight course for Lachland Islet from the Boy in the Boat
can not be steered on account of a shallow spit off Oyster Bay, which
stretches out into the channel, a depth of 23 feet being found at low
water, 94° distant 1.2 miles from Point des Galets, and 170° 2¼ miles
from Lachlan Islet.

After passing Lachlan Islet, continue to steer northward for 2
miles, taking care to avoid the shallow ground north of it and the
shoal water extending off Maria Island, and then make a straight
course for Spring Bay, which appears as a gap in the land. On this
track through the East Lachlan Channel to Spring Bay there are
not less than 4½ fathoms water at low water spring tides.

Westward, distant 1,600 yards from Soldier Point and northeastward of Lachlan Islet, is the end of a shallow spit, extending from
Maria Island, upon which there are depths of 6 to 18 feet at low
water springs.

Vessels drawing over 21 feet are recommended not to approach
Spring Bay by the south entrance, as the sand spits around
Lachlan Islet and off the shores of the channel have been known
to silt up and alter their position. No stranger in a craft over 6
feet draft should pass westward of Lachlan Island.

Tidal currents.—In the Lachlan channels the flood current runs
to the north, the ebb to the south.

The coast from Cape Bougainville curves in a northerly direction
5¼ miles to the southeast point of Grindstone Bay. Between 1 and
2¼ miles northward of the cape there are two small inlets, close
off which are some sunken rocks; but there are depths of 17 to 24
fathoms 1 to 2 miles from the coast between the cape and the bay.

Cape Bailly (lat. 42° 20′, long. 148° 03′).—From Grindstone Bay,
which is a small inlet trending to the westward, the coast trends
northwest 2 miles, and thence north-northeast 4 miles to Cape
Bailly, on the south side of which are some rocks above water,
with depths of 10 fathoms close outside them, and 16 to 14 fathoms
between Grindstone Bay and the cape. The land from Cape
Bougainville to Cape Bailly is less elevated, but still steep and
wooded.

Ile des Phoques, situated distant 7 miles eastward from Grindstone Bay, is a sterile rock from 400 to 600 yards in extent; with

depths of 12 fathoms water close to the southward of it, and 26 to 24 fathoms between it and the shore.

Schouten Island, which is about 1,500 feet high, forms the east side of the entrance of Fleurieu or Oyster Bay, is 4½ miles long, east and west, and 1 to 2 miles broad, with Cape Faure, its south-western extremity, 9¼ miles eastward from Cape Bailly. Cape Sonnerat, the southern extremity of the island, has groups of islets and rocks extending 2 miles to the southward, the southern being the Taillefer Islets, the largest of which is of a pyramidal shape when seen from the eastward and about 150 feet high; an islet also 1 mile off the northwestern extremity of Schouten Island.

Water.—There is a small stream of excellent water on the south part of Schouten Island, where a boat may easily land; and the inlet·at the east end of the island may possibly afford a landing place.

Geographe Strait or Schouten Passage, which separates Schouten Island from the south point of Freycinet Peninsula to the northward of it, is about ½ mile across at its narrowest part, with apparently no other detached danger than a small rock above water, close off the south point of the peninsula. H. M. S. *Pioneer* in 1906 passed through this strait; the least depth obtained was 12 fathoms.

Freycinet Peninsula is 6 miles long north and south and 3½ miles across its broadest part, whence it gradually narrows to Cape Dégerando (lat. 42° 17′, long. 148° 20′), its south point. The east side of the peninsula from its south point trends northward for 6 miles to its northeast point, Cape Forestier, whence the northern ends turns westward 2½ miles to the isthmus which connects this with another peninsula to the northward of it. The east side is partly bordered with rocks, and the southwest and northwest sides are slightly indented. There is some tin mining on the peninsula.

Cape Forestier—Light.—A group-flashing white light with red sectors, visible 20 and 10 miles, respectively, 265 feet above water, is exhibited from a white tower, 13 feet high, on the extremity of the cape.

The isthmus, which connects Freycinet Peninsula with a smaller one to the northward of it, is 1½ miles long, northwest and southeast, and ½ mile broad, the greater portion of it being occupied by a pond of fresh water supplied by the rains; it is separated from the bay on its west side by a barrier of sand about 50 yards broad; the other part of the isthmus is tolerably well wooded.

The peninsula to the northward of this isthmus extends 4 miles east and west, the isthmus which connects it with the land farther north being 1 mile broad.

Thouin and Sleepy Bays.—Thouin Bay, on the east side of the southern isthmus, is 1 mile broad north and south, ½ mile deep;

but exposed to the eastward. From the north point of this bay the east side of the northern peninsula trends northward 2½ miles to the head of Sleepy Bay, a small bight on the south side of Cape Tourville.

Cape Tourville.—This cape projects about ½ mile to the southeast, with a cluster of small islets or rocks extending from it about 1 mile to the northeast; these, together with the cape, probably protect Sleepy Bay from the northward, although it must be fully exposed to the southward and eastward.

Aspect.—Freycinet Peninsula, 2,014 feet high, and Schouten Island, also high, are steep, and sterile toward the sea, but low and wooded on the west side: Cape Tourville being also high, these alternate mountains and isthmuses give this part of the coast from seaward, the appearance of a chain of islands. View A. on H. O. Chart No. 3570.

Fleurieu or Oyster Bay is formed on the east side by Schouten Island and the peninsula to the northward of it, and on the west side by the coast extending northward from Cape Bailly. This bay is 9½ miles wide at its entrance, whence it extends northward 14 miles to its low north shore. The bay is reported to be clear of dangers.

Depths.—The depths in Fleurieu Bay are from 11 to 7 fathoms.

The west shore of Fleurieu Bay from Cape Bailly extends north-northwestward 2 miles, and then turns westward 1 mile to the entrance of Little Swan Port. A rock above water, with 11 fathoms close to the northward of it, lies near the shore 1 mile northward from Cape Bailly.

Little Swan Port does not appear to be more than 200 yards wide at its entrance, but the port extends thence 3 miles in a southwest direction, with the width of a mile; it is, however, only fit for boats. There are two small islets in the western part of Little Swan Port, between which and its north shore is the mouth of Little Swan Port River, an inconsiderable stream winding from the westward. Little Swan Port Mountain, 1,757 feet high, is situated 9 miles southwestward from Cape Bailly.

From Little Swan Port the west shore of Fleurieu Bay extends northward for 3½ miles to Buxton Point, and is intersected by two small streams, one at 1¼ miles, and the other at 2¾ miles north of the entrance of the port. The coast from Buxton Point, after turning about 1 mile to the northwest, trends north-northeast 5½ miles to Webber Point, between which and Waterloo Point 2½ miles northward from it, the coast forms a slight indentation, with a small stream flowing into it 1 mile northward of Webber Point, and an inlet close to the southward of Waterloo Point.

Swansea (lat. 42° 09′, long. 148° 06′).—On the left bank of the small stream is the settlement of Swansea, where there is a post and telegraph station. A wharf extends 1,000 feet into 11 feet of water. There is communication by coaches to Campbelltown on the main line and Sorrell on the branch line of railway. Population is about 300.

About 1 mile northwest of Waterloo Point is the mouth of a small stream flowing from the southwestward, whence the north shore of Fleurieu Bay curves east-northeastward 7 miles to the entrance of Great Swan Port. For about 5 miles westward from this opening the shore forms the south side of a low tongue of land, which separates Fleurieu Bay from Great Swan Port.

The east shore—Refuge Islet.—The east shore of Fleurieu Bay has already been described as far north as the isthmus northward of Freycinet Peninsula.

Anchorage.—The eastern side of Fleurieu Bay affords good anchorage, sheltered by Refuge Islet, which, with some rocks close to the southward, lies near the shore 1 mile from the northwest part of the isthmus.

Wood and water may be procured from this anchorage with facility, the latter from the pond on the isthmus.

Hepburn Point (lat. 42° 07′, long. 148° 16′).—The bay on the west side of the northern isthmus, abreast of Sleepy Bay, is 1¼ miles wide at its entrance between the west point of the northern peninsula, and Hepburn Point to the northward of it, and is about 2 miles deep. Although there are several rocks in this bay, it is said to afford good anchorage. From Hepburn Point the east shore of Fleurieu Bay trends northwestward 1½ miles to the entrance of Great Swan Port.

Soundings.—The middle of Fleurieu Bay has not been sounded, but there are depths of 12 to 6 fathoms from 1 mile off Buxton Point to close off Webber Point, whence, to within ½ mile of the north shore, there are from 7 to 5½ fathoms, with 4½ fathoms between the latter depth and the entrance of Great Swan Port. From ½ mile off the west point of the northern peninsula to the same distance off Hepburn Point there are depths of 7 to 4½ fathoms.

Great Swan Port.—From its entrance, northward of Hepburn Point, which appears to be not more than 200 yards wide, Great Swan Port trends westward 5 miles along the north side of the tongue of land before noticed, to the mouth of Swan River, which flows into the port from the northward and westward. The port from its entrance increases to 1 mile in width.

Moulting Lagoon.—About 1½ miles northeast of the mouth of Swan River is a narrow opening communicating with Moulton Lagoon, which extends thence 5 miles in a north-northeast direction, forming, by a projection of the northwest shore, two basins, the south-western being 2¼ miles and the northeastern 1¼ miles in extent. There are several islets, or rocks, in this lagoon, and a small stream flows into the northeastern basin from the northward.

The east coast of Tasmania from Cape Tourville extends 2 miles in a north direction to the southeast point of Blue Stone Bay, and receding thence ½ mile to the westward, it trends north-northwest 3 miles to a double headland, having on its west side a small inlet; northwest 1¼ miles from this is a larger opening. Between the latter and Moulting Lagoon, 2 miles to the westward of it, the land rises to Mount Peter.

Cape Lodi (lat. 41° 55′, long. 148° 20′).—From the northwest of these two inlets the coast trends northward 7 miles to Cape Lodi, 3 miles south of which a point projects ¼ mile to the southeast from the line of coast. There are depths of 14 fathoms within 1 mile of the southeast point of Blue Stone Bay and 9 fathoms close off Cape Lodi, with from 8 to 5 fathoms near the coast between these points, but very few soundings have been taken.

Peggy Point.—From Cape Lodi the coast curves northward for 2¾ miles to Peggy Point, close off which is a small islet. The coast from Peggy Point forms a bay 2 miles deep, extending north-north-westward a distance of 7½ miles to the south extreme of Long Point. The shores of this bay are intersected by several inlets and small streams, the largest of the former being an opening with a small islet in it, situated 3 miles northwestward from Peggy Point. About 1 mile from Peggy Point toward the opening is Diamond Islet close to the shore. The small township Bicheno is situated on this coast. Population is about 75.

Long Point is of a peninsular form, with its east face extending about 1 mile north and south, and forms a small bight on either side of the isthmus which connects it with the mainland to the westward of it.

Seymour Jetties.—These little bays have jetties, with coal chutes for the convenience of coasting vessels. The township is named Seymour.

Coal.—There are exports of coal, fire clay, etc., from Long Point, a seam of good coal having been opened, about ½ mile from Long Point, and worked by a company.

St. Patrick Head.—About 1 mile westward of the north part of Long Point is an inlet with a narrow entrance, trending north and south parallel with the shore. From this inlet the coast extends 9¾ miles in a northerly direction to St. Patrick Head, which, together with the coast for about 2 miles south of it, is bordered by a reef. There are depths of 10 fathoms close to the northward of the reef which projects but a short distance from the head in that direction.

Soundings.—From about 10 miles east of Cape Tourville to 7 miles northeastward of Long Point there are depths of 66 to 40 fathoms, with similar depths about 5 miles from the shore, but immediately outside those soundings there is no bottom at 89 fathoms.

The coast from St. Patrick Head trends west-northwestward 1¼ miles and thence north-northwest 3 miles to the entrance of a creek trending irregularly nearly 2 miles in a north-northwest direction. About 2 miles farther to the northward is the mouth of a small stream flowing from the northward and westward. From the mouth of this stream the coast extends northward 5 miles to the entrance of a creek, having a small arm trending to the westward and a larger one to the southward; 2 miles southward from this creek lies Paddy Islet, about half a mile off the coast. The coast, consisting of a slightly curved sandy beach, next trends northward for 6 miles to St. Helen's Point.

Falmouth, a post town and telegraph station, is situated at the northwest end of the creek, 5½ miles northward from St. Patrick Head. Population is about 75.

Signal station.—There is a signal station at Falmouth, and communication can be made by the international code; it is connected by telegraph.

Maurouard Isle (lat. 41° 21′, long. 148° 21′), which lies south 5 miles from St. Helen's Point, and a little more than a mile from the shore, is nearly 1,500 yards long, east and west, with a rocky reef extending from it to the southward. Some fresh water has been found on the isle, and in case of absolute necessity a vessel might anchor in 18 fathoms between it and the shore. Between Mourouard Isle and St. Helen's Point there are depths of 16, 11, and 5 fathoms within ½ mile of the shore.

Aspect.—From Cape Tourville to Cape Lodi the coast is rocky and barren, but toward St. Patrick Head it appears to be well wooded, and, rising higher near St. Helens Point, presents several remarkable points of a pyramidal shape in the interior, the three most worthy of notice within 10 miles of the coast being Lyne Sugarloaf, 1,777 feet high, 8 miles westward from Cape Lodi; Mount St. John, 2,550 feet high, 10 miles west-southwestward from Long Point; and Mount Nicholas, 2,812 feet high, about 10 miles to the westward of St. Patrick Head. There are more lofty mountains in the interior, St. Paul Dome, 11 miles westward from Mount St. John, being 3,368 feet, and Ben Lomond, about 11 miles northwest from the Dome, being 5,010 feet high.

St. Helens Point is the north end of a long and comparatively narrow tongue of land, with a continuous ridge of hills on it, extend-

175078°—20——25

ing in a north-northeasterly direction to **Bare Top Hill,** which at 1,340 yards within the northeastern extremity of the point, rises to the height of 250 feet. The point from Bare Top Hill to its northeastern extremity· is about 1,500 yards broad, and thickly fringed with rocks, none of which appear to extend beyond 200 yards from the shore, except from the southeastern extremity of the point, whence a reef of rocks extends about 2 miles to the southeast; there are depths of 60 fathoms a mile off the end of this reef, and 66 fathoms 4 miles farther to the eastward.

George Bay is an extensive harbor with a shallow entrance on the west side of the long tongue of land just noticed, with a wide, deep approach from the sea, between St. Helens Point and the southern part of Grant Point to the northwest of it, about 1¼ miles apart.

Depths.—The depths in George Bay are from 10 to 12 fathoms, but in the entrance there are only 8 to 9 feet on the outer bar and 5 to 7 feet on the inner bar.

Grant Point (lat. 41° 15', long. 148° 21'), the outer northwest point of the entrance of George Bay, is a rocky projection distant 2 miles, north-northwestward from Bare Top Hill. The land rises from Grant Point to the height of 1,203 feet at Mount Pearson, which lies westward, distant 4½ miles from the point. Elephant Rock, which lies a quarter of a mile northeastward from Grant Point, is the southeastern of a cluster of small islets and rocks, altogether not exceeding 400 yards in extent.

Entrance to George Bay.—From Grant Point the shore trends south-southeast nearly ½ mile to the inner northwest point of the entrance, which is 1¼ miles wide between this point and the northwestern extremity of St. Helens Point.

From the northern extremity of St. Helens Point a very broken rocky shore extends southwestward 1 mile, whence a low, smooth shore trends westward ⅓ mile to Blanche Point, which lies nearly 1 mile westward from Bare Top Hill.

Granite Rock (lat. 41° 16', long. 148° 20'), **Dora, and Clerk Points.**—From the inner northwest point of the entrance the irregular rocky west shore extends southward a little over 1 mile to Granite Rock Point, whence a more uniform shore trends southwestward nearly ¼ mile to Dora Point, and then ¼ mile farther in the same direction to Clerk Point, which lies ⅓ mile westward from Blanche Point.

South and Middle Shoals, which are practically one bank, occupy the greater part of the southeastern side of the entrance. They have irregular depths of 1 to 5 feet of water on them.

North Shoal.—The shore for about ⅓ mile northward from Granite Rock Point is fronted by North Shoal, which has 2 to 5 feet water, and extends ¼ mile from the shore, terminating in a narrow irregular spit, with 4 and 5 feet water 800 yards northeastward from Granite Rock Point.

Outer bar.—The channel between North and Middle Shoals, which is the principal passage into George Bay, is from 120 to 270 yards wide, with only 8 and 9 feet water on the bar across its entrance. This bar appears to be permanent, as its state when surveyed in 1862 agreed with a report of 1853 and with earlier records. From the bar the depth of water increases to 4¾ fathoms in the channel 200 yards southward of Granite Rock Point, but rapidly decreases toward the inner bar, off Dora Point.

Beacons and buoys.—Beacons are placed on Granite Rock Point which in line lead over the bar for those locally acquainted.

The channel within is marked by buoys or stakes, white on the west side and black on the east side, not to be depended on, and would be only available for small craft locally acquainted.

The east shore is low between Blanche Point and Atkins Point 1½ miles south-southwest from it, and forms two bights, separated by Pelican Point, a low sandy spit projecting 400 yards to the southwest and south. The northern bight is filled by shoals and sandbanks, with 1 to 8 feet water between them; and the southern bight is occupied by a sand and mud flat.

Horseshoe Bank, with 1 to 5 feet water, is a continuation of the South Shoal, extending ⅓ mile southwestward from Blanche Point, and is one of the great obstacles to the navigation of George Bay. It occupies nearly all the entrance opposite Blanche Point, leaving only a narrow channel between it and the northwestern shore.

Inner bar (lat. 41° 17′, long. 148° 20′).—The channel from Dora Point is uncertain; sometimes it takes a former direction, known as Glover Channel, at others by the west shore, and when surveyed in 1862 it passed between the northwest spit of Horseshoe Bank and some small patches close off Dora Point, the channel being there about 70 yards wide, with a bar on which there was from 5 to 7 feet of water.

Thence the channel gradually increased to 150 yards in width southward of Clerk Point, with depths of 1½ to 2¼ fathoms; but at 100 yards northward of this point a spit, with 4 feet water, projects from the west shore to within 60 yards of Horseshoe Bank.

On the east side of the northwest spit of Horseshoe Bank there was in 1862 a blind channel 150 yards wide, with depths of 4¼ to 1¼ fathoms, extending 400 yards southward into the bank, beyond which distance there were only from 2 to 4 feet water for more than 200 yards in that direction. This, however, in 1863 appears to have become the channel, as it then crossed the Horseshoe Bank.

The west shore of George Bay from Clerk Point forms a bight extending 1.4 miles southward to a projection ¼ mile westward of Atkins Point. This bight, which is 800 yards deep, with several ledges of rocks along its northwest shore, is filled with mud flats covered at half-flood, the east edge of which, from Clerk Point, trends southward for a distance of 1¼ miles to 200 yards off the south point of the bight, whence the outer edge of the west mud flats sweeps round to the shore, about ⅓ mile southwest of the point.

The outer edge of these flats is steep and irregular, except 300 yards northwest of Pelican Point, where a 5-foot spit projects about 100 yards. There are several long narrow ditches running nearly north and south through these flats.

The main channel into George Bay from Clerk Point is bounded on the west side by the edge of the mud flats just described, and on the east side by the Horseshoe Bank, the edge of the shoals northward of Pelican Point, and the sand and mud flats thence to Atkins Point. From Clerk Point the channel increases to 250 yards in width, abreast of the south Point of Horseshoe Bank, with from 13 to 7 and 9 feet water; between Pelican Point and the 5-foot spit to the northwest of it, the channel is nearly 200 yards wide, with from 2 to 4 fathoms water. From Pelican Point to Atkins Point the channel is generally about 250 yards wide, with an average depth of 3 fathoms in the fairway; there are depths of 3 fathoms within 80 yards of Pelican Point, and 4 fathoms close to Atkins Point.

The shores.—From a steep point on which stands the constable's house, 300 yards south-southwestward from Atkins Point (lat. 41° 19′, long. 148° 20′), the east shore trends southward for nearly ½ mile to the northern extremity of the southeast bight of George Bay, which extends thence south-southwestward 1¼ miles to a small islet, close to a projecting bend of the shore, nearly ½ mile southeast of which is a small lagoon. This bight is bordered by flats, which from its northeast point extends westward 1,600 yards; the western portion of these flats, for a distance of ¼ mile, being about 400 yards broad and divided by channels, with from 13 to 8 feet water.

The north edge of these flats is separated by a narrow channel from a bank from 100 to 200 yards broad, which, from 400 yards northwest of the northeast point of the bight, extends west-southwest 800 yards. Although the narrow channel between this bank, and the flats to the southward of it, has from 12 to 17 and 11 feet water, it appears too narrow at its western end to have any outlet in that direction. From 300 yards southward of the northeast point of the bight, the flat which borders the bight extends from 100 to 400 yards from the shore, projecting farthest from the middle of the bight.

From the projection ¼ mile westward of Atkins Point, the west shore sweeps round about 1 mile in a southwest direction to the east point of the entrance of Moulting Bay, between 600 yards and ½ mile eastward of which some rocks lie close to the shore. This forms the south shore of a hilly promontory, ¼ mile to 1 mile broad, projecting 1¼ miles from the northward, between Moulting Bay and the bay from Horseshoe Bank to Atkins Point.

The main channel from Atkins Point trends in a southwesterly direction 1¼ miles to its opening into George Bay, and is bounded on the northwest side by the south edge of the western mud flats and the shore, thence to the east point of the entrance of Moulting Bay, and on the opposite side by the bank and flats before noticed. The channel is 150 yards wide abreast of the constable's house, with a depth of 4 fathoms close to the point on which the house stands; thence it increases to ¼ mile in width ⅓ mile farther to the southwestward, between which and its opening into George Bay it varies from nearly 200 yards to ¼ mile in width; with irregular depths of 10 feet to 5 fathoms, the bottom being sand throughout the channel, from the entrance into the bay.

The east and central portion of this extensive landlocked harbor of George Bay, contains, independently of its southwest and north arms, an area of nearly 1 square mile, with regular depths of 5 to 12 fathoms, over a bottom of mud.

Southwest Arm.—From the small islet at the southwest extremity of the southeast bight of George Bay the southeast shore curves in a southwest direction nearly 1 mile to the south bight of the arm, which extends 1,800 yards east and west, and is ⅓ mile deep. About ½ mile southwest from the east point of the bight is a cliffy projection, on which are some farm buildings. From the west point of this bight the shore curves northwest ¼ mile to a projecting point, and thence nearly ½ mile in a westerly direction to Jasons Gates Bridge. Some rocks above water extend about 100 yards from the projecting point. From the islet to the bridge there are depths of 3 fathoms within 200 yards of the shore.

The land behind this shore is hilly, and ⅓ mile southeast of the east end of the south bight rises to the height of 180 feet; there are several small streams in the valleys between these hills, and some springs close to the beach, about ½ mile to the southwest of the small islet before mentioned.

McDonald Point (lat. 41° 19′, long. 148° 17′), the north point of the southwest arm of George Bay, is a sandy projection lying southwestward 1 mile from the east entrance point of Moulting Bay, and forms the southeast side of the mouth of George River.

The northwest shore of the southwest arm of George Bay, from McDonald Point, extends irregularly ½ mile in a southwesterly direction to a small islet in the mouth of a narrow creek, trending about

northwest 800 yards to the foot of a little ridge of hills which extend thence nearly 1 mile in a west-northwest direction.

St. Helens, on the north side of the southwest arm, is a post
town and telegraph station. Coaches run to St. Marys and Scottsdale railway stations; and a small steamer trades every alternate
week to Hobart or Launceston. A large amount of tin ore from
Blue Tier and Thomas Plain is shipped here. Population is
about 500.

Oyster Patch (lat. 41° 20′, long. 148° 16′), about 100 yards in
extent, with 6 feet water on it, lies 1 mile southward from McDonald
Point. There are from 2 to 5 fathoms water close round the patch,
and 3 and 4 fathoms between it and the flat 200 yards to the northward of it. There are some stones on the mud flat 670 yards to the
northeast of the patch.

The navigable water in the southwest arm of George Bay is ¼
mile wide at its entrance, whence it varies from 800 to 1,500 yards
in width to within 1,340 yards of the bridge. There are depths of
11 and 12 fathoms across the entrance, with similar depths up to
Oyster Patch, and from 9 to 3 fathoms within 200 yards of the south
shore and of the mud flats, the bottom throughout being mud. The
navigable water in the western corner of the arm for about 1,340
yards outside the bridge is 500 yards broad, with 7 and 8 fathoms
close off the rocks which project from the south shore, whence the
depths gradually decrease toward the bridge, with 3 fathoms within
100 yards of the south shore and of the mud flats on the north side.

George River is 200 yards wide at its entrance, between McDonald
Point and the low point to the northwest of it, and fronted by small
banks, extending ¼ mile to the northward. The river flows from
the northwestward to about 200 yards northward of the ridge before
mentioned, and thence trends eastward to the entrance. From the
entrance to 1,500 yards above it, where the river is only 50 yards
wide, the depth of water is from 1 to 3 feet.

Moulting Bay (lat. 41° 18′, long. 148° 18′), which is the north
arm of George Bay, is 1,500 yards across east and west at its
entrance, whence it extends northward 1½ miles. From the east
entrance point to another projection ½ mile north of it the shore is
steep with 2½ to 5 fathoms about 50 yards from it. But with this
exception the shores of the bay appear to be inaccessible, especially
to the northward and westward, on account of a continuous mudflat the edge of which, from the north point of the steep east shore
just mentioned, extends 400 yards from the shore 1,500 yards farther
to the northward. From the north and west shores the mud flat
extends from 200 to 500 yards, and from the west entrance point it
projects half way across toward the east shore, leaving an entrance

800 yards wide, with 2¼ to 9 fathoms water. Within the entrance there is a space 1 mile long, north and south, and 1,500 to 500 yards wide, with 5 to 2 fathoms water on a bottom of mud. The north and northwest shores of Moulting Bay are low, and intersected by several small streams. On the west side of the entrance the land is hilly and rises to a summit 700 feet high, about 1 mile westward from the west point of the entrance of the bay.

Directions.—Although there is a sufficient space in George Bay (lat. 41° 17', long. 148° 22'), for a fleet of the largest ships, it is only available for vessels of light draught on account of the narrow intricate channel leading into the bay from its outer entrance, and the bars which obstruct the channel.

As the outer entrance of George Bay is exposed to the northward and eastward, gales from between these points may naturally be expected to cause heavy breakers upon the outer bar, when it would appear unsafe for any craft to attempt to enter, and none other than those locally acquainted should attempt to enter without a pilot.

The directions which applied some years ago are as follows: In a small craft, however, adapted to the depth of water on the bar, having with smooth water and a commanding breeze, approached near enough to the entrance to clearly distinguish Granite Rock Point, bring the beacons on the point in range, which lead over the bar between Middle and North Shoals, then steer to pass about 100 yards southward of the point; thence, keep at the distance of about 100 yards along the west shore, between Granite Rock and Dora Points, passing between Dora Point and Horseshoe Bank. Having cleared the small patches close off the latter point, take the channel that may be the most practicable one, either along the west shore; or through Horseshoe Bank, which a stranger should ascertain before passing Dora Point. A portion of the channel is marked by white buoys or beacons on the starboard hand and black buoys or beacons on the port hand, but they are not to be depended on.

From the southern extremity of Horseshoe Bank steer a mid-channel course for Atkins Point, keeping midway between Pelican Point and the spit to the westward of it; and after passing close to Atkins Point, and that under the Constable's House to the southwest of it, steer in mid-channel between the north shore, and the shoals immediately to the southward of it, 1¼ miles, which will clear the channel into George Bay.

Tides.—It is high water, full and change, in George Bay at 9 h. 42 m.; springs rise 3 feet, neaps 2 feet.

The east coast of Tasmania from Grant Point (lat. 41° 15', long. 148° 21'), curves to the northwestward and northward about 3½ miles to a point close off which lies Sloop Rock. At 1,500 yards to the westward of Grant Point is the entrance of a lagoon which branches to the southwestward and westward.

Between the point abreast of Sloop Rock, and another projection 2¼ miles northward of it, the coast forms an indentation ¼ mile deep, with a small double inlet in its southwest corner, and a sunken rock close off its north point. From the north point of this bay the coast sweeps around north-northwestward 2 miles to a small inlet, and thence extends northward 5 miles to the entrance of Anson Lake in the southern part of the Bay of Fires.

The Gardens (lat. 41° 10′, long. 148° 18′) are some sunken patches which lie near the coast between the north point of the indentation just noticed and the small inlet 2 miles to the northward of it, but the outermost of these dangers does not appear to extend beyond a mile from the shore.

The entrance to Anson Lake is so small as not to be discernible from seaward, and scarcely permits boats to enter at high water. Landing can seldom be effected outside the entrance. At most times there is a dangerous, heavy surf rolling on to the beach.

Bay of Fires.—Between the entrance to Anson Lake and Eddystone Point, 4½ miles to the north-northeastward, is the Bay of Fires which affords good shelter when the wind is steady from the westward, but should be left immediately there is a lull, as the wind often at the termination of a gale shifts suddenly to the southeast. The anchorage is in 10 fathoms, near the center of the bay. The water shoals suddenly near the beach.

Eddystone Point and north coast of Tasmania have been described.

Soundings.—From a depth of 60 fathoms at 5 miles off St. Helens Point the soundings gradually decrease to 36 fathoms at 3 miles off Eddystone Point. Between this line of soundings and the shore, for about 5 miles southward from Eddystone Point, the soundings decrease to 15 fathoms at about 1 mile from the shore.

CHAPTER VIII.

AUSTRALIA—EAST COAST, CAPE HOWE TO PORT JACKSON.

NEW SOUTH WALES.

The coast.—From Telegraph Point (lat. 37° 33′, long. 149° 55′), ¼ mile northwest of the north point of Gabo Island, the coast, which consists of bare white sand hillocks, the highest being 143 feet high, trends northeastward 3½ miles to a sandy point, with a ledge of dry and sunken rocks extending ½ mile to the southward from it, on which the sea nearly always breaks; this point may be mistaken for Cape Howe, as its bare sandhills make it much more conspicuous than that cape. Thence the coast trends northeastward 1¼ miles to Cape Howe.

Reef.—At from ¾ to 1 mile south-southwestward from the above-mentioned sandy point is a rocky shoal with 19 feet water on its shoalest part. This reef is ¼ mile outside a line drawn from Cape Howe to Gabo Island Lighthouse and 2 miles from the former; it breaks in heavy weather. A red sector of the Gabo Island light shows over the reef.

Cape Howe is a low point, composed of stones and sand, covered with ti trees; the land to the westward is almost level for 4 miles to the foot of Howe Hill and the Howe range of mountains which extends nearly north-northwestward 5 miles from that hill.

At 3 miles to the north-northwest of the Howe Range is another range, called Table Hills, attaining an elevation of 1,786 feet, so that the whole aspect of the country about Cape Howe is that of a mountainous district.

A geodetic station on nearly the highest part of the point 200 yards inland marks the boundary between Victoria and New South Wales; is 148 feet above the level of the sea.

Howe Hill or **Cunawurræ**, 1,300 feet high, is conspicuous, rising abruptly from the adjacent lowland, its southern aspect exhibiting a steep fall and its summit being shaped like a haystack. From Howe Hill a range of round and flat-top hills extends in a northerly direction to Wonboyn River. On the southwest side of Howe Hill is a lagoon of fresh water.

389

Boundary.—About 2¼ miles to the northward of Howe Hill there is a deep cutting through the thick timber, open from the, eastward; this is the boundary line cut by the Government of Victoria, dividing that State from New South Wales.

The coast.—From Cape Howe a rocky coast trends north-north-westward 3¾ miles to Black Head, thence in the same direction 5 miles to some cliffs of granite and porphyry, which sweep round in a northerly direction 5 miles to the southern extremity of a long sandy beach in Disaster Bay. A barren heath extends from Cape Howe to the cliffs, but these are surmounted by steep grassy hills, bearing gum, oak, and other trees.

The coast from Cape Howe to the south-bluff of Disaster Bay consists of steep rocky points, with a few sandy indentations.

Disaster Bay, situated about 12 miles northward from Cape Howe, is 5 miles broad at its entrance, 3½ miles deep, terminating in a curved sandy beach 3¼ miles in length. In the northwest part of Disaster Bay, and about 3½ miles from the rocky bluff forming the south point of the bay, is Bay Cliff, the south head of Wonboyn River, which is only accessible to boats in fine weather; the narrow mouth of this river is sometimes fordable for cattle, but the sand is continually shifting.

Anchorage with northeast winds may be had in from 13 to 17 fathoms water near the northern shore, with no dangers and a cliffy coast.

Green or Bundooro Cape (lat. 37° 15′, long. 150° 04′), lying 15 miles northward from Cape Howe, is a smooth, low point, covered with grass, dotted with patches of small bushes, and sloping gradually to the eastward, from an elevation of 501 feet at 2½ miles inland; its coast to the northward is low and rocky; there are depths of 16 fathoms within ¼ mile from the cape, and no outlying dangers.

Light.—A flashing white light, 144 feet above water, visible 19 miles is shown from a white octagon lighthouse, 80 feet in height, on the extremity of Green Cape.

Signal station.—There is a signal station at Green Cape, and communication can be made by the International code, and by Morse lamp at night. It is connected by telegraph.

The coast.—From Green Cape the coast trends northwestward 3 miles to Bitangabee Creek (with 9 feet of water), which is a good harbor for small vessels, and has a jetty on its south side; thence north-northwest 4¼ miles to Mowwarry Point; and then northwestward 3½ miles to Red Point. The coast from Green Cape to Red Point is bold, with rocky points and small sandy beaches having depths of 15 to 20 fathoms within ½ mile of the shore, the land along it being generally barren heath with good grass on the points; the back country is hilly, and thickly wooded. Haycock Hill, 922 feet above the sea, and westward 2½ miles from Mowwarry point, is the highest of these hills; but the most elevated land in this locality is Mount Imlay, a remarkable and densely wooded peak, 2,910 feet in height, situated 16 miles to the westward of Green Cape.

Soundings.—There are depths of 50 fathoms about 6 miles off Green Cape, 48 fathoms 7 miles off Mowwarry Point, and 43 fathoms 8 miles northeastward of Red Point, with gradually decreasing depths toward the coast, along which there are from 31 to 22 fathoms at 1 mile from it.

Mowwarry Rock, 80 feet high and shaped like a haystack, is conspicuous from the southward; it lies close to Mowwarry Point.

Red or Burrowrajin Point (lat. 37° 06′, long. 149° 58′), the south head of Twofold Bay, lies 3½ miles northwestward from Mowwarry Rock, and may be known by a white stone tower named Boyds Tower on it 66 feet above the sea.

Twofold Bay is 2¾ miles wide between Red Point and Worang Point, the north point of the bay, and from a depth of 20 fathoms midway between the points the bay extends 4 miles westward to the head of Nullica Bay.

Depths.—The depths in Twofold Bay are 10 to 15 fathoms, shoaling gradually toward the shore.

The entrance is free from dangers, with the exception of a rock with 5 fathoms water, on which the sea breaks only in bad weather from the eastward, lying 800 yards northward from the white stone tower on Red Point; 300 yards from the tower in the same direction is a rock covered at high water, on which sea always breaks. Between the two rocks there are 7 fathoms of water. The west end of the long sandy beach at the entrance to Walker River kept in sight clear of Jews Head, bearing 257° leads north of the 5-fathom rock, in a depth of 14 fathoms.

Aspect.—The land about Twofold Bay appears more mountainous than the coast immediately north or south of it; the hills which are either round or sharp topped, lying in clusters, and gradually increasing in elevation to the westward. Mount Imlay is sometimes obscured, but when seen is an excellent mark for entering the bay.

The south shore of Twofold Bay, between Red Point and Honeysuckle Point, 1,500 yards to the westward of it, forms an exposed bay, having 4 fathoms water close to its points and 8 to 11 fathoms between them.

From Honeysuckle Point a bold cliffy shore extends west-southwestward nearly ⅓ mile to Jews Head, off the northwestern extremity of which and in a direction toward Lookout Point is a rock with 3 fathoms water, 300 yards from the shore; the white tower on Red

Point, bearing 127°, leads clear of the rock. From Jews Head the coast trends southwest 600 yards to **Manganoo** Point, the northeast Point of East Boyd Bay.

East Boyd Bay (lat. 37° 05', long. 149° 57'), which appears to afford the most sheltered anchorage for large vessels on the south side of Twofold Bay, extends from **Munganoo** Point southwest nearly 1 mile to Brierly Point, and is ¼ mile deep. A bank with 12 to 9 feet water on it extends about 200 yards from the shore around the bay to within ¼ mile of Brierly Point, from which the bank projects ⅓ mile to the northward.

East Boyd, on the east side of this bay, about ⅓ mile southward of Munganoo Point, was a whaling station, and contains a few houses, probably uninhabited, as the whaling industry is practically dead.

Anchorage.—Small vessels may anchor in East Boyd Bay in from 3½ to 2½ fathoms, sandy bottom, by bringing Worang Point and Munganoo Point in line about 16°; large vessels find shelter from south and southeast winds farther out in 5 or 6 fathoms, and smoother water with easterly winds than on the opposite shore in Snug Cove.

Kiyerr Inlet, at 300 yards to the southward of Brierly Point, is a shallow opening only a few yards wide, forming the mouth of a lagoon, separated from the south shore of Twofold Bay by a low narrow barrier, extending from Kiyerr Inlet to a point 800 yards southeastward of Torarago Point, on which is Mootries House.

The lagoon, which is full of low islets and shoals, forms the estuary of the Towamba or Walker River, an inconsiderable stream winding into Twofold Bay from the southward, having 6 feet at low water on the bar. This river is sometimes frequented by small craft to ship potatoes and other produce; whale oil used to be shipped here, the boiling-down establishment being situated in a little bight south of Brierly Point, named Kiyerr, beyond which the river is only navigable for boats.

Between Brierly Point and Whale Spit, which dries at low water and projects about 550 yards northeastward from Torarago Point, a bank having from 6 to 12 feet water on it, extends 800 to 400 yards from the low narrow barrier just noticed, and about 150 yards from Whale Spit. Red Point open north of Jews Head leads clear of Whale Spit.

Nullica Bay, which forms the western bight of Twofold Bay, is 1¼ miles wide between Torarago and Oman Points, with 4 to 5 fathoms water between them; the bay shoals gradually to 12 feet for nearly a mile to the westward, 300 yards from the beach.

Between Whale Spit and the mouth of Myruial Creek, situated about 1 mile westward of Torarago Point, is a sandy bay, ¼ mile deep, close behind which are the remains of the township of West Boyd, consisting of a few deserted houses.

A flat, with from 6 to 18 feet water on it, extends 400 to 800 yards from the shore between Whale Spit and Myruial Creek.

From Myruial Creek the northwest shore curves around northward and eastward about 1¼ miles to Oman Point. A ledge of rocks extends 300 yards from the north point of the mouth of the creek; thence to Oman Point, shoals with from 6 to 13 feet water on them extend 200 to 400 yards from the shore, except at ¼ mile northward of the creek, where there are 3 fathoms close to the land.

The shore from Oman Point trends northeastward nearly ½ mile to Cocora Point, which forms the western extremity of Snug Cove. On the east side of Oman Point a shoal, with 12 feet water on it, projects about 300 yards from the shore; but there are depths of 4 fathoms between the shoal and Cocora Point.

Lookout Point is a rocky peninsula ⅛ mile broad, with a few stunted trees on its summit, and presenting a steep cliffy aspect to the southward and eastward; the point is situated about midway on the north shore of Twofold Bay, from which it projects toward the southeast about 1,340 yards, and is connected with the mainland by a low narrow spit of sand, 100 yards wide, forming two small bights or coves, the southern of which is Snug Cove.

Light (lat. 37° 04', long. 149° 56').—A fixed red light with white sector visible 7 miles, 144 feet above water is shown from a white octagonal tower on the southern extremity of Lookout Point.

Signal station.—There is a signal station at the point, connected by telegraph. Communication can be made by the international code.

Storm signals are shown from a mast near the lighthouse.

Rocks.—About 300 yards eastward from the lighthouse is a rock awash at low water, and 120° distant about 470 yards from the lighthouse is a sunken pinnacle rock, with 3¼ fathoms water over it; on the former the sea nearly always breaks, but on the latter only in bad weather.

Eden (lat. 37° 04', long. 149° 56') is a small settlement situated at the back of Lookout Point. The town is built on the slopes and valleys between two hills which jut out into the bay, dividing it into two parts. A road from Eden gives access to the Monaro district and passes through Cooma, the terminus of the southern railway, which is 108 miles from Eden. The principal trade consists in the shipment of live stock to Hobart, pigs and bacon to Melbourne, and wool and hides to Sydney. The Illawarra Co.'s steamers trade between Eden and Sydney, calling at intermediate ports. The population is about 700. There is a telegraph station here.

Light.—A light is shown from the outer end of the jetty, off which is a warping buoy.

Lifeboat.—There is a lifeboat and a life-saving rocket apparatus at Eden.

Town.—The principal part of the town is from ½ mile to 1 mile from the jetty and a hill intervenes so that the lights of the township are not visible from Snug Cove.

Snug Cove, the anchorage off the township of Eden, extends nearly 1,340 yards east and west between Cocora Point and the lighthouse on Lookout Point, but is not more than ¼ mile deep, and is bordered by a shoal, which extends from 100 to 200 yards from the shore. There are depths of 5 to 3 fathoms in the cove, bottom soft clay and sand, where two or three small craft can lie landlocked off the jetty, by shutting in Red Point with the south extremity of Lookout Point.

Larger vessels anchor in 6 fathoms, about ½ mile westward of the lighthouse, or in 4 fathoms about 300 yards farther to the northwestward, with the lighthouse bearing 100°, and Eden Jetty, on the east side of the cove, 55°. This anchorage is, however, exposed to the heavy swell of an east or southeast gale, only partially broken by the heads, but during northeast winds it is snug.

Yallumgo Cove is a small inlet on the northeast side of the isthmus which connects Lookout Point with the mainland; some dry and sunken rocks extend across the entrance of the cove, and a reef borders the shore immediately to the northward of it.

All the points which project into Twofold Bay are the terminations of thickly timbered ranges of hills, with numerous creeks and lagoons between them, most of which have salt or brackish water.

Calle-Calle Bay (lat. 37° 04′, long. 149° 56′), the exposed northern bight of Twofold Bay, is 1½ miles wide, northeast and southwest, between Lookout and Worang Points, and is nearly 1¼ miles deep, with from 13 to 12 fathoms across its entrance, from the middle of which the depths gradually decrease to 3 fathoms at the head of the bay, close off the mouth of Curalo Lagoon, a narrow shallow opening, at times apparently blocked up.

Calle-Calle Bay affords shelter from the northeast winds, but it is not a desirable anchorage, being open to southeast and southerly winds, and almost always disturbed by a swell.

Curalo Lagoon is an extensive sheet of salt water which from its entrance trends, with gradually increasing width, about 1 mile to the southwestward, where it is ½ mile wide, with a branch extending into the thickly timbered land to the northwestward. This lagoon, abounding with excellent fish, is only separated from the northwestern shore of Calle-Calle Bay by a low narrow tongue of land.

Pilots.—On a vessel off Twofold Bay making the usual signal she will be boarded by a pilot as soon as practicable.

The pilot, who is also harbor master, and a boat's crew are stationed in the bay.

Supplies.—Wood in abundance can be procured in all parts of Twofold Bay. A small supply of brackish water may be obtained from a well about 200 yards from Eden Jetty. The pump can supply about 7 tons a day. This water is unfit for drinking and should only be used for boilers in case of urgent necessity. Fresh provisions can be obtained at reasonable prices. The ponds and lagoons, which are at the back of most of the beaches are frequented by ducks, teal, herons, redbills, and some small flights of curlew and plover, but these are rapidly being trapped and shot for the Sydney market; the bay appears to be well stocked with fish.

Directions.—**Twofold Bay** (lat. 37° 04′, long. 149° 56′) is so open to seaward and so free from detached dangers that there is very little difficulty in entering it; Mount Imlay, bearing 238°, leads midway between the entrance points. On approaching the bay, take care to avoid the sunken rocks which lie to the northward of Red Point, and having distinctly made out the lighthouse and other objects, Torarago Point kept on a 246° bearing leads in through the middle of the bay, when anchor either in Snug Cove off the township of Eden or in either of the anchorages off East or West Boyd, according to the prevailing wind, or as most convenient.

When running for Twofold Bay in bad or thick weather after dark the light must not be depended on for making the place, as it is difficult to distinguish in such weather and the lighthouse should not be approached within ¼′ mile.

In entering Snug Cove with a southerly wind, care must be taken to shorten sail in good time and to drop the anchor in 6 or 5 fathoms before Red Point comes on with the southern extremity of Lookout Point, and in veering cable the lead should be hove over the stern of the vessel.

In rounding Lookout Point and entering Snug Cove with much sea on, the red tower amongst the trees at the back of the West Boyd town, kept open north of Torarago Point, 238°, leads in 11 fathoms ¼ mile south of the rocks.

For shelter in Twofold Bay from a southeasterly gale the anchorage off East or West Boyd, on the south side of the bay, is far preferable to Snug Cove, and it is by no means certain that it is not so even with an easterly gale. The southern part of Nullica Bay, off West Boyd, is a very convenient anchorage and was the constant resort of coasters.

Tides.—It is high water, full and change, in Twofold Bay at 8 h. 15 m.; rise 5 to 7 feet.

The coast.—Mewstone is a small rock 20 feet high lying 200 yards southeast of Worang Point; it is steep-to, but there is no passage inside. From this rock the coast trends north-northwestward 2¼ miles to a point having close in front of it Bullara or Lennard Island, which is flat, with a reef projecting a short distance from its northern extremity. From this island the coast curves northwestward 2 miles to the red Quoraburagun Cliffs and thence northward 2¼ miles to Ioala Point, which is connected by a reef of rocks with Haystack Rock, a remarkable round-shaped bowlder, 50 feet high, lying close off the point. A succession of rocky points from Haystack Rock sweeps round northwestward and westward 1 mile to the entrance of Panbula River. The most elevated land between Twofold Bay and this river is Mount Robinson, a long hill, 1,127 feet high, at 4 miles northwestward of Worang Point. The land is everywhere thickly wooded and rises gradually to Mount Robinson.

Merimbula Bay is a sandy indentation lying between Ioala and Merimbula Points; it is about 1¾ miles deep, with 16 to 17 fathoms water, shoaling gradually to 8 fathoms within ¼ mile of the beach.

Hunter Rock (lat. 36° 56′, long. 149° 57′), with a depth of 3½ fathoms, on which the sea seldom breaks, lies north, ½ mile from Haystack Rock. The small sandy beach between two conspicuous bluffs on the west side of the entrance to Panbula River, kept open of the northern extremity of Ioala Point, leads to northward of Hunter Rock; the extremities of Ioala Point and Ballara Island in range lead to the westward; and Haystack Rock on with Quondolo Red Cliff in the sandy bight south, leads to the eastward of it. There is a channel between Hunter Rock and Ioala Point, and the northern of the two bluffs well shut in by Ioala Point leads through it.

Panbula River discharges itself into the southwest corner of Merimbula Bay; and is accessible only for boats or small craft immediately after floods, which sweep the bar away. The river is about 200 to 400 yards wide and trends southwestward nearly 2 miles into Panbula Lake, which is about 1½ miles in extent, with several small streams flowing into it. The village of Panbula, at about 2½ miles to the westward of the entrance of the river, is situated near the Walker branch, which flows into the lake from the westward, between Melton Hill to the northward and Mowbray Range to the southward of it. The population of Panbula is about 600. There is a telegraph station here.

Anchorage.—Good anchorage sheltered from southwest and southerly winds may be obtained off the entrance of Panbula River in 6 fathoms, with the north part of Ioala Point bearing 100°, and Merimbula Point 21°.

Tides.—It is high water, full and change, at Panbula River at 9 h. 00 m.; springs rise, 4 to 6 feet.

Merimbula Creek and Lake.—From the bluff forming the west head of Panbula River the coast is a long sandy beach, curving in a northerly direction for 3 miles to Merimbula Creek, which extends from the lake to the northwest corner of the bay; there are at times a depth of 7 feet at high water on the bar of this creek. Merimbula Lake is somewhat of a triangular form and about 1½ miles in extent, its east side being separated from the sea by a narrow sandy flat covered with scrub.

No directions can be given for entering Merimbula Creek; it can only be recommended for boats.

Merimbula is a township situated on Merimbula Lake, with a population of about 150; there are several storehouses and a pier situated nearly at the mouth of the lake; good roads extend back into the country, along which wool and other produce is carted down for export to Sydney. There is a telegraph station here, and biweekly communication with Sydney by steamer from the jetty.

Merimbula Point (lat. 36° 54′, long. 149° 57′), projecting about a mile in a southeast direction from the north side of the entrance of Merimbula Creek, is a steep, cliffy headland, affording at a quarter of a mile off shore shelter from northeast winds, in about 6 fathoms, sand; but on the appearance of a southerly wind it must be left, as a heavy sea rolls in.

Jetty.—There is a jetty on the south side of Merimbula Point. Head and quarter mooring buoys are in position off the jetty.

The coast.—About 1 mile northward from Merimbula Point is Panbula Inlet, whence the coast extends north-northeastward nearly 2 miles to Tura Head, between which and Turingal Point, 4 miles northward from the head, is a bay 1½ miles deep, divided midway by Bournda Island, close behind which is a salt-water pool.

Wallagoot Lagoon, which lies nearly 1 mile to the westward of Turingal Point, is also salt, and is only separated from the shore of the bay by a narrow ridge, without any apparent opening.

Wolumla or Massey Peak.—From Panbula Inlet to Bournda Island the coast consists of sandstone and pipeclay cliffs, with grassy headlands and low scrubby ranges behind. The most remarkable hill behind this part of the coast appears to be Massey Peak, a thickly wooded mountain 2,660 feet high, distant 11 miles westward from Tura Head.

Tathra Head (lat. 36° 43′, long. 150° 00′).—From Turingal Point, an uneven line of granite and pipe-clay cliffs, with grassy land over them, extends 4 miles northward to Tathra Head, between

175078°—20——26

which and Wajurda Point, 2 miles northward from the head, is an exposed bay about 1,500 yards deep.

Tathra.—A small pier extends from the south shore of this bay, with moorings off it, laid down by the Government; it is visited by the steam vessel on her way to Merimbula, and when the wind is off the land, or the weather fine, by small schooners; most of the Bega district market commodities are exported from this place. Communication with the town of Bega, 10 miles inland, is by coach.

Mogareka Inlet and Bega River.—About 1,500 yards to the southwestward of Wajurda Point is Mogareka Inlet, the mouth of the Bega River, which is sometimes open, with 6 feet water on the bar. Close within its mouth, where there is a small islet, this inlet forms three branches, two of which trend to the southward, whilst the main branch winds southwest and southeast about' 2½ miles to two small islets, above which the river flows between the. ranges of hills, from the westward. Between Tathra Head and Mogareka Inlet the shore is low and sandy; but between the inlet and Wajurda Point the land is more elevated, with some rocks near the shore.

Baronda Head and Inlet.—Baronda Head is a rocky projection nearly ½ mile to the northward of Wajurda Point, and forms the north side of the mouth of Baronda Inlet, which is dangerous even for boats, being very narrow, with sunken rocks on either side of it. From Baronda head a beach extends northward 2 miles to Tanya Lagoon, and at 1½ miles northward from this lagoon is Bithry Inlet, which is not fordable.

Bunga Head.—From Bithry Inlet the coast consists of a series of small projecting rocky points trending irregularly north-northeast 3½ miles to Bunga Head, which is a steep, cliffy headland, forming the most prominent projection from the coast when seen from the northward or southward; the cliff is 200 feet high, having a peaked summit 400 feet high; several detached dry and sunken rocks fringe its base from 100 to 200 yards distant. About ¼ mile southward from the head lies Mimosa Rock. Hence the coast trends northward 1½ miles to Goalen Head, a green, smooth, sloping point with dark, rocky shores; about 1 mile to the northwestward of which is Erungona Creek. The coast from Bithry Inlet to Erungona Creek is closely bordered with dry and sunken rocks, except at about 1½ miles to the southwestward of Bunga Head, where there is a sandy beach nearly ½ mile long.

Thubbul Inlet and River.—From Erungona Creek a low sandy beach curves to the northward 1 mile to Thubbul Inlet, the estuary of the river of the same name. This inlet is narrow and fordable at its mouth, within which it is 400 to 600 yards wide, and trends about westward 1,340 yards to where the river winds into it from the northwestward. At about 2½ miles above the mouth of the inlet the water is said to be fresh.

Aspect.—The land between Mogareka Inlet and Thubbul River is generally poor, with high scrubby hills, destitute of grass. Mount Townsend or Mumbulla, 2,630 feet high, lies westward 10 miles from Bunga Head, and is the summit of a high, thickly timbered range of mountains, rising in gradations toward it from north and south; it appears round topped from some views, whilst from others it appears sharp with a nipple top; there are several peaks 1,500 to 2,100 feet high around it within a radius of 2 miles. From about 2 miles southward of this mountain, one ridge trends in a westerly direction, whilst others branch off to the southeastward and eastward, terminating at Wajurda Point and at other points of the coast between Tanya Lagoon and Erungona Creek.

Soundings.—There are depths of 65 fathoms, sand, at about 15 miles off Twofold Bay, and no bottom at 100 fathoms at the same distance off Bunga Head, from which depth the soundings decrease with some regularity toward the land. From 10 miles eastward of Twofold Bay to about the same distance to the eastward of Thubbul River, there are from 50 to 70 fathoms at an average distance of 10 miles from the shore.

Baragga Point (lat. 36° 30′, long. 150° 04′)—**Burragat Rocks.**—Baragga Point is the central of a series of small rocky points, bordered with sunken rocks, which, from Thubbul Inlet, sweep round in a northerly direction 2½ miles to a salt lagoon close to the shore, whence a sandy beach trends northward nearly 2 miles to Jerimbut Point, which has a reef of sunken rocks projecting from it, and is fronted by the three Burragat Rocks, above water.

Bermaguey Inlet.—The coast from Jerimbut Point extends north nearly 2¼ miles to a rocky projection, at 1,500 yards to the westward of which is Bermaguey Inlet, across the narrow entrance of which is a 6-foot bar, with apparently some sunken rocks close off the east point of the entrance. This inlet appears to be much encumbered by two islets or banks, lying in it, one being close within the entrance, and the other at about ½ mile farther to the westward. The township of Bermaguey has postal and telegraphic communication; population about 500. A steamer calls bi-weekly.

Jetty.—There is a jetty, with mooring buoy off it, 400 yards westward from the point.

Anchorage.—Small vessels can obtain anchorage, protected from southerly winds, under the head to the eastward of Bermaguey Inlet, in 5 fathoms water, near the jetty.

Tides.—It is high water, full and change, on Bermaguey Bar at 9 h. 20 m.; rise 5 feet.

The coast.—For the first 2 miles northward of Thubbul Inlet the country is good for cattle, but thence to Bermaguey Inlet there are thick scrub and forest. From Bermaguey Inlet a low sandy beach, backed by a swamp, curves to the northward 2½ miles to the south part of Murunna Point. Close behind the beach, ¼ mile to the westward of the north part of Murunna Point, is Walluga Lake, the water of which is salt. Thence a sandy beach, backed by good pasture, with plenty of fresh water, extends north-northeastward nearly 3 miles to a double point, and 1½ miles beyond is Cape Dromedary. Shoal water extends over ½ mile off shore at 1¼ miles northward of Murunna Point.

Mount Dromedary, the most remarkable object on this part of the coast, and visible in clear weather from a distance of 60 miles, is a double mountain 2,706 feet high, which, from its figure, was named Mount Dromedary; it stands 4 miles back from the coast, with Ajungagua Hill, 702 feet high, between the mount and the cape of that name.

Cape Dromedary (lat. 36° 18′, long. 150° 09′), which lies 5¼ miles eastward from the mountain of the same name, is the eastern of a series of granite and ironstone points, extending from 1½ miles south-southwestward of the cape to Barbunga Lagoon, situated at 1¼ miles north-northwest from it. Several rocks lie along these points, and between Cape Dromedary and Barbunga Lagoon reefs of dry and sunken rocks project some distance from the shore.

Nugget Point.—From Barbunga Lagoon a sandy shore, with some sunken rocks close to it, trends north-northeastward 1¼ miles to Nugget Point, from which a succession of granite and ironstone points and small sandy bays extend north-northwest nearly 3½ miles to Wagonga Inlet. Nugget Point and the other projections between it and Wagonga Inlet are bordered by reefs. Between Barbunga Lagoon and Wagonga Inlet there is good grazing country along the headlands, but it is intersected by salt and brackish lagoons, and some parts are thickly wooded.

Montagu or Barunguba Island (lat. 36° 15′, long. 150° 14′), situated east-northeastward 3¾ miles from Nugget Point, may be called two islets, being divided near the center by a deep rocky chasm, through which the sea breaks with heavy easterly winds; it is about 1 mile long in a north and south direction and ⅓ mile broad.

The south part of the island, 210 feet high near its center, is of granite formation, with long rank grass and scrub growing on it, abounding with rabbits; the soil appears to be of a rich quality; its coast is rocky, and from 300 to 500 yards from the south extreme of the island, which is low and bare, there are a number of large granite bowlders.

The north part of this island, some 20 or 30 feet lower than the south, is of volcanic formation, masses of conglomerate lying about it; it is also covered with long grass.

A ridge of rocks with from 5 to 9 fathoms water on it extends from the southwestern extremity of Montagu Island southward a little more than a mile; on its south end are depths of 6 fathoms, on which the sea frequently breaks in bad weather. The ledge is steep on both sides, dropping suddenly into 13 and 15 fathoms to the westward, and 15 to 20 fathoms to the eastward. It should be avoided by vessels, particularly small coasters, in heavy weather, as there is a confused sea on the ridge.

A rock about 30 feet long, and with a depth of 15 feet over it, is reported by fishermen to exist at a distance of 1½ miles 150° from Montagu Island Lighthouse. Vessels are advised to give the position a good berth in passing.

Light.—A fixed and flashing white light, 262 feet above water, visible 20 miles, is shown from a gray, circular, granite lighthouse, 53 feet high, on the summit of Montagu Island.

Signal station.—There is a signal station at Montagu Island Lighthouse. and communication can be made by the international code, and at night by Morse lamp. It is not connected by telegraph.

Tides.—It is high water, full and change, at Montagu Island at 8 h. 30 m.; springs rise 5 to 7 feet.

Anchorage.—Small vessels sometimes anchor with easterly and southerly winds in a small bight on the west coast formed by the two parts of the island, but it can not be recommended for large vessels. Small craft unable to get off the land or fetch a safer anchorage would find tolerable shelter in this little cove by getting as close in as possible with the gap open 94°, 200 yards offshore. The bottom is irregular and rocky.

Directions.—In navigating this part of the coast, steam vessels, and sailing vessels having a fair wind, bound northward, are recommended to keep inside Montagu Island, and about 2 miles off the mainland all the way to the northward, to avoid the southerly current usually found outside; at this distance from the coast an eddy sometimes runs to the northward.

Soundings.—From 5 miles off Thubbul River to about 1 mile westward of Montagu Island the soundings range from 51 to 17 fathoms, on a sandy bottom; but at 7 miles southeastward of the island there is no bottom at 100 fathoms.

Coast.—Wagonga Inlet has a narrow entrance, sometimes accessible to small vessels, but there is generally a heavy break across it. Within the entrance this inlet extends about 1½ miles to the southwestward, with several creeks branching to the northward and southward, and an islet, or bank in the middle of it, between which and the southeastern bight of the inlet is the anchorage for such small vessels as may enter.

Wagonga is a small telegraph and postal township; the population of the district is about 150. There are two sawmills. In the vicinity of Mount Dromedary are extensive gold reefs.

The port of Narooma is situated at Wagonga Inlet. The population is about 400. A weekly steamer service is made between this port and Sydney. There is a telegraph station.

Yellow Head (lat. 36° 10′, long. 150° 09′)—**Marka Point.**— From Wagonga Inlet the coast, consisting of sandy beaches and rocky points mostly bordered by reefs, extends northward 2½ miles to Yellow Head, on the north side of which is Minmuga Lake, a salt lagoon, about ¼ mile wide, trending to the westward. A low sandy beach from Minmuga Lake extends north 1¾ miles to Burra Lake, which is about ¾ mile in extent, and continues its northerly direction for 1½ miles to a smaller lagoon, north-northeastward 1,500 yards from which is Marka Point; the coast from this lagoon to Marka Point and for about a mile to the northwestward of it is bordered with sunken rocks.

Turos (or Boogon) Inlet and Turos River.—At about 1,500 yards to the northwest of Marka Point is a small hilly projection, whence a low narrow tongue of land extends northward 1,500 yards to the mouth of Turos Inlet, which does not appear more than 200 yards wide and is sometimes closed; but after heavy rains it is open, and only fordable at low water. There is a sunken rock about 400 yards off the mouth of the inlet. Within its entrance Turos Inlet forms a labyrinth of points, creeks, and islets, extending about 3 miles north and south, and about the same distance east and west. At about 3 miles westward of its entrance Turos Inlet receives the waters of Turos River, a considerable stream winding from the southwestward, through a good cattle country, over which there are several stations.

Binge-Binge Point (lat. 36° 01′, long. 150° 10′).—From the opening into Kialy Lagoon at 1 mile to the northward of Turos inlet, a sandy beach, with scrubby land behind it, trends northward 2 miles to a rocky projection, about half a mile to the northward of which is Binge-Binge Point; both points have reefs of rocks projecting from them. From Binge-Binge Point a succession of small bays and rocky points curve round northward for 1¼ miles to Mullinburra Point, and thence northward 2¼ miles to Congar Creek, a narrow inlet, with sunken rocks close to its mouth.

Petro Head and Black Rock.—From Mullinburra Point the coast trends to the northwestward 1,500 yards to Petro Head, with

Black Rock lying about 400 yards off it, from which a rocky ledge extends northward to abreast of Congar Point, which lies 1¾ miles northward from Petro Head; there are depths of only 3½ fathoms on some parts of this ledge.

The sea always breaks on the rock, and with strong southeasterly gales it breaks heavily on some patches of the ledge, which would be dangerous to a small deeply laden vessel, or might cause a large one to strike the ground in the hollow of the sea, which runs here in a heavy southeast gale. From Congar Point the beach continues in a northerly direction 1 mile to Yowaga Point, thence in the same direction for 1½ miles to the north base of Toragy Point, the headland forming the south side of the entrance of Moruya River.

Between Binge-Binge and Toragy Points there are good grassy headlands, with salt lagoons and scrub between them; near the latter point there are some forest gum and swamp oak trees, besides scrub.

Toragy Point and Moruya River.—Toragy Point is the northeastern extremity of a rock peninsula, with some grassy slopes on its north side; extending nearly ½ mile from east-northeast to west-southwest there are some rocks above water close off the northeastern extremity within 100 yards of which there are from 5 to 6 fathoms water.

Pilot station.—On the west point or inner south head of this peninsula there is a pilot station, with a signal staff.

Two rocky patches, with 5 fathoms water over them, on which the sea breaks in bad weather, lie off the entrance of Moruya River. The southern patch lies 83° 1 mile from the signal staff and the other 600 to 800 yards to the northward of the southern patch. Nearly ½ mile from these patches to the eastward the bottom is broken and rocky, with 6 to 9 fathoms. There are from 8 to 9 fathoms, sandy bottom, between the patches and the south head. From 200 yards off Toragy Point to ½ mile off Congar Point the bottom is rocky and irregular, varying in depth from 2½ to 7 and 8 fathoms. and the sea breaks from 1,000 to 1,500 yards off shore in detached patches for the whole distance.

Moruya River (lat. 35° 55′, long. 150° 10′) forms a bar harbor, of which the narrowest part of the entrance lies between the inner south head of Toragy Point and a low point about 400 yards to the northwestward of it; but the channel is contracted to barely 100 yards in width by the North Spit, which projects to the southward and eastward from the low northwestern point to within 150 yards of the rocks at the inner south head.

From depths of 6 fathoms at 200 yards off the northeast point of the peninsula the water decreases to 8 feet at about 200 yards from the shore midway between the northeast and west points. Thence to the narrow part of the entrance, between the signal staff and the North Spit, there are irregular depths of 7 to 14 feet at about 200 to 100 yards from the shore. At 100 yards northward from the signal staff is a small rock above water, close outside which there is only 3 feet of water.

Two breakwaters have been constructed at the river entrance, and other extensive improvements are being carried out; a sand pump is working with great success, and a granite training wall is being erected.

The channel of Moruya River, from its entrance between the signal staff and the North Spit, is bounded to the southward by the edge of the western of two flats, and thence by the rocky south shore. And the channel is bounded to the northward by the inner edge of the North Spit, by the north shore thence to a small sand spit and then to the edge of the mud flat which trends to the westward from the sand spit.

The channel between the southern edge of the North Spit and the flat to the southward of it is about 100 yards wide, with from 7 to 10 feet water in the fairway, but farther in there is only a depth of 4 feet.

Although Moruya River is only adapted to steam and other small vessels of light draft, it promises to become a place of considerable importance, being the only outlet by water for the produce of the Araluen and Braidwood districts, with their gold fields.

Moruya (lat. 35° 55′, long. 155° 07′), on the south shore of Moruya River, and 3 miles from its entrance, has a population of about 1,000. There is a telegraph station; and regular communication by coach and steam vessels is maintained. The district is agricultural and mining.

Signal station.—There is a signal station at Moruya, and communication can be made by the International Code. It is connected by telegraph.

Directions.—Small craft must always enter Moruya River with the flood tide, more especially when there are freshets in the river; for then the ebb current runs out through the narrow mouth at the rate of 7 knots, forming eddies that would prevent any vessel from steering, and place her in great danger of being set on the rocks to the southward, or on the sand spit to the northward. The channel across the bar is continually shifting, the average depth being about 7 feet at high water springs.

The bar is seldom smooth, and during southerly gales breaks for a considerable distance to the eastward of the heads.

When over the bar (lat. 35° 55′, long. 150° 10′), a mariner must be guided by circumstances, as the sand banks near the entrance of the river change every tide.

Broulee, 4 miles to the northward, affords good anchorage for vessels awaiting tide or otherwise.

The tides probably differ very little from those in Bateman Bay, 10 miles farther to the northward. It may therefore be assumed that it is high water, full and change, on Moruya Bar, at 8 h. 0 m.; rise 4 to 6 feet.

The usual tidal signals are shown on the flagstaff on the outer south head.

Aspect.—Peak Alone, 3,130 feet above the sea, situated 12 miles westward of Mount Dromedary, although a solitary mountain, may be considered as the northeasternmost of Maneroo Range. The land adjacent to the coast between Mount Dromedary and the entrance to Moruya River is low, level, and thickly timbered; receding to the westward it maintains the same characteristic features, broken only by a few undulating ranges of 300 to 400 feet in height till it meets the base of the high coast range of mountains (some 10 or 12 miles inland), extending from the southern extremity of Challenger Range in a northern direction for 20 miles to a cluster of high conspicuous peaks named Horns, the highest of which, Evening Peak, lies 14 miles westward of Toragy Point.

This range, however, is not altogether uniform and uninterrupted, being broken about its center, or at Mount Lambert, which rises to an elevation of 3,200 feet; it here loses its general direction, and forms deep gorges, gullies, and isolated hills, but from seaward these features are not perceptible, and a high unbroken range of mountains with sharp peaks will be seen for a distance of 20 miles north and south.

Moruya River, running through the Honoria Valley, divides this range of mountains from the Duke of Edinburgh Range, which, commencing at Mount Haig, 3,381 feet above the sea, extends in a northern direction for 24 miles to Mount Fane, near the head of Clyde River, where it becomes broken into detached steep table-topped mountains. Some of the higher mountains in this range are conspicuous from seaward, Collaribbee, Budawang, and Curroebilly, being respectively 3,424, 3,630, and 3,619 feet high. A road through a pass over 2,000 feet high has been cut near Budawang, connecting the Braidwood district to the westward of the range with the coast.

The coast, from the northwest point of the entrance of Moruya River, consists of a sandy beach extending northward 2¼ miles to a small stream between which and Broulee Island, at east-northeast-ward 1¼ miles from it, are two points, with reefs projecting from them.

Broulee Island (lat. 35° 51′, long. 150° 12′), about ¼ mile east-ward from the coast, is inclosed by a reef of dry and covered rocks, close outside which there are depths of 4 fathoms. This island forms the southwest point of a bay which thence extends northeast 2½ miles to Burrewarra Point, and is nearly 1½ miles deep; the irregular west and north shores of the bay are bordered by dry and covered reefs.

Anchorage.—Iñ the northwest part of the bay between Burrewarra Point and Broulee Island there is good anchorage, sheltered from north and northeast winds. Northward and westward of Broulee Island small vessels find shelter from southerly winds in about 3 fathoms, with the north point of the island bearing 123°, and Toragy Point in range with the point of the mainland immediately west of Broulee Island.

Soundings.—The 100-fathom edge of the bank of soundings, from 5½ miles to the eastward of Montagu Island, extends to about 15 miles east of Burewarra Point, and between Montagu Island and Burrewarra Point there are depths of 50 fathoms between 4 and 5 miles, 30 fathoms at 2½ miles from the coast, and thence regularly decreasing soundings toward the land, the bottom being everywhere sand.

Burrewarra Point is a rocky headland, 182 feet high, projecting about ½ mile from the coast line; it is closely fringed by a reef of dry and sunken rocks, and there is a sunken patch close to the northwestward of it. Between this headland and a double rocky point at 1¾ miles north of it, are two little bights separated by a small prominent point, connected by a ledge of dry rocks with a rocky islet lying ¼ mile to the eastward of the point. From the double point a bay, partly bordered by a reef, extends northward 1,500 yards to South Head, which has a rocky reef projecting from it and forms the south point of Bateman Bay.

Bateman Bay (lat. 35° 44', long. 150° 13') extends from South Head 4¼ miles northward to North Head, and runs in a northwesterly direction for a distance of 3¼ miles from Tollgate Islets in the middle of the entrance to the bar of Clyde River.

The swamps and lakes on the north shore of Bateman Bay abound with swan, duck, teal, etc.; snapper fishing will be found off the rocky points.

Bateman is a village on the Clyde, near its entrance into the bay. The district is noted for its fine timber, and there are several sawmills. There is a telegraph station at the village. Population is about 300.

Depths.—The depths in Bateman Bay decrease gradually from 9 fathoms at the entrance to 4 fathoms at the head of the bay 800 yards off Square Head.

Black Rock, 32 feet high, lies 1 mile northeastward from South Head, and is about 200 yards in extent, with from 6 to 10 fathoms water close around it. Between this rock and Tollgate Islets there is a channel 1¼ miles wide, having from 10 to 15 fathoms water, on a sandy bottom.

Coast.—The southwest shore of Bateman Bay, from the South Head, extends north-northwest 1¼ miles, and thence northwest 2¾ miles, to Observation Head, and consists of a series of rocky points and small sandy bays. From a point at ⅜ mile northward of South Head a reef of dry and covered projects ⅛ mile toward Black Rock. All the other projections of the southwest shore of the bay are also bordered by reefs of a similar kind, but none of them extend beyond 300 yards from the points.

Trennant Rock, a small pinnacle rock with 3½ fathoms water on it and 12 to 13 fathoms close-to, lies southward 600 yards from the southern extremity of Tollgate Islets. There is a passage between Trennant Rock and Tollgate Islets with 8 to 10 fathoms by keeping within 200 yards of the southwest islet.

North Head, in range with the east end of Tollgate Islets, leads directly to the rock, and, seen open either side, leads clear.

Tollgate Islets (lat. 35° 45′, long. 150° 16′), which are two in number, are connected by a ledge of rocks and reefs, and extend together nearly ½ mile northeast and southwest; the southwestern islet is 157 feet high, and both are closely fringed with rocks, having 9 to 6 fathoms water at about 300 yards outside them.

Southwest Tollgate Islet is infested with snakes.

From the Tollgate Islets a ledge of rocks with 12 to 15 feet water extends northwestward ⅓ of a mile; 300 yards north and northeast of the ledge the bottom is rocky, with 3 to 6½ fathoms, deepening suddenly to 9 and 10 fathoms, sand.

A small vessel may take shelter under the lee of these islets; but it would be imprudent for a stranger to do so, except in a case of absolute necessity.

There is a channel 1 mile wide, with 7 to 10 fathoms between Tollgate Islets and Three Islet Reef, which lies southwestward ¾ mile from North Head.

Observation Head, 50 feet high, is inclosed by a reef of dry and sunken rocks. Snapper Islet, at ⅓ mile northeastward from the head, is 75 feet high, about 100 yards in extent, and is the north western and larger of two islets lying west-northwest and south-southeast 200 yards from each other. Both islets have reefs about them; the eastern islet having a reef which extends 300 yards to the southward. From 1¼ miles northwestward of South Head to within 1 mile of Observation Head there are from 12 to 8 fathoms at about ⅔ of a mile from the shore; but from 1 mile southward of Observation Head to Snapper Islet the shore is fronted by a shoal, with from 3 to 1 fathom water on it.

North shore.—North Head and Three Islet Point are both fringed with dry and covered rocks; but at the head of the little bay between these points there appears to be a sandy beach. Three Islet Point derives its name from three islets lying close together and extending 400 yards southward from the point. Three Islet Reef, before noticed, which extends 300 yards southward from the outermost islet, has a dry rock on it, with 7 fathoms water, 400 yards to the westward of it.

Reef Point is a small projecting headland at ⅜ mile westward of Three Islet Point, with a reef of dry and sunken rocks extending about 300 yards from it.

Acheron Ledge (lat. 35° 43', long. 150° 16'), which lies ¼ mile southwestward from Reef Point, is about 300 yards long, northwest and southeast, with a rock above water on either end, and a bank, having 1½ to 4 fathoms water on it, extending 600 yards to the southward from it. There are from 6 to 7 fathoms water between Three Islet Reef and this bank; and there is the same depth of water on the west side of the bank.

Chain Bay extends from Reef Point northwest ¾ mile to a point, from which ledges of dry and covered rocks project 400 yards to the southward; the east shore is also bordered by a reef; but between this and the ledges which project from the northwest point of the bay there is a sandy beach, ⅓ mile long, with 1½ to 2 fathoms within 200 yards of the shore. Immediately behind Chain Bay there is some cultivated land, with buildings near it.

White Cliff—Square Head.—From the northwestern point of Chain Bay the north shore trends northwestward 600 yards to White Cliff, and is bordered with dry and covered rocks, which project 200 yards from the cliff, and extend along shore 600 yards to the westward. Off White Cliff reefs extend 600 yards in a west-southwest direction. From White Cliff a smooth shore curves a little more than a mile in a west-southwest direction to the inner fall of Square Head, which is 400 yards broad, and projects ½ mile southward from the low land behind it, to 1 mile northward of Observation Head.

Directions from the southward.—Approaching Bateman Bay from the southward, give Burrewarra Point a good berth, and do not haul into the bay until Black Rock bears 257°, as there are dangerous rollers along the coast from the point to the rock. From about ½ mile northward of Black Rock steer 333° for Square Head, which leads about 600 yards to the eastward of the Snapper Islet Reefs.

From the northward.—Entering Bateman Bay from the northward, steer about 224° for Tollgate Islets, passing the North Head at the distance of about ½ mile; and when Square Head bears 300°, steer for it on this bearing, which leads ⅓ mile south of Three Islet Reef and ¼ mile southward of Acheron Ledge.

Between Tollgate Islets and Three Islet Reef there are depths of 7 to 10 fathoms rocky bottom, and with easterly gales, during the ebb stream out of Clyde River, there is a heavy confused sea all the way to the mouth of the river, with occasionally a heavy break in the bay.

Anchorages.—Tollgate Islets (lat. 35° 45′, long. 150° 17′) afford shelter in case of necessity, with, winds from east-southeast to south-west. The best anchorage is in 8 fathoms sandy bottom, with the center of the southwest islet bearing 156°, distant ½ mile, when, if the wind should shift to the northward, vessels can get under way and pass between the islets and Black Rock. In weighing from this anchorage and taking the southern passage, to pass westward of the 16-feet ledge, keep the North Head Summit (which is the highest hill over North Head) in line over the first little sandy beach immediately within, and westward of Three Islet Point, bearing about 16°. This leads in 7 fathoms water 100 yards west of the danger.

Small craft can anchor much closer in, with the southwest islet on the same bearing in 5 to 6 fathoms.

The anchorage recommended in Bateman Bay for large vessels is in 5 or 6 fathoms, sand, at about ½ mile westward of Acheron Ledge; and for vessels of about 10 feet draft, is 300 yards to the westward of the large Snapper Islet, (which can be passed on the north side close-to in 21 feet), in from 12 to 15 feet, with the center part of Tollgate Islets in range with the edge of the north cliff of Snapper Islet. A vessel will ride easy at anchor, though a heavy ground swell is experienced; on the ebb current setting out of Clyde River a kedge should be run out to the westward from the stern, to prevent being brought broadside on to the swell.

Although the anchorages in Bateman Bay appear much exposed to seaward, a vessel with good ground tackle, may lie here with comparative safety, almost at any time, if her berth be well chosen.

Tides.—It is high water, full and change, at Observation Head, Bateman Bay at 8 h. 00 m.; springs rise 4 to 6 feet.

Clyde or **Bundoo River** (lat. 35° 43′, long. 150° 12′).—The entrance of this river may be considered to lie between Observation and square Heads, where the greatest depth of water in mid-channel is 3½ fathoms.

The bar.—The channel over the bar which extends across the entrance had, in 1917, six feet of water over it at low water springs. It is advisable to ascertain the depth from the Officer in Charge before attempting to cross it.

Two red buoys mark the southern edge of the spit on the north side of the entrance.

Range lights—Front light.—A light is shown from a position 1,750 yards northwestward from the extremity of Observation Head.

Rear light.—A light is shown from a position on the southern shore 2,040 yards west-northwestward from the extremity of Observation Head. These range lights are moved as necessary to meet the changes in the channel over the bar.

Light.—A light is shown from the ferry jetty southeastward from Smoke Point.

Training wall.—A training wall extends from a position 320 yards east-northeastward from the present position of the front range light, 600 yards west-northwestward.

Depths.—Within the bar are depths of 9 feet, gradually increasing to 3 and 4 fathoms above the training wall.

Nelligen (lat. 34° 38', long. 150° 08'), a small postal township, is situated 7 or 8 miles up, on the right bank of Clyde River, and has a population of about 200. The district is principally agricultural and pastoral. About 15 miles further up the Clyde is the settlement of Brooman, from which a large amount of produce is carried. There is a telegraph station at Nelligen.

Pilot.—There is no pilot at this river, and only local steam vessels and small coasters well acquainted with the place frequent it.

Aspect.—The land about Bateman Bay is low and thickly wooded, receding from each shore to an elevation of 400 to 600 feet. Further inland the country is mountainous; Mount Oldrey, a conspicuous round summit, 2,212 feet above the sea, is the highest of Clyde Range, and lies 12½ miles westward of the north head of Bateman Bay. This range extends in a southeast direction from Duke of Edinburgh Range, and is separated from Belmore Range, which lies parallel to it further south, by Macleay River. Mount Collaribbee, 3,385 feet high, lies 5½ miles west-northwestward from Mount Oldrey.

Belmore range is separated from Mount Haig by a deep gorge, and extends from Duke of Edinburgh Range, in a southeast direction along the north bank of Moruya River 7 or 8 miles, terminating in a remarkable semidetached mountain, Mount Wandera, 1,945 feet high, also named, from its shape, Camels Hump. This range is made up of five distinct and peculiar summits, mostly anvil-shaped, with valleys or gorges between each.

From seaward Honoria Valley is remarkable when seen on a northwest bearing, apparently dividing the barrier of high mountains, and separating the coast from the inland ranges.

Flat Rock—Wasp Islet.—The coast from North Head of Bateman Bay extends 5½ miles northward to Point Upright, and consists of a series of small points and sandy beaches. At 2¼ miles northward of North Head lies Flat Rock, 1 mile to the northward of which lies Wasp Islet, 40 feet high, thence to Point Upright the land recedes nearly 1 mile, forming a bay about 2 miles long, terminating in a sandy beach. At 1 mile to the west-southwest of Point Upright is the mouth of a narrow inlet, winding 1½ miles southwestward into a lagoon 2 miles long north and south and ½ mile wide.

Beagle Bay.—At the south end of the beach above mentioned is the township of Beagle Bay, off which mooring buoys are laid for steamers loading timber.

Point Upright (lat. 35° 38', long. 150° 21') is the termination of a ridge of hills extending from the westward, and was so named from its perpendicular cliffs.

Grasshopper Islet, 40 feet high, is situated on a reef which projects above ½ mile to the northeastward from Point Upright, and about ¼ mile northeastward from Grasshopper Islet lies a ledge of sunken rocks, on which the sea breaks heavily.

Pebbly Beach township is about 1 mile north of the islet. Mooring buoys are laid for steamers loading timber.

Dawson Islets.—Between Point Upright and a rocky projection 2 miles northward of it the coast forms a bay, of which the southern half is a sandy beach, with some sunken rocks along it. From the north point of this bay the coast continues north-northeastward 1½ miles to a point at the base of Mount O'Hara, which rises close behind, to the height of 1,110 feet. A reef, on which are the two Dawson Islets, extends about ½ mile eastward from this point.

O'Hara Islet and Head.—O'Hara Head is 2¼ miles northeastward from Dawson Islets, and O'Hara Islet, 15 feet high, lies near the shore at ½ mile northward of Dawson Islets. Between O'Hara Head and First Sandy Point, at 2¼ miles north-northeast from it, the coast consists of a sandy beach, with rocky points extending ¼ mile northwestward of the head, and a rocky point with a sunken rock about 300 yards off it, nearly in the middle of the beach. Near a small islet close to the northward of O'Hara Head there is anchorage for coasters, but it is not recommended.

Kiola township is near the above anchorage. Mooring buoys are laid for steamers loading timber.

Brush Island (lat. 35° 32', long. 150° 26'), which lies 200 yards off First Sandy Point, is about ½ mile across and 140 feet high, the sand hills being covered with scrub and abounding with rabbits. About 800 yards northeast from the eastern extremity of Brush Island, which lies nearly 1 mile from First Sandy Point, is a dangerous sunken rock on which the sea only breaks with a heavy swell.

Vessels bound northward and keeping inshore to avoid the current should be careful when passing this island to haul out 1 mile to seaward.

Anchorage.—On the northwest side of Brush Island anchorage may be obtained during southerly winds in 6 fathoms, sandy bottom, with O'Hara Head in range with First Sandy Point, and the northern extremity of the island bearing 100°.

Brawley Point is the name of a small township near the above anchorage. Mooring buoys are laid for steamers loading timber.

Stokes Islet.—From First Sandy Point a beach, having a building, Murramarang House, on it ⅓ mile from the point, curves north-northwest 1¼ miles to a projection, rather over ¼ mile westward from which is the entrance to a lagoon; from the projection the coast extends northwest 1¼ miles, in and out, to the narrow mouth of the creek trending to the westward. The north side of the mouth of this inlet is formed by a small peninsular headland, between which and a point at 1½ miles northward of it is a bay having a small opening at 1 mile northwest of the head, forming the mouth of a lagoon about ¼ mile in extent. Stokes Islet, which lies ½ mile northeast of this opening, is surrounded by reefs apparently connected with the shore to the northward of the islet.

Crampton Islet, situated 1 mile north-northeast from Stokes Islet, is situated on a reef which extends across the mouth of a narrow inlet trending northward 1¾ miles, and separated from the sea by a low narrow tongue of land.

Lagoon Head.—Between Crampton Islet and Lagoon Head, at 2½ miles north-northeast from it, is a sandy bay nearly ½ mile deep. From Lagoon Head a rock coast, bordered by a reef, trends in a northerly direction 1 mile to a narrow opening trending northwestward 1¼ miles into a lagoon about 2 miles long, north and south, and ½ mile across its widest part, with a small stream flowing into its northern end from the westward. This lagoon is separated from the coast to the southward and eastward of it by ranges of hills. From the mouth of the lagoon a sandy beach extends northeastward 1,500 yards to a prominent rocky point off which a narrow ledge of rocks projects southeastward 1,500 yards. The outer rock of this ledge is 6 feet above high water, and there is a depth of 21 fathoms ¼ mile eastward of it. The sea breaks all the way between the outer rock and the shore. From this ledge rocky points extend northeast 2 miles to Warden Head, which has a reef projecting nearly ½ mile from its southeastern extremity. The entrance to Ulladulla Harbor is about ⅓ mile north-northwest of Warden Head.

Warden Head—Light (lat. 5° 22′, long. 150° 31′).—A group flashing white light, 112 feet above water, visible 12 miles, is shown from a circular iron white lighthouse, 32 feet high, on Warden Head.

Sullivan Reef, lying ¼ mile northeastward from the northeast part of Warden Head, is a rocky patch (on which the sea nearly always breaks), extending north and south little more than 200 yards, and about 50 yards wide; its center is dry at low water springs, with 10 feet on its north end, 12 feet on the south, and 4 to 6 fathoms close-to all round. This reef, lying nearly across the fairway of the entrance to Ulladulla Harbor, forms a natural breakwater, and tends considerably to break the heavy seas rolling in toward the artificial breakwater at the head of the harbor during easterly winds.

There is a passage both north and south of this reef, the northern one is to be preferred, and is about ½ mile wide between the reef and North Head, with from 7 to 8 fathoms water near the middle.

Rock.—A rock, with 2 fathoms water on it, lies nearly ½ mile northeastward of North Head. The rock is believed to be about 200 yards in extent.

Ulladulla Harbor is 800 yards wide northwest and southeast between the rocky shelf which projects ·100 yards from Warden Head, the southeast point, and the detached rocks which extend nearly the same distance from North Head, the northwest point of entrance. From the middle of the entrance, Ulladulla Harbor trends westward ½ mile and is ¼ mile wide, except at its western end, where a sandy bay forms the inner harbor, which extends north-northwest 400 yards, and is 200 yards deep.

The north and south shores of all but the inner harbor are bordered by shelves of rock extending farthest from the south shore, from which, at ¼ mile westward of Warden Head, the rocky shelf projects 300 yards to the northward.

Depths.—In mid-channel, between the entrance heads, there are depths of 30 to 36 feet, sand, which shoal gradually toward the pier end, between which and the rocky ledge extending from the point northward of it, a distance of nearly 200 yards across, there are depths of 19 to 20 feet.

Inner Harbor.—The little bay which forms Ulladulla Inner Harbor affords anchorage for small craft in from 13 to 15 feet water; it is sheltered from the eastward by a curved pier, which extends from the south point of the bay over the western end of the sunken ledge, 27°, 160 yards, and thence 331°, 80 yards, to the pierhead, where there is a depth of 17 feet at low water.

Between the pierhead and the rocky shelf which projects east-southeastward nearly 100 yards from the north point of the Inner Harbor, the entrance is nearly 200 yards wide, with from 18 to 21 feet water in the fairway, whence the depths gradually decrease to 5 feet, close to the northern ledge; but there is deeper water on the south side, there being 11 feet close along the back of the outer part of the pier. From 17 feet at the pierhead the depth gradually decreases to 3 feet within 30 yards of the shore, except on the ballast in the south corner of the harbor, toward which a jetty projects from the inner south end of the pier into 2 feet water. On the southwest side of the Inner Harbor there is a slip, between which and the jetty two streams run through the beach; a small stream also flows into the northwestern part of the harbor.

Range lights.—Two lights are exhibited from the beach at the head of the bay; these lights in range bearing 272°, lead into Ulladulla Harbor to abreast the pierhead.

Ulladulla (lat. 35° 21′, long. 150° 30′) has a population of about 224 persons. Dairy farming is the principal industry in the surrounding district, and there is plenty of good timber, giving employment for sawmills. There is a telegraph station here, six mails a week, and also steam communication three times a week with Sydney.

Directions.—When approaching Ulladulla from the northeast and eastward the heads are difficult to distinguish, being low, and the points adjacent resembling them. From the southward, when abreast of Brush Island, the northernmost low point seen projecting from the shore is Warden Head; on a nearer approach it may be identified by a deep gap cut through the trees, a short distance inland, and there is also the lighthouse on it. Cooks Pigeon House, a remarkable conical mountain, bearing 272°, leads to the entrance of the harbor, and when within 5 or 6 miles the white houses at the head of the harbor and the sandy beach under them will be distinguished.

Cooks Pigeon House Mountain kept over the center of this sandy beach, on a 272° bearing, leads into the harbor in mid-channel, northward of Sullivan Reef; when within the reef the Pigeon House Mount will be lost sight of. Sullivan Reef generally breaks, and should be given a berth of about 400 yards. North Head should not be approached too close, as there is a depth of 14 feet about 70 yards southward of the low-water rocks extending off it. When nearing the pier keep more toward the north shore to avoid the flat extending 200 yards east-northeastward from the pier end, with from 8 to 12 feet on its northern extremity; then haul in as the wind will allow, passing the pier at a convenient distance. Small vessels anchor inside the pier in from 13 to 15 feet of water 150 yards from the beach.

The bottom is sand over rock, and if likely to blow hard from the westward it is necessary to run out a warp to a tree on shore, or bury a kedge in the sand, to prevent the possibility of dragging and tailing on to the pier. Small steam vessels discharge alongside the pier.

During summer, sailing vessels when leaving are recommended in the early morning to tow or warp out toward North Head, during the calm preceding the northeast wind, when a long tack can be made toward Sullivan Reef, and the next tack will clear the North Head, or almost a fair wind may be made by taking the channel between the reef and the south shore; but this should not be attempted unless in fine weather, as it is narrow with a rocky, irregular bottom, and in bad weather the sea breaks across the channel. When Lagoon Head is seen open to the eastward of Warden Head, bearing 224°, a vessel is outside Sullivan Reef and in about 16 fathoms.

At night.—To enter the harbor (lat. 35° 21′, long. 150° 30′) steer in with the range lights in line, bearing 272°, taking care to be nothing to the southward whilst in the vicinity of Sullivan Reef. These lights lead in clear of danger to small vessels until abreast the pierhead.

Current.—Vessels bound northward find little or no current by keeping inshore between Brush Island and Cape St. George.

The current from the northward striking the bluff headlands about Jervise Bay appears to be diverted from its general direction, and strikes the coast again above Brush Island, leaving in the space inshore of this limit comparatively slack water.

Tides.—It is high water, full and change, in Ulladulla Harbor at 8 h. 30 m.; springs rise 6 feet.

Coast—Aspect.—From the North Head of Bateman Bay to the back of Ulladulla Harbor a range of hills from 600 to 1,370 feet high follows the trend of the coast a little distance inland from it. The most conspicuous and first from the southward is Mount O'Hara, which is flat, 1,100 feet high, 1 mile from the coast and 3½ miles northward of Point Upright. From Mount O'Hara the range is somewhat lower for 5 miles to Wason Heights, consisting of three distinct round-topped hills, lying about 3 or 4 miles from the coast; thence the range continues in a north-northeast direction, gradually decreasing in height till lost in the flat, rich agricultural lands between Ulladulla and Cook's Pigeon House, which bears westward, distant 11 miles from the entrance of Ulladulla Harbor. Cook's Pigeon House, 2,398 feet high; Table Hill, which lies between 4 and 7 miles northeastward of it; and Mount Sidney, 2,496 feet high, at 4 miles northward of the Pigeon House, form a conspicuous group from seaward off this part of the coast. From the northern extremity of Table Hill a ridge descends in an easterly direction toward the coast, while a lofty range extends 15 miles west-southwestward from the hill to Mount Campbell, and then turns southward 11 miles to Budawang Hill, which is 3,630 feet high; hence the range trends south-southwestward to within 2 miles of Mount Collaribbee, before mentioned.

Nurrawherre Inlet (lat. 35° 18′, long. 150° 30′).—Between the North Head of Ulladulla Harbor and a projecting point 1¼ miles to the northward of it, the coast forms a sandy bay, with small reefs projecting from both its points and also from the southern part of the beach. From the north point of this bay the coast curves north-northwestward 1¼ miles to Nurrawherre Inlet, the northern half being a sandy beach, and thence north-northeastward 2¼ miles to a low point forming the southeast side of the opening of Cuhudjuhrong Lake, close in front of which is Green Islet. Reef extends from the south head of Nurrawherre Inlet to Preservation Rock at 400 yards off it, and also a short distance from the foot of a hill at ½ mile to the northward of it.

Green Islet and Cuhudjuhrong Lake.—Green Islet is fringed by a reef apparently connected with the bar across the mouth of Cuhudjuhrong Lake behind it, which trends westward 4 miles, and is ⅓ mile wide. At about a mile within the mouth of this lake a cluster of islets begins, immediately to the northward of which an arm of the lake branches to the northward; the islets cover a distance of 1¼ miles.

Rocky patch.—There is a small rocky patch, with 4 fathoms water on it and steep-to, distant 1¼ miles eastward from the center of Green Islet.

Red Point.—From the northeast point of the mouth of Cuhudjuhrong Lake the coast trends northeastward 2 miles to Red Point, which projects 1,340 yards eastward from the coast-line; there are two small beaches between the inlet and the point with hilly land behind. Between Red Point and the narrow mouth of Swan Lake, at 3 miles northward from it, is a bay 1 mile deep, the irregular shore of which is intersected by two small streams. Cadmurrah Beach lies between ¾ mile to the southward and 1¼ miles northeastward of the mouth of Swan Lake, and reefs project from the points at each end of it.

Swan Lake.—From its mouth Swan Lake continues very narrow for about 1,340 yards in a northerly direction, within which it forms a lagoon about 1 mile in extent. From the entrance to Swan Lake a sandy beach trends east-northeastward 1¼ miles to the west point of Wreck Bay, which point forms the south side of the narrow mouth of Sussex Inlet.

Sussex Inlet and St. George Basin.—Sussex Inlet is·a narrow channel trending in a northerly direction, 3 miles into St. George Basin, which is 5 miles in length, east and west, with a breadth of about 3 miles; it is separated from Wreck Bay by a low tongue of land, 1 mile broad, extending 4 miles to the eastward from Sussex inlet. On the west shore of the basin are several large creeks, into the northern of which flows a small stream.

Wreck Bay—St. George Head (lat. 35° 12′, long. 150° 43′).—Wreck Bay is a dangerous bight for sailing craft. Between Sussex Inlet and St. George Head it is 5 miles in length, and is 2¼ miles deep in the eastern part. The north shore of the bay, from the mouth of Sussex Inlet, consists of a sandy beach fronted by rocks, extending east-northeastward as far as a rocky point 2½ miles northwestward of St. George Head; but between this point and the head the eastern end of the bay is a rocky bight bordered with reefs.

Caution.—When navigating near this part of the coast during bad weather with easterly and southeasterly winds guard against being set into Wreck Bay. At night by keeping Perpendicular Point Light in sight, vessels will be well to the eastward of Wreck Bay.

Anchorage.—During the summer, anchorage may be obtained in Wreck Bay, 1 mile off shore, in 7 fathoms, sandy bottom, with St. George Head bearing 145°.

Aspect.—The land from Ulladulla Harbor to St. George Head is mostly low and thickly wooded, with ridges of hills extending inland from the coast between Red Point and Wreck Bay. From Table Hill the main range takes an irregular semicircular direction to the northward and northeastward, and, after rounding St. George Basin, terminates at St. George Head.

Soundings.—From 2 miles southeastward of South Head to the same distance eastward of the north head of Bateman Bay the soundings range from 25 to 30 fathoms; they increase to 50 fathoms at 4 miles eastward of Point Upright, between which and 4 miles southward of St. George Head the depths range between 50 and 60 fathoms, decreasing to 30 fathoms at 3 miles east-northeastward of the head.

Cape St. George.—From St. George Head the coast trends northeastward 3¼ miles to Cape St. George; there is a small exposed bay midway between these two headlands, and the cape has some sunken rocks close about it.

From Cape St. George a cliffy coast, with from 27 to 29 fathoms water at 1 mile off it, winds northward 2¾ miles to Governor Head.

Jervis Bay (lat. 35° 07′, long. 150° 47′).—**Bowen Isle,** which forms the southwest point of the entrance of Jervis Bay, lies close off Governor Head, from which it is only separated by a breach 300 yards across, appearing as if the cliff had been torn to pieces, and leaving here and there a straggling rock above water, and where there is a passage for small craft, in very calm weather, which requires local knowledge to use. The isle is 1,200 yards long north and south, 800 yards broad, and 133 feet high. Shoal water extends 200 yards north of the isle and about 600 yards west of it, where it nearly meets the shoal water extending northwestward from Governor Head.

Bowen Isle, which for situation, soil, scenery, and fresh water seems the most desirable spot in Jervis Bay, is moderately wooded and has much clear ground, covered only with long grass. Its sea front is formed of high vertical cliffs, in many places deeply rent. From these cliffs the isle slopes gradually but irregularly toward the bay, and that side is low and formed of sand intermixed with rocks.

Water.—The largest and most convenient stream of fresh water lies directly at the back of a little sandy bight on the west side of Bowen Isle, where boats can easily load in fine weather.

Depths.—Within its entrance Jervis Bay extends 8 miles north and south, and from 3 to 5½ miles east and west, with regular depths, gradually decreasing from 78 and 17 fathoms in the entrance to 9 and 6 fathoms within ½ mile of the greater portion of the shores of the bay.

Middle Ground, a rocky patch about 100 yards across, with 8 fathoms water on it, lies about 1,400 yards northward from the north point of Bowen Isle, in the fairway of the entrance to Jervis Bay.

Perpendicular Head, situated 2 miles northeastward from the north point of Bowen Isle, is a bold cliffy headland, 263 feet high, forming the northeast point of the entrance of Jervis Bay. This point from its rising perpendicularly to a flat surface, without tree or shrub, is a most conspicuous feature of the coast, but there is an inner north head to the entrance, formed by Longnose and Dart Points, which lie 1¾ and 1⅝ miles westward from Perpendicular Head.

Light (lat. 35° 05′, long. 150° 50′).—A group flashing white light, 304 feet above water, visible 24 miles, is exhibited from a white circular tower 56 feet high, on Perpendicular Head.

Vessels navigating the coast at night will be well to seaward of Wreck Bay to the southward and Crookhaven Bight to the northward by keeping this light in sight.

Signal station.—There is a signal station at Perpendicular Head Lighthouse, and communication can be made by the international code, and by Morse lamp at night. It is connected by telegraph.

Storm signals.—Notice of existing storms on any part of the coast are signaled to passing vessels.

Landing can be effected at a place situated ¼ mile northwest from the lighthouse, marked by a small white patch above it, but it is not available with a southerly or southeasterly swell. An alternative landing place is in Blackboat Bay, where the lighthouse boat is kept, situated 1 mile northwest of the lighthouse.

Bumbora Rock.—The shore between Longnose and Dart Points forms an irregular bight, with a reef projecting nearly ¼ mile from Longnose Point, with Bumbora Rock, about 100 yards in extent with 12 feet water over it, on which the sea breaks in bad weather, lying

700 yards southward from the point. Between the reef off Longnose Point and Bumbora Rock there is a depth of 6 fathoms with 11 fathoms close to the rock. With the exception of this reef and rock and the Middle´ Ground, the entrance into Jervis Bay is free from all dangers, with depths of 15 to 20 fathoms, sand.

Anchorage.—The whole of the west side of the bay is exposed to the heavy swell thrown in by southeast gales, and is consequently unsafe for anchorage; the sea breaking on it may be heard at a considerable distance.

There is anchorage either in Darling Road, to the westward of Bowen Isle, or in Montagu Road, to the northward of Dart Point, but Darling Road is the preferable of the two.

Plantation Point.—**Darling Road,** the southern bight of Jervis Bay, is situated between Bowen Isle and Captain Point. About 3 miles northward of Captain Point is Plantation Point; some sunken rocks lie close to the extremity of that point, one which breaks being 400 yards eastward of the point and a 3-fathom rocky patch, which has been seen to break, at $\frac{1}{4}$ mile eastward from it. Vessels should not pass between this patch and the point, and inside the 2-fathom line the ground is foul. A sand spit with from $1\frac{1}{2}$ to 3 fathoms water, projects $\frac{1}{4}$ mile northward from the northeastern part of Darling Road, and adds to the security of the anchorage by breaking the swell which sets in; with this exception the shore may be approached within $\frac{1}{4}$ mile in depths of $5\frac{1}{2}$ to 7 fathoms. The shore from 1 to 2 miles southward of Plantation Point is only separated from the eastern end of St. Georges Basin by an isthmus 2 miles broad.

Naval college.—A naval college has been established at Captain Point consisting of numerous buildings necessary for the training of naval officers.

Anchorage (lat. 35° 08′, long. 150° 45′).—Darling Road affords good anchorage in 6 to 8 fathoms, with the south end of Bowen Isle in line with the east point of Darling Road, and Hole-in-the-Wall (a white cliffy projection with a hole through it) bearing 111°, or as convenient.

Fish.—Good seine, also hook and line fishing, is to be had off the points and beaches.

The western bight of Jervis Bay, from Plantation Point to Flora Point, nearly 4 miles to the northward of it, is $1\frac{1}{2}$ miles deep.

Dent rock, a pinnacle rock, with 2 feet on it at low water, lies 300 yards off shore, and 1,600 yards westward of the eastern extremity of Plantation Point. At $1\frac{1}{2}$ miles northwestward of Plantation Point is the mouth of Moona Moona Creek, from which a ledge of sunken rocks extends 200 yards.

Carrambeen Creek—Huskisson.—At about 1,600 yards northward of Moona Moona Creek is Carrambeen Creek, with the shipbuilding village of Huskisson just inside it, where there is a telegraph and a post office. Shoal water extends about ½ mile off these creeks, and a patch of rocks nearly covered at high water lies about 600 yards off Carrambeen. A small beacon surmounted by a cage marks the extremity of the rocks at the south side of the entrance to Carrambeen Creek. The population of Huskisson is about 150.

Flora Point, with the shore for ½ mile to the westward and ¼ mile northward of it, is bordered by rocks, with 3 fathoms water about 300 yards outside them.

Eastern shore.—Montagu Road, on the eastern side of Jervis Bay, affords secure anchorage; it extends from the south point of the road 1½ miles northward to Montagu Point, and is about ¼ mile deep, with depths of 5 to 6 fathoms at ¼ mile offshore.

The anchorage in Montagu Road is in 6 to 7 fathoms, stiff ground. Although large vessels may lie here landlocked, they are exposed to a heavy fetch of the sea from the southward, to which also every other part of the bay is subjected; it is therefore indispensable that the ground tackle be good.

Pier.—A pier is situated about ¼ mile to the northward of the south point of Montagu Road.

Green Point (lat. 35° 01', long. 150° 46') is nearly 1 mile northwest of Montagu Point, the shore between being rocky, with a depth of 5 fathoms at 400 yards off it; at 200 yards southward of Green Point a reef extends 400 yards off, with Green Islet on it; a small patch of 2½ fathoms is situated nearly 600 yards southward from Green Islet.

Hare Bay, the northern bight of Jervis Bay, is 2¼ miles wide between Green Point and Flora Point, 1¼ miles deep, and is divided into two bights by Red Point, nearly equidistant from Green and Flora Points. Between the rocks off Green Point and the 3-fathom curve at nearly ½ mile off Flora Point, there are depths of 4 to 6 fathoms across the bay. The eastern and larger of the two bights into which Hare Bay is divided is mostly occupied by the shallow sand and mud flats, extending off Cararma, or Cabbage Tree Creek. The western bight has 3 fathoms at about 800 yards from the shore, and there is a 3-fathom rocky patch at 1,500 yards eastward from Flora Point, while the depths inside the 3-fathom curve are irregular.

Directions—Jervis Bay.—Outside Jervis Bay, and along the coast to the northward as far as Beecroft Head, a heavy, confused sea is experienced during even moderately bad weather; and perhaps along the coast of New South Wales there is no place where greater care is required in managing small vessels in bad weather than in the vicinity of this steep peninsula. This is in consequence of the agitated state of the sea, which may be attributed to the current setting out of Shoalhaven Bight meeting at Beecroft Head another branch of the stream in a disturbed state from crossing Young Banks, and the currents being much stronger near projecting headlands than elsewhere.

When bound into Darling Road, the southern bight of Jervis Bay, steer for the north point of Bowen Isle and round it not nearer than 400 yards, afterwards keeping Perpendicular Head well open northwestward of the low northwest point of Bowen Isle, bearing 66°, until Governor Head bears 111°, which clears the shallow flat extending westward of Bowen Isle, when stand in for the anchorage. Although the isle may be closely approached on its east and north sides, in light winds sailing craft should keep without the influence of the swell.

From the northward and eastward give Perpendicular Head a good berth, for, although it is bold and clear close-to, a fresh southeast wind does not blow home, but becomes light and baffling, while the swell sets heavily upon it; and as there is no anchorage near the head, except at a great depth, it should be carefully avoided by sailing craft in light winds. After passing Perpendicular Head the chief danger in the approach to Montagu Road is the reef and Bumbora Rock off Longnose Point, to clear which steer in rather north of mid-channel between Bowen Isle and Longnose Point, till Green Point is seen well open of the bluff northward of Dart Point, before steering for the anchorage. The vicinity of the Middle Ground should be avoided.

Tides.—It is high water, full and change, in Jervis Bay at 8 h. 30 m.; springs rise 5 feet.

Supplies.—Fresh water can be obtained at the head of Darling Road, and ¼ mile along the beach, to the southward from Hole-in-the-Wall, is a small stream always running. With the exception of a few fishermen, who are migratory, the village of Huskisson, in the northwest corner, and the Naval College at Captain's Point, the shores of Jervis Bay are generally uninhabited. Fresh meat, vegetables, and bread can be obtained at Huskisson. The nearest town is Nowra, 9 miles northwest from Huskisson.

Coast—Crocodile and Beecroft Heads.—From Perpendicular Point a line of cliffs trends northeastward 1¾ miles to Crocodile Head (lat. 35° 04′, long. 150° 51′), 356 feet high, and thence northward 1½ miles to a bight ½ mile wide, with three small islands, named the Drum and Drumsticks, in the middle of it. From the north point of this bight a continuation of the line of cliffs extends northward 2 miles to Beecroft Head, the eastern extremity of Crookhaven

Bight.   There are depths of 30 to 32 fathoms within 1 mile of the shore between Perpendicular Head and Beecroft Head.

Young Banks are two shoal rocky patches lying 38°, distant 3 miles, and 44°, 5½ miles from Beecroft Head.   The Southwestern patch is ⅓ mile long northeast and southwest, and ¼ mile wide, with depths of 7 to 10 fathoms over it, 16 to 17 on its north side, with a small patch of 10 fathoms lying about 1,400 yards northwestward of it, and 15 fathoms close-to on the south side; the sea has been seen to break on them in heavy gales.

Between these banks and Beecroft Head there is a channel, with depths of from 11 to 20 fathoms; between the southwestern patch and the northeastern one the depths are from 14 to 20 fathoms, rocky bottom; 12 fathoms is the least water on the northeastern patch, deepening suddenly to 43 fathoms on its east side, and 14 to 20 fathoms ½ mile to the westward of it.

The vicinity of these banks should be avoided, as the current when strong causes heavy overfalls even during smooth water.

Clearing marks.—Perpendicular Point Lighthouse kept open of Crocodile Head, leads to the eastward; and Cambewarra, the southern and highest mountain of Shoalhaven Range, just seen open to the southward of Mount Berry, a remarkable isolated conical hill, on the lowland about Shoalhaven River, bearing 294° leads about 1 mile to the northward.

At night by keeping Perpendicular Head Light in sight vessels will be clear eastward of these banks.

Crookhaven Bight (lat. 35° 00′, long. 150° 50′).—From Beecroft Head, the coast trends northwestward for ⅓ mile, thence around low rocky points, and two small bights to Crookhaven Bight, where good anchorage may be obtained in Abraham's Bosom, in about 6 or 7 fathoms water, with the northern extremity of Beecroft Head bearing 78° distant ½ mile.

In rounding this north point, which is low and projecting, care must be taken to avoid a sunken rock with 2 feet water on it and on which the sea breaks occasionally, lying 300 yards to the northward of it, with 9 fathoms close-to, and 6 fathoms between it and the shore.   At the northern extremity of Crookhaven Bight, near some outlying rocks off the beach, there is an isthmus, about 200 yards across to a fordable creek running into the head of Jervis Bay.

Shipwrecked mariners in this vicinity can reach Huskisson by walking across this neck and skirting the northwest beaches of Jervis Bay to the settlement.

Shoalhaven Bight.—Vessels should during a prevalence of bad weather keep outside this bight, which extends from Beecroft Head to Black Point, a distance of 13 miles, as heavy tide rips exist, caused probably by the strong ebb currents from the Crookhaven and Shoalhaven Rivers (which are much accelerated by rains) discharging into the bight, directly across the course of the ocean current running to the southward.

From the west end of Crookhaven Bight a smooth shore curves in a northerly direction 3½ miles to Kinghorn Head (a low projecting point), between ¼ mile and 2 miles to the southward of which is a lagoon 2 miles long, north and south, and 1 mile wide, only separated from the sea by a very narrow ridge. From Kinghorn Head a sandy beach curves north-northwestward 2 miles to a narrow rocky point, forming the east head of the entrance of the Crookhaven River.

Shoalhaven Rivers, between which and Sydney there is communication by steam vessel three times a week, are separated from each other by Comerong Island.

Caution.—No one should attempt to enter these rivers unless thoroughly acquainted with them, more particularly that of the North Shoalhaven, without having first procured a pilot, who will board vessels upon the usual signal being made.

Bar—Beacons.—The approach to Crookhaven River—the southern Shoalhaven River—is over a bar, with seldom less than 10 feet at low water close to the western side of the east head. From here the channel, between the rocky shore on the eastern and the sandy shoal on the western side, trends south-southwestward about ½ mile to the entrance of the river, which is between a rocky point projecting from the south shore and the sand head at 150 yards to the westward of it; the channel here is about 100 yards wide, with about 20 feet water. From the north edge of this sand head a narrow ledge of rocks, above water, extends nearly 400 yards to the northward, outside of which there are breakers over the sandy shoal on the west side of the channel from the bar to the entrance of the river.

A white beacon is placed on the northeastern extremity of the reef extending about 350 yards from the east head. There are three other white beacons on the south side of the channel.

During south or southwest winds sailing vessels should not attempt to enter if there is much sea on, but go into Jervis Bay, or to the anchorage known as Abraham's Bosom, in Crookhaven Bight.

Light (lat. 34° 54′, long. 150° 47′).—A fixed red light, 72 feet high, visible 8 miles, is shown from a white brick lighthouse 23 feet high, at the south entrance point of Crookhaven River.

Lifeboat.—There is a lifeboat and life-saving rocket apparatus at the entrance of Crookhaven River.

Comerong Island.—The coast forming the east side of Comerong Island, after trending in a north-northwest direction 2½ miles from the entrance to Crookhaven River, terminates at the north point of the island, which forms the south side of the mouth of the Northern Shoalhaven River, and may be distinguished by a small hillock on it. On the west side, at ¼ mile southward from the point, on the northern end of Comerong Island, is the narrow mouth of a lagoon, whence the north and northwest coasts of the island trend westward ¼ mile, and southwestward nearly a mile to the northwestern entrance of the navigable channel, which communicates between the two Shoalhaven Rivers. With the exception of the hillock on its north point, Comerong Island appears to be low and flat, with its south coast mostly lined with mangroves, covered at high tides. There is some forest and bush land on the southeast part of the island.

Northern Shoalhaven River.—The mouth of this river lies between the north point of Comerong Island and the sand head at ¼ mile to the northward of it, and is crossed by a bar, on which the sea breaks. The bar is of a shifting nature and the channel liable to change. The breakers on the south spit of the entrance extend about 1¼ miles offshore. Within its entrance the river expands to more than ½ mile in width, but again narrows to 400 yards between the northwestern extremity of Comerong Island and a point of the opposite shore. Close to the northwestward of the entrance of the navigable channel is an islet, whence the Northern Shoalhaven River extends westward 5 miles to Pig Islet, with a width of nearly ¼ mile. Above Pig Islet the river winds in a westerly direction along the south side of Cambewarra Range.

Nowra, a postal township on the right bank of the river about 1½ miles above Pig Islet, has a population of about 2,000. There is a telegraph station here.

Nowra has direct steam communication three times a week and daily communication by railway with Sydney.

Signal station.—At North Shoalhaven River is a signal station, and communication can be made by the International Code. It is connected by telegraph.

Tides.—It is high water, full and change, in Shoalhaven Rivers at 8h. 30m.; springs rise 9 feet, neaps 6 feet.

Aspect.—Mount Cambewarra, 9 miles west-northwestward from the entrance of the northern Shoalhaven River, is the most remarkable summit of a range of mountains extending from the mount southwestward about 8 miles, and northeastward 4 miles to some table-land. But the highest land near this part of the coast is Mount Berry, which, at 2 miles northwestward of the entrance of the river, rises from its north shore to the height of 1,000 feet, and serves as a good guide.

Coast—Black Point (lat. 34° 47′, long. 150° 51′).—The coast from the entrance of the northern Shoalhaven River forms a sandy beach, extending, with a slight curve north-northeast 5¾ miles to a small double creek, whence the coast trends east-southeastward ¼ mile to Black Point and Black Rock, with a reef extending south-southwestward ¼ mile from them. Between Black Point and the south point of Geering Bay at 2 miles northward from it, are two small bights, separated from each other by a point with a projecting reef.

Geering Bay extends 1 mile north and south and is ⅓ mile deep, with a sandy beach, at the north end of which is a small double inlet. The south point of the bay and the beach for some distance to the northward of it have reefs extending along them.

Gerringong is a village on the south coast road, in the southern part of Geering Bay. There are railway and telegraph stations, and a steam vessel to and from Sydney calls here. The population is about 500.

Caution.—Immediately on the appearance of bad weather from south-southeast, shifting easterly, vessels at anchor off Geeringang should proceed to sea, as during bad weather from seaward a heavy sea sets in.

From Geering Bay a succession of rocky points and small bights extends northward 4 miles to the south head of Kiama Harbor. The points which project from the northern and greater portion of this coast have ledges of sunken rocks projecting from them.

Kiama Harbor (lat. 34° 40′, long. 150° 53′) is a cove, available only for vessels of light draft, sheltered from the southward and eastward by a peninsula, which, together with the rocky shelf about it, extends 800 yards in an easterly direction from the mainland. This peninsula is 400 yards broad and rises at the center to a hill, with a flagstaff on it, close to which is Blowhole Rock. There is a beacon, 46 feet above the sea, on the eastern rock off Kiama Peninsula. There are two detached rocks above water near the south side, and one close to the north point of the peninsula, the latter lying 250 yards northward from the flagstaff. From this rock the entrance of the harbor is 600 yards across in a northwest direction.

Depths.—From 150 yards northward to about the same distance northwestward of the rock which forms the east point of the entrance, there are depths of 4 to 6 fathoms, irregular toward the shore; thence the depths decrease somewhat irregularly, up the harbor, in a southwesterly direction, to 6 feet water at about 150 yards from the shore. The channel leading to the basin is about 60 yards wide, with depths of from 15 to 23 feet.

The harbor basin, formed by jetties or breakwaters, extending to the westward from the east side, and to the north-northwest from the south side of the harbor, is 150 yards long, about 80 yards wide, and there are depths of 15 to 17 feet all over it. There is a red buoy moored in 14 feet water off the entrance on the south side of a 12-foot patch.

Kiama, with a population of about 1,500, lies at the head of the harbor, westward of the basin; but the shore being mostly rock, with shoal water off it, there is no convenient landing place in front of the town. Although small Kiama Harbor is of considerable importance in the beautiful Illawarra district. There are railway and telegraph stations at the town.

The west side of Kiama Harbor trends northward ¼ mile to the northwest point of the entrance, and consists of perpendicular rock, 40 to 50 feet high, bordered by rocky shelves extending from 50 to 150 yards from the shore, and projecting farthest toward the entrance of the basin on the opposite side, the intermediate space in the harbor being about 120 yards wide, with from 10 to 17 feet water.

Light (lat. 34° 40', long. 150° 53').—A fixed green light, 119 feet above water, visible 9 miles, is exhibited from a white circular lighthouse, 40 feet high, on the hill above Blow-hole Rock.

Range lights.—Two range lights, one on the south end of the wharf and one on shore, 142 yards apart, range bearing 221°, lead into the harbor.

Directions.—As the harbor is difficult of access to a sailing vessel it should only be attempted under favorable conditions, and strangers are recommended to hoist the pilot signal, approach the harbor with the end of the breakwater on a southwest bearing and to wait for instructions from officer in charge at a safe distance off the shore. Any swell off the coast sweeps round the bay and through the mouth of the harbor into the basin, rendering it impossible for vessels to lie alongside.

A life-saving rocket apparatus is installed at this port.

Tides.—It is high water, full and change, in Kiama Harbor at 9 h. 00 m.; springs rise 4½ feet.

Aspect.—The most elevated summit of Flinders Ridge, to the southwestward of Kiama, appears to be the Nipple, distant 6½ miles west-southwestward from Kiama Head. It is 2,240 feet high, with one ridge extending west-southwest from it, and another, the Crown Mountains, 10 miles northwestward to the southwestern termination of Reliance Range. The Fall, 2,106 feet high, and Broughton Head, 1,800 feet high, are two other heights between Flinders Ridge and the coast, the former at 2° miles northeastward, and the latter 3 miles east-northeastward, from the Nipple. From the northwestern trend of Flinders Ridge, ridges descend northeastward toward the coast.

Coast—Bass Point.—The coast from Kiama Harbor trends irregularly northward, 1 mile to a projecting head, and thence north-northwest 2 miles to Minumurra River, close off which is Stack Islet surrounded by a reef having 16 fathoms water at ½ mile to the southeastward of it. Reefs also extend from the projecting head into the bay to the northwestward. From Minumurra River two small sandy bays extend 1¼ miles in a northeasterly direction, whence a line of cliffs trends in the same direction 1¾ miles to Bass Point.

Windang Islet (lat. 34° 33′, long. 150° 54′).—Between Bass Point and Red Point, distant 6 miles northward from it, the low coast forms an exposed bay nearly 2 miles deep, the sandy beaches, of which its shore consists, being separated by three rocky points, the northernmost being 2⅜ miles from Bass Point. At 1 mile northward from this point lies Windang Islet close to the beach.

Shell Harbor, between Windang Islet and Bass Point, has a narrow entrance only fit for small craft with local knowledge. The district is principally agricultural. The population is about 500.

Communication.—There are railway and telegraph stations here, and also communication with Sydney by steam vessel.

Illawarra Lake.—The beach immediately behind Windang Islet forms the only barrier between the sea and the southeast corner of Illawarra Lake, which, from 5½ miles west-northwestward of Bass Point, extends northeastward about 5 miles to nearly 2 miles westward of Red Point, and is 3 miles wide.

Macquarie River flows into the southwest corner of the lake and Mullet River into its northwestern part. The lake is shallow.

A considerable quantity of fish are caught in this lake and sent to Sydney market.

Coast.—Red Point, so named from the dull red color of the cliffs and rocks of which it is composed, has four hillocks on it, which present the form of a saddle. Red Point may also be readily known by Mount Kembla, a remarkable hill 1,786 feet high, at about 6 miles west-northwest from it, which, from its form, was first named Hat Hill. From the back of this hill Reliance Range trends southwestward 12 miles, and north-northeastward 10 miles, descending in the latter direction from Mount Keira, 1,573 feet high, to the coast.

From Red Point, cliffs trend northwestward 1 mile to the south point of a sandy bay, within which is Port Kembla.

Red Point Islets and Tom Thumb Islands.—The former are three low rocky islets extending, nearly in line, 1 mile eastward from Red Point, the western and largest being 70 feet high; and Tom Thumb Islands, which are two in number, the southeastern being 20 feet high and the northwestern 15 feet, are also rocky, and lie respectively 2 miles north-northeastward, and 2¼ miles northward from Red Point. They are known also as the Five Islands.

Port Kembla (lat. 34° 29′, long. 150° 56′).—This port is situated about 1 mile to the northward of Red Point.

Depths.—The depths in Port Kembla are from 8 fathoms at the entrance to 4 fathoms at 200 yards from the shore.

Breakwaters.—The eastern breakwater, starting from the eastern point of the south side of the port, extends 800 yards to the northward. The northern breakwater extends 700 yards eastward from the western shore of the port. The outer ends of the breakwaters bear 311° and 131°, from each other, distant 900 yards.

Anchorage.—The port is well sheltered from all winds except those from between northeast and north-northwest, and the holding ground is good in depths of 40 feet and less.

Range lights are shown from two white towers, 670 feet apart, situated between the coal jetties; in range bearing 206°, they lead between the breakwaters.

A flagstaff stands on a mound 46 feet in height, about 300 yards southwestward from the inner end of the eastern breakwater.

Jetties.—The Southern Coal Co.'s jetty is 480 yards long, with depths of 24 to 30 feet near its extremity. Vessels of 8,000 tons have loaded coal at it. Mount Kembla Coal Company's jetty is 300 yards long with a depth of 20 feet at its end. The low-level jetty, between the eastern breakwater and the Mount Kembla Coal Company's jetty, is about 360 yards long in a northerly direction. There is a depth of 30 feet at its outer end. The coal-loading jetty, about 300 yards northwestward from the Southern Coal Company's jetty, extends from the shore about 330 yards in a northeasterly direction. There is a depth of about 35 feet at the outer end. All four jetties are connected by rail with the coal mines.

Buoys.—There are warping buoys off the Southern Coal Co.'s and Mount Kembal Coal Co.'s jetties.

Tides.—It is high water, full and change, at Port Kembla at 8h. 45m.; springs rise 5½ feet, neaps 3½ feet.

Shipping.—Considerable shipping enters and clears Port Kembla.

Tom Thumb Lagoon.—From Port Kembla the coast trends . northwards about 3 miles to Wollongong Head, forming a slight indentation. At 1¼ miles westward from the northwestern Tom Thumb Islet the beach which forms this bay is intersected by the narrow and shallow mouth of Tom Thumb Lagoon, which is 1¼ miles long, north and south, 1 mile wide, and is separated from the seashore by a narrow tongue of land extending 1 mile from the northward to the mouth of the lagoon.

Soundings.—From 30 fathoms at 1¼ miles off Beecroft Head the depth of water decreases to 10 fathoms at 3 miles north-northeastward of the head, whence the soundings increase with some regularity to 42 fathoms at 5 miles off Black Point, between which and 2½ miles off Bass Point the soundings range from 30 to 44 fathoms, and then again decrease to 33 fathoms at 2½ miles southeast Red Point. The 100-fathom curve from 15 miles eastward of Black Point extends northward to about 17 miles off Wollongong Head; at about 20 miles eastward of Kiama Head there are depths of 280 fathoms, fine dark sand.

Wollongong Head (lat. 34° 26′, long. 150° 56′) is a rocky peninsula projecting eastward 400 yards from the lower land to the westward of it; it is 300 yards across from its south side to its north point, and rises to a hill 59 feet high, on which is a signal station, situated 100 yards within the southeastern extremity of the head. Some rocky ledges, which dry, lie about 200 yards from the southeast point of the head.

Wollongong Harbor is situated immediately westward of Wollongong Head. It is about 400 yards in length east and west, with depths of 2 to 3 fathoms in the fairway and protected by a breakwater. On the western side of the entrance to the harbor rocky ledges extend about 100 yards offshore, with a sandy beach southward of them. From a depth of 5 fathoms at 300 yards to the northward of the north point of Wollongong Head the depth of water decreases regularly to about 3 fathoms within about 100 yards of the detached rocks which project from the northwest point of the harbor.

Breakwater.—Wollongong Harbor is protected by a breakwater 230 yards in length, extending west-northwestward from the northwest point of Wollongong Head.

Basin—Piers.—On the northeast side of the harbor, within the breakwater, extensive artificial works have been constructed, forming a basin 200 yards in length by 50 yards in breadth, having an area of 3 acres, with depth of 11 to 15 feet, and wharfage accommodations of about 600 yards. There is also a pier between the breakwater and the basin, with a depth of about 18 feet at its face. There is a crane on it for discharging cargo. There is a timber jetty on the west side of entrance to the basin.

Para Reef.—The only detached danger in Wollongong Bay is Para Reef, lying east-southeastward about ½ mile from the entrance to Para Creek, on the edge of the 3-fathom curve fronting the shore. It has a depth of 9 feet, and when there is any easterly swell the sea breaks on it.

Lights.—A group flashing white light with red sectors, 56 feet above water, visible 10 miles, is shown from a circular iron lighthouse, 56 feet above high water, near the end of the breakwater. The light is visible inside the breakwater to guide vessels into the basin.

A light is shown in the lower part of the northeast side of the lighthouse when it is dangerous for vessels to enter the harbor.

Two range lights are shown on the south shore of the harbor. When southward of Para Reef in line 189°, they lead in to the off end of the breakwater in not less than 15 feet at low water.

Buoy.—A black buoy is placed abreast the breakwater lighthouse, about 15 yards eastward of the line of the range, to indicate the end of the rubble mound extending from the breakwater.

Signal station.—There is a signal station at Wollongong Lighthouse, and communication can be made by the International code. It is connected by telegraph. There is also a flagstaff on the head.

Pilot.—A pilot with boat and crew is stationed at Wollongong, who will render service and assistance to vessels when required, but cannot be depended on for going outside to bring vessels in.

Wollongong town is situated at the base of one of the highest peaks of the Illawarra range. It is gradually increasing in importance, and has about 5,000 inhabitants. The coal mines, about 3 miles distant from the port, are connected with the basin by locomotive and horse traction railways. A large quantity of coal is shipped.

Passenger steam vessels ply frequently to and from Sydney. It is connected with Sydney by rail.

Exports of produce and imports of merchandise amount to about 60,000 tons annually; besides coal, the exports consist of pigs, calves, poultry, cheese, bacon, and butter.

A life-saving rocket apparatus is installed at this port.

Directions.—If from the southward with a strong southerly wind, sail should be shortened and the vessel hove to, 2 miles southward of Flagstaff Hill, to enable the pilot to get aboard; if from the northward sail should be shortened in time and the vessel hove-to off the port, for the same purpose.

From the southward, after passing outside Tom Thumb Islets, steer about 336°, then round Wollongong Head at a prudent distance; when northward of the flagstaff, haul up to 200 yards off the rocks, and work up for the end of the breakwater, taking care to avoid Para Reef, which lies nearly ¼ mile to the northward of it.

From the northward.—When running along the shore from the northward for Wollongong, do not, when within 4 miles of it, bring the extremity of Wollongong Head to bear to the southward of 212°, until Mark Hill, a long double summit hill, barren at each end and thickly wooded in the center, is in range with the center of Flagstaff Bluff on Wollongong Head, bearing 218°, which leads to the eastward of Bellambi Reef in 11 fathoms water, then steer for the harbor. keeping a good lookout for and avoiding Para Reef.

At night.—Wollongong Light shows red over Bellambi Reef. The entrance is approached in the white sector of the light, passing into the red sector when rounding the breakwater. The lights on the south shore of the harbor in range, 189°, lead in from southward of Para Reef.

Caution.—Vessels are frequently detained for days owing to the heavy surf which breaks across the entrance.

Tides.—It is high water, full and change, in Wollongong Harbor at 8 h. 45 m.; springs rise 4¾ feet.

Coast—Towradgi Point (lat. 34° 23', long. 150° 56').—From 400 yards northward of the ledges on the northwest shore of Wollongong Harbor a sandy beach extends northward, with a slight curve for 1½ miles to Towradgi Point, which is formed of bluestone bowlders; a sunken reef extends from it. From Para Creek, behind the ledge of rock at the south end of this beach, a coast range of low sand hills extends close behind the beach to Towradgi Point; these hills are covered with coarse grass and scrub, with dense bush behind them.

Sandspit Point.—Between Towradgi Point and Sandspit Point, at 1,500 yards north-northeast from it, is a sandy bay, with ranges of low sand hills close along the beach, separated by two creeks, the mouths of which are barred across by the beach. Sandspit Point, which has some rocks close to its extremity, is inclosed by a reef, which is always covered.

Bellambi Point and Reef (lat. 34° 22', long. 150° 57').—From Sandspit Point a sandy beach, apparently bordered by a shoal, winds in and out, ½ mile in a north-northeast direction to Bellambi nearly east-southeastward ½ mile from the east side of Bellambi Reef, which partially dries at low water and always breaks, extends nearly south-southeastward ½ mile from the east side of Bellambi Point, and a rock about 200 yards in extent, lies 800 yards southeastward from the point. There are depths of 3 fathoms at 200 yards to the southeastward, and 11 feet water at 50 yards to the northwestward of the rock, with a boat channel nearly midway between it and the shore.

Clearing mark.—Mark Hill and Flagstaff Bluff on Wollongong Bluff in line 218° lead eastward of the reef.

Wollongong light shows red over Bellambi Reef.

Bellambi Bay extends northward from Bellambi Point for 1¼ miles to a point with a ledge of rock projecting about 200 yards from it, and is ⅓ mile deep. The southern part consists, like Bellambi Point, of rock with a sandy surface, extending from the point ½ mile in a westerly direction to a small creek close behind it. Hence to the northwest point of the bay the shore consists of a smooth sandy beach.

Lights.—An occulting white light with red sector, 54 feet above water, visible 8 miles, is shown from an iron framework tower on the summit of the foreshore at the head of the bay between the old and new jetties. It is privately maintained.

A fixed white light, 54 feet above water, is shown from a flagstaff on the head of the South Bulli Coal Co.'s jetty in Bellambi Bay.

Bellambi village—Depths.—From 150 yards northward to ¼ mile northwestward of Bellambi Point there are depths of 3 to 6 fathoms water, from which the depths decrease, somewhat irregularly, to 2 fathoms to within 100 yards of the shore in a little bight extending ¼ mile westward from Bellambi Point, and forming the sea frontage of Bellambi village, the population of which is about 800. Here is a railway station.

Jetty.—In the eastern corner of this bight is a jetty 166 yards in length, with a depth of 22 feet at its end at low water. There is also a new jetty.

Coal.—Vessels drawing up to 19 feet (according to the tide) take in their cargoes under the coal staith from the railway trucks at the end of the jetty. A tramway, 3 miles long, leads to the mines, where a superior steam coal is worked in a seam some 9 feet thick.

Directions.—Approaching Bellambi Bay from the southward, there is to the northward of the northernmost Hat Peak a broken point in the mountain range, named Coorimal or Broken Nose, which bearing 257° leads into the bay clear of the reefs.

From the northward.—From 4 miles off Hacking Point steer 224° until a white sandy point, forming the east extremity of the bay, is seen ahead. While steering for this point an iron-roofed store will come in sight, which, bearing 212°, leads to the jetty.

At night.—Vessels from the northward should keep in the white sector of the Bellambi Light; those from the southward should keep in the white sector of the Wollongong Light until they open the white sector of the Bellambi occulting light.

Coast—Waniora Point (lat. 34° 20′, long. 150° 57′).—From the northwest point of Bellambi Bay the coast extends northward 1 mile to Waniora Point, which has a reef projecting from it, and separates a small bight to the southward, from a sandy bay extending northward 1 mile from the point.

Half a mile southward of Waniora Point a reef extends about 800 yards to the eastward.

Bulli.—Within the bight to the northwest of Waniora Point is Bulli coal station, where a wooden pier 208 yards in length projects east-northeastward over the rocks into 22 feet of water. This position is slightly protected from the southward by a reef of rocks, dry at low water, extending southeastward 300 yards from Waniora Point. Steam vessels lie as close as prudent to the pier, and load in ordinary weather, as they are able to get away on the appearance of a shift of wind to seaward, but it is a dangerous place for sailing vessels to load at, even under favorable circumstances.

Bulli is a postal township with a railway and telegraph station, and a population of about 1,500 persons. It is the headquarters of the Bulli Mining Co.; the mine is about 400 feet above sea level, worked by a tunnel about 2 miles long, the coal being taken by railway locomotives to the jetty about 1½ miles distant. There are several farms in the neighborhood.

Communication with Sydney is either by rail ór by the Wollongong steam vessel.

Coal Cliff—Stanfield Bay.—From Bulli Point the coast trends north-northeastward 5¾ miles in a direct line to Coal Cliff; for the first 2 miles it appears to be low and bordered by reefs, but the remaining portion consists of a line of cliffs. Stanfield Bay is merely the northern of two small bights lying between Coal Cliff and a point 1½ miles northward from it.

Clifton, a mining township, with post, telegraph, and railway stations, is situated about 300 feet above the sea on Coal Cliff. A jetty extends 500 feet from the mouth of the coal mine into 20 feet water, whence steam colliers ship the coal. Population is about 600.

Wata Mooli.—The coast from Stanfield Bay extends, with a little indentation, northeastward 8 miles to Wata Mooli, a boat harbor, where water may be procured, with shelter for large boats, from all winds except those from the southward and eastward. This coast consists of a line of cliffs, except where it is broken for about 1,500 yards by a breach nearly midway between Stanfield Bay and Wata Mooli.

Aspect.—Ranges of hills extend close along the coast from Bellambi Bay to Hacking Point; Bulli Hill within Bulgo Point rises to 1,048 feet, but these hills are not very remarkable; that most worthy of notice being Table Hillock, 484 feet high, at 1½ miles southwestward from Hacking Point.

Soundings.—From about 15 miles off Red Point the 100-fathom curve extends north-northeastward to about the same distance off Hacking Point, with depths of from 75 to 40 fathoms between 11 and 5 miles from the shore. But there are depths of 56 fathoms, mud, at 3 miles off Wata Mooli, and from 24 to 16 fathoms at 1 mile from the shore from Wata Mooli to Hacking Point, the bottom being mostly sand.

Hacking Point (lat. 34° 04′, long. 151° 11′).—From Wata Mooli the coast trends irregularly, in and out, 4¼ miles in a north-northeast direction to Hacking Point, and is mostly fringed with dry and covered rocks, not extending far from the shore. Hacking Point protects Port Hacking from southerly or southeasterly gales.

Jibbon Bumbora, situated about ⅓ mile to the southeastward of Hacking Point, is a detached patch of rocks from 400 to 600 yards in extent, with 15 fathoms water at ¼ mile to the eastward, and 7 fathoms close to the northward of them. The sea nearly always breaks on these rocks.

Port Hacking is a small anchorage, suitable for coasters, within Hacking Point. The entrance, which lies between Hacking Point and Glaisher Point, 1,340 yards to the westward of it, is ½ mile wide, with from 4 to 5 fathoms water between the ledges of rocks, which project from both points of the entrance. From Hacking point the port extends nearly 1 mile in a west-southwest direction to a bar with from 3 to 6 feet water on it, stretching across an opening between two rocky points, lying north and south about 800 yards from each other and forming an inner entrance leading from Port Hacking into the shallow but extensive inlet to the westward.

Port Hacking is nearly ½ mile wide, with two small sandy bights on the south side, divided by a rocky projection, close to which there is a depth of 12 feet. The north shore, which extends southwestward barely ¼ mile from Glaisher Point to the north point of the inner entrance forms the south end of a hilly promontory ¼ to ½ mile broad, extending 1¼ miles from the northward. There are depths from 11 to 6 fathoms from 1 mile northeastward of the port to the entrance of Port Hacking, but from 5 fathoms in the entrance the depth of water inward decreases rapidly to 3 and 2 fathoms.

Although apparently a good anchorage during southerly winds, the swell which sweeps around Bate Bay into Port Hacking renders it an undesirable anchorage.

Pier.—A pier with a depth of 9 feet at its extremity extends in a northwesterly direction from the rocky ledge that lies eastward about 800 yards from Pulpit Point. Two conical red mooring buoys are situated off the outer end of the pier.

Port Hacking Inlet.—From the inner entrance of Port Hacking the inlet trends westward for about 4 miles. It is very irregular in shape and is navigable only by small boats. The boat channel is on the northern side of the inlet.

Tides.—It is high water, full and change, in Port Hacking at 7h. 45m.; springs rise 7½ feet.

Bate Bay (lat. 34° 03′, long. 151° 11′), an exposed bight, 1¼ miles deep, between Port Hacking and Botany Bay, extends from Glaisher Point northeastward 3½ miles to Potter Point, close to the northwestward of which is Botany Cone, 180 feet high, with some dry and sunken rocks extending to the southward from its base. At 1,500 yards westward from Potter Point is another rocky projection, from which a reef, which breaks in places, extends 1,500 yards to the southwestward, with patches of 4 fathoms outlying it. The west shore of

Bate Bay trends irregularly nearly 1 mile northward from Glaisher Point, and is cliffy to a point fringed by a reef on the northwest side of which is the southern extremity of Curranulla Beach.

A reef, dry in places at low water, extends 400 yards east-south-eastward of the point mentioned, with depths less than 5 fathoms for 600 yards.

Curranulla or **Cronulla Beach** commences within the point mentioned, and trends along the foot of bare sandy ridges to the rocky projection 1,500 yards westward of Potter Point. There are depths of 10 fathoms in the middle of the entrance of the bay, between which and 9 fathoms at $\frac{1}{4}$ mile offshore abreast of it, there are irregular depths of 7 to 11 fathoms. About 1,500 yards westward of Potter Point there is a boat harbor having a depth of 6 feet in the center.

Cape Solander.—From Potter Point a cliffy coast, closely fringed with rocks, trends northeastward nearly $\frac{1}{2}$ mile to Cape Baily, whence a more elevated line of cliffs extends about $1\frac{3}{4}$ miles northward to Cape Solander, the southwest point of the entrance of Botany Bay. A sand hill rising from a bluff $\frac{1}{2}$ mile northward of Cape Baily, is the only hill worthy of notice over the cliffs between Capes Baily and Solander.

Botany Bay (lat. 34° 00′, long. 151° 14′) has an historic interest arising from it being the place where Capt. Cook anchored H. M. S. *Endeavour* on Saturday, April 28, 1770, and afterwards took possession of New South Wales in the name of the British Crown. The site of his landing on the southern shore of the bay, within the point named by him Sutherland Point but now known as Inscription Point, is marked by a monument.

The outer heads of Botany Bay are Cape Solander and Cape Banks northeast $1\frac{1}{4}$ miles from it, the latter being a small peninsula projecting a little more than 200 yards from the cliffy land to the northward of it. There are depths of 6 fathoms close to Cape Solander, and 7 fathoms at 200 yards from Cape Banks, with 10 to 17 fathoms between them. From Cape Solander the southwestern side of the entrance is a continuation of cliff trending northwestward, 1,500 yards to Inscription Point, and is bordered by a rocky ledge, not extending beyond 100 yards from the shore; this point and the shore for about $\frac{1}{3}$ mile to the southwestward of it are fringed by a narrow reef.

Depths.—The western half of the bay is shallow, less than 3 fathoms, but the eastern half has from 4 to 6 fathoms between the entrance and a mile southward of the pier, but it is open to the southeastward. This space of comparatively deep water is 1 mile wide, close within the entrance, whence it gradually narrows to the northwestward. There are depths of 6 to 9 fathoms across the entrance between Inscription Point and La Perouse Monument, whence the depth decreases gradually to 4 fathoms at 1,500 yards from the jetty.

At 200 yards to the northwestward of Inscription Point there are depths of 5½ fathoms, whence the 3-fathom edge of the shoal water which occupies the greater portion of the bay trends half a mile to the westward, and after winding 2½ miles in a northwest direction turns southeastward to about 800 yards off the jetty. From this 3-fathom edge the water gradually shoals toward the south and west shores and Cook River, the most shallow part being apparently off the low peninsula and the lagoon to the eastward of it, on the south side of the bay, where there are only depths of 6 feet at about 1,500 yards from the shore. From 800 yards off the jetty the 3-fathom edge of the shoal water which projects from the northeast shore of the bay trends in an east-southeast direction to the first point to the northward of La Perouse Monument, with very shallow water toward the shore.

The northeastern side of the entrance of Botany Bay, from Cape Banks, trends irregularly about ½ mile westward to Henry Head, between which and the northeast inner head of the entrance is Congwong Bay, a shallow bight. The northeast inner head, which lies northward 1,500 yards from Inscription Point, has a tower on it, close to the west side of which is a monument to the memory of the unfortunate French navigator, La Perouse. Bear Islet extends 300 yards from the south point of this head, with which it is connected by narrow fringing reefs.

Bumbora Shoal.—At 300 yards to the southward of Bear Islet is Bumbora, a rocky patch with 4 feet of water on it, which breaks, and between which and the southwestern side of the entrance there is a channel nearly ½ mile wide, with regular depths of 8 to 10 fathoms, and a clear approach from seaward.

Within its entrance Botany Bay forms nearly an equilateral triangle, of which each side is about 4 miles long; the shores are low and wooded, with very shoal water extending a considerable distance from them, except for about 1 mile within the northeast inner head, where some low hilly ranges terminate in two points, one at 800 yards and the other nearly 1 mile northwestward from the monument, each point being closely fringed by a reef, with a depth of 3¼ fathoms close outside the former and 2½ fathoms at 200 yards off the latter point. Between the monument and the northwestern point there are two shallow sandy coves.

The south shore of Botany Bay from Inscription Point (lat. 30° 00′, long. 151° 13′) sweeps round ½ mile in a southwesterly direction, whence a sandy beach extends westward 1¼ miles to Bonna Point, between ½ mile and 1¼ miles to the westward, of which the south shore of the bay is formed by the northern end of a low, flat peninsula named Towra Point, extending from the southward and separating a shallow lagoon named Weeney Bay, on its east side, from the estuary of George River on its west side. The lagoon within its entrance includes Weeney and Quibray Bays, extending nearly 2 miles in an east and west direction and 1 mile south from its entrance.

Shoal Patch is a rock with 8 feet water over it at low water spring tides and lies 550 yards west-northwestward from Inscription Point. Eastward of it the depths increase quickly to 28 feet and there are depths of 12 to 20 feet at about 400 yards westward and southwestward of the rock.

Botany township is situated on the north side of the bay, and has a population of over 10,000. It is much frequented by pleasure parties from Sydney, with which place it is connected by a tramway. There is a pier, with a few feet water at its extremity, situated a short distance southeastward of Sir Joseph Banks Hotel and Pleasure Gardens. There is a post and telegraph office here.

George River.—The estuary of this river is nearly 1 mile wide, between Towra Point and Dolls Point, a low broad point to the westward of it, whence a bight, 1 mile wide, extends 2 miles to the southward. Killigalere and Shell Points give the southern part of this bight the form of a lagoon, which is named Woolooware Bay, and is about the same size as that to the eastward of it. Between Sans Souci and Commons Point the estuary of George River is about ¼ mile wide, whence its main course, between the numerous creeks on either side of it, trends 3 miles west-northwestward to the main river, where a narrow creek joins it from the southward; to this junction the river flows from the northward and westward.

Cook River.—From the west side of the estuary of George River the low sandy shore, Lady Robinsons Beach, which forms the west side of Botany Bay, curves north-northeastward nearly 4 miles to Cook River, a considerable stream flowing from the northwestward, and which formerly supplied the city of Sydney with fresh water.

Light (lat. 33° 57′, long. 151° 12′).—A small light is exhibited from the end of the pier at the entrance to Cook River, as a guide to vessels navigating that river.

North shore.—From the mouth of Cook River, which is about 800 yards wide, the northeast shore of Botany Bay trends east-southeastward 1½ miles to the foot of a hillock, between which and the first point to the southeastward of it is a shallow sandy bay, 1 mile across and nearly ½ mile deep.

Jetty.—At ½ mile westward of the hillock is a small jetty, close behind which is the hotel.

Directions—Anchorage.—To enter Botany Bay (lat. 34° 00′, long. 151° 14′) keep in about mid-channel between the outer heads; Cape Banks just open of Henry Head to the westward of it leads southward of Bumbora Patch. Having passed this danger, haul in toward the northeast shore, and anchor in about 6 fathoms, with La Perouse Monument bearing about 122°. A vessel seeking shelter in Botany Bay from a southerly gale will probably find as good, if not better, anchorage in 3 or 4 fathoms ½ mile to the northwestward of Inscription Point, but care must be taken to avoid Shoal Patch. Although the anchorage in the eastern part of Botany Bay is of considerable extent where vessels may lie in from 7 to 4 fathoms water, there is no shelter from southeasterly winds, and when they blow hard from that quarter a heavy sea rolls in to the bay.

Caution.—Masters of vessels, fishing parties, and others are cautioned against anchoring, creeping, dredging, or fishing near the position of the telegraph cables connecting Botany Bay with New Zealand.

Tides.—It is high water, full and change, in Botany Bay at 8 h. 10 m.; springs rise 7½ feet. The currents set in and out the entrance at the rate of 1½ knots an hour.

Coast—Long Bay.—From Cape Banks a line of cliffs extends in a northerly direction 1½ miles to the southwest point of Long Bay, which is ¼ mile wide at its entrance, whence it runs in 1,340 yards, and terminates to the northward in a narrow cove. Some sunken rocks outlie at a short distance the northeast head of the bay.

Coogee Bay.—From the projecting northeast head of Long Bay the cliffy coast trends northward nearly 1,340 yards to the south point of Marubra Bay, which is 1 mile wide north and south, and nearly ½ mile deep. Between the cliffy headland which forms the north point of this bay and a projecting point at 1¾ miles to the northward of it is Coogee Bay, ¼ mile deep in its northern part, where there are two small inlets, the southwestern of which has some sunken rocks close off it.

Eclipse Bluff.—From the north point of Coogee Bay—which projects nearly 1,340 yards southeastward from the coast line, and has some rocks close to the shore at ½ mile northward of it—a double bight extends north-northeastward 1½ miles to Eclipse Bluff. A point at ½ mile southwestward of the bluff separates Grama Grama Bay on its southwest side from Bondi Bay northeast of it. A sewer outfall has been constructed, with a tall conspicuous chimney on the top of Eclipse Bluff. From Eclipse Bluff a line of coarse sandstone cliffs extends, with a slight curve, northward for 2½ miles to the Outer South head of Port Jackson.

Soundings.—At nearly 14 miles eastward from Hacking Point there are depths of 90 fathoms, dark sand, from close outside of

which the 100-fathom curve extends north and northeastward to a
position at 19 miles eastward from the Outer South head of Port
Jackson. From this 100-fathom curve, the soundings decrease with
regularity toward the shore, which, from 4 miles southward of Hacking Point to the entrance of Port Jackson, may be generally approached to the distance of about 1 mile, in 20 to 30 fathoms, the
bottom being everywhere sand. There are no detached dangers
beyond ¼ mile of this part of the coast.

Port Jackson (lat. 33° 52′, long. 151° 12′), independently of
being the port of the metropolis of New South Wales, is the most
commodious and secure harbor on the east coast of Australia; and
although some vessels, in former times have been wrecked in attempting to enter, these disasters, in most cases, may be attributed
rather to want of judgment and common prudence than to any real
difficulty in making or entering the port.

Aspect.—In approaching Port Jackson from the eastward the
summit of the northern or the two Sydney Heads will, in clear
weather, be first seen, from its being considerably higher than the
adjacent coast. As the port is neared it will be easily identified
by the lighthouse and signal station on Outer South Head and the
bold, perpendicular profile of Outer North Head. The water tower
on Bellevue Hill, southward of the entrance, is also a conspicuous
object.

The characteristic features of the coast to the northward and
southward of Port Jackson assume somewhat different aspects; for,
although North Head, with its immediate vicinity, presents a high,
table-topped precipitous appearance, yet the high, undulating hills,
thickly covered with trees, which rise from the coast farther to the
northward, are strikingly in contrast with the sterile table-topped
cliffs which extend southward of the port; and these hills or cliffs
would, even if the lighthouse did not present a conspicuous feature,
point out whether the land seen is to the northward or southward
of the entrance of Port Jackson.

Depths.—The port is available for all classes of vessels; the least
depth from sea to the city by the Eastern and deepest channel is
40 feet, and by the Western Channel 21 feet. The Western Channel
is being dredged to 40 feet.

Outer South Head.—Outer South Head (lat. 33° 51′, long. 151°
17′) is a precipitous projection of the coast, which here consists
of coarse sandstone cliffs, of a light reddish color; the summit of
the head is 300 feet above the sea.

Light.—A flashing white light, 344 feet above water, visible 25
miles, is shown from a white circular stone building 76 feet high,
standing near the edge of the cliff ¼ mile to the southward of Outer
South Head.

Signal station.—Near the edge of the cliff, about ¼ mile to the northward of Macquarie Lighthouse, is a signal station and telegraph office connected with Sydney. Communication can be made by the International Code, and at night by Morse Lamp. Storm signals are shown.

. **The Gap.**—From Outer South Head the cliffy coast line trends northward 1 mile to Inner South Head, which forms the rounding point on the south side of the entrance to Port Jackson.

Midway between Outer and Inner South Head Lighthouse the profile of the cliffs breaks down to a deep hollow and indentation of the coast, known as the Gap, which is so remarkable that it has on a dark night even been mistaken for the entrance of Port Jackson.

Gap Bluff, immediately to the northward of the Gap, is 170 feet in height.

The water is deep along the coast between Outer and Inner South Heads; there are depths of 14 fathoms at ¼ mile, and from 4 to 9 fathoms 200 yards from the shore; the cliffs are so precipitous as to afford no refuge in case of shipwreck.

Inner South Head.—From Gap Bluff the ridge gradually descends to Inner South Head, which is 60 feet high, and has a lighthouse upon its extremity.

Light.—A fixed white light, 90 feet above water, visible 14 miles, is shown from a tower 30 feet high, painted red and white in vertical stripes, on the edge of the cliff of Inner South Head.

Caution.—Artillery practice takes place occasionally to seaward from South Head. Mariners are warned that when a red flag is hoisted at any of the forts or rifle ranges between North Head and Botany Bay, it indicates that target practice with shot or shell, or rifle practice, is taking place seaward from these positions.

Masters of vessels of every description shall, when the red flag is flying, navigate at a distance of at least 3 miles from the coast, and shall not remain anywhere in the vicinity of the targets.

Additional signals will be shown at South Head and Fort Denison consisting of a ball with the International Code Signal H. J. Y. underneath, to warn vessels that artillery practice is in progress.

The Department of Defense will give seven days' notice of the dates on which practice with shot and shell will take place from the forts.

South Reef is a ledge of rocks extending nearly 200 yards to the northward from Inner South Head; it is easily seen in the daytime by the sea constantly breaking upon it.

Clearing marks.—On the western shore southward of Middle Head and above Obelisk Bay are two white obelisks, each 30 feet high; the eastern and lower obelisk is at the edge of an elbow of the coast. These obelisks in range 274° lead northward of South Reef. At night the red sector of Grotto Point Light covers the South Reef and a little more than 200 yards northward of it.

Outer North Head, on the north side of the entrance of Port Jackson, is a flat-topped perpendicular cliff, 240 feet high.

Inner North Head, nearly 1,500 yards within Outer North Head, is a continuation of the cliffy coast from Outer North Head but it is lower.

The entrance of Port Jackson (lat. 33° 50′, long. 151° 18′) is 2,200 yards wide, between Outer North and Inner South Heads; the narrowest part between Inner North and South Heads is little more than 1,500 yards across from cliff to cliff; and this breadth is reduced by rocky spits extending from both points. The entrance is clear of danger, and the depths are regular, gradually decreasing from about 20 fathoms southward of Outer North Head. Although there is a depth of 10 fathoms within 200 yards of the northern shore, the sea generally rolls in and breaks heavily upon the cliff.

When a heavy sea is running between Sydney Heads, the following signal will be shown at South Head Signal Station, and will also be hoisted at the masthead of the northern flagstaff at the Fort Phillip Signal Station, and at the masthead of the flagstaff on the northern end of Middle Head: A square red flag with a diamond shape below.

Range Lights.—Two range lights have been established on the west side of the Sound to lead into the fairway of the entrance at night.

The front light is exhibited from a white tower, at an elevation of 61 feet, situated on the southwestern extremity of Grotto Point.

The rear light is exhibited, at an elevation of 142 feet, from a white tower on the high land south of the spit, Middle Harbor, at a distance of 1,760 yards 295° from the front light.

These in range lead midway between Inner North Head and South Reef.

Northern shore.—**The sound** is immediately within the entrance, and branches off into Spring Cove, North, and Middle Harbors. Although the Sound occupies an area of nearly 1½ square miles with regular depths of 8 to 9 fathoms, it is too much exposed to the ocean swell to afford safe anchorage, except with off-shore winds. Vessels may wait here for a tug or a favorable opportunity for entering the port.

Examination anchorage.—This anchorage is in the northern part of the Sound northward of a line joining Inner North Head and Grotto Point, and southward of a line drawn 268° from the south side of the point between North Harbor and Spring Cove.

Spring Cove and quarantine establishment.—From Inner North Head the cliffs recede to the northwestward for about ½ mile, terminating at a hummocky point, which forms the sheltering point of Spring Cove, where four or five vessels, in moderate weather, may obtain safe anchorage to ride out quarantine. Vessels must moor inside the line of mark buoys.

The Quarantine establishment and burial ground are situated at about ¼ mile from the cliff, between Spring Cove and Inner North Head.

North Harbor, a bight northwestward of Spring Cove, with depths of from 6 to 8 fathoms, although not apparently open to the fury of southeast gales, is a treacherous anchorage; but after running for Spring Cove, and finding the limited space so filled by vessels as to prevent taking up a berth, anchorage may be had in 6 fathoms, about ¼ mile northward of the north point of the cove, at 200 yards from the eastern shore, and in some measure sheltered from the sea which southeast gales send into the middle and western portions of North Harbor.

Jetties.—There are three jetties in the northeastern portion of North Harbor, chiefly used by steamers of the Port Jackson Steamship Co.

Buoy.—A spit of 2 to 2½ fathoms extends about 300 yards to the southward from Dobroyd Point, the west point of the entrance to North Harbor. The extremity is marked by a spar buoy with "Danger" on a cross arm.

Hunter Bay, the western part of the sound, is 800 yards wide between Middle Head and Grotto Point. No vessel should enter Hunter Bay when blowing hard from the eastward, as it is then a sheet of broken water which, although it has depths of 3½ to 4½ fathoms, would defy any ground tackle and swamp a laden craft if her draft prevented her crossing the 9-foot bar between Hunter Bay and Middle Harbor.

Middle Harbor, which trends to the northwestward from Hunter Bay, has about 9 feet of water on its bar and carries from 13 to 5 fathoms water for about 1,500 yards above the bar, where, after narrowing to 200 yards in width abreast of Hillery Spit, it turns to the westward with a depth of 15 fathoms and branches into deep creeks leading to residences and pleasure resorts and powder hulks in Bantry Bay.

Middle Head, situated 1,340 yards westward from Inner South Head, is a lofty, precipitous, bold bluff of whitish freestone immediately facing the entrance of Port Jackson. As it is exposed to the ocean swell, the sea breaks upon it with great violence during easterly gales.

George Head (lat. 33° 51′, long. 151° 15′), 1,500 yards southwestward from Middle Head, is 209 feet high. A shoal of less than 3 fathoms extends from George Head ¼ mile toward Middle Head.

Bradley Head, the southernmost projection of the north shore of Port Jackson, lies nearly 1¼ miles southwestward from George Head. Shoal water extends 200 yards eastward from the head and 150 yards southward of it.

Beacon.—A pillar stands 25 yards off the southwestern extremity.

Light.—From a white tower situated 20 yards eastward from the southern extremity of Bradley Head, a light is exhibited at an elevation of 22 feet above high water.

A fogsignal is sounded at the lighthouse.

Between George and Bradley Heads are Chowder and Taylor Bays, separated by Chowder Head.

Chowder Bay has a large hotel and a fenced-in bathing place, with large pleasure grounds attached. It is now known as Clifton Gardens.

Robertson Point.—Robertson or Cremorne Point is situated 1,400 yards northwestward of Bradley Head. Between these points are Mossmans Bay and Athol Bight. Shallow water extends about 80 yards southward from the point.

Light.—A fixed white light, visible 5 miles, 24 feet above water, is exhibited from a white circular tower 28 feet in height, on the edge of the drying reef eastward of Robertson Point.

Kirribilli Point (lat. 33° 51′, long. 151° 13′), the most prominent projection from the north shore of Port Jackson westward of Bradley Head, is 1½ miles westward from Bradley Head. A rocky spit extends a short distance from the point, which, with the shoal water at Fort Macquarie Point, reduces this part of the harbor to less than ¼ mile in breadth. There is a landing place at Kirribilli Point, and a wharf a short distance westward of it, for the convenience of the large stores adjoining the grounds of Admiralty House.

Light beacon.—A pile beacon marks the extremity of the spit extending a short distance from this point.

A light is exhibited from the beacon.

A patch of 30 feet, with from 7 to 9 fathoms around, lies 200 yards southwestward from Kirribilli Point beacon. There is a patch of 31 feet 100 yards northwest of the 30-foot patch.

On the northern shore, within Kirribilli Point, are Milsons Point, Blues Point, and Balls Head, with Lavender Bay and Berry Bay between them, for which see chart.

Southern shore—Watson Bay.—From Inner South Head the southern shore of Port Jackson trends southward for about ½ mile to Green Point, the northern extremity of Watson Bay, upon which, at

high-water mark, is a white obelisk about 12 feet high. Parsley
and Vaucluse Bays, which are separated by Vaucluse Point, are two
bights forming a continuation of Watson Bay. Both points of the
common entrance are fringed with sunken rocks; and from Bottle
and Glass Spit, the southern point, foul ground borders the shore
for nearly ¼ mile to the southwestward, terminating at Shark Point.
There is a white obelisk, about 25 feet high, on the southwest shore
of Parsley Bay.

Range lights.—A fixed red light, visible 5 miles, exhibited from
a white towner in Vaucluse Bay at an elevation of 52 feet, is the front
light.

A fixed red light, visible 5 miles, is exhibited from a white tower,
at a height of 275 feet, ¼ mile 186° from the front light.

These lights in range 186° lead through the Sow and Pigs Eastern
Channel.

Lifeboat.—Watson Bay is the lifeboat and pilot station, and there
is smooth anchorage in 6 to 7 fathoms water. There are two piers,
extending about 120 and 300 feet, respectively, from the shore of
the bay.

Shark Island is small, 30 feet high, and thickly wooded. It lies
1,200 yards off Bradley Head and about the same distance southwestward of Shark Point; a spit of foul ground extends nearly 200
yards from its northern end.

Buoy.—The shoal ground, with depths under 3 fathoms, extending 150 yards southeast of the island, is marked by a black buoy.

Light.—From a white tower, in 11½ feet water, on the northern
extremity of the spit off Shark Island, a light is shown at an elevation of 40 feet above high water.

Clarke Island, 1,700 yards above Shark Island, is 42 feet high,
and similar in aspect to that island, but smaller; the edge of the
5-fathom curve lies at 100 yards northeast and southwestward of it,
and less in other places around it.

Beacon.—A black beacon, with cage, is erected on the edge of the
reef extending southward of the island.

Darling Point—Rushcutter Bay.—A flat with depths under 3
fathoms extends ¼ mile northwestward of the western extremity of
Darling Point, two-thirds across the entrance to Rushcutter Bay; its
western and northern extremities are marked by cage beacons in
about 4 fathoms of water. A small wooden beacon marks the extremity of the 1-fathom part of the flat.

Garden Island, which lies nearly 1 mile southwestward of Bradley Head, is considerably larger and higher than the others. The
island is appropriated as a depot for the Australian Navy and has
coal and other wharves on the west side. On the island are store-

houses, engineers' shops, a sawmill, boathouse, boat slip, barracks, heavy sheers, etc.

Dolphin.—A dolphin, with a cage top, marks the edge of the shoal water off the northeast point, in about 5 fathoms.

Shallow water extends about 70 yards from its southwest side; and between the island and Potts Point, 300 yards to the southward, is a ridge 200 yards wide, with depths of about 4¼ fathoms over it, except near the shores, where it shoals gradually.

A telegraph cable and water pipe are laid from the south end of the island to Potts Point. The island is also in telephonic communication with Admiralty House on Kirribilli Point.

Fort Denison—Light.—Fort Denison, formerly called Pinchgut Islet, lies ¼ mile northwestward of the north end of Garden Island. This islet is a mass of bare rock and masonry, with a white tower, surmounted by a lantern, on its northeastern extremity, from which a light is exhibited, at a height of 61 feet. The islet should not be approached within 100 yards, as spits extend out a short distance.

Fogsignal.—In thick or foggy weather a fogsignal is sounded at Fort Denison.

Benelong Point—Light beacon.—Tram sheds are situated near the northern extremity of Benelong Point, which separates Farm any Sydney Coves. It is connected by submarine telegraph cable with Fort Denison and with Bradley Head lighthouse. Shallow water extends 100 yards from the point, the northwestern extremity being marked by a pile beacon from which a light is exhibited.

A small light is exhibited from the northeast corner of Fort Macquarie Horse ferry jetty.

Sydney Cove—Light.—A light is shown from a red beacon with cage topmark erected on the eastern edge of the shoal extending southward from Dawes Point.

Dawes Point Light.—A light is shown from a pole at the northwest corner of a large cargo shed on Dawes Point.

Fogsignal.—In thick or foggy weather a fogsignal is sounded on the northeastern extremity of the point on the end of the Horse Ferry Wharf.

Goat Island—Range lights.—Two lights are exhibited at the northeast end of Goat Island; the western and rear light, 32 feet above high water, is shown from a pole with a black square daymark at the east end of the jetty; the eastern and front light, 12 feet above high water, is shown from a pile beacon with white diamond off the northeast end of the island. When in range bearing 279° they lead in the fairway past Kirribilli and Dawes Points.

Millers Point—Light.—From the northwest corner of the wharf on Millers Point a light is exhibited from a white steel tower at an

175078°—20——29

elevation of 69 feet. This light facilitates the approach to Darling Harbor at night.

Entrance channels—Depths.—The bar, with Sow and Pigs Shoal, formerly extended nearly across Port Jackson, between Inner South and George Heads. East channel, on the eastern side of these shoals, has been dredged to a low-water depth of 40 feet, with a width of 700 feet. The dredging of the western channel to the same depth and width is in progress, but some time will elapse before it is completed; the average width (1919), carrying 40 feet and upward, is about 200 feet, the remainder carrying a minimum depth of 22½ feet.

It is intended that the eastern channel shall be uséd only by vessels leaving the port and the western channel by those entering it when the latter is sufficiently dredged.

On the northern portion of the Sow and Pigs Flats is a group of rocks, showing at half tide, marked by an iron rod beacon surmounted by an open-hoop ball; the beacon lies nearly midway between the two shores. Foul ground, with depths under 3 fathoms, extends 150 yards northward of the beacon. An irregularly shaped flat of rough ground, with depths of from 2 to 3 fathoms, lies to the southward of the beacon; it is about 600 yards in length, north and south, and 400 yards wide; its southern extremity lies 700 yards southward from the beacon.

At 150 yards to the westward of this flat is a 3-fathom patch 150 yards in extent, the southwestern extremity of which lies 700 yards southwestward from the beacon.

Caution.—At times a heavy swell sets into the harbor, which requires an allowance of a fathom for scend in the entrance; this must be duly considered by vessels of heavy draft.

Lightbuoys.—The dangers are marked as follows:

Sow and Pigs Lightbuoy, a black conical buoy, exhibiting, at a height of 22 feet, a white flashing light, is moored about 250 yards north-northwestward of the beacon on Sow and Pigs Shoal.

Eastern Channel Lightbuoy, black and showing a flashing red light, is moored on the northern extremity of the 5-fathom curve, east side of Eastern Channel, at nearly 400 yards westward of Hornby Lighthouse.

Western Channel Lightbuoy, black and showing a flashing red light, is moored in about 6½ fathoms close westward of the 3-fathom patch of 650 yards southwestward of Sow and Pigs Shoal Beacon, on the east side of Western Channel.

Eastern Channel Light Beacon.—A white tower, showing a light at an elevation of 31 feet, is situated in Eastern Channel, in 28 feet of water, off the southeastern extremity of Sow and Pigs Shoal ¼ mile west-southwestward from the obelisk on Green Point.

Pilots.—The pilot station at Watson Bay is within ½ mile of the signal station on Outer South Head, and the lookout is kept at the signal tower, from which night signals of vessels requiring pilots are answered.

A Government steam vessel is stationed in Watson Bay to take pilots off to vessels making the port. If the state of the sea will not admit of a pilot being put on board from the steam vessel, she will lead the way into smoother water between the heads, where tugs will be in attendance.

Pilotage regulations.—The pilot service of New South Wales is under the control of the department of navigation. Port charges, harbor and light dues, removal charges, etc., will be found in the Official Handbook of the port of Sydney. Masters of vessels should obtain a copy for their guidance. The master of every merchant vessel not by law exempt from the necessity of accepting the services of a pilot is to place her in charge of the first licensed pilot that may come alongside, and such master is not to enter the harbor or proceed to sea, or quit his anchorage without having a licensed pilot on board, under penalty equal to the amount of pilotage to which he would have been subject if a pilot had been employed.

Steam tugs may be summoned by signal when required; usually one or more are in waiting outside.

Speed.—Vessels other than ferryboats must not proceed at a speed exceeding 8 knots when west of a line drawn between Bradley Head and Shark Island and not exceeding 6 knots when west of a line drawn north and south through Fort Denison.

The speed of any steamer navigating the port shall be reduced while passing the wharves at Millers Point or any works of the commissioners in progress or any dredge or other vessels employed in the execution of such works, so that such works or dredge or other vessel shall not be interrupted or damaged.

Sound signals.—A steam whistle, or bell, or foghorn shall not be sounded on any vessel navigating the waters of the port for the purpose of summoning or signaling to workmen or passengers or to any person whatsoever, but shall only be sounded for such purposes as relate to navigation.

Pratique.—Vessels from foreign ports are forbidden by the port regulations to pass an imaginary line drawn from Bradley Head to Shark Point until boarded by the health officer and other authorities.

Western channel, between the Sow and Pigs Shoal and the western shore, is being dredged to a depth of 40 feet.

Range.—St. John's Church at Darlinghurst just open of Bradley Head, bearing 222°, leads through West Channel, passing 200 yards eastward of George Head.

Eastern channel has been dredged to a depth of 40 feet over a breadth of 700 feet, as before mentioned.

Shark Bay Lightbuoy, painted black, off Shark Beach, showing a white flashing light, is moored on the 5-fathom patch at about 400 yards northward from Shark Point, eastern side of eastern channel; the lightbuoys seaward of it are mentioned with the Sow and Pigs.

Range lights for east channel and cross mark for the southern edge of Sow and Pigs Shoal:

The two white lighthouses in the vicinity of Vaucluse Bay in range, bearing 186°, lead through East Channel in not less than 40 feet at low water springs.

The Eastern Channel light-beacon, on the southeastern elbow of the Sow and Pigs shoal, marks the rounding point from the East Channel.

The white flashing light buoy off Shark Bay should be left on the port hand by vessels bound up the harbor.

Macquarie or Outer South Head Lighthouse, open southward of the red and white checkered obelisk upon the eastern wooded slope of Parsley Bay, bearing 123°, clears the southwestern or inner edge of the Sow and Pigs Flats.

Above the bar—Sow and Pigs.—Port Jackson, above the bar, is so free from dangers that the chart is a sufficient guide.

General directions for approaching Port Jackson.—Steam vessels have no difficulty in entering at all times both by night and by day, provided the weather is clear, being guided by the buoys and lights as charted.

Pilots are available by those requiring them. Pilotage is compulsory to all who are not exempt by law.

The most unfavorable times for sailing vessels entering the port are in easterly gales, southerly gales, and light variable winds, with a ground swell rolling in upon the heads. Sailing vessels of, say, over 500 tons should not attempt to enter the port under canvas unless the wind is somewhere between east and north. Tugs are always available.

Easterly gales sometimes blow very hard, causing a heavy sea upon this coast, which not only breaks with great violence upon Sydney Heads, but occasionally right across the entrance and directly home to Middle Head.

Easterly gales are frequently attended by haze banks, which might prevent the lights being seen at night, until too late for a vessel to claw off the land; sailing vessels should therefore, day or night, keep the sea rather than bear up for Port Jackson in a gale from the eastward, and should not approach the coast within 10 miles.

If a sailing vessel is making for Port Jackson in bad weather she should endeavor to pass the South Reef as close as is prudent and round up to the southward and anchor until picked up by a tug. If unable to do this she should run for the Sound and pick up an anchorage there as near Spring Cove as possible without encroaching on the Quarantine Reserve.

It must be borne in mind, when getting an offing, that the weather gauge will be to the northeastward as the gale expends itself, and that in standing to the northward the vessel is safe as long as Outer South Head (Macquarie) Light is open of Outer North Head, within which line no vessel should approach the coast.

The southerly gales are strong squally winds, which rush down the harbor, and frequently embarrass sailing vessels when working up between the heads, sometimes taking them aback, and exposing them to destruction against the North Head Cliffs; vessels should therefore wait outside until the wind becomes more steady, unless in very good working order and the flood stream is running.

Sailing vessels should not attempt to enter between the heads with light variable winds, as, under such circumstances, they frequently become unmanageable, and, being left to the mercy of the ground swell, may be set upon either of the heads; therefore, it is advisable to anchor and wait for a steady breeze, or summon a steam tug, before getting too near the heads.

If bound to Port Jackson in heavy weather, and from want of observations, uncertain of the latitude, and the land is fallen in with either to the southward or the northward of it, shelter may be found in Botany Bay, to the southward, or Broken Bay or Port Stephens to the northward, according to circumstances.

Botany Bay lies about 10 miles southward, and Broken Bay lies 16 miles northward of Port Jackson; and it is of the utmost consequence that those in charge of such vessels as may happen to be in bad condition, and unable to keep offshore, should be aware of these useful places of refuge.

Sailing vessels approaching Port Jackson at night, with southerly or westerly winds, should keep the sea until daylight, but with winds from the northward or eastward, and favorable weather, they may safely enter.

The depths have been accurately ascertained within the range of Macquarie or Outer South Head light, and are a valuable assistance when near the land in thick weather. East of the entrance of Port Jackson, at 18 miles offshore, the depth is 100 fathoms, light green sand, from which it shoals regularly to 20 fathoms, close in with the land and with the entrance. Northward of the port the 100-fathom curve is farther offshore, and on the contrary, to the southward, this depth does not extend off more than 14 miles.

From the southward.—Vessels approaching Port Jackson from the southward would most probably sight the entrance to Botany Bay and also the buildings to the northward of it, namely, Little Bay hospital and the suburbs of Coogee and Bondi, the latter showing numerous lights at night.

At night time care should be taken not to get Macquarie Light on a bearing inside of 359°, as the light can be seen over the land when close to Port Hacking.

Eclipse Bluff, 2¼ miles from Macquarie Light, forms the commencement of a line of cliffs which continues to the entrance of Port Jackson.

A small tower, in connection with the sewerage outfall, is a conspicuous feature near Eclipse Bluff.

Coming from the southward vessels should steer for an offing of from 1,500 yards to a mile off Macquarie Light.

If the weather be dark or thick preserve a good offing until Sydney Heads or Outer South Head lighthouse be seen, in order to clear the projection of the coast about Botany Bay, where it is comparatively low, and where the current sometimes sets southwest, toward the shore.

A lookout should be kept on the signal station. If the following signals appear it will be an indication to vessels approaching the port that a vessel is proceeding outward through the east channel:

By day: The red burgee with a black shape, 2 feet in diameter underneath.

By night: A red light above a white light.

Steer for Outer North Head, about 0°, until the range marks or lights on Grotto Point and the Spit (in Middle Harbor) are in line 295°; then round-in, keeping along the north shore about a quarter of a mile off, until nearly abreast of Inner North Head, when the vessel can be gradually brought on to the range for the east channel or the west channel, as desired.

From the eastward make the land in a latitude of 33° 50′, and on approaching the heads enter in the white sector of the Grotto Point Light, keeping the red range light open to the northward and proceeding as before directed.

Entering by east channel.—Steer through the east channel with the range lights at Vaucluse in line 186° until approaching eastern channel white light beacon tower marking the southeast corner of the Sow and Pigs Shoal; round this about 100 yards off and then steer to pass Bradley Head at the distance of about 400 yards.

On approaching Bradley Head remember that George Head, in range with Middle Head, clears the shoal water eastward of it, and at night Macquarie Light, in range with Shark Island Light, leads

300 yards southward of the shoal water south of Bradley Head. When Fort Denison opens clear of Bradley Head commence to round it, keeping at least 200 yards off until Dawes Point is about midway between Fort Denison and Kirribilli Point. Here speed must be reduced to 8 knots and a sharp lookout kept for vessels leaving the northern anchorage and also for ferry steamers coming out of Athol Bight.

Pass to the northward of Fort Denison about 100 yards, and the same distance southward of Kirribilli; afterwards, keeping on the starboard side of the fairway and proceed to destination.

The lights on Goat Island, in range 279°, mark the middle of the fairway between Kirribilli and Bennelong Points, and also between Milsons and Dawes Points.

Entering by west channel.—Enter the heads as before directed and steer for Grotto Point until Bradley Head light tower is brought midway between the western channel lightbuoy and George Head, or nearly in range with St. John's Church at Darlinghurst, 222°.

Proceed through the channel, keeping Bradley Head light tower as above directed until the Macquarie Lighthouse opens to the southward of the checkered obelisk on Vaucluse Slope, or until the western channel lightbuoy has been passed, when the shoal water will have been passed. Keep Bradley Head half a point on the starboard bow and proceed as before directed.

Outward—By east channel.—Keep on the starboard side of the fairway and keep a good lookout for ferry traffic, especially when passing circular quay. Steer to pass about 100 yards southward of Fort Denison; whence steer. for the south end of Shark Island. This will take the ship about 600 yards off Bradley Head, and when the lightbuoys on the western side of the Sow and Pigs open clear of Bradley Head commence to alter course to port, maintaining a distance of at least 600 yards from Bradley Head until the Outer North Head is in range with the eastern chanel white light beacon tower on the southeastern corner of the Sow and Pigs Shoal. Keep the obelisk on Green Point ahead, passing to the northward of the Shark Bay lightbuoy and rounding the eastern channel white light beacon tower about 200 yards off until getting the east channel range in line astern. For a range ahead in going out of east channel, keep Manly Pier open of Kihi Point.

Proceed through the channel and, passing the red sector of the Grotto Point Light, steer out through the heads with the entrance range open to the southward.

When the Macquarie Light opens clear of Gap Bluff the vessel will be clear of the eastern extremity of the South Reef.

Outward by West Channel.—As at present the port regulations provide for vessels under 20 feet draft proceeding to sea by the

West Channel, such vessels should round Bradley Head as above directed, and when abreast of Shark Island Light proceed for the West Channel, keeping a good lookout for vessels inward bound by the East Channel.

The Cockatoo Island.—As in the moving of vessels in the upper parts of the harbor it is compulsory to engage a harbor pilot, directions are unnecessary. The least water between Ball's Head and Longnose Point is 30 feet. Vessels of large draft may proceed to the Sutherland and Woolwich docks on a straight course. Ships may proceed to Sutherland Dock on either side of Cockatoo Island according to wind and tide. The southern passage is a little deeper than the northern.

Working into Port Jackson.—A westerly wind blows right out of the entrance, but there is ample working room for a well-handled vessel between the heads, the shortest tack being ½ mile, between South Reef and Inner North Head. Tugs are available.

Caution.—To insure success in working in, and to avoid mishap, smart working and readiness with both anchors is absolutely necessary to cope with flaws and gusts of wind, as well as the ground swell, which perplex even those who frequent Port Jackson.

Tides.—At North Head it is high water, full and change, at 8 h. 15 m., mean springs rise 6 feet; and at Sydney at 8 h. 38 m., springs rise 5¼ feet and neaps 4 feet. At Cockatoo Island it is high water 20 minutes later than at Fort Denison.

From April to October night tides are higher than day tides; in June and July this sometimes amounts to 2 feet, and the reverse occurs in December and January. From October to April day tides are the higher, and it is stated that the highest tides are with westerly winds.

The usual sequence of the tides is from lower low water to higher high water.

Tidal currents.—In the offing, within the line of the current, the ebb current sets to the southward and the flood to the northward. Outside the bar, as before stated, the ebb sets across the sound toward Inner North Head and then about 123° close along shore in the direction of Outer North Head, leaving the space between the line of the Outer Heads and Inner South Head in slack water during the ebb. The ebb and flood currents set fairly across the bar northeast and southwest and up as far as Shark Island. Above Bradley Head the ebb current runs eastward and the flood westward, the maximum velocity of the ebb being 2 and of the flood 1½ knots.

The tidal current, especially at spring tides, is interrupted by the numerous headlands and bays causing eddies to shoot off the points partly into the adjoining bays and partly off the points into the stream.

In the vicinity of Cockatoo Island the tidal currents change at approximately the times of high and low water on the shore.

The flood current sets to the westward between Cockatoo Island and Hunters Hill, and also on the southern side of the passage between that island and Balmain;- between Cockatoo Island and Spectacle Island it sets to the northwestward.

The ebb current in each case sets in the opposite direction.

Off the entrance to Fitzroy Dock both flood and ebb currents form an eddy setting to the northeastward, which has its greatest strength during the flood.

. Anchorages—Woolloomooloo Bay and Farm Cove (lat. 33° 52', long. 151° 13').—Farm cove and the northern part of Woolloomooloo Bay are reserved for the anchorage of vessels of war, and several sets of moorings are laid here for the use of H. M. ships. Man-of-war anchorage is included in the space bounded by a line drawn from Fort Macquarie, passing about 100 yards south of Fort Denison, to the northern extremity of Garden Island, and a line from the southern extremity of Garden Island, westward to the land; the space extending 600 yards eastward of the island and about the same distance wide from north to south, is also included.

Merchant vessels can find good anchorage north of a line joining Bradley Head and Kirribilli and south of a line drawn from the northern end of Garden Island to the northern end of Clarke Island (excepting 600 yards from Garden Island), and thence to Shark Island Light.

A sector of white light is shown over the fairway of Woolloomooloo Bay from Fort Denison Light.

Oil wharf is on the western side of Garden Island, extending 90 feet to the northeastward from the coal wharf. The depths along its northwestern face has been dredged to 33 feet.

Water.—At the boat camber inside a jetty on the west side of Farm Cove, water of excellent quality is at all times of tide supplied, free of charge, to Government boats. Water is also supplied to shipping at all wharves in the port, and to ships in the stream by water boats.

The charges for water supplied through hoses supplied by the commissioners is 2s. and in other cases 1s. 6d. per 1,000 gallons.

Sydney Cove lies westward of Fort Macquarie. Navigation is prohibited in this cove between the hours of 8 a. m. and 9.30 a. m., and the hours of 5 p. m. and midnight, and on Saturdays between the additional hours of noon and 1.30 p. m., except for certain mail steamers on the latter day. Further details will be found in the Harbor Regulations.

The fairway of Port Jackson, within the limits of which vessels are not allowed to remain at anchor, is bounded on the north side

by an imaginary line extending from the distance of 300 yards eastward of Middle Head to about 70 yards southward of Bradley Head,
Kirribilli, and Blues Points; and on the south side by a line passing
at the distance of about 70 yards westward of Inner South Head
and Green Point, northward of Shark Island Light, Clarke and
Garden Islands, and Dawes Point.

Prohibited anchorage.—A vessel shall not anchor

(a) Near any of the following lines:

    (1) A line between Potts Point and the south end of Garden
        Island.

    (2) A line between Bradley Head and Fort Denison.

    (3) A line between Fort Denison and Fort Macquarie.

    (4) A line between Millers Point and the southeast end of
        Goat Island.

    (5) A line between Simmons Point and the southwest end of
        Goat Island.

    (6) A line between Whitehorse Point and the Fitzroy Dock.

(b) Within or near the area bounded by the following lines:

    (1) A line joining the eastern extremity of Milsons Point and
        the horse ferry at Dawes Point.

    (2) A line joining the northwest corner of Dawes Point and
        McMahons Point.

    (3) A line drawn east and west through McMahons Point.

(c) In or near any of the following positions:

    (1) South of a line drawn between Dawes Point and Millers
        Point.

    (2) Within 250 feet of Pyrmont Bridge, or Glebe Island
        Bridge.

    (3) The entrance to Five Dock Bay.

    (4) The entrance to Lane Cove River, Greenwich.

    (5) Also in any part of the port in which a cable, telephone
        wire, or water pipe is laid.

    (6) Within 600 feet of any wharf, or in such a position as
        shall obstruct the approach thereto.

It is not here deemed necessary to enter into a minute description
of Port Jackson above Fort Macquarie, as a vessel having arrived
thus far will be berthed by the Portmaster's directions; other details
of the shore will be best understood by reference to the chart.

Submarine cable.—Persons are cautioned against anchoring,
creeping, dredging, or fishing near the submarine cables between
Dawes Point and Milsons and Blues Points. Fresh-water pipes are
laid between the first two of these points.

Sydney (lat. 33° 52', long. 151° 12'), the capital and seat of government of New South Wales, is situated on the south side of the
port, about 4 miles within the entrance. The shores in all directions

are broken up by steep jutting points, of moderate elevation, forming bays and coves which are harbors of themselves, and allow the heart of the city to be readily reached from sea.

The limits of greatest commercial activity extend, with the exception of Farm Cove, from Wooloomooloo Bay to Darling Harbor; the great natural advantage of deep water generally continuing to the shore being fully utilized and artificially improved, so that from Fort Macquarie westward the frontage is skirted by an almost unbroken line of wharves and quays, amply provided with covered sheds and warehouses.

Extensive works have been completed in Wooloomooloo Bay and afford excellent facilities for dealing with modern shipping. A large jetty, 1,140 feet long by 203 feet wide, has been constructed in the center of the bay. Double-decked cargo sheds have been erected on this jetty. The wharves on the eastern side of the bay have been reconstructed and fitted with modern accommodation.

Circular Quay, or Sydney Cove, originally the first settlement, has been remodeled within the last 10 years. Mail and other ocean steamships are accommodated here on the east and west sides, the depths being up to 30 feet. The south side is reserved for ferry accommodation. The customhouse faces Circular Quay.

At the head of Darling Harbor is the terminus for goods traffic of the railway system of New South Wales. In Pyrmont Bridge across the upper part of Darling Harbor is a swing opening of 70 feet in width.

Within the city there are numerous and extensive factories and foundries. Steam ferryboats ply between the city and its transmarine suburbs.

Trade.—In the year ending June, 1918, the total number of vessels entering the port was 7,538, with a gross registered tonnage of 9,058,568 and net tonnage of 5,320,400. The goods imported during the year amounted to 3,447,653 tons. The value of oversea, interstate, and State imports was £55,371,749; overseas exports, exclusive of gold, were valued at £36,288,589.

The estimated population of the city and immediate suburbs is 800,000.

Telegraph.—The telegraph office at Sydney is always open, both day and night.

Radio station.—A radio station has been established in the Pennant Hills in latitude 33° 40′, longitude 151° 00′. It is open to the public from 7 a. m. to 2 a. m. The call letters are V I S. Weather forecasts are repeated to vessels at 8 p. m. and 9.30 p. m.

Berthage accommodation.—There are nine berths, 600 to 700 feet long, with depths of 30 to 35 feet, and two with depths of 40 to 45 feet; 28 berths, 500 to 600 feet long, with depths of 28 to 30 feet;

and numerous smaller berths. The total berthage is 57,556 feet, and
there are 5,836 feet under construction.

Time ball.—A ball is dropped from the top of the observatory at
Sydney at 1 h. 00 m. 00 s. p m., New South Wales standard time,
equivalent to 15 h. 00 m. 00 s. Greenwich mean time. The ball is
hoisted half way as preparatory at five minutes before the signal.
When the signal fails in accuracy the ball is at once hoisted half way
and kept up for one hour. The amount of error is published in the
local daily papers.

Coaling facilities.—Coal for steaming purposes can be obtained
at Sydney in any quantity. Vessels are supplied from hulks and
lighters or direct from the colliers, and loading facilities are established at Pyrmont Wharf.

Docks and slips.—Every facility is to be obtained at Sydney for
repairing vessels of any size or description, with abundant supplies
and stores of every kind. See Appendix I.

The Commonwealth Government do no undertake the construction
or repair of vessels or machinery for private owners in competition
with private firms. The docks and slips of the Commonwealth Government are not available for docking privately owned vessels unless
the circumstances are exceptional. The use of the workshops and
machinery belonging to the establishment may be hired.

Vessels in the service of the British Admiralty or the Governments
of the Commonwealth and the States and also foreign men-of-war
are exempted from dock dues, but are required to pay all expenses.
No vessel is permitted to enter the docks with gunpowder or explosives of any kind on board.

Building and repairs.—At Mort's Dock & Engineering Co.
(Ltd.) steel vessels of 1,200 tons can be built; engines of 1,500 indicated horsepower made and repaired. Casting to 40 tons; cylinders
bored to 120 inches. Boilers constructed up to 16 feet diameter and
with shell plates up to 1⅜ inches. Shafting forged and turned to 24
inches diameter and 38 feet long. Pipes of 36 inches diameter brazed.
Repairs of every kind executed. Masts made and boats built. The
sheers at H. M. dockyard lift 160 tons; others, private, up to 60 tons.
At Mort's establishment there is a machine for testing cables up to
300 tons.

Other engineering and shipbuilding firms in Sydney are Poole &
Steele, in Johnsons Bay; Chapman (Ltd.), Waterview Bay; Wood-
ley's (Ltd.), Gore Bay; Morrison & Sinclair, Johnsons Bay; David
Drake, Balmain.

Passenger steamship companies having establishments at Sydney are: Peninsular & Oriental, Orient, Anglo-Australasian, Australian & American, British India, Canadian Australian, China Nav-

igation, Eastern & Australian, New Zealand Shipping, Messageries Maritimes, etc.

Explosives.—Vessels having explosives on board shall not at any time lie or be to the westward of a line between the pillar on the southern extremity of Bradley Head and the eastern extremity of Clarke Island, or to the southward of a line between the northern extremity of Clarke Island and the northern extremity of Woollahra Point, or within 300 yards of the shore of the port. International flag B must be shown at the foremast head.

The penalty incurred by a breach of this regulation is a sum not exceeding £50.

This regulation does not apply to ships of war and Government vessels, nor to vessels having on board small or necessary quantities of explosives properly shipped and stored in magazines. See Regulations.

Port regulations.—Masters of vessels should procure a copy of the Port Regulations on first arrival.

Signals.—The following shall be the signals in use in the port:

1. Explosives on board_____Flag B at the foremast head.
2. Pilot exemption_____Square white flag and numerical flags of Port Jackson code of signals denoting port from which arrived.
3. Want custom boat_____Jack at peak.
4. Quarantine_____Q at foremast head.
5. Cholera, yellow fever, or
   plague_____International Code L.
   Signal for (4) and (5)
   from sunset to sun-
   rise: Three l i g h t s
   (two red and one
   white) p l a c e d   a s
   nearly as practicable
   amidships, at d i s -
   tances of 6 feet apart,
   in the form of an
   equilateral triangle
   with the apex (the
   white light) above.
6. Mails on board_____International Code T until mails are landed.
7. Want harbor pilot_____International Code S at foremast head.
8. Want sea pilot_____Jack at foremast head.
9. Want medical assistance___International Code H at peak.
10. Want water police_____International Code N Y at mainmast head.
11. Want steam tug_____International Code Y P at mainmast head.
12. Want water boat_____International Code G U J at mainmast head.
13. Want boarding officer_____Numerical flags of Port Jackson code of signals denoting port from which arrived, or International Code signal.
14. Oil vessels_____Red flag at a masthead; red light at a masthead at night.

Ballast, etc.—No ballast, ashes, or other material, nor any dead animal or putrefying matter is to be thrown into the waters of the port.

Proceeding to sea.—1. Every vessel proceeding to sea and intending to pass through the eastern channel shall, when to the eastward of a line joining Bradley Head and Clarke Island, exhibit at the foremast head the following signal:

By day: A red burgee with a black shape 2 feet in diameter underneath, the distance apart to be 6 feet.

At night: A red light above a white light, the distance apart to be 6 feet.

Such signals shall be kept displayed until the vessel is to the eastward of a line joining the North Head and the Inner South Head. When repeated at the South Head Signal Station these signals shall be an indication to vessels approaching the port that a vessel is proceeding outward through the eastern channel.

2. Every vessel entering or leaving the port in tow shall, when the draft of such vessel permits, be navigated through the western channel.

For continuation of the coast to the northward see Australian Pilot, Vol. III.—H. O. publication No. 169.

APPENDIX No. I.

Particulars of dry docks, patent slips, etc.

Port.	Name of dock.	Length.		Breadth of entrance.	Depth at M. H. W. S.		Springs rise.	Lifting power.	Date built.	Remarks.
		On blocks.	Over all.		On sill.	On blocks.				
		Feet.	*Feet.*	*Feet.*	*Feet.*	*Feet.*	*Feet.*	*Tons.*		
Melbourne	G. S. Duke & Son	510	520	65	21	23½–20	2¾	....	1875	
	Wright, Orr & Co	415	428	50	21	23½–19½	2¾	....	1880	
Melbourne, Williamstown.	Alfred	450	470	77	27	24¾–23¾	2¼	....	1874	The Alfred Dock has taken a vessel 486 feet long between perpendiculars, 507 feet over all, and 23½ feet draft.
	Floating Dock	....	216	36¾	14	12	2¼	900	1896	
	Patent Slips:									
	Williamstown Patent Slip Co	[1] 165	....	....	....	[2] 8 [3] 12	2¼	400	1860	
	Melbourne Coal Shipping & Engineering Co.	[1] 66	....	....	....	....	2¼	....	....	
Sydney	Sutherland Government	608	602	83½	32	32	5¼	....	1890	With caisson in outer stop 30 feet longer. The depth in this dock is considerably affected by the wind.
	Fitzroy Government	477	480	56	21	19½	5¼	....	1857	With caisson in outer stop 22½ feet longer.
	Mort's, New	700	700	82	28	28½	5¼	....	....	To be increased to 775 feet. At Woolwich.
	Atlas Pontoon	196	....	56½	....	14	5¼	1,400	1888	
	Mort's Balmain	640	640	69	18	....	5¼	....	1899	At Balmain.
	Jubilee Floating	....	320	44	....	15	5¼	1,200	1887	At Balmain.
	Rowntree Floating	....	161½	42	....	15	5¼	400	1864	At Balmain.
	Goodall's Floating	100	....	21	7½	....	5¼	120	....	In Johnsons Bay.
	Drake's Floating	....	150	47	....	7½	5¼	100	....	In White Bay.
	Patent Slips:									
	Mort's No. 1	[1] 270	....	....	....	[2] 10 [3] 16	5¼	1,500	1885	At Balmain.
	Mort's No. 2	[1] 200	....	....	....	[2] 9 [3] 14	5¼	800	1868	At Balmain.
	Mort's No. 3	[1] 58	....	....	....	[2] 5 [3] 10	5¼	40	....	At Balmain.
	Waterview Bay	[1] 210	....	....	....	[2] 8 [3] 18	5¼	1,000	1914	
	Cockatoo Island	[1] 105	....	....	9	....	5¼	300	....	
Hobart	Patent Slip, Domain	[1] 150	....	....	....	[2] 8 [3] 13	4½	350	1854	
Launceston	Floating	....	172	37	12½	....	12½	1,200	1894	
Devonport	Patent Slip	[1] 112	....	....	....	[2] 7 [3] 12	11	300	....	

[1] Cradle. [2] Forward. [3] Aft.

APPENDIX No. II.

Particulars of wet docks, basins, locks, etc.

Port.	Name of dock, basin, etc.	Area.	Length.	Breadth.	Depth.	Entrance.		Lock.			Springs rise.	Quay-age.	Date built.	Remarks.
						Depth at M.H.W.S.	Width.	Length.	Breadth.	Depth on sill a C.D.L.[1]				
		Acres.	*Feet.*	*Feet.*	*Feet.*	*Feet.*	*Feet.*	*Feet.*	*Feet.*	*Feet.*	*Feet.*	*Feet.*		
Melbourne...	Victoria dock......	95	400	100	26	......	160	......	......	......	2¾	8,900	......	
	Spencer Street dock	......	......	......	15	......	......	......	......	......	2¾	......	......	
Hobart......	Constitution dock...	......	......	......	10½	......	......	......	......	......	4¾	1,074	......	
	Victoria dock	......	......	......	10½	......	......	......	......	......	4¾	1,378	......	

C. D. L.—Chart datum level.

APPENDIX No. III.

List of principal ports, showing particulars of depths, etc.

Port.	Depth at M. L. W. S. in channel of approach.	Depth at M. L. W. S. in anchorage.	Rise of tide.		Remarks.
			Springs.	Neaps.	
Devonport	16 feet.	14 to 18 feet at wharf	11	8	20 to 28 feet at wharves.
Geelong	25 feet.	4½ fathoms	3	2½	30 feet to Port Melbourne piers (being deepened to 35 feet); 26 feet in Yarra River.
...ne	33 feet in south channel, Port Phillip.	3 to 5 fathoms	2½	......	
Port Jackson	40 feet east channel, 21 feet west channel (being deepened).	7 fathoms	5½	4	28 to 50 feet at wharves.
Port Kembla	8 fathoms.	6 fathoms	5½	3½	
Port Phillip	40 feet.	6 to 9 fathoms	7	5½	It is dangerous for vessels drawing more than 29 feet to enter.
for affiar:					
Port Dalrymple	10 fathoms.	8 fathoms	10	7½	
Beauty Point Harbor	8 fathoms.	40 feet at wharf	10	......	
Launceston	9 feet.	12 to 16 feet at wharves.	12½	......	

INDEX.

A.

175078°—20——31

AGENTS FOR THE SALE OF HYDROGRAPHIC OFFICE PUBLICATIONS

IN THE UNITED STATES AND ISLANDS.

ABERDEEN, WASH.—The Evans Drug Co.

ASTORIA, OREG.—The Beebe Co., Astoria Branch.

BALBOA HEIGHTS, CANAL ZONE.—The Captain of the Port.

BALTIMORE, MD.—John E. Hand & Sons Co., 17 South Gay Street.

BELLINGHAM, WASH.—E. T. Mathes Book Co., 110 West Holly Street.

BOSTON, MASS.—Charles C. Hutchinson, 154 State Street.

    W. E. Hadlock & Co., 152 State Street.

    Kelvin & Wilfrid O. White, 112 State Street.

CHARLESTON, S. C.—Henry B. Kirk, 10 Broad Street.

CHICAGO, ILL.—A. C. McClurg, 330 East Ohio Street.

CLEVELAND, OHIO.—Upson Walton Co., 1294–1310 West Eleventh Street.

CRISTOBAL, CANAL ZONE.—The Captain of the Port.

DULUTH, MINN.—Joseph Vanderyacht.

EASTPORT, ME.—S. L. Wadsworth & Son, 5–8 Central Wharf.

GALVESTON, TEX.—Fred C. Trube, 2415 Market Street.

    Purdy Brothers, 2217 Market Street.

GLOUCESTER, MASS.—Geo. H. Bibber, 161 Main Street.

GULFPORT, MISS.—Southern Stationery Co., 2504 Fourteenth Street.

HONOLULU, HAWAII.—Hawaiian News Co.

JACKSONVILLE, FLA.—H. & W. B. Drew Co., 45 West Bay Street.

KETCHIKAN, ALASKA.—Ryus Drug Co.

KEY WEST, FLA.—Alfred Brost.

MANILA, PHILIPPINES.—Luzon Stevedoring Co.

MOBILE, ALA.—Cowles Ship Supply Co., 13–19 Dauphin Street.

NEW ORLEANS, LA.—Woodward, Wight & Co., Howard Avenue and Constance Street.

    Rolf Seeberg Ship Chandlery Co., P. O. Box 1230.

    J. S. Sareussen, 210 Tchoupitoulas Street.

NEWPORT, R. I.—W. H. Tibbetts, 185 Thames Street.

NEWPORT NEWS, VA.—W. L. Shumate & Co., 133 Twenty-fifth Street.

    John E. Hand & Sons Co., 2310 West Avenue.

NEW YORK, N. Y.—T. S. & J. D. Negus, 140 Water Street.

    John Bliss & Co., 128 Front Street.

    Michael Rupp & Co., 112 Broad Street.

    C. S. Hammond & Co., 30 Church Street.

    Kelvin & Wilfrid O. White, 38 Water Street.

NORFOLK, VA.—William Freeman & Son, 243 Granby Street.

PENSACOLA, FLA.—McKenzie Oerting & Co., 603 South Palafox Street.

PHILADELPHIA, PA.—Riggs & Bro., 310 Market Street.

    John E. Hand & Sons Co., 208 Chestnut Street.

PORTLAND, ME.—Wm. Senter & Co., 51 Exchange Street.

PORTLAND, OREG.—The Beebe Co., First and Washington Streets.

    The J. K. Gill Co., Third and Alder Streets.

PORT ARTHUR, TEX.—N. M. Nielson.
PORT TOWNSEND, WASH.—W. J. Fritz, 320 Water Street.
ROCKLAND, ME.—Huston Tuttle Book Co., 405 Main Street.
ST. THOMAS, VIRGIN ISLANDS.—S. Fischer, Harbor Master.
SAN DIEGO, CALIF.—Arey-Jones Co., 933 Fourth Street.
SAN FRANCISCO, CALIF.—Geo. E. Butler, Alaska Commercial Building.
    Louis Weule Co., 6 California Street.
    A. Lietz Co., 61 Post Street.
SAN JUAN, PORTO RICO.—Superintendent Lighthouse Service.
SAN PEDRO, CALIF.—Marine Hardware Co., 509 Bacon Street.
SAVANNAH, GA.—Savannah Ship Chandlery & Supply Co., 25 East Bay Street.
SEATTLE, WASH.—Lowman & Hanford Co., 616–620 First Avenue.
    Max Kuner Co., 804 First Avenue.
TACOMA, WASH.—Cole-Martin Co., 926 Pacific Avenue.
TAMPA, FLA.—Tampa Book & Stationery Co., 513 Franklin Street.
WASHINGTON, D. C.—W. H. Lowdermilk & Co., 1418 F Street NW.
    Brentanos, F and Twelfth Streets NW.
WILMINGTON, N. C.—Thos. F. Wood, 1–5 Princess Street.

IN FOREIGN COUNTRIES.

BELIZE, BRITISH HONDURAS.—A. E. Morlan.
BUENOS AIRES, ARGENTINA.—Rodolfo Boesenberg, 824 Victoria Street.
CANSO, N. S.—A. N. Whitman & Son.
HABANA, CUBA.—Eduardo Mencló, 10 Mercaderes.
HALIFAX, N. S.—Phillips & Marshall, 29 Bedford Row.
MANZANILLO, CUBA.—Enrique Lauten, Marti 44.
MONTREAL, CANADA.—Harrison & Co., 53 Metcalfe Street.
PORT HAWKESBURY, C. B. I., N. S.—Alexander Bain.
PRINCE RUPERT, B. C., CANADA.—McRae Bros., Ltd., P. O. Drawer 1690.
QUEBEC, CANADA.—T. J. Moore & Co., 118–120 Mountain Hill.
ST. JOHN, N. B.—J. & A. McMillan, 98 Prince William Street.
SHANGHAI, CHINA.—Capt. W. I. Eisler, care of American Post Office.

BRANCH HYDROGRAPHIC OFFICES.

These offices are located as follows:

BOSTON_____Fourteenth Floor, Customhouse.
NEW YORK_____Rooms 301–302, Maritime Exchange, 78–80 Broad Street.
PHILADELPHIA_____Main Floor, The Bourse Building.
BALTIMORE_____Room 123, Customhouse.
NORFOLK_____Room 2, Customhouse.
SAVANNAH_____Second Floor, Customhouse.
NEW ORLEANS_____Room 215, Customhouse.
GALVESTON_____Room 301, Customhouse.
SAN FRANCISCO_____Merchants' Exchange.
PORTLAND, OREG._____Room 407, Customhouse.
SEATTLE_____Room 408, Lowman Building.
DULUTH_____Room 1000, Torrey Building.
SAULT SAINTE MARIE_____Room 10, Federal Building.

NOTE.—By authority of the Governor of the Panama Canal some of the duties of Branch Hydrographic Offices are performed by the Captain of the Port at Cristobal and the Captain of the Port at Balboa. A set of reference charts and sailing directions may be consulted there; and ship masters may receive the Pilot Charts, Notice to Mariners, and Hydrographic Bulletins in return for marine data and weather reports. Observers' blanks and comparisons of navigational instruments may be obtained at the same time.

The Branch Offices do not sell any publications, but issue the Pilot Charts, Hydrographic Bulletins, Notices to Mariners, and Reprints to cooperating observers.

They are supplied with the latest information and publications pertaining to navigation, and masters and officers of vessels are cordially invited to visit them, and consult freely the officers in charge. Office hours, 9 a. m. to 4.30 p. m.

O.

Lightning Source UK Ltd.
Milton Keynes UK
UKHW012131290119
336431UK00008B/380/P